Quality of Telephone-Based Spoken Dialogue Systems

QUALITY OF TELEPHONE-BASED SPOKEN DIALOGUE SYSTEMS

SEBASTIAN MÖLLER
Institut für Kommunikationsakustik (IKA)
Ruhr-Universität Bochum
Germany

Sebastian Möller
Institut für Kommunikationsakustik (IKA)
Ruhr-Universität Bochum
Germany

Möller, Sebastian.
Quality of telephone-based spoken dialogue systems / Sebastian Möller.
p. cm.
Includes bibliographical references and index.
ISBN 0-387-23190-0
1. Automatic speech recognition. 2. Telephone systems--Quality control. I. Title.

TK7895.S65M65 2004
006.4'54--dc22

2004059514

ISBN 0-387-23190-0 e-ISBN 0-387-23186-2 Printed on acid-free paper.

Printed in the United States of America.

9 8 7 6 5 4 3 2 1 SPIN 11307129

springeronline.com

Contents

Preface

An increasing number of telephone services are offered in a fully automatic way with the help of speech technology. The underlying systems, called spoken dialogue systems (SDSs), possess speech recognition, speech understanding, dialogue management, and speech generation capabilities, and enable a more-or-less natural spoken interaction with the human user. Nevertheless, the principles underlying this type of interaction are different from the ones which govern telephone conversations between humans, because of the limitations of the machine interaction partner. Users are normally able to cope with the limitations and to reach the goal of the interaction, provided that both interlocutors behave in a cooperative way.

The present book gives a systematic overview of assessment, evaluation, and prediction methods for the quality of these innovative services. On the basis of cooperativity considerations, a new taxonomy of quality of service (QoS) aspects is developed. It identifies four types of factors influencing the quality aspects perceived by the user: Environmental factors resulting from the physical situation of use (transmission channels, ambient noise); factors directly related to the machine interaction partner; task factors covering the interaction goal; and non-physical contextual factors like the access conditions and the involved costs. These factors are shown to be in a complex relationship to different categories of perceived quality, like cooperativity, efficiency, usability, user satisfaction, and acceptability. The taxonomy highlights the relationships between the different factors and aspects. It is a very useful tool for classifying assessment and evaluation methods, for planning and interpreting evaluation experiments, and for estimating quality on the basis of system characteristics.

Quality is the result of a perception and a judgment process. Consequently, assessment and evaluation methods involving human test subjects are necessary in order to quantify the impact of system characteristics on perceived quality. The system characteristics can be described with the help of interaction parameters, i.e. parameters which are measured instrumentally or on the basis of expert annotations. A number of parameters and evaluation methods are defined, both on a system component level and for the fully integrated system. It is shown that technology-centered component assessment has to go hand in hand with user-centric evaluation, because both provide different types of information for the system developer. The resulting information about quality is needed in all phases of system specification, design, implementation, and operation, in order to efficiently set up systems which offer a high quality to their users.

Three new experimental investigations illustrate the relationships between system characteristics on the one side, and component performance or perceived quality on the other. First, the effect of the transmission channel on speech recognition and speech output is analyzed with the help of a network simulation model. The results are compared to human communication scenarios, and quality or performance estimations are obtained on the basis of system characteristics, using quality prediction models. In a second step, interaction experiments with a fully integrated system are carried out, and interaction parameters as well as user quality judgments are collected. The analysis of the obtained data shows that the correlation between both types of metrics is relatively low. This is a proof for the hypothesis that quality models for the overall interaction with the SDS can cover only a part of the factors influencing perceived quality. With the help of the QoS taxonomy, alternative modelling approaches are proposed. Still, the predictive power is too limited to avoid resource-demanding experiments with human test subjects. The reasons for this finding are discussed, and necessary research directions to overcome the limitations are pointed out.

The assessment, evaluation and prediction of quality requires knowledge from a number of disciplines which do not always share a common ground of information. Although being written from the perspective of an engineer in telecommunications, the book is directed towards a wide audience, from experts in telecommunications and signal processing, communication acoustics, computational linguistics, speech and language sciences, up to psychophysics, human factor design and ergonomics. It is hoped that this – admittedly very ambitious – goal can at least partially be reached, and that the book may provide useful information for designing systems and services which ultimately satisfy the needs of their *human* users.

Bochum

SEBASTIAN MÖLLER

Acknowledgments

The present work was performed during my occupation at the Institut für Kommunikationsakustik (IKA), Ruhr-Universität Bochum. A number of persons contributed in different ways to its finalization. Especially, I would like to thank the following:

- my colleague PD Dr. phil. Ute Jekosch for supporting this work over the years, and for providing the scientific basis of quality assessment,
- the former head of the institute, Prof. Dr.-Ing. Dr. techn. h.c. Jens Blauert, for providing a scientific home, and for enabling and supporting the work with interest and advice,
- Prof. Dr.-Ing. Ulrich Heute (Christian-Albrechts Universität zu Kiel, Germany) and Prof. Dr. Rolf Carlsson (KTH Stockholm, Sweden) for their interest in my work,
- my colleagues Alexander Raake and Jan Krebber for taking over some of my duties so that I had the time for writing, and for numerous fruitful discussions,
- the student Janto Skowronek for the huge amount of work performed during his diploma thesis and his later occupation at the institute,
- the students Christine Dudda (now Pellegrini) and Andreea Niculescu for their experimental work on dialogue system evaluation contributing to Chapter 6,
- the student co-workers Sven Bergmann, Sven Dyrbusch, Marc Hanisch, Marius Hilckmann, Anders Krosch, Jörn Opretzka, Rosa Pegam, Sebastian Rehmann and Joachim Riedel for their countless contributions during the last years,
- Dr. Ergina Kavallieratou for her work on speech recognition contributing to Chapter 4,

- Stefan Schaden and many other colleagues at IKA for discussions and suggestions,
- James Taylor and Dr.-Ing. Volker Kraft for reviewing and correcting the manuscript,
- Prof. Dr. Hervé Bourlard and his colleagues from the Institut dalle Molle d'Intelligence Artificielle Perceptive (IDIAP) in Martigny, Switzerland, for providing a scientific basis in early spring 2000,
- Dr. Martin Rajman and his colleagues Alex Trutnev and Florian Seydoux at Ecole Polytechnique Fédérale de Lausanne (EPFL), Switzerland, for their support in developing the Swiss-French prototype of the BoRIS system,
- Dr.-Ing. Jens Berger for his support with signal-based quality prediction models,
- numerous colleagues in Study Group 12 of the International Telecommunication Union (ITU-T) following and supporting my work with interest,
- the system administrators of the institute's computer network and the members of the office for providing and maintaining their resources, and
- my family and friends for strongly supporting me during the past five years.

A part of the work was supported by the EC-funded IST project INSPIRE ("INfotainment management with SPeech Interaction via REmote-microphones and telephone interfaces", IST-2001-32746).

Definitions and Abbreviations

Definitions

# ASR REJECTIONS	average number of speech recognizer rejections in a dialogue
# BARGE-INS	average number of barge-in attempts from the user in a dialogue
# CANCEL ATTEMPTS	average number of cancel attempts from the user in a dialogue
# HELP REQUESTS	average number of help requests from the user in a dialogue
# SYSTEM ERROR MESS.	average number of system error messages in a dialogue
# SYSTEM QUESTIONS	average number of system questions in a dialogue
# SYSTEM TURNS	average number of system turns in a dialogue
# SYSTEM WORDS	average number of system words uttered in a dialogue
# TIME-OUT PROMPTS	average number of time-out prompts in a dialogue
# TURNS	average number of turns in a dialogue
# USER QUESTIONS	average number of user questions in a dialogue
# USER TURNS	average number of user turns in a dialogue
# USER WORDS	average number of user words uttered in a dialogue
# USER WORDS IV	average number of in-vocabulary user words in a dialogue
# WORDS	average number of words uttered in a dialogue
α, α_i, $\overline{\alpha}$	false speaker rejection rate
$\alpha(i)$	scaling factor for $C_x(i,j)$
A	expectation factor
$a_0(i,j)$, $a_1(i,j)$, $a_2(i,j)$	coefficients for $C_x(i,j)$
$AN{:}CO$	number of correct system answers
$AN{:}FA$	number of failed system answers
$AN{:}IC$	number of incorrect system answers
$AN{:}PA$	number of partially correct system answers
β, $\beta_{\bar{i}}$, $\overline{\beta}$	false speaker acceptance rate
Bpl	packet loss robustness factor
γ, γ_i, $\overline{\gamma}$	speaker misclassification rate
$C_{ref}(i,j)$	recognition confusion matrix corresponding to E_{ref}
$C_x(i,j)$	recognition confusion matrix corresponding to E_x
c_{AVP}	number of correctly recognized attribute-value pairs
c_i	cost measures of the PARADISE model
c_s	number of correctly recognized sentences
c_w	number of correctly recognized words
$CA{:}AI$	percentage of system answers judged to be appropriate and inappropriate by different experts (%)
$CA{:}AP$	percentage of appropriate system answers (%)
$CA{:}IA$	percentage of inappropriate system answers (%)
$CA{:}IC$	percentage of incomprehensible system answers (%)
$CA{:}TF$	percentage of completely failed system answers (%)

CE	concept efficiency (%)
CER	concept error rate (%)
$COMP$	subjective evaluation of task completion
CR	correction rate (%)
D	frequency-weighted difference in sensitivity between the direct and the diffuse sound (dB)
d_{AVP}	number of deleted attribute-value pairs
d_s	number of deleted sentences
d_w	number of deleted words
$DARPA_s$	DARPA score
$DARPA_e$	DARPA error
$DARPA_{me}$	modified DARPA error
DD	dialogue duration (s)
Dr	D-value of the handset telephone, receive side (dB)
Ds	D-value of the handset telephone, send side (dB)
E_{ref}	reference recognition rate of the simulated recognizer
E_x	target recognition rate of the simulated recognizer
G	signal-to-equivalent-continuous-circuit-noise ratio (dB)
$\%GoB$	percentage of users rating a connection good or better (%)
$HC_{U1}, HC_{U2}, HC_{S1}, HC_{S2}$	recognition success metrics according to Kamm et al. (1997a)
I	impairment factor
i_{AVP}	number of inserted attribute-value pairs
i_s	number of inserted sentences
i_w	number of inserted words
IC	information content (%)
Id	impairment factor for impairments occurring delayed with respect to the speech signal
Ie	equipment impairment factor
$Ie{,}eff$	effective equipment impairment factor, including transmission errors
Iq	impairment factor for quantizing distortion
Iqo	recognizer-specific impairment factor
IR	implicit recovery (%)
Is	impairment factor for impairments occurring simultaneously with the speech signal
κ, κ_{conf}, κ_{dia}	kappa coefficient (per configuration, per dialogue)
Le	frequency-dependent loss of the talker echo path (dB)
Lst	frequency-dependent loss of the sidetone path (dB)
$LSTR$	listener sidetone rating (dB)
$M(i,j)$	understanding error confusion matrix
MRS	mean recognition score
N_{AVP}	total number of attribute-value pairs in an utterance
n_{AVP}	number of correctly not set attribute-value pairs
$N_c(i)$	total number of concepts in the ith dialogue
N_d	total number of dialogues
$N_q(i)$	total number of user queries in the ith dialogue
$N_u(i)$	number of unique concepts newly understood by the system in the ith dialogue

Nc	circuit noise level (dBm0p)
NES	number of errors per sentence
$\overline{NES}$, $\overline{NES_{iso}}$	mean number of errors per sentence (isolated word recognition)
$Nfor$	noise floor level (dBmp)
No	total equivalent circuit noise level (dBm0p)
Nor	equivalent circuit noise caused by room noise at receive side (dBm0p)
Nos	equivalent circuit noise caused by room noise at send side (dBm0p)
Nro	recognizer-specific noise parameter (dBm0p)
OLR	overall loudness rating between mouth and ear reference points (dB)
p	probability for rejection of the null hypothesis
$P(A)$	actual agreement rate
$P(E)$	chance agreement rate
$PA{:}CO$	number of correctly parsed user utterances
$PA{:}FA$	number of user utterances which failed parsing
$PA{:}PA$	number of partially parsed user utterances
$perf_{max}$	maximum performance value
$perf_{min}$	minimum performance value
$\%PoW$	percentage of users rating a connection poor or worse (%)
Ppl	random packet loss probability (%)
Pr	A-weighted sound pressure level of room noise at receive side (dB(A))
PRE	proportion reduction in error
Ps	A-weighted sound pressure level of room noise at send side (dB(A))
Q	signal-to-quantizing-noise ratio (dB)
$QComb$	system performance measure (Bonneau-Maynard et al., 2000)
QD	query density (%)
qdu	quantizing distortion unit
Qo	recognizer-specific robustness factor (dB)
ρ	Spearman rank order correlation coefficient
R, R_n	(normalized) transmission rating
R^2	mean amount of variance covered by the regression analysis
r	Pearson correlation coefficient
RD	response delay
RLR	receive loudness rating between the 0 dBr point in the network and the ear reference point (dB)
$RLRset$	receive loudness rating of the telephone handset (dB)
Ro	basic signal-to-noise transmission rating factor
S	total number of sentences in the reference
s_{AVP}	number of substituted attribute-value pairs
s_s	number of substituted sentences
s_w	number of substituted words
SA	sentence accuracy (%)
SAT	mean overall system performance rating (Bonneau-Maynard et al., 2000)
SCR	system correction rate (%)
SCT	average number of system correction turns
SER	sentence error rate (%)
SLR	send loudness rating between the mouth reference point and the 0 dBr point in the network (dB)
$SLRset$	send loudness rating of the telephone handset (dB)

SRD	system response delay (s)
STD	system turn duration (s)
$STMR$	sidetone masking rating (dB)
T	mean one-way talker echo path delay (ms)
Ta	overall delay between the mouth reference point of the talker and the ear reference point of the listener (ms)
TD	turn duration (s)
$TELR$	talker echo loudness rating (dB)
$topline$	topline performance value
$topline_R$	topline transmission rating value
Tr	round-trip delay for listener echo (ms)
TS, TS_{bin}, TS_{ord}	task success measures (%)
UA	understanding accuracy (%)
UCR	user correction rate (%)
UCT	average number of user correction turns
URD	user response delay (s)
US_w	user satisfaction rating according to Walker et al. (1998a)
$\widehat{US_w}$	estimation of US_w
UTD	user turn duration (s)
W	total number of words in the reference
w_i	weighting coefficients for $\widehat{US_w}$
WA	word accuracy (%)
WA_{iso}	word accuracy for isolated word recognition (%)
$WEPL$	weighted echo path loss for listener echo (dB)
WER	word error rate (%)
WER_{iso}	word error rate for isolated word recognition (%)
WES	word error per sentence
$\overline{WES}, \overline{WES_{iso}}$	mean word error per sentence (isolated word recognition)
$WPST$	average number of words per system turn
$WPUT$	average number of words per user turn

Abbreviations

ACR	absolute category rating
ADPCM	adaptive differential pulse code modulation
AM	amplitude modulation
AMR	adaptive multi-rate
ANN	artificial neutral network
ANOVA	analysis of variance
AoS	acceptability of service
ASR	automatic speech recognition
ATIS	air travel information system
AVM	attribute-value matrix
AVP	attribute-value pair

BP	bandpass
BT	British Telecom
CART	classification and regression tree
CELP	code-excited linear prediction
CLID	cluster identification
CNET	Centre National d'Etudes des Télécommunications
CNRS	Centre National de la Recherche Scientifique
CP	cooperativity principle
CSELT	Centro Studi e Laboratori Telecommunicazioni
CSLU	Center for Spoken Language Understanding
CSR	continuous speech recognition
CVC	consonant-vowel-consonant
DARPA	Defense Advanced Research Projects Agency
DDL	dialogue description language
DP	dynamic programming
DR	design rationale
DRT	diagnostic rhyme test
DSD	design space development
DTMF	dual tone multiple frequency
EAGLES	European Advisory Group on Language Engineering Standards
EER	equal error rate
ELDA	Evaluation and Language Resources Distribution Agency
ELRA	European Language Resources Association
EPFL	Ecole Polytechnique Fédérale de Lausanne
ETSI	European Telecommunications Standards Institute
EURESCOM	European Institute for Research and Strategic Studies in Telecommunications
FUB	Fondazione Ugo Bordoni
GG	generic cooperativity guideline
GSM	global system for mobile communication
GSM-EFR	GSM enhanced full-rate
GSM-FR	GSM full-rate
GSM-HR	GSM half-rate
GUI	graphical user interface
HENR	human equivalent noise ratio
HHI	human-to-human interaction
HMI	human-machine interaction
HMM	hidden Markov model
HTK	hidden Markov model toolkit
HTML	hypertext markup language
IDIAP	Institut dalle Molle d'Intelligence Artificielle Perceptive
IEC	International Electrotechnical Commission
IKA	Institut für Kommunikationsakustik
INT, INT_n	mean (normalized) rating on an intelligibility scale
IP	internet protocol
IRS	intermediate reference system
IRU	informally redundant utterance
ISCA	International Speech Communication Association
ISDN	integrated services digital network

ISO	International Organization for Standardization
ITU	International Telecommunication Union
ITU-T	Telecommunication Standardization Sector of the ITU
IVR	interactive voice response
LD-CELP	low-delay code-excited linear prediction
LDC	Linguistic Data Consortium
LPC	linear predictive coding
MAPSSWE	Matched-Pair-Sentence-Segment-Word-Error test
MFCC	mel-frequency cepstral coefficient
MIT	Massachusetts Institute of Technology
MLP	maximum likelihood process
MN	McNemar test
MNRU	modulated noise reference unit
MOS, $\mathrm{MOS_n}$	mean opinion score (normalized)
$\mathrm{MOS_{LE}}$, $\mathrm{MOS_{LE,n}}$	mean opinion score on a listening-effort scale (normalized)
MRT	modified rhyme test
NIST	National Institute of Standards and Technology
NLP	natural language processing
OOV	out-of-vocabulary
PARADISE	paradigm for dialogue system evaluation
PBX	private branch exchange
PCM	pulse code modulation
PESQ	perceptual evaluation of speech quality
PIN	personal identification number
PLP	perceptual linear predictive
PNAMBIC	pay no attention to the man behind the curtain (see WoZ)
PSOLA	pitch-synchronous overlap and add
PSTN	public switched telephone network
QOC	questions-options-criteria rationale
QoS	quality of service
RAD	rapid application developer
RAMOS	recognizer assessment by manipulation of speech
RASTA	relative spectral
ROC	receiver operating curves
RPE-LTP	regular pulse excitation long term prediction
SALT	speech application language tags
SASSI	subjective assessment of speech system interfaces
SDL	specification and description language
SDS	spoken dialogue system
SG	specific cooperativity guideline
SI	speaker identification
SLDS	spoken language dialogue system
SNR	signal-to-noise ratio
SPL	sound pressure level
SPSS	Statistical Package for the Social Sciences
SQL	structured query language
SUS	semantically unpredictable sentence
SV	speaker verification

TC STQ	Technical Committee Speech Processing, Transmission and Quality Aspects
Tcl/Tk	tool command language and tool kit
TCP/IP	transmission control protocol / internet protocol
TCU	turn-constructional unit
TDMA	time division multiple access
TFW	time-and-frequency warping
TIPHON	Telecommunications and Internet Protocol Harmonization Over Networks
TMF	time-frequency-modulation
TOSQA	Telekom objective speech quality assessment
TRP	transition-relevant place
TTS	text-to-speech
UMTS	universal mobile telecommunications system
VoiceXML	voice extensible markup language
VoIP	voice over internet protocol
VSELP	vector sum excited linear prediction
VUI	voice user interface
WoZ	Wizard-of-Oz
WSR	Wilcoxon signed rank test
XML	extensible markup language

Chapter 1

MOTIVATION AND INTRODUCTION

Modern telecommunication networks promise to provide ubiquitous access to multimedia information and communication services. In order to increase the number of their users, telephone network operators create new speech interaction services for communication, information, transaction and E-commerce, via an interconnected global network of wireline and mobile trunks. For mobile network operators, speech-based services are a key feature to being different from other operators. Other companies are cutting costs by automating call centers and customer-service operations, and can improve internal operations via web- and telephone-based services, especially for mobile workers. The Gartner group expects 2003 about one third of the automatized telephone lines to be equipped with automatic speech recognition capabilities (Thyfault, 1999).

Apart from the significant advances which have been made in speech and language technology during the last twenty years, the possible economical benefit for the service operators has been a key driving force for this development. Following the argumentation of Whittaker and Attwater (1995), speech-based systems help to

- enable market differentiation,
- exploit revenue opportunities,
- improve the quality of existing services,
- improve the accessibility of services,
- reduce the cost of service provision, and
- free-up people to concentrate on high-value tasks.

These reasons can be decisive for companies and service operators to integrate speech and language technology into their services. Railway information can be seen as a typical example: Based on a study of 130 information offices in six countries (Billi and Lamel, 1997), over 100 million calls were handled per year, with at least another 10 million calls remaining unanswered. About 91% of the callers solely asked for information, and only 9% performed a reservation task. It was estimated that over 90% of the calls could be handled by an automatic system with a recognition capability of 400 city names, and over 95% by a system with a 500 city names capability. Thus, automatic services seem to be a very economic solution for handling such tasks. They help to reduce waiting time and extend opening hours. The negative impact on the employment situation should however not be disregarded.

Amongst all potential application areas of spoken dialogue systems, it is the telecommunication sector which has provided the most powerful impetus for research on practical systems to date (Fraser and Dalsgaard, 1996). From a telecom operator's point of view, the new services differ in three relevant aspects from traditional ones (Kamm et al., 1997b). On the service side, traditional voice telephony was amended by the integrated transmission of voice, audio, image, video, text and data, in fixed and mobile application situations. On the transmission technology side, analogue narrow-band wireline transmission has been replaced by a mix of wireline and wireless networks, using analogue or digital representations, different transmission bandwidths, and different media such as copper, fiber, radio cells, satellite or power lines. On the communication side, the model changed from a purely human-to-human communication to an interaction partly between humans, and partly between humans and machines. These changes have consequences for the developers of spoken dialogue systems, for transmission network operators, and of course for the end users.

Interactive speech systems are "computer systems with which humans interact on a turn-by-turn basis" (Fraser, 1997, p. 564). They enable and support the communication of information between two parties (agents), mostly between a human user and a machine agent. Here, only those systems will be addressed in which spoken language plays a significant role as an interaction modality. According to Dybkjær and Bernsen (2000), p. 244, the most advanced commercial systems

> "have a vocabulary of several thousand words; understand speaker-independent spontaneous speech; do complex linguistic processing of the user's input; handle shifts in initiative; have quite complex dialogue management abilities including, e.g. reasoning based on the user's input, consultation of the recorded history of the dialogue so far, and graceful degradation of the dialogue when faced with users who are difficult to understand; carry out linguistic processing of the output to be generated; solve several tasks, and not just one; and robustly carry out medium-length dialogues to provide the user with, for instance, train timetable information on the departures and arrivals of trains between hundreds of cities".

Whereas not all of these characteristics need to be satisfied, the focus will be set in the following investigations on systems which accept continuously spoken input from different speakers, allow initiative to be taken from both the user and the system, and which are capable of reasoning, correction, meta-communication (communication about communication), anticipation, and prediction. These systems are called 'spoken dialogue systems' (SDSs), in some literature also 'spoken language dialogue systems' (SLDSs). They have to be differentiated from systems with more restricted capabilities, e.g. command systems or systems accepting only dialling tones as an input. A categorization of interactive speech systems will be given in Section 2.1.3.

Most of the currently available systems enable a task-orientated dialogue, i.e. the goal of the interaction is fixed to a specific task which can only be reached if both interaction partners cooperate. This type of interaction is obviously very restricted, and it should not be confused with a normal communication situation between humans. In task-orientated dialogues, the structure of the task was shown to carry a significant influence on the structure of the dialogue (Grosz, 1977), and this structure is a prerequisite for systems whose speech recognition and understanding capabilities are still very limited. In practical cases, this restriction is however not too severe, because task-orientated dialogues are highly relevant for commercial applications.

Spoken dialogue systems can be seen as speech-based user interfaces (so-called voice user interfaces, VUIs) to application system back-ends, and they will thus compete with other types of interfaces, namely with graphical user interfaces (GUIs). GUIs have the advantage of providing immediate feedback, reversible operations, incrementality, and of supporting rapid scanning and browsing of information. Because the visual information may easily and immediately indicate all options which are available at a specific point in the interaction, GUIs are relatively easy to use for novices. Spoken language interfaces, on the other hand, show the inherent limitations of the sequential channel for delivering information, of requiring the user to learn the language the system can understand, of hiding available command options, and of leading to unrealistic expectations as to their capabilities (Walker et al., 1998a). Speech is perceptually transient rather than static. This implies that the user has to pick up the information provided by the system immediately, or he/she will miss it completely.

These arguments against SDSs are however only valid when such systems just mimic GUIs. Human-to-human interaction via spoken dialogue shows that humans are usually able to cope with the modality limitations very well. Even better, spoken language is able to surpass several weaknesses which are inherent to direct manipulation interfaces like GUIs (Cohen, 1992, p. 144):

> "Merely allowing the users to select currently displayed entities provides them little support for identifying objects not on the screen, for specifying temporal relations, for identifying and operating on large sets and subsets of entities, and for using the context of interaction. What is missing is a way for users to *describe* entities, by which it is

meant the use of an expression in a language (natural or artificial) to *denote* or *pick out* an object, set, time period, and so forth."

It seems that the limitations of speech-based interfaces can and have to be addressed by an appropriate system design, and that in this way interfaces offering a high utility and quality to their users can be set up. Some general design principles are already well understood for GUIs, and should also be taken into account in SDS design, e.g. to represent objects and actions continuously, or to allow rapid, incremental, reversible operations on objects which are immediately acknowledged (Shneiderman, 1992; Kamm and Walker, 1997). These principles reflect to some extent the limitations of the human memory and cognitive and sensory processing. In SDSs, a continuous representation can be reached by using consistent vocabulary throughout the dialogue, or by providing additional help information in case of time-outs. Immediate feedback can be provided by explicit or implicit confirmation, and by allowing barge-in. Summarization might be necessary at some points in the interaction in order to respect the human auditory memory limitations.

Before developing a spoken dialogue system, it has to be decided whether speech is the right modality for the application under consideration, and for the individual tasks to be carried out. For example, users will not like to say their PIN code out aloud to a cash machine in the street, and long timetable lists are better displayed visually. The decision on an appropriate modality can be taken in a systematic way by using modality properties, as it was proposed by Bernsen (Bernsen, 1997; Bernsen et al., 1998). If speech is not sufficient as a unique modality, multimodal systems may be a better solution (Fellbaum and Ketzmerick, 2002). Such systems are able to handle several input and output devices addressing different media in parallel. A user may interact with the system using different modalities of input and output, and combinations of modalities are possible. For example, a user may point to a touchscreen device and ask "How can I get there?". Or a system may display a route on the screen and inform the user: "You have to turn right at this point!". Cohen (1992), p. 143, pointed out that a major advantage of multimodal user interfaces is "to use the strengths of one modality to overcome weaknesses of another".

Still, the speech modality will remain an essential element in multimedia communication services. This fact is underlined by the strong persistence of unimodal narrow-band telephone services even in networks which would allow for wideband and audio-visual services. Remote access to information is highly desirable (in privacy, but also for mobile workers), and the telephone is a lightweight and ubiquitous form of access which is available to nearly everyone. Speech is also the only modality to address devices which are physically very small, or which are desired to be invisible. Thus, it can be expected that speech will continue to persist as the main modality in human-to-human communication and develop for human-machine interaction as well, besides other multimodal forms.

In order to design spoken-dialogue-based services to be as efficient, usable and acceptable as possible, both the underlying technologies (speech transmission, speech recognition, language understanding, dialogue management and speech synthesis), as well as the human factors which make human-machine interfaces to be usable, have to be considered. The perception of a service by its users will depend on the underlying technology, but the link between both is particularly complex, because it involves a human partner which cannot easily be described and modelled via algorithms. It would be wrong to assume that the notable progress which has been reached in the last years for SDSs would be grounded on a thorough theoretical basis, at least not on one for the human interaction partner. A solid theoretical basis can however only be built when the underlying characteristics of the human-machine interaction via spoken language are well understood. The optimal way to advance our knowledge in this respect is to analyze the human-machine interaction situation, and to assess and evaluate the characteristics of the systems under consideration from the human user's point of view. A user-centric evaluation will help to identify, describe and measure the interaction problems which can be observed, and is thus a prerequisite for setting up better theories and systems.

The development of spoken dialogue systems requires not only a change in the focus of speech and language research activities, namely from typed text to spoken language input, from read to spontaneous speech recognition, and from simple recognition to interpretation and understanding of speech input (Maier et al., 1997; Furui, 2001b). In addition, it increases the need for carrying out subjective[1] interaction experiments with human users in order to determine the quality of the developed systems, and the resulting satisfaction of their users. As a wide range of novice users is the target group of current state-of-the-art systems and services, the need for subjective assessment and evaluation of quality is increasing (Hone and Graham, 2001, p. 2083):

> "In the past speech input/output technology was successful only in a limited number of specialised domains. Now speech technology is increasingly seen as a means of widening access to information services. Many see speech as an ideal gateway to the mobile internet. Others see it as a way of encouraging more widespread use of information technology, particularly by previously excluded groups, such as older people. The eventual success of speech as a means of broadening access in this way is very heavily dependent on the perceived ease of use of the resulting systems."

[1] The expression "subjective" is used in the following to indicate that a measurement process involves the direct perception and judgment of a human subject, which acts as a measuring organ. It is not contrasted to "objective", and carries no indication of intra- or inter-subject validity, reliability, or universality.

Unfortunately, the need for a systematic evaluation still seems to be underestimated. Despite the efforts for the development of assessment and evaluation criteria made both in the US and the EU during roughly the last decade (e.g. the DARPA programs and the EAGLES initiative), the dimensions of quality of a spoken-dialogue-based service which are perceived by its users are still not well understood. One obvious reason is that only few real interactive systems have been available so far. However, there is also a lack of stable reference criteria which an evaluation or assessment could be based on. Whereas for speech recognizers the target is relatively clear (namely to achieve a high word accuracy), for other components there is no basic categorization available, e.g. for dialogue in terms of dialogue acts, grammars, etc. When the components are integrated to form a working system, the impetus of each part on the quality of the whole has to be estimated. This task is particularly difficult because so far no analytic and generic approaches to quality exist, neither for the analysis of the users' quality percepts, nor for the system elements which are responsible for these percepts.

It is the aim of the present work to contribute to the closing of this gap. A particular type of service will be addressed, namely the interaction with a spoken dialogue system over a telephone network. For such a service, the transmission channel and the acoustic environment carry a severe impact on the performance of the dialogue system. Modern transmission networks like PSTN (public switched telephone network), ISDN (integrated services digital network), mobile networks, or IP-based networks introduce a number of different impairments on the transmitted speech signal. Whereas the effects of these impairments on a human interaction partner can partly be quantified, it is still unclear how they will impact the interaction with a spoken dialogue system. It has to be assumed that their impact on the interaction quality may be considerable, and thus has to be taken into account in the system development and evaluation phases. The problem has to be addressed jointly by transmission network planners and by speech technology experts. As a consequence, this book is directed towards a wide audience in telecommunication engineering, speech signal processing, communication acoustics, speech and natural language technology, communication science, as well as human factors and ergonomics. It will provide useful background information from the involved fields, present new theoretical and experimental analyses, and serve as a basis for best practice system design and evaluation.

Chapter 2 describes the covered interaction situations from a global point of view, following the way which is taken by the information from the source to the sink, and vice versa. It involves the acoustic user interface, the transmission channel, the speech recognizer, the speech understanding component, the dialogue manager, the underlying application program, the response generation, the speech synthesis, the transmission channel, and the acoustic interface to the human user. An overview of the most important elements of this chain will be given in this chapter. Humans interacting with a spoken dialogue system usually behave in a different way than can be observed in human-to-human communication scenarios. This behavior will be addressed in the following section, indicating aspects which are important for the quality perception of the user. On the basis of this analysis, a new taxonomy of all relevant aspects of quality will be developed. It shows the relationship between the relevant elements of the system or service and the user's quality percepts which are organized on different levels (efficiency, usability, user satisfaction, acceptability). For human-to-human interaction over the phone, several of these relationships can already be quantitatively described, using quality prediction models. For spoken dialogue systems accessed over the phone, modelling approaches are still very limited, and system designers have to rely on intuition and on simulation experiments. Apparently, there is a lack of assessment and evaluation data which would be useful for quantifying the effects of system elements on perceived quality.

Methods and methodologies for quality assessment and evaluation will be discussed in Chapter 3. The individual elements of a spoken dialogue system will be addressed first individually, and commonly used methods and developments will be pointed out. However, the quality of the whole system and of the service visible (audible) to the user will not be just a sum of the individual components. It is therefore necessary to quantify the contribution of the individual components to the quality perception of the whole. A way in this direction is to collect interaction parameters which relate to individual aspects of the system, and to relate them to quality features perceived by the user. Thus, quality assessment and evaluation requires the collection of subjective quality judgments from users who are interacting with the service under consideration. The collection can be largely facilitated by simulation environments as long as not all system components are available. A list of interaction parameters and judgment aspects will be compiled by the end of that chapter. It will form a basis for new evaluation experiments, and can serve as a source of information for system developers.

On the theoretical basis for quality description and analysis, experimental data will be presented in Chapters 4 to 6. The experiments address three different parts of the communication chain, namely the recognition of telephone-impaired speech signals (Chapter 4), the quality of synthesized speech when transmitted over the telephone network (Chapter 5), and quality aspects of the interaction with a fully working system (Chapter 6). Each problem is addressed in an analytical way, using simulation environments in order to gain control over the elements potentially influencing the performance of the system, and consequently the user's quality percepts. Relationships between system or interaction parameters on the one hand, and quality judgments on the other, are established with the help of quality prediction models. For the transmission channel impact, signal-based or parametric models are used. They have to be partly extended in order to obtain reasonable predictions. For a fully working service, the relationship between interaction parameters and quality aspects is addressed with the help of linear regression models. Although the predictions for the whole service are far from perfect, the taxonomy of quality aspects developed in Chapter 2 proved to form a solid basis for quality modelling approaches which aim at being generic, and applicable to a variety of other systems.

The interaction scenario which is addressed here is of course limited. However, it is expected that the structured approach to quality of telephone-based spoken dialogue systems can be transferred to other types of systems. Examples are systems which are operated directly in different acoustic environments (car navigation systems, smart home systems), or multimodal systems. Such interactive systems will become increasingly important in the near future, and their success and acceptance will depend to a large extent on the level of quality they offer to their users.

Chapter 2

QUALITY OF HUMAN-MACHINE INTERACTION OVER THE PHONE

Telephone services which rely on spoken dialogue systems are now being introduced at a large scale for information retrieval and transaction tasks. For the human user, when dialing the number, it is often not completely clear that the agent on the other side will be a machine, and not a human operator. The uncertainty is supported by the fact that the user interface (e.g. a telephone handset) is identical in both cases. As a consequence, comparisons will automatically be drawn to the quality of human-to-human communication over the same channel, for carrying out the same task with a human operator. Thus, it is useful to investigate both scenarios in parallel. While acknowledging the differences in behavior from both – human and machine – sides, it seems justifiable to take the human-to-human telephone interaction (here short 'human-to-human interaction', HHI) as *one* reference for telephone-based human-machine interaction (short 'human-machine interaction', HMI). Depending on the task, another reference may be a web site for online timetable consultation, or a TV news-ticker with stock rates. The references have to be taken into account when the quality of a telecommunication service, the quality of the dialogic interaction with a machine agent, or the quality of transmitted speech are to be determined.

The quality of transmitted speech has been a topic of investigations for a long time in traditional telephony. Its importance is still increasing with the advent of mobile phones and packetized speech transmission (e.g. Voice over Internet Protocol, VoIP), and new assessment methods and prediction models are currently being developed. When the interaction partner on the other side of the transmission channel is a machine instead of a human being, the question arises of how the performance of speech technology, in particular of speech recognition and speech understanding, but in a second step also of speech synthesis, is influenced by the transmission channel. Without doubt, the quality of transmitted speech will be linked in some way to the performance of speech technology devices which are operated over the transmission channel. Both entities should however not be confused, because the requirements of the human and the machine interaction partner are different. Depending on how well both requirements are fulfilled, the dialogue will be more or less successful, resulting in a higher or lower interaction quality for the user. The interaction quality largely determines the quality of the whole telecommunication service which is based on an SDS.

Whereas structured approaches have been documented on how to design spoken dialogue systems so that they adequately meet the requirements of their users (e.g. by Bernsen et al., 1998), the quality which is perceived when interacting with SDSs is often addressed in an intuitive way. Hone and Graham (2001) describe efforts to determine the dimensions underlying the user's quality judgments, by performing a multidimensional analysis on subjective ratings obtained on a large number of different scales. The problem obviously turned out to be multidimensional. Nevertheless, many other researchers still try to estimate "overall system quality", "usability" or "user satisfaction" by simply calculating the arithmetic mean over several user ratings on topics as different as perceived synthesized speech quality, perceived system understanding, and expected future use of the system. The reason is the lack of an adequate description of quality dimensions, both with respect to the system design and with respect to the perception of the user.

The quality of the interaction will depend not only on the characteristics of the machine interaction partner itself, but also on the transmission channel and the acoustic situation in the environment of the user. In the past, the impact of transmission impairments in HHI has been analyzed in detail, and appropriate modelling approaches already allow it to be quantified in a predictive way (Möller and Raake, 2002). Unfortunately, no such detailed analysis exists for the transmission channel impact on the interaction of a human user with a spoken dialogue system over the phone. This gap has to be filled, because modern telecommunication networks will have to guarantee both – a high speech communication quality between humans, and a robust and successful interaction between humans and machines[1]. Apparently, adequate planning and evaluation of quality are as important for the designer of transmission networks as they are for the designers of spoken dialogue systems.

In this chapter, an attempt is made to close the gap. The starting point is a description of communication scenarios in which a human user interacts with a spoken dialogue system over some type of speech transmission network, see Section 2.1. It takes into account the source and the sink of information, as well as the transmission channel. Different types of networks will be briefly discussed, and the main modules of a spoken dialogue system will be presented. The human interaction with an SDS in these scenarios is described in

[1]The author admits that this is an ambitious goal. Usually, telecommunication networks are designed for HHI only, and speech technology devices have to cope with the resulting limitations.

Section 2.2, on the basis of a theory which has successfully been used for the definition of design guidelines for spoken dialogue systems. The guidelines encompass general principles of cooperative behavior in HMI, and form one aspect of interaction quality.

A more general picture of interaction quality and of the quality of services offered via SDSs is presented in Section 2.3. A new taxonomy is developed which allows quality aspects to be classified, and methods for their measurement to be defined. To the author's knowledge, this taxonomy is the first one capturing the majority of quality aspects which are relevant for task-orientated HMI over the phone. It can be helpful in three respects: (1) System elements (both of the transmission channel and of the spoken dialogue system) which are in the hands of developers, and responsible for specific user perceptions, can be identified; (2) the dimensions underlying the overall impression of the user can be described, together with adequate (subjective) measurement methods; and (3) prediction models can be developed to estimate quality – as it would be perceived by the user – from instrumentally or expert-derived interaction parameters. The taxonomy will be compared to definitions of quality on different levels (efficiency, usability, user satisfaction, acceptability) which can be found in the literature. Practical experiences with the taxonomy for analyzing and predicting quality are presented in Chapter 6.

An adequate definition of quality aspects is necessary in order to successfully build spoken dialogue systems and telecommunication networks. The specification, design and evaluation process is illustrated in Section 2.4, both for the transmission network and for the spoken dialogue system. It is shown that quality aspects should be taken into account already in the early phases of system specification and design in order to meet the requirements of the user, and consequently to build systems which are acceptable on the market. The congruence between system properties and user requirements can be measured by carrying out assessment and evaluation experiments, and an overview of the respective methods is given in Chapter 3. On the basis of experimental test data, it becomes possible to anticipate quality judgments of future users, and to take design decisions which help to optimize the usability, user satisfaction, and acceptability of the system. Chapters 4 and 5 show how quality prediction models can be used to estimate the transmission channel and environmental impact on speech recognition performance and synthesized speech, respectively, and Chapter 6 presents first steps towards structured quality prediction methods for the overall human-machine interaction.

2.1 Interaction Scenarios Involving Speech Transmission

In this book, quality for a specific class of human-machine-interaction will be addressed, namely the interaction of a human user with a spoken dialogue

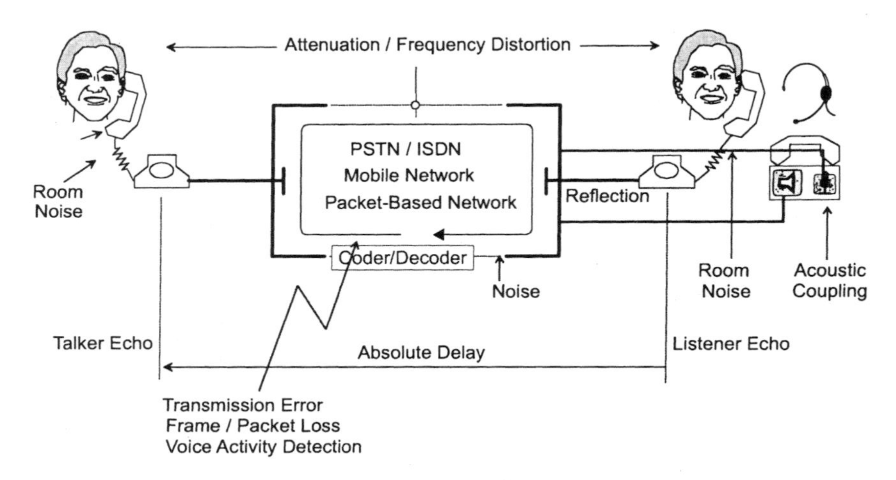

Figure 2.1. Human-to-human telephone conversation over an impaired transmission channel.

system via some type of speech transmission network, in order to carry out a specific task. Whereas the scenario is similar to normal human-to-human communication over the phone, it has to be emphasized that fundamental differences exist, resulting from both the machine agent and from the behavior of the human user (cf. Section 2.2). Nevertheless, the scenarios are similar in their physical set-up, and it has been stated that the quality of HHI over the phone will represent one reference for the quality of HMI with a spoken dialogue system.

The two scenarios are depicted in Figures 2.1 and 2.2. In both cases, the human user carries out a dialogic interaction via some type of telecommunication network. The network will introduce a number of transmission impairments which are roughly indicated in the pictures, and which will impact the quality of transmitted speech (when perceived by a human communication partner) as well as the performance of a speech recognizer (and subsequent speech and natural language technology components in the spoken dialogue system). On its way back to the human user, the transmission channel will also degrade the speech signal generated by the dialogue system. Because telecommunication networks will be confronted with both scenarios, it is important to consider the requirements of both the human user and the speech technology device. The requirements will obviously differ, because the perceptive features influencing the user's judgment on quality are not identical to the characteristics of a speech technology device, e.g. of an automatic speech recognizer (ASR).

The human user carries out the interaction via some type of user interface, e.g. a telephone handset, a hands-free terminal, or a computer headset. The acoustic

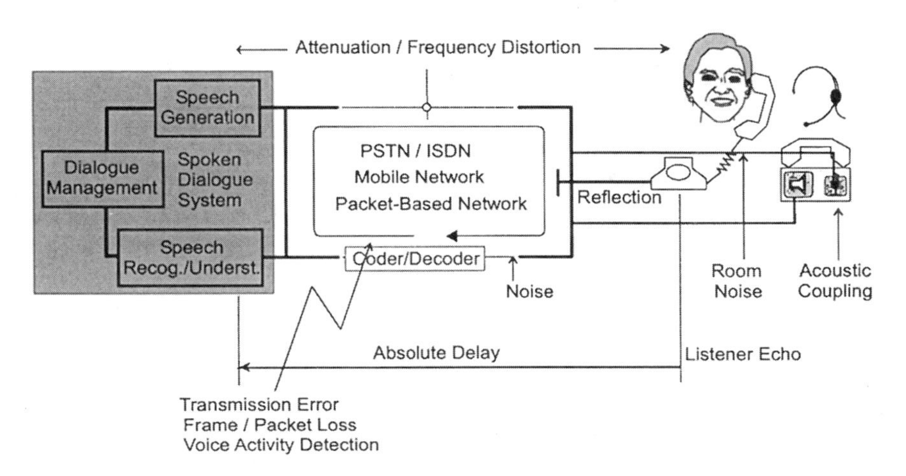

Figure 2.2. Interaction of a human user with a spoken dialogue system over an impaired transmission channel.

characteristics of the mentioned interfaces are very diverse, and so is their sensitivity to room acoustic phenomena occurring in the talking and listening environment of the user. For example, ambient noise may significantly impact the intelligibility of speech signals transmitted through a hands-free terminal, and it also carries an influence of the talking behavior of the user. As a result, such an environmental factor will have to be taken into account for the overall quality of the interaction, be it with a human user or with a machine agent. When the interaction partner is a machine, the acoustic characteristics on the machine side can be neglected, the interface being set up in a purely electric way via a 4-wire connection to the telecommunication network.

In the following sections, the characteristics of the transmission system – including the room acoustics at the user's side – and of the spoken dialogue system components will be discussed in more detail. The description is quite generic in character, as it refers to a large number of transmission networks (wirelines, wireless and IP-based) and of components which are used in nearly all types of spoken dialogue systems. It will therefore be valid for a large number of current state-of-the-art and future services which will be offered via telecommunication networks.

2.1.1 Speech Transmission Systems

Telephone speech quality, in the times of telephone networks administered and operated at the national level, was closely linked to a standard analogue or digital transmission channel of 300-3400 Hz bandwidth, terminated at both sides by conventionally shaped wirebound handsets. Most national and international connections featured these characteristics until the 1980s. Common impairments were transmission loss, linear distortions, continuous circuit noise, as well as quantizing noise associated with waveform PCM coding processes. These features were usually described in a simplified way in terms of a signal-to-noise ratio, SNR. Due to the low variability of the physical channel characteristics, users' expectations largely reflected their experiences with such connections over the years – a relatively stable reference for judging quality was achieved.

This situation completely changed with the advent of new coding and transmission technology, new terminal equipment, and with the establishment of mobile and IP-based networks on a large scale. Telephone speech quality is no longer necessarily linked to a specific transmission channel nor to a specific user interface. Rather, a specific transmission channel may be accessed through different types of user interfaces (e.g. handset phones, hands-free terminals, headsets), or one specific user interface serves as a gate to different transmission channels (wireline or mobile telephony, IP-based telephony). The services which are accessible to the human user now span from the standard human-to-human telephone service to a large variety of HMI services, e.g. for timetable information, stock exchange rates, or hotel reservation. Kamm et al. (1997b) state that an ever increasing percentage of the traffic in such modern networks is between humans and machines. The variety of transmission channels and user interfaces has severe consequences for the quality of transmitted speech, and consequently also for the quality of services which are accessed through the networks.

The underlying reason for this change is an integration of different types of networks. In the past, two types of networks have evolved mainly in parallel: On the one hand the connection-orientated, narrow-band telephone network, which is implemented in a mixed analogue (Public Switched Telephone Network, PSTN) and digital way (Integrated Services Digital Network, ISDN), and which has been augmented by cellular wireless telephone networks (e.g. the Global System for Mobile communication, GSM); and on the other hand a packet-based network which makes use of the Internet Protocol (IP), the internet.

Networks of the PSTN/ISDN type are connection-orientated, i.e. they allocate a specific transmission channel for the whole duration of a connection. Voice transmission is generally limited to a bandwidth of around 300-3400 Hz (lower frequencies with ISDN), which corresponds to a standard digital transmission bit-rate of 64 kbit/s in order to reach an SNR of roughly 40 dB (nearly independent of the signal level due to a non-linear quantization). This bit-rate may be reduced by making use of medium- to low-rate speech coders, or an extended wideband transmission channel (50-7000 Hz) may be offered in ISDNs. Signalling is performed through a parallel data channel. Mobile telephone networks mainly follow the same principle, but because of the limited bandwidth, speech coders operating at bit-rates around 13.6-6.8 kbit/s have to be used. Multi-path propagation and obstacles in the wireless transmission path severely impact the quality of the received signal and make channel coding and error protection or recovery techniques indispensable.

IP-based networks, on the other hand, are packet-switched and connectionless. Routing and switching is performed by data packets, using the standard transmission protocol TCP/IP (Transmission Control Protocol/Internet Protocol). The information which is to be transmitted is divided into packets consisting of a header (source and address of the packet) and a payload (voice, audio, video, data, etc.). Packet-based networks are designed to handle bursty transmission demands like data transfer, but they are not optimally designed for synchronous tasks like voice, audio or video transmission. Nevertheless, it is often desirable to install only one network which is able to handle a multitude of different transmission requirements in an integrated way. In such a case, the transmission of on-line speech signals over an IP-based network may be an economic alternative, and huge efforts have been invested into the respective technology and quality requirements in recent years, cf. the TIPHON project and the "Technical Committee Speech Processing, Transmission and Quality Aspects" (TC STQ) initiated by the European Telecommunications Standards Institute, ETSI.

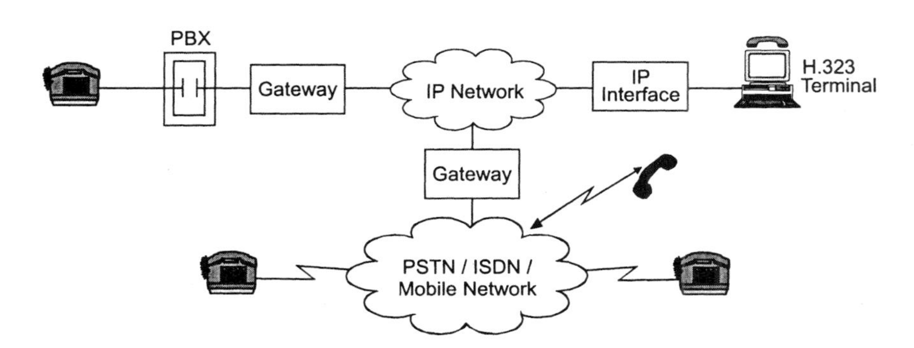

Figure 2.3. Interconnection of a mixed PSTN/ISDN/mobile network with an IP-based network, terminated with wireless or wireline telephones, an H.323 terminal, and a PBX, see ITU-T Rec. G.108 (1999).

Although in principle both types of networks allow the transmission of speech signals and data, PSTN/ISDN networks are mainly used for the transmission of time-critical information like speech signals in an on-line communication, whereas IP networks usually transmit non time-critical data information. For more than a decade, these two types of networks have tended to be integrated, forming one interconnected network where traffic can be routed through different sub-networks. The interconnection between connection-orientated and IP-based networks is generally performed by so-called gateways. Figure 2.3 gives an example of such an interconnected network, namely a mixed PSTN/ISDN connected to an IP-based network, and terminated with wireless or wireline telephones, an H.323 terminal, and a private branch exchange (PBX). In addition, mobile networks of the new generation (e.g. the Universal Mobile Telecommunications System, UMTS) base their normal voice service on IP transmission technology. Interconnected networks may also provide multimodal services, e.g. a spoken dialogue web interface or an audio-visual teleconference service.

From a physical point of view, speech transmission networks consist of terminal elements, connection elements, and transmission elements (ITU-T Rec. G.108, 1999):

- *Terminal elements*: All types of analogue or digital telephone sets, wired/cordless or mobile, including the acoustic interface to the user. They can be characterized by the frequency responses of the relevant transmission paths (send direction, receive direction, electrical coupling of the talker's voice in the telephone set), or in a simplified way using the so-called 'loudness ratings' (see the discussion in Section 2.4.1). For wireless terminals, additional degradations are caused by delay, codec and digital signal processing distortions, and time-variant behavior as a result of echo cancellers or voice activity detectors integrated in the terminals.

- *Connection elements*: All types of switching elements, e.g. analogue or digital private branch exchanges (PBX), mobile switching centers, or international switching centers. They may be implemented in an analogue or digital way. Analogue connection elements can be characterized by loss and noise, digital ones by the delay and quantizing distortion they introduce. In the case of 2-wire/4-wire interfaces (hybrids), signal reflections may occur which result in echoes due to a non-zero transmission delay.

- *Transmission elements*: Physical media including cables, fibres, or radio channels. The signal form may be analogue or digital. Analogue media are characterized by their propagation time, loss, frequency distortion, and noise; digital ones by their propagation time, codec delay, and signal distortions.

The list shows the sources of most of the degradations which can be observed in current speech transmission networks. A separation into three types of elements is however not always advantageous, because the boundary between user interface and transmission network is blurred. Modern networks make it necessary to take the whole transmission channel mouth-to-ear into account, in an integrative way. The degradations which occur on this channel are quantified with respect to their influence on the acoustic signal reaching the user or the spoken dialogue system, irrespective of the source they originate from. Such a point of view is taken in Section 2.4.1 where the individual degradations are discussed in more detail, and system parameters for a quantitative description are defined.

2.1.2 Room Acoustics

Speech communication via the telephone usually takes place from locations which are not shielded against ambient noise, concurrent talkers, or reverberation. Thus, in nearly all practically relevant cases the acoustic environment in which an SDS-based service is used has to be taken into account. Room acoustic influences are particularly important for services accessed through the mobile network, as the acoustic situation is usually worse than in locations with fixed installed telephone sets. An example for a critical situation is a hands-free terminal mounted in a moving car.

Ambient room noise is picked up by the microphone in all types of user interfaces simultaneously to the desired speech signal. However, user interfaces differ in their sensitivity for the mostly diffuse ambient noise compared to the directed speech sound. In the presence of a diffuse sound field, generated for example in a highly reverberant room, the transmission characteristic of the sending microphone towards ambient noise can be determined. The sensitivity towards a directed speech sound can be measured with the help of a head and torso simulator, as it is specified in the respective ITU-T Recommendations, e.g. in ITU-T Rec. P.310 (2003) for digital handset telephones, and in ITU-T Rec. P.340 (2000) for hands-free terminals. For handset telephones, the weighted average difference in sensitivity between direct and diffuse sound can be expressed by a one-dimensional scalar factor, the so-called D-factor of the handset under consideration.

The disturbing effect of ambient noise on the user is usually characterized by a frequency weighted (A-weighted) average sound pressure level which can be measured using a sound level meter. However, it has been shown that the specific spectral and temporal characteristics of the noise, and the meaning which is associated to it by the human listener, may also carry a significant influence on how loud and how annoying it is perceived (Bodden and Jekosch, 1996; Hellbrück et al., 2002). On the telephone connection, the A-weighted noise power level can be transformed into an equivalent level of circuit noise. This transformation is current practice for some network planning models, see Section 2.4.1.1.

Apart from the direct influence on the transmitted speech signal, ambient noise leads to a change in talking behavior. This 'Lombard reflex' (Lombard, 1911; Lane et al., 1961, 1970) affects the loudness of the produced speech signal, the speaking rate, and the articulation of the talker. Several authors have shown that the Lombard reflex significantly influences the performance of speech recognizers, and consequently has to be taken into account when evaluating spoken dialogue systems. In standard telephone handsets, a part of the speech produced by the talker is coupled back to its own ear, in order to compensate for the shielding effect of the handset, and to give a feedback on the proper work of the device. This so-called 'sidetone' path will also loop back a part of the ambient noise to the user's ear.

The room acoustic situation is particularly important when a service is accessed from a hands-free terminal. Such user interfaces are prone to coloration and reverberation resulting from early and late reflections in the talker's environment (Brüggen, 2001). Hands-free terminals are also very sensitive to the ambient noise, as the talker is usually located at an unknown distance and direction with respect to the microphone. Due to the physical set-up, microphone and loudspeaker are located very closely compared to the talker/listener, and are usually not decoupled from each other. Thus, level switching or echo cancelling devices have to be integrated in the user interface. These devices introduce time-variant degradations on the speech signal (front-end clipping, signal distortions, residual echo) and during the pauses, effects which are partly masked by inserting so-called 'comfort noise'. Whereas echo cancellers may help to prevent spoken dialogue systems from loosing their barge-in[2] capability, they nevertheless introduce degradations on the speech signal which impact the performance of speech recognizers.

2.1.3 Spoken Dialogue Systems

Spoken dialogue systems can be seen as an interface between a human user and an application system which uses speech as the interaction modality (Fraser, 1997). This interface must be able to process two types of information: The one coming from and going to the user through the speech-technology-based interface (voice user interface, VUI), and the one coming from and going to the application system through a specialized (e.g. SQL-based) interface. The connection between the user and the application system is an indirect one: The SDS must achieve a number of actions in order to be able to give a response, and the response will depend on the internal state of the system, or on the context of the interaction. This situation is the most common one found in practical

[2]Barge-in is defined as the ability for the human user to speak over the system prompt (Gibbon et al., 2000, p. 382). Two types of barge-in may be distinguished: One in which the user can interrupt the system without being understood, and one where the user can stop the system output and the speech is understood.

applications so far. Another situation exists, namely a system which supports – in one way or another – human-to-human communication. Examples for the latter are multilingual translation systems like the one set up in the German VerbMobil project. They will mainly be disregarded in the following chapters, although several considerations (e.g. the experiments described in Chapters 4 and 5) also refer to this type of SDS.

Seen from the outside, the task of the SDS is to enable and support the spoken interaction between the human user and the service offered by the application system. This task leads to a number of internal sub-tasks which have to be handled by the system: The coherence of the user input has to be verified, taking into account linguistic and task- or domain-related knowledge; communicative and task goals have to be negotiated with the user, and problems occurring during the interaction have to be resolved; references like anaphora or ellipses in the user's utterances have to be resolved; inferences which are reasonable in the communicative and task context have to be drawn, and the most probable user reaction has to be predicted; and appropriate and relevant responses to the user have to be generated.

Interactive dialogue systems have been defined as "computer systems with which humans interact on a turn-by-turn basis" (Fraser, 1997, p. 564). Depending on the complexity of the dialogic interaction, four types of systems can be differentiated:

- *Command systems*: They are characterized by a direct and deterministic interaction. To each stimulus from one agent corresponds a unique response from the other. The response is independent of the state or context of each agent. This type of interaction is normally not considered as a dialogue, and is called a "tool metaphor". Example: Pressing a key on the keyboard results in a character appearing on the screen.

- *Menu dialogue systems*: To this class belong simple question-answer user interfaces, where dialogue and task models are merged. The interaction is mainly system-directed, permitting only very little user initiative (e.g. barge-in). In contrast to command systems, several exchanges may be necessary in order to provoke one action of the application system. On the other hand, one user input can provoke different responses, depending on the internal state of the system, e.g. the current level in the menu structure. Example: So-called "Interactive Voice Response" (IVR) systems which enable an interaction via Dual Tone Multiple Frequency (DTMF) or keyword recognition.

- *Spoken dialogue systems (SDSs)*: This narrow class of systems disposes of distinct and independent models for task, user, system, and dialogue. Context information is taken into account using a particular knowledge base or dialogue history. Multiple types of references can be processed. An SDS may be capable of reasoning, of error or incoherence detection, internal correction, anticipation, and prediction. Examples of such systems are given in Section 2.1.3.7.

- *Multimodal dialogue systems including speech*: Systems of this class show the same characteristics as SDSs do, but in addition they are able to process and synchronize different modality information. Appropriate modalities for the individual interactions have to be selected, both at the system input and output side. Examples can also be found in Section 2.1.3.7.

The investigations in this book focus on the third class, in which spoken language is the primary means of interaction.

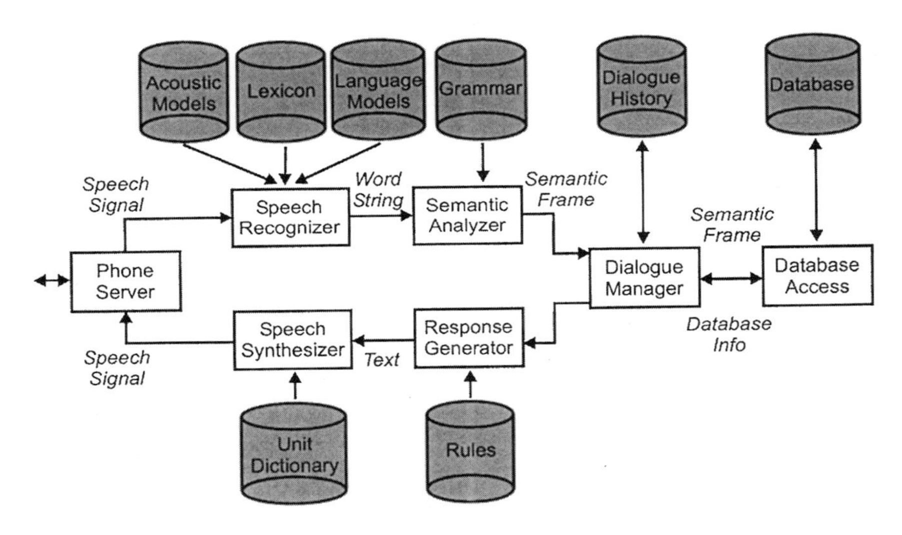

Figure 2.4. Pipelined structure of a telephone-based spoken dialogue system, similar to Lamel et al. (2000b).

The functionality of such a system, operated over the telephone network, can be best displayed in a sequential or pipelined structure. An example of such a structure is depicted in Figure 2.4. It consists of six major components which are accessed by the user via a phone server interface. The speech signal from the user is first processed to the speech recognizer. During the recognition process, it is transformed into a word string or a word hypothesis graph which is then semantically analyzed. The output is a semantic frame representing what has been "understood" from the user's utterance. It is the task of the dialogue manager to interpret the semantic frame in the context of the dialogue and the task, and to keep track of the dialogue history. When all relevant information has been collected from the user, a query to the underlying application (in this example a database) can be launched. The information originating from the application program, as well as other communicative goals of the dialogue manager, have to be transformed into a response for the user. This is the task of the response generation module. It generates a response in text form, which is then transformed by the speech synthesizer into a speech signal which is transmitted to the human user.

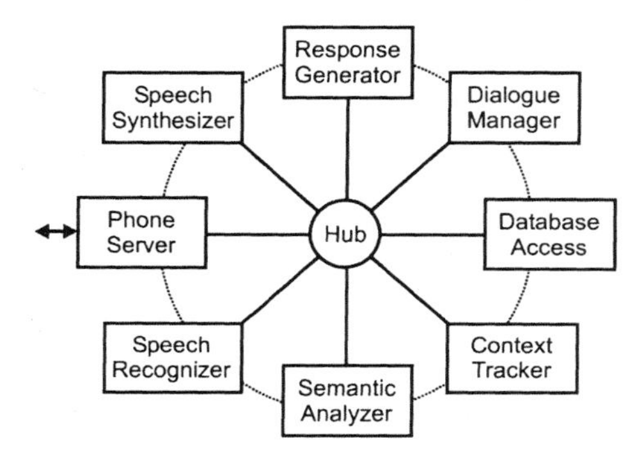

Figure 2.5. Hub structure of a telephone-based spoken dialogue system, similar to Seneff (1998) and Zue et al. (2000).

This principle structure may be implemented in different ways, see the overview given by Antoniol et al. (1998) as an example. One popular way is the so-called "hub architecture" (Seneff, 1998; Zue et al., 2000) which is used in the DARPA Communicator project, see Figure 2.5. It consists of a central programmable hub which communicates with the different modules (phone server, speech recognizer, semantic analyzer, context tracker, database access, dialogue manager, response generator, and speech synthesizer) in specific scripting languages, and thus controls the flow through each dialogue. This architecture facilitates resource-sharing between participating sites, and replacing single modules in case new versions become available. Polifroni and Seneff (2000) show how this hub architecture can also be used in the system evaluation phase, by connecting a batchmode server to automatically re-process previously stored interactions, and an evaluation server to accumulate performance statistics. Another possibility is to divide the components into a speech platform (containing the phone server, the speech recognizer and the speech synthesizer), a voice application server (similar to a web application server, and containing the semantic analysis, the dialogue management and the response generation), and the application back-end (e.g. the database). This structure is similar to the one used for web services, and the interface between the parts is usually implemented in an extensible markup language (XML) in order to be compatible with existing web services.

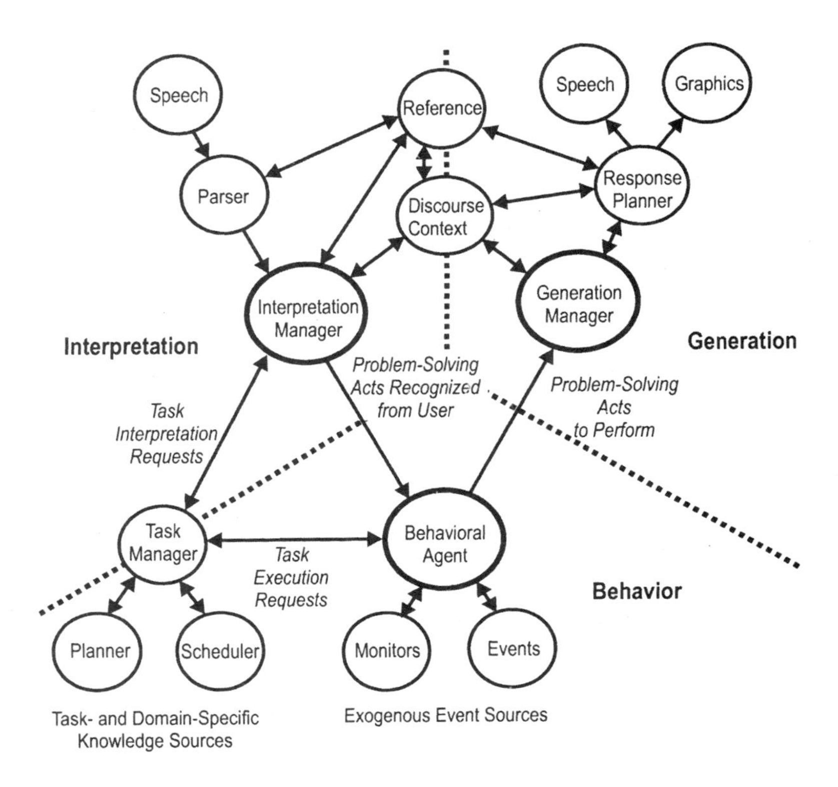

Figure 2.6. Asynchronous TRIPS architecture according to Allen et al. (2001).

In principle, the hub architecture also administers the flow of information in a pipelined way. In contrast to this, it is also possible to set up systems where the different modules operate truly asynchronously. An example is the TRIPS architecture which has been described by Allen et al. (2001) and Blaylock et al. (2002). It allows asynchronous working of interpretation, behavior (reasoning and acting) and generation parts, without disposing of a dialogue manager. This architecture is displayed in Figure 2.6.

The TRIPS system is able to take task initiative in a principled way, without necessarily being driven by the user input. Each component is implemented as a separate process. Information is shared by broadcasting messages between the modules, through a central hub which supports message logging, syntax checking and broadcasting. The interpretation manager interprets the user's input and interacts with the reference module to resolve referring expressions, and with the task manager to perform plan and intention recognition. It broadcasts recognized speech acts and their interpretation to the other modules. The behavioral agent plans system behavior based on its own goals and obligations, on the user's utterances and actions, and on changes in the (world's) context state. Actions that require task- and domain-dependent processing are performed by the task manager. Actions that involve communication and collaboration with the user are sent to the generation manager in the form of communicative acts. The generation manager then plans the specific contents of utterances and produces the speech output. Its behavior is driven by discourse obligations from the discourse context, and by the directives it receives from the behavioral agent.

In the following sections, a brief description of the most important modules which are commonly part of SDS architectures is given. In addition to the ones depicted in Figure 2.4, also a speaker recognition module is discussed, because it is a key module for all types of tasks which operate on person-critical information (e.g. home banking, personal data file access). Additional modules which may be necessary for specific systems are omitted because of their specific character, e.g. a language identification component in multi-lingual systems, or a text verification component for systems which process written text. More details on the common modules can be found in McTear (2002).

2.1.3.1 Automatic Speech Recognition (ASR)

The task of an automatic speech recognizer is to transcribe an acoustic signal into a list of words. It makes use of an acoustic model describing potential acoustic signals, a lexicon containing all potential words (vocabulary), and a language model (grammar) indicating the ways words may be ordered within an utterance. The aim of the recognition algorithm is to determine the word sequence with the highest likelihood for generating the observed signal, given the acoustic model, the lexicon, and the language model as constraints. The output of the recognizer is the most likely word sequence, sometimes augmented with a lattice of hypothetical word sequences and their corresponding probabilities, or a word graph as a compact representation of a large number of hypotheses (Souvignier et al., 2000). The task is complicated by the linguistic variability, the variability due to the speaker, and the one due to the transmission channel and to the talker's environment.

Typical recognizers used in SDSs are able to recognize continuously spoken utterances from different speakers (speaker-independent CSR). The first step is to extract spectral features characterizing the content of the speech signal. A feature vector typically contains the signal energy, 12-14 mel-scale cepstral coefficients, and their first or second order derivatives. Sometimes, more robust features like RASTA (Hermansky and Morgan, 1994), PLP (Hermansky, 1990), or J-RASTA (Koehler et al., 1994) are used in order to reduce the impact of additive and/or convolutional noise. The second step is the determination of (locally) optimally matching fundamental speech units, e.g. phonemes, which can be associated to the observed feature vectors. This task is performed with the help of the acoustic model. During the training process, prototypical features are computed and stored in the acoustic model. Most state-of-the-art recognizers use continuous-mixture-density or discrete-density Hidden Markov Models (HMMs) or neural networks (e.g. multi-layer perceptron networks) for the representation of the fundamental speech units (Bourlard and Wellekens, 1992). The third step is the decoding process, i.e. the determination of the (globally) best matching sequence of words, given the probabilities for the fundamental speech units. This step makes use of the lexicon and the language model. Both local and global matching require powerful search algorithms. An overview of current methods and algorithms can be found e.g. in Rabiner et al. (1996), Furui (2001a), Boite et al. (2000), or Juang (1991).

Language models are mainly statistical in nature, indicating the probability that a sequence of words has been observed in the training corpus (n-gram backoff language models). Together with the vocabulary they are estimated on the basis of a working dialogue system with a similar task, or via Wizard-of-Oz experiments (see Section 3.8). Such a corpus is of course limited, and several techniques have been used in order to obtain better models of the data which will be observed with the fully operational system. One method is to use grammatical classes for specific entities like cities, days or times, which are not easy to fully capture in the training corpus (Lamel et al., 1997). Such category grammars can be collected by parsing with a simple context-free grammar, and counting the occurrences of each category in the parsed sentences, see Meng et al. (1996). An alternative method is to artificially increase the amount of training material by inverting the parsing process and creating sentences from the grammars, see Souvignier et al. (2000) for an example. Because the language of the human user will depend on the dialogue context, language models which are specific for each dialogue state have shown to yield performance improvements, see Gerbino et al. (1995), Popovici and Baggia (1997), or Xu and Rudnicky (2000) for examples.

The incorporation of prosodic information can yield performance improvements in speech recognition and speech understanding, as it helps to disambiguate meanings. Speakers make use of prosody when starting a new discourse segment, and accentual focussing is commonly used in several languages to signal new information. A lack of a strategy to capture this information may be inefficient and lead to misunderstandings (Bruce et al., 1995). Prosody can also be used to automatically detect turns when the user first becomes aware that the system made an error (Litman et al., 2001), to automatically detect recognition errors (Hirschberg et al., 2000), or to enhance segmentation and parsing in speech understanding (Nöth et al., 2002). For crucial information, recurrence is sometimes made to a spelling module, because recognition accuracy for spelled words is usually higher than for spoken words (Meyer and Hild, 1997).

2.1.3.2 Speaker Recognition

When the speaker-related information instead of the linguistic information in the speech signal is of interest, a speaker recognition component can be used to compare the features of a user utterance to those stored in a speaker model database. Two cases have to be distinguished: Speaker verification decides between two alternatives (the identity of a known speaker is correct or wrong), and speaker identification performs an identification of one speaker out of all possible speakers in the database. In order to protect against pre-recorded speech, speaker verification can be performed text-dependent (the speaker has to utter a fixed text) or text-prompted (a random utterance is generated and the speaker has to read it immediately) instead of text-independent. In those cases, the speaker recognizer is combined with a speech recognizer in order to ensure that the prompted text is spoken. Speaker recognizers may be necessary to restrict access to critical information, e.g. for telephone banking, calling card services, specific directory assistance, voice dialling, or unified messaging. An overview of commonly used methods is given by Furui (1996, 2001a).

Speaker recognition involves a training step for modelling the speech of the speaker population, and a comparison step. The models for each speaker must contain the average parameter values and the according covariances, because speech as a biometric variable carries an inherent variability. The quality of a speaker model depends on the reliability of the estimates, as well as on the mismatch between training and use conditions. During the comparison step, a threshold decision is taken as to whether the observed and the stored model parameters are identical or not. The threshold has to be set as a compromise between the false acceptance and the false rejection cases, e.g. in terms of an Equal Error Rate (EER) or of Receiver Operating Curves (ROC), see Section 3.6. Often, however, the effects of these two errors are not identical. The thresholds then have to be adapted according to the expected effects of false acceptances or rejections for the specific task, and according to the prior probability of break-in attempts.

Boves and den Os (1998) and van Leeuwen (2003) consider the security impression which can be reached via speaker recognition. In general, the "security level" of a system is expressed as the amount of effort it takes to have a successful break-in attempt. Different human- and technology-originated weaknesses of speaker verification facilitate such attempts (van Leeuwen, 2003). Customers have to reach a feeling of security in order to trust in the service – otherwise the whole service might appear useless to them. However, security is not an aspect which could be easily added to existing services. Rather, a service has to be designed for security from the very beginning. In order to reach an acceptable level of security, different biometric and other measures are often combined. These measures can (and should) be integrated into the remaining dialogue. For example, when the identification confidence is low but the user attempts to access the service in a usual way (doing his/her usual transactions), this might be accepted nonetheless. Thus, a combination of speaker recognition with other constituents of a user model is desirable in most cases.

2.1.3.3 Language Understanding

On the basis of the word string produced by the speech recognizer, a language understanding module tries to extract the semantic information and to produce a representation of the meaning that can be used by the dialogue management module. This process usually consists of a syntactic analysis (to determine the constituent structure of the recognized word list), a semantic analysis (to determine the meanings of the constituents), and a contextual analysis.

The syntactical and semantical analysis is performed with the help of a grammar and involves a parser, i.e. a program that diagrams sentences of the language used, supplying a correct grammatical analysis, identifying its constituents, labelling them, identifying the part of speech of every word in the sentence, and usually offering additional information such as semantic classes or functional classes of each word or constituent (Black, 1997). The output of the parser is then used for instantiating the slots of a semantic frame which can be used by the dialogue manager. A subsequent contextual understanding consists in interpreting the utterance in the context of the current dialogue state, taking into account common sense and task domain knowledge. For example, if no month is specified in the user utterance indicating a date, then the current month is taken as the default. Expressions like "in the morning" have to be interpreted as well, e.g. to mean "between 6 and 12 o'clock".

Conversational speech, however, often escapes a complete syntactic and semantic analysis. Fortunately, the pragmatic context restricts the semantic content of the user utterances. As a consequence, in simple cases utterances can be understood without a deep semantic analysis, e.g. using keyword-spotting techniques. Other systems perform a caseframe analysis, without attempting to carry out a complete syntactic analysis (Lamel et al., 1997). In fact, it has been shown that a complete parsing strategy is often less successful in practical applications, because of the incomplete and interrupted nature of conversational speech (Goodine et al., 1992). In that case, robust partial parsing often provides better results (Baggia and Rullent, 1993). Another important method to improve understanding accuracy is to incorporate database constraints in the interpretation of the best sentence. This can be performed, for example, by re-scoring each semantic hypothesis with the a-priori distribution in a test database.

Because the output of a recognizer may include a number of ranked word sequence hypotheses, not all of which can be meaningfully analyzed, it is useful to provide some interaction between the speech recognition and the language understanding modules. For example, the output of the language understanding module may furnish an additional knowledge source to constrain the output of the recognizer. In this way, the recognition and understanding process can be optimized in an integrative way, making the most of the information contained in the user utterance.

2.1.3.4 Dialogue Management

An interaction with an SDS is usually called a dialogue, although it does not strictly follow the rules of communication between humans. In general, a dialogue consists of an opening formality, the main dialogue, and a closing formality. Dialogues may be structured in a hierarchy of sub-dialogues with a particular functional value: Sub-dialogues concerning the task are generally application-dependent (request, response, precision, explanation), sub-dialogues concerning the dialogue are application-independent (opening and closing formalities). Meta-communication sub-dialogues relate to the dialogue itself and how the information is handled, e.g. reformulation, confirmation, hold-on, and restart.

It is the task of the dialogue manager to guarantee the smooth course of the dialogue, so that it is coherent with the task, the domain, the history of the interaction, with general knowledge of the 'world' and of conversational competence, and with the user. A dialogue management component is always needed when the requirements set by the user to fulfill the task are spread over more than one input utterance. Core functions which have to be provided by the dialogue manager are

- the collection of all information from the user which is needed for the task,
- the distribution of dialogue initiative,
- the provision of feedback and verification of information understood by the system,
- the provision of help to the user,
- the correction of errors and misunderstandings,
- the interpretation of complex discourse phenomena like ellipses and anaphoric references, and
- the organization of information output to the user.

Apart from these core functions, a dialogue manager can also serve as a type of service controller which administers the flow of information between the different modules (ASR, language understanding, speech generation, and the application program).

These functions can be provided in different ways. According to Churcher et al. (1997a) three main approaches can be distinguished which are not mutually exclusive and may be combined:

- *Dialogue grammars*: This is a top-down approach, using a graph or a finite-state-machine, or a set of declarative grammar rules. Graphs consist of a series of linked nodes, each of which represents a system prompt, and of a limited choice of transition possibilities between the nodes. Transitions between the nodes are driven by the semantic interpretation of the user's answer, and by a context-free grammar which specifies what can be recognized in each node. Prompts can be of different nature: closed questions by the system, open questions, "audible quoting" indicating the choices for the user answers in a different voice (Basson et al., 1996), explanations, the required information, etc. The advantages of the dialogue grammar approach is that it leads to simple, restricted dialogues which are relatively robust and provide user guidance. It is suitable for well-structured tasks. Disadvantages include a lack of flexibility, and a very close relation or mixture of task and dialogue models. Dialogue grammars are not suitable for ill-structured tasks, and they are not appropriate for complex transactions. The lack of flexibility and the mainly system-driven dialogue structure can be compensated by frame-based approaches, where frames represent the needs of the application (e.g. the slots to be filled in) in a hierarchical way, cf. the discussion in McTear (2002). An example of a finite-state dialogue manager is depicted in Appendix C.

- *Plan-based approaches*: They try to model communicative goals, including potential sub-goals. These goals may be implemented by a set of plan operators which parse the dialogue structure for underlying goals. Plan-based approaches can handle indirect speech acts, but they are usually more complex than dialogue grammars. It is important that the plans of the human and the machine agent match; otherwise, the dialogue may head in the completely wrong direction. Mixtures of dialogue grammars and plan-based approaches have been proposed, e.g. the implementation of the "Conversational Games Theory" (Williams, 1996).

- *Collaborative approaches*: Instead of concentrating on the structure of the task (as in plan-based approaches), collaborative approaches try to capture the motivation behind a dialogue, and the dialogue mechanisms themselves. The dialogue manager tries to model both participants' beliefs of the conversation (accepted goals become shared beliefs), using combinations of techniques from agent theory, plan-based approaches, and dialogue grammars. Collaborative approaches try to capture the generic properties of the dialogue (opposed to plan-based approaches or dialogue grammars). However, because the dialogue is less restricted, the chances are higher that the human participant uses speech in an unanticipated way, and the approaches generally require more sophisticated natural language understanding and interpretation capabilities.

A similar (but partly different) categorization is given by McTear (2002), who defines the three categories finite-state-based systems, frame-based systems, and agent-based systems.

In order to provide the mentioned functionality, a dialogue manager makes use of a number of knowledge sources which are sometimes subsumed under the terms "dialogue model" and "task model" (McTear, 2002). They include

- *Dialogue history*: A record of propositions made and entities mentioned during the course of the interaction.

- *Task record*: A representation of the task information to be gathered in the dialogue.

- *World knowledge model*: A representation of general background information in the context the task takes place in, e.g. a calender, etc.

- *Domain model*: A specific representation of the domain, e.g. with respect to flights and fares.

- *Conversation model*: A generic model of conversational competence.

- *User model*: A representation of the user's preferences, goals, beliefs, intentions, etc.

Depending on the type of dialogue managing approach, the knowledge bases will be more or less explicit and separated from the dialogue structure. For example, in finite-state-based systems they may be represented in the dialogue states, while a frame-based system requires an explicit task model in order to determine which questions are to be asked. Agent-based systems generally require more refined models for the discourse structure, the dialogue goals, the beliefs, and the intentions.

A very popular method for separating the task from the dialogue strategy is a representation of the task in terms of slots (attributes) which have to be filled with values during the interaction. For example, a travel information may consist of a departure city, a destination city, a date and a time of departure, and an identifier for the means of transportation (train or flight number). Depending on the information given by the user and by the database, the slots are filled with values during the interaction, and erroneous values are corrected after a successful clarification dialogue. The slot-filling idea allows to efficiently separate the task described by the slots from the dialogue strategy, i.e. the order in which the slots are filled, the grounding of slot values, etc. In this way, parts of the dialogue may be re-used for new domains by simply specifying new slots together with their semantics. The drawback of this representation is a rather strict and simple underlying dialogue model (system question – user answer). In real-life situations, people tend to ask questions which refer to more than one slot, to give over-informative answers, or to introduce topics which they think would be relevant for the task but they weren't asked for (Veldhuijzen van Zanten, 1998).

A main characteristic of the conversation model is the distribution of initiative[3] between the system and the user. In principle, three types of initiative handling are possible: system-initiative where the system asks questions which have to be answered by the user, user-initiative where the user asks questions, or mixed-initiative offering both possibilities. It may appear obvious that users would prefer a more flexible interaction style, thus mixed-initiative dialogues. However, mixed-initiative dialogues are generally more complex, in that they require more knowledge on the part of the user about the system capabilities. The possibility to take the initiative leads to longer and more complex user queries which are more difficult to recognize and interpret. Consequently, more errors and correction dialogues might impact the user's overall impression of a mixed-initiative system. This observation has been made in the evaluation of the ELVIS E-mail reader system by Walker et al. (1998a), where the mixed-initiative system version – although being more efficient in terms of the number of user turns and the elapsed time to complete a task – was less preferred by the users against a system-initiative version. It was assumed that the additional flexibility caused confusion for the users about the possible options, and lead to lower recognition rates.

The choice of the right initiative strategy may depend on additional factors. Veldhuijzen van Zanten (1998) found that the distribution of initiative in the dialogue is closely related to the "granularity" of the information that the user is asked for, i.e. whether the questions are very specific or not. The right granularity depends on the predictability of the dialogue and on the prior knowledge of the user. When the user knows what to do, he/she can give all relevant information in one turn. This behavior, however, makes the dialogue less predictable, and decreases the chances for a correct speech recognition. In such cases, recurrence to lower-level questions can be made when high-level questions fail.

[3]There seems to be no clear definition of the term 'initiative' in the literature on dialogue analysis. Doran et al. (2001) use the term to mean that "control rests with the participant who is moving a conversation ahead at a given point, or selecting new topics for conversation."

Apart from the initiative, a second characteristic of the conversation model is the confirmation (verification) strategy. Common strategies are explicit confirmation where the user is explicitly asked whether the understood piece of information is correct or not (yes/no question), implicit confirmation where the understood piece of information is included in the next system question on a different topic, "echo" confirmation where the understood piece of information is repeated before asking the next question, or summarizing confirmation at the end of the information-gathering part of the dialogue. In general, explicit confirmation increases the number of turns, and thus the dialogue duration. However, implicit confirmation carries the risk that the user does not pay attention to the items being confirmed, and consequently does not necessarily correct the wrongly captured items (Sturm et al., 1999; Sanderman et al., 1998). Shin et al. (2002) observed that users discovering errors through implicit confirmation were less likely to succeed and took a longer time in doing so than through other forms of error discovery such as system rejections and re-prompts. Summarizing confirmation has the advantage that the dialogue flow is only minimally disturbed, but it is not very effective because of the limited cognitive capability of the user. It is particularly complicated when more than one slot contains an error. Confidence measures can fruitfully be used to determine an adequate confirmation strategy, making it dependent on the reliability of the recognized attribute.

The dialogue strategy does not necessarily have to be static, but can be adapted towards the needs of the current interaction situation, and towards the user in general. For example, a system may be more or less explicit in the information which is given to the user, as a function of the expected user expertise (user model), see e.g. Whittaker et al. (2003). In addition, a system can adapt its level of initiative in order to facilitate an effective interaction with users of different degree of expertise and experience, see Smith and Gordon (1997) for an investigation on their circuit-fix-it-shop system, or Litman and Pan (1999) for a comparison between an adaptive and a non-adaptive version of a train timetable information system. Relaño Gil et al. (1999) suggest that different control strategies should be available, depending on the characteristics of the user, and on the current ASR performance. Confidence measures of ASR performance can be used to determine the degree of system adaptation.

A prerequisite for an efficient adaptation is the user model. Modelling different typical user interactions can provide guidance for constraint relaxation, for efficient dialogue history management, for selecting adequate confirmation strategies, or for correcting recognition errors (Bennacef et al., 1996). In a slot-filling approach, the individual slots can be labelled with flags indicating whether the user knows which information is relevant for a slot, which values are accepted, and how these values can be expressed (Veldhuijzen van Zanten, 1999). Depending on the value of each label adequate system guidance can be provided. Whittaker et al. (2002) proposed to adapt the database access and the response generation depending on the user model. For example, the user's general preferences can be taken into account in searching for an adequate answer in the database, and the most frequently chosen information – which is potentially more relevant for this particular user – can then be presented first. Stent et al. (2002) showed that a user model for language generation can fruitfully be used to select appropriate information presentation strategies. General information about the set-up of user models is given in Wahlster and Kobsa (1989). Abe et al. (2000) propose to use two finite-state-automata, the first one for describing the system state, and the second one for describing the user state.

2.1.3.5 Communication with the Application System

In principle, an SDS provides an interface between the human user and the application system. For both spoken and written language processing, two application areas seem to be (and have been since the 1960s and 1970s in written language processing) of highest financial, operational, and commercial importance: Database interfaces and machine translation. As it has already been pointed out, the focus here will be on the HMI case, opposed to the human-machine-human interaction in spoken language translation. Instead of a database, the application system may also contain a knowledge base (for systems that support cooperative problem solving), or provide planning support (for systems that support reasoning about goals, plans and actions, and which are not limited to pre-defined plans, thus involving plan recognition). All application systems may provide transaction capabilities, as it is common practice in telephone banking, call routing, booking and reservation services, remote control of home appliances, etc.

Obtaining the desired information or action from the application system is not always a straightforward task, and sometimes complex actions or mediations have to be performed (McTear, 2002). For all application systems, it has to be ensured that the language used by the dialogue manager matches the one of the application program, and that the dialogue manager does not make false assumptions about the contents and the possibilities of the application program. The first point may be facilitated by inserting an additional "information manager" module which performs the mapping between the dialogue manager and the application system language (Whittaker and Attwater, 1996). The latter point may be particularly critical in cases that the application system functionality or the database is not static, but has to be extracted from other data sources. An example is a weather forecast service where the underlying information is extracted periodically from specific web sites, namely the MIT JUPITER system (Zue et al., 2000).

Another requirement for a successful communication with the application system is that the output it furnishes is unambiguous. In case of ambiguities either from the user or from the application system side, the dialogue manager may not be able to cope with the situation. Usually, interaction problems arise in such cases, e.g. because of ill-formed user queries (e.g. due to misconceptions about the application program), because of an ambiguous or indeterminate date (both from the user or form the application program), or because of missing or inappropriate constraint relaxation.

2.1.3.6 Speech Generation

This section addresses the two remaining modules of the structure depicted in Figure 2.4, namely the response generator and the speech synthesizer. They are described together, because the strict separation into a component which generates a textual version of the output for the user (response generation) and another one which generates an acoustic signal from the text (speech synthesizer) is not always appropriate. For example, pre-recorded messages (so called "canned speech") can be used in cases where the system messages are static, or the acoustic signal may be generated from concepts, using different types of information (textual, prosodic, etc.). In a stricter definition, one may speak of "speech output" as a module which produces signals that are intended to be functionally equivalent to speech produced by humans (van Bezooijen and van Heuven, 1997).

Response generation involves decisions about what information should be given to the user, how this information should be structured, and about the form of the message (words, syntax). It can be implemented e.g. as a formal grammar (Lamel et al., 1997) or in terms of simple templates. On a lower level, the response generator builds a template sentence at each dialogue act, filling gaps from the content of the current semantic frame, the dialogue history, and the result of the database query. Top-level generation rules may consist in restricting the number of information items to be included into one output utterance, or in structuring the output when the number of information items is too high. The dialogue history enables the system to provide responses which are consistent and coherent with the preceding dialogue, e.g. using anaphora or potentially pronouns. Response generation should also respect the user model, e.g. with respect to his/her expected domain knowledge and experience.

The speech output module translates the message constructed by the response generation into a spoken form. In limited-domain systems, a template-filling strategy is often used: template sentences are taken as a basis for the fixed parts of the sentences, and they are filled with synthesis from concatenation of shorter units (diphones, etc.), or with other pre-recorded expressions. However, when the system has to be flexible and provide previously unknown information (e.g. E-mail reading), a full Text-To-Speech (TTS) synthesis is necessary. TTS systems have to rely on the input text in order to reconstruct the prosody which reflects – amongst other things – the communicative intentions of the system utterance. This reconstruction is often paid with a loss of prosodic information, and therefore the integration of other information sources for generating prosody is desirable.

Full TTS synthesis consists of three steps. The first one is the symbolic processing of the input text: Orthographic text is converted into a string of phones, involving text segmentation, normalization, abbreviation and number resolution, a syntactical and a morphological analysis, and a grapheme-to-phoneme conversion. The second step is to generate intonation patterns for words and phrases, phone durations, as well as fundamental frequency (F_0) and intensity contours for the signal. The third and final step is the generation of an acoustic signal from the previously gained information, the synthesis in the proper sense of the word.

Speech synthesis can be performed using an underlying model of human speech production (parametric synthesis), namely with a source-filter model (formant synthesis) or with detailed models of articulatory movements (articulatory synthesis). An alternative is to concatenate pre-recorded speech units of different length, e.g. using a pitch-synchronous overlap-and-add algorithm, PSOLA (Moulines and Charpentier, 1990), or by selecting units of a large inventory. In recent years, the trend has been obviously in favor of unit-selection synthesis with longer units (sometimes phrases or sentences) which are available in a large unit database, and in several prosodic variants. The selection of units is then based on the prosodic structure as well. Other approaches make use of Hidden Markov Models or stochastic Markov graphs for selecting speech parameters (MFCCs, fundamental frequency, energy, derivations of these) describing the phonetic and prosodic contents of the speech to synthesize, see e.g. Masuko et al. (1996), Eichner et al. (2001), or Tamura et al. (2001). An overview of different speech synthesis approaches is given by Dutoit (1997) or van Santen et al. (1997).

Whereas synthesized speech is often still lacking in prosodic quality compared to naturally produced, pre-recorded speech provides high intelligibility and naturalness. This is particularly true when recordings a made with a professional speaker. The disadvantage is a severe limitation in flexibility. Recent unit-selection synthesis methods try to bridge the gap between pre-recorded and synthesized speech, in that they permit unrestricted vocabulary to be spoken, while using long segments of speech which are concatenated. The quality will in this case strongly depend on the coverage of the specific text material in the unit database, and perceptually new effects are introduced by concatenating units of unequal length.

The question arises which requirements are the most important ones when acoustic signals have to be generated in an SDS. Tatham and Morton (1995) try to formulate general and dialogue-specific requirements in this context. General requirements are that (1) the threshold of good intelligibility has to be passed, taking into account both the segmental and supra-segmental generation and the synthesizer itself; and (2) that a reasonable naturalness of the speech has to be reached, in the sense that the speech resembles (or can be confused) with the one from a human, that the voice has an appropriate "tone" for what is being said, that the "tone" changes according to the content of the conveyed message, and that the synthesized speaker seems to understand the message he/she is saying. The second statement may however be disputed, because a degraded naturalness may be an indication of the system's limited conversational capabilities, and thus lead to higher interaction performance due to changes in the user's behavior. Dialogue-specific requirements include that the "tone" of the voice should suite the dialogue type, that the synthesized speaker should appear confident, that the speaking rate is appropriate, and that the "tone" varies according to the message, and according to the changes in attitude with respect to the human user. Additional requirements may be defined by the application system and by the conversation situation. They may lead to speaker adaptation, and to the generation of speaking styles for specific situations (Köster, 2003; Kruschke, 2001). The respect of these requirements may lead to increased intelligibility, naturalness, and to an increased impact and credibility of the information conveyed by the system.

2.1.3.7 SDS Examples

In the following section, references are listed to descriptions of spoken dialogue systems which have been set up in (roughly) the last decade. Most of these systems are research prototypes. They are sorted according to their functionality, and a brief section of multimodal systems has been added. The list is not complete, but will give an impression about functionalities which have already been addressed, and will provide guidance for further reading. Overviews over the most important European and US projects and system have been compiled by Fraser and Dalsgaard (1996) and by Minker (2002).

Travel Information and Reservation Tasks:

- *General systems addressing several tasks*: SUNDIAL system providing multi-lingual access to computer-based information services over the phone. Languages: English, French, German and Italian. Domains: Intercity train timetables (German, Italian), flight enquiries and reservation (English, French), hotel database (Italian), see Peckham (1991) and Peckham and

Fraser (1994). DARPA Communicator system for travel-related services including flight, hotel and car arrangements, see e.g. Levin et al. (2000).

- *Systems for train timetable information*: VODIS (Voice Operated Database Inquiry System), see Peckham (1989) and Cookson (1988); Philips system, see Aust et al. (1995); RailTel and Dialogos system at CSELT, RailTel system at CNET, see Billi and Lamel (1997) and Billi et al. (1996); TOOT system at AT&T, see Litman et al. (1998); TRAINS system, see Sikorski and Allen (1997); ARISE system at CSELT and CNET, see Sanderman et al. (1998), Baggia et al. (1998), Lamel et al. (1998b), Lamel et al. (2000a), and Baggia et al. (2000); Spanish Basurde[lite] system, see Trias-Sanz and Mariño (2002).

- *Systems for flight information*: ATIS systems developed under the US DARPA/ARPA program (Price, 1990; Goodine et al., 1992), e.g. the PEGASUS system from MIT (Zue et al., 1994), the CMU system (Issar and Ward, 1993), or the BBN system (Bates et al., 1993); Danish Dialogue System, see e.g. Bernsen et al. (1998), Dalsgaard and Baekgaard (1994), or Baekgaard et al. (1995).

- *Systems for bus travel information*: Norwegian TABOR system, see Johnsen et al. (2000).

Phone Directory, Call-Routing, and Messaging Tasks:

- *Systems for phone directory, call routing, switchboard, and messaging*: Experimental phone directory system at FUB, see Delogu et al. (1993); Annie system at AT&T, see Kamm et al. (1997a); Operetta^{TM} system from Vocalis, see Fraser et al. (1996); VATEX system from KDD, see Naito et al. (1995); PADIS/PADIS-XL systems from Philips, see Kellner et al. (1997) and Seide and Kellner (1997); Telecom Italia directory assistance, see Billi et al. (1998); AT&T directory assistance, see Buntschuh et al. (1998); ADAS Plus automated directory assistance system from NORTEL, see Gupta et al. (1998); automatic call routing based on users responses to the prompt "How may I help you?", see Gorin et al. (1996, 1997); AT&T TTS help desk, see di Fabbrizio et al. (2002).

- *Systems for E-mail access over the phone*: ELVIS from AT&T, see Walker et al. (1998a); CSELT system developed as part of the SUNDIAL project, see Gerbino et al. (1993); E-MATTER system developed in the EU IST program, see Bel et al. (2002); Nokia EVOS system, see Oria and Koskinen (2002).

- *Systems for other telephone services*: Telephone service order, disconnect and billing inquiry systems, see Mazor and Zeigler (1995).

Other Information and Reservation Tasks:

- *Systems for workshop/conference services*: Prototype system from AT&T, see Rahim et al. (2000).
- *Systems for weather information*: JUPITER at MIT, see Polifroni et al. (1998).
- *Systems for tourist information*: PARIS-SITI, see Devillers and Bonneau-Maynard (1998); Czech system InfoCity, see Nouza and Holada (1998).
- *Systems for restaurant information*: Swiss MaRP and German BoRIS systems, see Möller and Bourlard (2002) and Chapter 6.
- *Systems for automobile classifieds*: WHEELS, see Meng et al. (1996).
- *Systems for cinema ticket reservation*: Experimental Austrian system, see Pirker et al. (1999).
- *Systems for home-banking*: OVID project for phone banking, see Jack and Lefèvre (1997); Nuance demonstrator system, see McTear (2002).
- *Systems for postal rate information*: Austrian system, see Erbach (2000).
- *Systems for general information retrieval over the internet*: Japanese system, see Fujisaki et al. (1997).

Problem-Solving and Decision-Taking Tasks:

- *Systems for cooperative problem-solving*: Experimental Circuit-Fix-It-Shop system, see Smith and Gordon (1997).
- *Systems for decision-taking*: ComPASS system for error diagnosis supporting CNC machine operators, see Marzi and John (2001).

Other Specialized Tasks:

- *Census systems*: Voice-response questionnaire for the US census, see Cole et al. (1994).
- *Translation systems*: VerbMobil for appointment scheduling situations, see Wahlster (2000) or Bub and Schwinn (1996); JANUS system, see Lavie et al. (1996) or Zhan et al. (1996).

Multimodal Systems:

- MASK kiosk for train inquiry, combining speech and tactile input and visual/speech output, see Lamel et al. (1998a, 2002).

- Swedish AUGUST system providing tourist information on Stockholm, using an animated agent communicating with the user via synthetic speech, facial expression, head movements, thought balloons, maps and tables (Gustafson et al., 1999).

- Dutch MATIS system for train timetable information, providing speech and pointing input and spoken and visual output, see Sturm et al. (2002b).

- SmartKom system for travel information, car and pedestrian navigation, and a home portal to information services, combining speech, gesture and mimic inputs and outputs, see Wahlster et al. (2001) or Portele et al. (2003).

2.2 Interaction with Spoken Dialogue Systems

It has been argued that the phone interaction between humans can be seen as one reference for the interaction of a human with an SDS over the phone. However, there are a number of differences between both types of interaction. They become obvious when the capabilities of the interlocutors in the interaction are compared.

Bernsen et al. (1998) identified the following capabilities of the human interaction partners in a task-orientated HHI:

- Recognition of spontaneous speech, including the ability to recognize words and intonational patterns, generalizing across differences in gender, age, dialect, ambient noise level, signal strength, etc.

- Very large vocabulary of words from widely different domains.

- Syntactic-semantic parsing capability of complex, prosodic, non-fully-sentential grammar of spoken language, including the characteristics of spontaneous speech input.

- Resolution capability of discourse phenomena such as anaphora and ellipses, and tracking of discourse structure including discourse focus and discourse history.

- Inferential capabilities ranging over knowledge of the domain, the world, social life, the shared situation, and the participants themselves.

- Planning and execution capability of domain tasks and meta-communication tasks.

- Dialogue turn-taking according to clues, semantics, plans, etc., the interlocutor reacting in real-time while the speaker is still speaking.

- Generation of language characterized by a complex semantic expressiveness and a style adapted to the situation, message, and to the interlocutor.

- Speech generation including phenomena such as stress and intonation.

These capabilities have to be compared to the ones of a machine agent observed in a task-orientated HMI, e.g. a phone-based interaction with an SDS (Niculescu, 2002):

- Limited recognition of continuous (partly spontaneous) task-related utterances, depending on the articulation characteristics of the speaker, and on the acoustic environment.
- Limited domain- and meta-communication-related vocabulary.
- Limited syntactic-semantic parsing capability; especially when confronted with spontaneous speech only partial parsing will be possible.
- Limited resolution capability of discourse phenomena and references. Limited discourse tracking capability via a dialogue history. Limited dialogue focus recognition capability.
- Planning and execution capability of domain tasks and meta-communication tasks. Capability to apply meta-communicative strategies (corrections, clarifications, repetitions, etc.) in case of misunderstandings.
- Dialogue turn-taking according to pre-defined rules, potentially with barge-in capability.
- Limited language generation capability according to rules.
- Unlimited vocabulary speech generation with limited intonational phenomena (stress, intonation).

It becomes obvious that the communicative capabilities of the interaction partners are not balanced. This imbalance will have an impact on the quality experienced by the human in the HMI.

In view of the limitations of the machine interaction partner, the question arises whether the term "conversation" makes sense in the context of HMI, and a debate was started about this point already more than a decade ago, see e.g. Luff et al. (1990). The question has some practical value, because in the case that HMI can be seen as a kind of conversation, then rules and descriptive models of conversation which have been derived by (human-to-human) conversation analysis might be useful for implementing spoken dialogue systems as well. Button (1990) argues that – although acknowledging the potential usefulness of the findings of conversational analysis for system development – such rules are often of a different quality than those required to implement a computer program. Simple rules can often only provide a rough indication about how communication works, and one cannot ignore the very details which are highly important for a successful conversation. Another key difference is that people are social agents, whereas computers are not. Citing Gilbert et al. (1990), "the meaning of an expression is relative to such contextual matters as who says it, to whom it is said, where and on what kind of occasion it is said, the social relations between speaker and hearer, and so forth". Thus, the correspondence between phenomena and descriptors ("indexicality") is complicated in a way which makes it very difficult (if not impossible) to be applied to set up computer programs. Nevertheless, it is clear that findings from computational analysis – although they cannot straightforwardly be implemented in computer programs – can fruitfully be used in the design and the evaluation of HMI. An example of this fact are the design guidelines for cooperative HMI defined by Bernsen et al. (1996) which will be discussed in Section 2.2.3.

It has been pointed out that only a specific class of HMI will be addressed in the following. This class can be characterized as follows (see also Bernsen et al., 1998):

- The interaction is task-orientated, and limited to certain application domains.
- It is mediated by a speech transmission network, and limited to the speech modality.
- The types of communication which can be carried out are the domain communication, a limited "social communication" (greetings, excuses, etc.), and meta-communication (communication about the interaction itself).
- The system offers a "service" to human users, e.g. to obtain information, or to perform a transaction. Note: Because of this fact, the *quality of service* is a right entity to characterize the interaction with an SDS, see definitions given in Section 2.3.
- The interaction requires a certain degree of cooperativity in order to be successful.
- The interaction has rarely any social function, at least for the computer.

In the following section, some of the consequences for the user behavior which result from the imbalance between both interaction partners will be illustrated. Then, the behavior of the machine interaction partner will be analyzed by a theory developed by Bernsen et al. (1998), see Section 2.2.2. This theory helps to identify the components of the machine agent which are responsible for its behavior. A key characteristic of the interaction is the notion of cooperativity. Design guidelines for cooperative system behavior will be described in Section 2.2.3, and they will form a basis for a more general definition of quality in Section 2.3.

2.2.1 Language and Dialogue Structure in HMI

The language and the dialogue structure of an interaction is influenced by a number of dimensions which characterize the interaction situation. Dahlbäck (1995, 1997), in his presentation of first steps towards a dialogue taxonomy, identified the following ones:

- Type of agent (human or computer): mainly carries an influence on the language used.
- Type of medium (e.g. spoken or written): influences the dialogue structure.
- Involvement of the interaction partners (monologue vs. dialogue).
- Spatial and temporal commonality (context).
- Task structure: dialogue-task distance (connection between task and dialogue structures, which is characterized by the need of understanding the underlying non-linguistic task, and by the availability of linguistic information required for doing so), and the number of different tasks.
- Kinds of shared knowledge between the dialogue participants: perceptual, linguistic, and cultural (also factual) knowledge.

Several investigations are reported in the literature which address the effect of one or several of these dimensions. In general, speech which is directed to a computer has been described as "formal" (Grosz, 1977), "telegraphic" (Guindon et al., 1987), "baby talk" (Guindon et al., 1986), and "computerese" (Reilly, 1987). Krause and Hitzenberger (1992) proved the existence of a language register which they called "computer talk". Kennedy et al. (1988) showed that the utterances in HMI are shorter, the lexical variation is smaller, and use of pronouns is minimized. Pirker et al. (1999) report that subjects abandoned politeness markers (e.g. "please") during the interaction with a very slow reservation system.

Such observations may result from the nature of the interaction partner (type of agent) which is more or less apparent to the user. Richards and Underwood (1984) found that the style and the content of users' utterances were significantly affected by the attributed nature of the system (human operator vs. computer), the computer being simulated by disguising a human voice and by instructing the subjects that they were speaking to a computer. In front of the "computer", subjects spoke more slowly, used a more restricted vocabulary, tended to use less potentially ambiguous pronouns, and asked questions in a more direct manner, see the discussion given by Fraser and Gilbert (1991b). This may be an indication that the human interaction partner takes the *assumed* linguistic (and perhaps task) knowledge of the machine agent into account when formulating his/her utterances.

In a different investigation (Fraser and Gilbert, 1991a), the same authors found that HHI utterances contained more words, more distinct forms, and more unfinished words than HMI utterances. In HMI, speakers produced fewer ellipses and fewer relative clauses than in HHI, and there was fewer overlapping speech. The authors attribute the observed differences to the influence of the system voice which was natural in the HHI case and synthesized in the HMI case. The system voice was also found to influence user behavior by Delogu et al. (1993). In her study, subjects were reported to repeat more often the same questions for synthesized prompts than for naturally produced system prompts. However, Sutton et al. (1995) reported that synthesized prompts did not lead to an increased number of adequate user responses for their automated spoken questionnaire. Thus, using synthesized speech does not necessarily influence the user's language in a way that it is more understandable to the system. This effect may however happen, as it was observed in the evaluation of the VODIS system (Cookson, 1988). Subjects "learned" to use simple-structured answers, often not more than one or two words, because this style of interaction was more successful in reaching the user's goals.

The mentioned observations are, however, not without contradictions. Amalberti et al. (1993) confirmed the cited results in that subjects talking to a computer tend to control and simplify their language, but made additional findings which are in contradiction to them: Subjects were observed to produce more utterances when talking to a computer, and no differences were observed with respect to the structural and pragmatic complexity of the utterances. The observed differences were ascribed to differences in representations of interlocutor ability (type of knowledge), which was implemented by a restricted behavior of the (simulated) computer. Analyzing typed dialogues, Dahlbäck (1995) reported nearly no differences between HHI and HMI. He supposes that the communication channel and the kind of task have a stronger influence on the dialogue than the perceived characteristics of the interlocutor. Dybkjær et al. (1993) reported that the number and linguistic diversity of speech produced by the subjects depended mainly on the subjects' professional background. Namely, secretaries produced less and less diverse tokens than linguists in the same situation. Such a person-specific factor may be dominant in describing the behavior of humans in the HMI.

Apart from the language used in the individual utterances, also the dialogue structure and the initiative seem to be different. Guindon (1988) showed that the dialogue structure was simpler in HMI dialogues. Although many system developers claim their systems to be mixed-initiative, Doran et al. (2001) found that their system massively dominated in taking the initiative. This is a difference to the interaction with a human expert where users and experts share the initiative relatively equitably. The fact may, however, not necessarily have an influence on user satisfaction. Users might prefer the situation of being asked by the system, because this provides better interaction guidance in an unknown situation. The system was generally more verbose than human experts (more words per turn), and used longer and more confirmations than the user did. In HHI, confirmations were observed to be shorter and more equally balanced between expert and user. The system tried to put more dialogue acts into a single turn than human experts did.

Turn-taking conventions are also different between HHI and HMI. Structured approaches exist for describing turn-taking, e.g. from Fox (1987). In her notion, turns are constructed of "turn-constructional units", TCUs (e.g. words or phrases), and each TCU is allocated to a specific speaker. Changes can – but need not – occur at the end points of TCUs, called "transition-relevance places", TRPs. In HMI, TRPs occur because either the system is silent, or because the user's response is completed. This is only a subset of the naturally occurring TRPs, and more complex turn-taking phenomena like double talk, overlap, and silences of specific length are currently not implemented in most SDSs.

2.2.2 Interactive Speech Theory

The described behavior of the human interaction partner is provoked by a number of elements of the machine agent. Bernsen et al. (1998) developed a theory which can be used to characterize the behavior of machine agents, e.g. when the performance of systems has to be compared. The theory is limited to the properties of current state-of-the-art SDSs, however, with the possibility to include novel interaction elements when they come up. It incorporates results from existing theories of HHI whenever they were believed to be useful and applicable to HMI, and captures the structure, contents and dynamics of the behavior of an SDS. The theory is bottom-up, with the later possibility to predict machine behavior, or at least to support the design of interaction models for HMI. It focusses on those elements which are directly in the hands of system developers, and gives indications on the influence they carry on system performance. In contrast to the interaction scenario depicted in Figure 2.2, it is limited to the "software" elements of the SDS (speech processing and dialogue implementation), and does not capture the "hardware" of the transmission channel and the physical user environment.

According to this theory, the elements of an SDS are organized in five layers which often correspond to the logical architecture of the system (the performance layer being replaced by the human user in that case). This structure is depicted in Figure 2.7. The lower four layers mainly reflect the quality elements of the SDS which can be optimized by the system designer, whereas the upper performance layer represents the features perceived by the human user

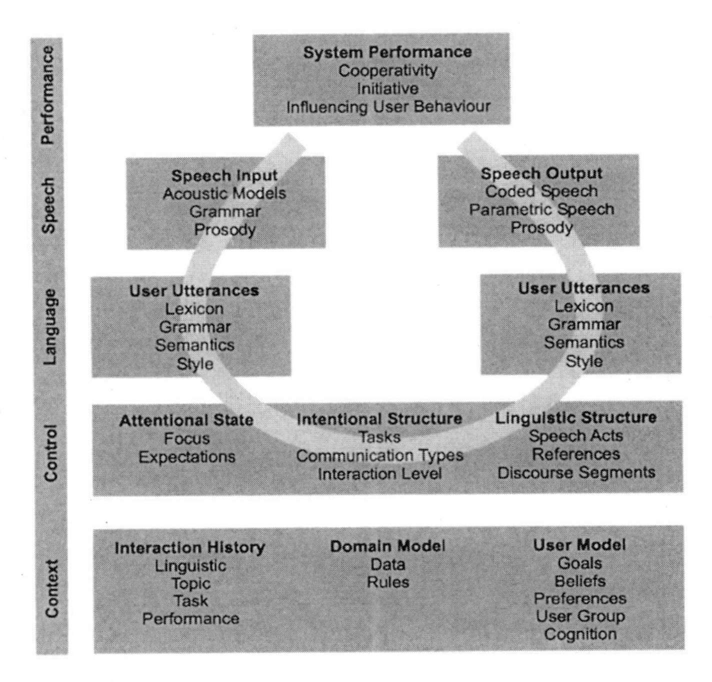

Figure 2.7. Elements of an interactive speech theory, taken from Bernsen et al. (1998). Element types are shown in bold type. The gray band and the gray boxes reflect the logical architecture of spoken dialogue systems.

in the interaction. A detailed description of each layer is given in Bernsen et al. (1998).

The lowest layer is the *context layer*. It contains all elements which are of crucial importance for language understanding and generation but which are not directly included in the lexicon and the grammar. Instead, the elements of this layer provide constraints on the lexicon and the grammar, e.g. for speech act interpretation, reference resolution, system focus and expectations, system reasoning, communication planning, and task execution. The layer contains the interaction history (selective record of information which has been exchanged during the interaction; relevant for the discourse and dynamically changing), the domain model (the aspects of the "world" about which the system is able to communicate), and the user model.

On top of the context layer, the *interaction control layer* determines which actions have to be taken at what point of the interaction. The decisions are taken on the basis of structures which have been determined at the development time of the SDS, but which are continuously updated at run-time. According to Grosz and Sidner (1986), three elements are important for the interaction control:

- The attentional state contains elements which concern what is going on at a certain point in time in the dialogue. It helps to constrain the search space and to resolve ellipses. It is determined by the set of topics which can be treated at a certain point in the dialogue.
- The intentional structure describes the purposes of the interaction. It subsumes elements which concern tasks and communication forms. For a task-orientated cooperative dialogue, intentions coincide with the task goals. Tasks can be structured into subtasks which may be interdependent, and which have to be solved in a certain time sequence. The intentional structure is not always stereotypical, e.g. for ill-structured tasks. The communication forms may be domain communication, meta-communication for repair and clarification, and other communication types like greetings, information about the system, etc. The interaction level describes the constraints on user communication at a certain stage of the dialogue. It may be adapted to the user's needs during the dialogue.
- The linguistic structure subsumes high-level structures in the input and output discourse. It includes speech acts (Searle, 1969), co-references, and discourse segments. Although there is no universally agreed-upon taxonomy of speech acts, they are thought to be important for speech understanding. Speech acts may be indirect, i.e. not disclosing what their actual intention is ("Do you have a match?"), or direct (apparently showing their intention), and indirect speech act identification causes problems for speech understanding. The resolution of co-references is another unsolved problem, and because of the lack of co-reference resolution, many SDSs perform robust partial parsing, or even keyword-spotting, instead of full parsing. Discourse segments are supra-sentential structures in the discourse which can be regarded as the linguistic counterparts of the task structure.

On top of the interaction control layer, the *language layer* describes the linguistic aspects of the spoken interaction. Spoken language is very different from written language (people do not follow rigid syntactic and morphological constraints in spoken dialogue), thus this adds some difficulty especially on the input side. Elements of the language layer are the lexicon (vocabulary) used by the system, the grammar describing how the words of the lexicon may be combined, the semantic representation of the words and phrases, and the language style, the latter being influenced by the grammar and lexicon. The user input style may be influenced through instruction and examples given by the system, or generally through the system's output style. The system's output style may be focussed or unfocussed (narrow or open questions), and feedback may be implicit or explicit, and immediately or summarizing, cf. the examples given above.

The *speech layer* describes the relationship between the acoustic speech signal on the one side, and a lexical string (e.g. enriched text) on the other. On the speech input side, speech recognition provides the mapping of the acoustic input signal to a repertoire of acoustic models, which are passed to the linguistic processing component in order to find the best matching lexical representation. On the output side, speech may be generated using pre-recorded utterances, carrier speech (templates), text-to-speech, or concept-to-speech. On both sides, the information stored in the system (acoustic model and grammar on the input side, unit inventory or rules on the output side) can be seen as a system element which may be optimized to reach high performance.

According to Bernsen et al. (1998), the final *performance layer* describes the observable behavior of the system during the interaction. It consists of the "elements" cooperativity, initiative, and influencing user behavior. Cooperativity has already been defined as a key requirement for the limited task-orientated HMI which is possible with current state-of-the-art speech technology. It will be discussed in more detail in the following section. Initiative depends on the speech acts performed by both interlocutors, and rules can be derived from speech acts for controlling initiative (Whittaker and Stenton, 1988). In a broad classification, initiative can be divided into system-initiative, user-initiative, and different levels of mixed-initiative. The behavior of the system also carries an influence on the behavior of the user. For example, the user behavior may be influenced by explicit systems instructions provided during the introduction or elsewhere in the interaction, via implicit system instructions (through system speech output), or via explicit developer instructions given to the users prior to the use of the system.

The classification of system elements helps to identify the sources of specific system behavior, and thus also the sources of quality features perceived by the user of a system. Two of the three elements in the performance layer are well reflected in the taxonomy of quality aspects which is developed in Section 2.3. Elements of the speech and the language layer can often be assessed directly or indirectly via questions to the user, or via parameters determined during the course of the interaction. Elements of the context and of the control layer are more difficult to identify in a specific interaction. Often, they become detectable in the case of interaction problems. A profound knowledge of the system architecture is then necessary to identify the exact source of the problem.

Apart from this theory, other models and theories for HMI exist. For example, Veldhuijzen van Zanten (1999) categorizes the elements of a dialogue manager into five layers: (1) intention (system and user goals); (2) attention (coherence of discourse); (3) guidance given to the user; (4) strategies for grounding information (verification, acknowledgement, etc., see Traum (1994)); and (5) utterances (speech act, word and speech level). Layer (1) contains some of the elements of the context layer in Bernsen's theory. Layers (2), (3) and (4) all comprise elements which are located on the interaction control layer. Layer (5) comprises both the elements of the language and the speech layer, plus a part of the linguistic structure element types. Layers (1) and (2) are discussed in more detail by Grosz and Sidner (1986). The theory was used to design adaptive dialogue management strategies, see Veldhuijzen van Zanten (1999).

2.2.3 Cooperativity Guidelines

Cooperativity has turned out to be a pre-requisite for a successful HMI, given the limited capacity of the current-state machine agents. Bernsen et al. (1998), p. 89, indicate: "A key to successful interaction design, we claim, is to ensure adequate cooperativity on the part of the system during interaction [...] This is a crucial interaction design goal in order to facilitate smooth interaction in domain communication, meta-communication and other types of communication".

Principles for cooperative behavior in HHI have already been defined by Grice (1975). In his definition, communication is cooperative action which requires that both parties have a minimal common purpose, or at least a mutually accepted direction. In order to act cooperatively in a conversation, people are expected to respect the Cooperativity Principle (CP): "Make your conversational contribution such as is required, at the stage at which it occurs, by the accepted purpose or direction of the talk exchange in which you are engaged" (Grice, 1975, p. 45). In situations where the meaning of an utterance (the implicature) is not identical with what has actually been said, interpretation methods (implications) are usually used by both participants which help to make it meaningful. Grice investigated situations in which the listener has to implicate what was meant, and set up four categories of underlying maxims which have to be assumed to be fulfilled by both participants in order to derive an implication:

- *Quantity* of information: Make your contribution as informative as required (for the current purposes of the exchange); do not make your contribution more informative than is required.
- *Quality*: Try to make your contribution one that is true; do not say what you believe to be false; do not say that for which you lack adequate evidence.
- *Relation*: Be relevant.
- *Manner*: Be perspicuous; avoid obscurity of expression; avoid ambiguity; be brief (avoid unnecessary prolixity); be orderly.

The maxims are not claimed to be jointly exhaustive. Other maxims may exist (e.g. aesthetic, social or moral in character) which are also normally observed by participants in talk exchanges, and these may also generate (non-conventional) implicatures. The conversational maxims are stated in a way as if the purpose were to have maximally effective exchanges. This idea is, however, too narrow, and the maxims have to be understood as to generally influencing or directing the actions or interpretations of others.

It is important to note that many dialogues are not strictly cooperative (Lee, 1999). For example, humans often answer in an indirect way to a question in order to convey conflicting information. Example: "Is there any direct train?" – "That will take much longer than the one with intermediate changes." Such indirect answers happen when a conversation partner wishes to achieve several communicative goals at once, be they conjunctive goals (i.e. an additional goal to the one being recognized by both agents) or avoidance goals (avoiding a certain state). Lee (1999) therefore differentiates between cooperative (shared beliefs and shared goals), collaborate (contradictory beliefs and shared goals) and conflicting (contradictory beliefs and goals) dialogues. Especially in HMI the assumption of mutual beliefs and shared goals might often not be satisfied, and the asymmetry between the interaction partners makes it very difficult to detect conjunctive or avoidance goals.

Although Grice's maxims have been developed in the observation of HHI, they have fruitfully been used for addressing the problem of cooperativity in HMI as well. A common assumption in both cases is that any particular conversation serves, to some extent, a common purpose or a set of purposes. The purpose may be more or less definite, and be either fixed beforehand or evolve during the conversation. In such conversations, interlocutors pursue the shared goals most efficiently – a goal which is congruent with most of the task-orientated interactions supported by current-state SDSs. The idea underlying the maxims is however different in both cases. They have been developed to analyze inferences which humans have when the interlocutor in a HHI deliberately violates one of the maxims. In a HMI, the non-deliberate violations are of interest. In the case that they can be avoided, the need for clarification and meta-communication dialogues, which are often difficult to handle, may be reduced. Thus, the respect of the maxims may help to prevent unwanted spoken interaction behavior, and may reduce communication errors and task failure.

On the basis of Grice's maxims, Bernsen et al. (1998) propose a set of guidelines which capture most of the interaction problems which have been observed in the interaction with a prototype SDS, namely the Danish system for flight information inquiry. The guidelines represent a first approximation to an operational definition of system cooperativity in task-orientated, shared-goal HMI. When a guideline is violated, it is likely that mis-communication occurs, which in turn may seriously damage the user's task performance.

The guidelines are grouped along seven interaction aspects, see Table 2.1. Four of them (informativeness, truth and evidence, relevance, manner) are identical to Grice's maxims. Three aspects have been added which are particularly

Table 2.1. Guidelines for cooperative system interaction, according to Bernsen et al. (1996) and Bernsen et al. (1998). GG: generic guideline; SG: specific guideline; *: guideline which is roughly equivalent to a Grice maxim.

Interaction aspect	*No.*	*Generic or specific guideline*
Aspect 1: Informativeness	GG1	* Make your contribution as informative as is required (for the current purposes of the exchange).
	SG1	Be fully explicit in communicating to users the commitments they have made.
	SG2	Provide feedback on each piece of information provided by the user.
	GG2	* Do not make your contribution more informative than is required.
Aspect 2: Truth and evidence	GG3	* Do not say what you believe to be false.
	GG4	* Do not say for which you lack adequate evidence.
Aspect 3: Relevance	GG5	* Be relevant, i.e. be appropriate to the immediate needs at each stage of the transactions.
Aspect 4: Manner	GG6	* Avoid obscurity of expression.
	GG7	* Avoid ambiguity.
	SG3	Provide same formulation of the same question (or address) to users everywhere in the system's interaction turns.
	GG8	* Be brief (avoid unnecessary prolixity).
	GG9	* Be orderly.
Aspect 5: Partner asymmetry	GG10	Inform the users of important non-normal characteristics which they should take into account in order to behave cooperatively in spoken interaction. Ensure the feasibility of what is required from them.
	SG4	Provide clear and comprehensible communication of what the system can and cannot do.
	SG5	Provide clear and sufficient instructions to users on how to interact with the system.
Aspect 6: Background knowledge	GG11	Take partner's relevant background knowledge into account.
	SG6	Take into account possible (and possibly erroneous) user inferences by analogy from related task domains.
	SG7	Separate whenever possible between the needs of novice and expert users (user-adaptive interaction).
	GG12	Take into account legitimate partner expectations as to your own background knowledge.
	SG8	Provide sufficient task domain knowledge and inference.
Aspect 7: Repair and clarification	GG13	Enable repair and clarification meta-communication in case of communication failure.
	SG9	Initiate repair meta-communication if system understanding has failed.
	SG10	Initiate clarification meta-communication in case of inconsistent user input.
	SG11	Initiate clarification meta-communication in case of ambiguous user input.

important in HMI with limited interlocutor capabilities (see above): Interaction partner asymmetry (because the machine is not a normal partner in the interaction, and users are partly aware of this fact and behave accordingly), background knowledge (which significantly differs between the two interlocutors), and the need for meta-communication, for repair, and for clarification (important because of the limited recognition, understanding and reasoning capabilities of the machine agent). The guidelines are further classified into Generic Guidelines (GG) which are important in both HHI and HMI, and Specific Guidelines (SG) tailored to specific aspects of the interaction with a spoken dialogue system.

The guidelines have first been developed on a Wizard-of-Oz corpus, then compared with Grice's maxims of cooperative HHI, tested on a user test corpus, and finally consolidated in the form given above (Bernsen et al., 1998). They reflect the experiences made with a specific flight inquiry system and helped to improve the system during the development phase. Independent from their development, Niculescu (2002) analyzes cases where the guidelines are violated in the case of a restaurant information system (see Section 6). Such an analysis, which has to be carried out manually, helps in identifying weaknesses of dialogue management implementations, and can provide solutions for better dialogue management design. Unfortunately, the schematic does not indicate any weighting of different dimensions. In case of conflicting guidelines, the system developer has to search for a compromise on its own.

The interaction aspects addressed by the guidelines can be usefully applied to system quality evaluation. For this purpose, Niculescu (2002) identifies 11 aspects of quality which are relevant for cooperative HMI, and which are classified in three levels. The levels and a part of the aspects have been renamed here in order to be congruent with the rest of the book, however without changing the underlying ideas. This results in a reduction to 9 aspects, the help capability and the dialogue structure aspect being subsumed under the transparency aspect:

- *Utterance level (question-answer level)*: Includes the aspects relevance and informativeness (amount and completeness of information), manner (intelligibility, comprehensibility, speech understanding capability), and meta-communication handling (repetition, confirmation of user input, confirmation of pauses).

- *Functional level (system capabilities)*: Includes transparency (ease of use, functional limits, help capabilities), congruence with the user's background knowledge and user expectations, initiative and interaction control, speed and smoothness (processing speed, dialogue interruptions).

- *Satisfaction level*: Includes perceived task success and the perception of the machine interaction partner (comparability with a human partner, trustworthiness, user's mood).

Niculescu (2002) uses the three-layered taxonomy to set up the questionnaire for the evaluation of the BoRIS restaurant information server, see experiment 6.2 described in Section 6.1.

2.3 Quality of Telephone Services Based on Spoken Dialogue Systems

In the past sections, two terms have been mainly used to describe how well a system or service fulfills the requirements of the user, or the ones of the designer or operator: *quality* and *performance*. The term *performance* describes the ability of a system to provide the function it has been designed for. Performance can easily be quantified when a measure[4] exists which is closely linked to the function under consideration. For example, the performance of a speech recognizer describes the degree to which it is able to transcribe spoken input into a sequence of words, and the quantifiable measure is e.g. the word accuracy or the word error rate. The term performance is, however, often used for units for which the function is not completely clearly defined, or for which no "natural" quantifiable measure exists. Network performance, for example, is defined as "the ability of a network or network portion to provide the functions related to communications between users" (ITU-T Rec. E.800, 1994, p. 4). However, no exact definition of the "functions related to communications" is given. Thus, the term performance has to be used with care, in particular when it describes the outcome of an assessment or evaluation experiment[5].

Spoken language systems are relatively complex systems which offer a number of different (and ill-defined) functions. The functions of the individual parts of the system (technical functions) have to be differentiated from the functions of the system as a whole (operational functions). Whereas the former can sometimes be defined in a general, application-independent way and allow performance measures to be computed, the latter always depend on the application a system has been designed for. Moore (1997, 1998) illustrates this dichotomy in a model which shows the relationship between technical system characteristics (called "features" in his terminology), technical system requirements and capabilities, operational requirements and capabilities, and the operational benefit. This model in depicted in Figure 2.8. Moore points out that a meaningful definition of the general suitability of a given technology for a particular application is dependent of a multi-factorial assessment along both technical and operational dimensions. The operational assessment has to investigate the re-

[4]The noun "measure" is the number or category assigned to an attribute of an entity by making a measurement, see ISO Standard ISO/IEC 9126-1 (2001).

[5]The International Telecommunication Union has now recognized the need for defining measures related to the performance of specific services, and proposals have been made for web hosting, e-mail or streaming media applications in ITU-T Delayed Contribution D.108 (2003).

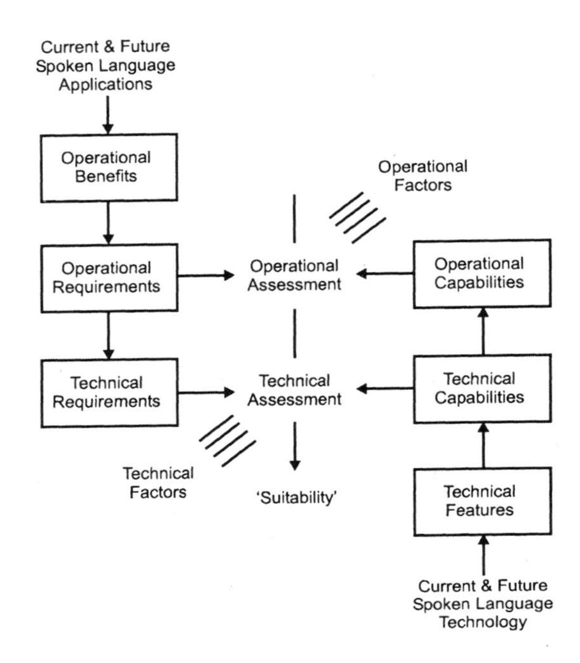

Figure 2.8. Model of the relationship between the applications of spoken language systems and the underlying technologies (Moore, 1997, 1998).

lationship (matching) between operational requirements and capabilities, and the technical assessment the relationship between technical requirements and capabilities. Operational benefits according to Moore include cost savings, manpower savings, increased operational effectiveness, increased productivity, workload reduction, increased security, increased safety, increased functionality, space savings, bandwidths savings, and improved quality of life. Thus, both the benefits for the system or service operator as well as the benefits for the user are addressed under this term. They are in a complex relationship with the operational requirements, and depend on operational factors such as accessibility, inter-operability, portability, mobility, reconfigurability, etc. Performance can be defined along each of these dimensions, and the assessment of a system as a whole should consequently take both aspects into account (Moore, 1998).

Whereas the functions of a speech recognizer are relatively well defined, the functions of a dialogue manager are numerous and mainly ill-defined. As a consequence, the performance of a dialogue manager can be measured with respect to different targets. For example, Danieli and Gerbino (1995) investigate the following operational functions of a dialogue manager for an information inquiry system:

- The ability to drive the user to find the required information (measures are e.g. task success or the average number of turns).
- The ability to generate appropriate answers to user queries (a measure is contextual appropriateness, see Section 3.8.5).
- The ability to maintain an acceptable level of interaction with the user, also when other modules have partial or total breakdowns (measures are e.g. the number of user and system correction turns, or the implicit recovery measure, see Section 3.8.5).

On the other hand, the performance of a dialogue manager with respect to technical functions can be expressed in terms of its coverage of natural language phenomena. Churcher et al. (1997b) list the following phenomena which a dialogue manager should be able to handle in order to be generic, and in order to provide a certain degree of "naturalness":

- Anaphora.
- Ellipses.
- Over-informative user utterances.
- Recovery from errors or misunderstandings.
- Language of different perplexity (related to the user's freedom of expression).
- Operational functions of different perplexity (density of topic changes which can be handled in a single dialogue, or the number of distinct tasks that can be performed).
- Type of interaction strategies (user-initiative, system-initiative, mixed-initiative).

It becomes obvious that an exact specification of technical or operational functions is a key requirement for making meaningful statements about the performance of a system.

Humans are the users of SDS-based services which are offered over the phone. Thus, human factors have mandatorily to be taken into account when the functions of a system/service and the degree of their fulfilment are determined. This necessity is not only important from a user's, but also from a system designer's point of view: System designers may specify functions which are not perceived, expected or desired by the user, and which are thus irrelevant in the application scenario. On the other hand, functions which are important for the user may not have been taken into account by the system designer, and lead to a degradation of quality as it is perceived by the user.

Because it is the user who perceives and judges, it is obvious that *quality* is not an entity which could be measured in an easy way, e.g. using a technical instrument. The quality of a service results from the perceptions of its user, in relation to what they expect or desire from the service. In the following, the definition of quality developed by Jekosch (2000) will be used:

> "Quality: Result of appraisal of perceived composition of a unit with respect to its desired composition."

The unit to be judged upon in this case is the service the user interacts with through the telephone network, and which is based on a spoken dialogue system. Its quality is a compromise between what he/she expects or desires, and the characteristics he/she perceives while using the service. The "perceived composition" contains the totality of features, i.e. recognizable and nameable characteristics of the service. These features are the result of a perception and a reflection process which happens in the perceiving subject (the user), and which is influenced by potential anticipation in the case that similar percepts are known or can be deduced (Jekosch, 2000). The processes are highly dependent on the situation in which the perception and judgment take place[6]. In order to reflect this fact, one could speak of a "quality event" (Jekosch, 2000), similar to the notion of an "auditory event" introduced by Blauert (1997).

The perceived composition is – of course – influenced by the physical composition, namely the set-up of the service and the system it relies on. Both are linked and separated by a perception process. At this point, it is useful to differentiate between *quality elements* and *quality features*, as it is further proposed by Jekosch. Whereas the former are system or service characteristics which are in the hands of the designer (and thus can be optimized to reach high quality), the latter are perceptive dimensions forming the overall picture in the mind of the user. Generally, no stable relationship which would be valid for all types of services, users and situations can be established between the two. However, it is possible to establish such relationships for particular systems in a particular usage scenario. Quality prediction models for HHI over the phone can be taken as an example, see Section 2.4.1.1. They establish a relationship between instrumentally measurable signals or parameters of the system on the one side, and estimations of user judgments on the other. Their scope is limited to narrow-band telephone networks with handset user interfaces, and they are useful for the design and set-up of such networks. However, it is generally not possible to obtain valid quality predictions for systems and scenarios which have not been taken into account at the time of model development.

[6]It should be noted that both the perception and the judgment process are influenced by non-auditory factors. This situation is similar to the assessment of product sound quality, as it has been described by Blauert and Jekosch (1996), Fig. 2.

For telephone services which are based on SDSs, the notion of quality elements is not as simple as for a telephone network alone. In such services, the user takes an active part in the interaction, and his characteristics and behavior may be decisive for the fulfillment of the operational and technical functions. The system and service characteristics will therefore be largely influenced by the user. In this case, quality elements can tentatively be defined as measures related to the system and the service which can be collected during the interaction. These measures will be called *interaction parameters* in the following chapters. Interaction parameters may, but need not be instrumentally measurable. Examples include speech levels or the duration of a dialogue which are instrumentally measurable, or a word error rate and a task success measure which can only be determined with the help of human experts. An overview of interaction parameters will be given in Section 3.8.5. Also for HMI over the phone, prediction models can be established transforming interaction parameters into estimations of user quality judgments. Such approaches are discussed in more detail in Section 6.3. They are usually designed for a specific service and domain, but a moderate degree of cross-domain validity can be reached under certain conditions.

In principle, the *Quality of Service (QoS)* can be addressed from two different points of view[7]. The first is the one of the service provider. He/she is mainly interested in the effects of individual elements of the service, and how they relate to the user's degree of satisfaction or acceptability. Service providers make use of a definition of QoS given in ITU-T Rec. E.800 (1994), p. 3:

> "Quality of Service (QoS): The collective effect of service performance which determine the degree of satisfaction of a user of the service."

More precisely, QoS comprises four factors, namely service support, service operability, serveability, and service security. Service support performance is the ability of the service provider to provide the service and assist in its utilization. Service operability performance is the ability of the service to be successfully and easily operated by a user. Serveability performance is the ability of the service to be obtained when requested by the user, and continue to be provided without excessive impairment for a requested duration. It includes accessibility (the ability of the service to be obtained), retainability (the probability that a service, once obtained, will continue to be provided under given conditions for a given time duration), and integrity (the degree to which a service is provided

[7]ITU-T Rec. G.1000 (2001) even defines four different points of view: The customer's requirements for QoS, the service provider's offering of QoS, the QoS achieved or delivered by the provider, and the QoS perceived by the customer. The following discussion will however be limited to the two main points of view, namely the one of the user and the one of the service provider, both allowing to set requirements and to define assessment and evaluation methods.

without excessive impairments) of the service. Service security performance is the protection provided against unauthorized monitoring, fraudulent use, malicious impairment, misuse, human error, and natural disaster.

The second point of view is the one of the user. He/she perceives and reflects on characteristics (features) of the service, compares the percepts with some type of internal reference, and judges them according to whether they fulfill his/her expectations or desires. Quality turns out as the result of the perception and judgment process. When investigating the quality of a service, it is important to take both points of view into account. Relationships have to be established between what the user expects or desires, his percepts, and the network and system characteristics which are responsible for evoking the percepts.

This double-sided concept of quality shows some similarities to a quality model which is standardized for software products, see ISO Standard ISO/IEC 9126-1 (2001). In that model, a separation is made between internal quality (the totality of characteristics from an internal point of view; comparable to a glass-box description of a system), external quality (totality of characteristics from an external point of view; comparable to a black-box description in a laboratory setting), and quality in use (user's view of the quality of the product when it is used in a specific environment and in a specific context of use). Achieving quality in use depends on achieving the necessary external quality, which depends on achieving the necessary internal quality. However, internal quality is not sufficient for external quality, which is in turn not sufficient for quality in use. For example, the level of quality in the user's environment may be different from that in the developer's environment, e.g. because of the differences between the needs and capabilities of different users, because of the differences in the environment, or because of different contextual factors. Thus, system developers normally require measures at all three levels.

Following the ISO definition, internal and external quality are defined by six system and service characteristics which are important for developers:

- Functionality: suitability, accuracy, interoperability, security.
- Reliability: maturity, fault tolerance, recoverability.
- Usability: understandability, learnability, operability, attractiveness.
- Efficiency: time behavior, resource utilization.
- Maintainability: analysability, changeability, stability, testability.
- Portability: adaptability, installability, co-existence, replaceability.

The quality in use, on the other hand, is defined by the four characteristics:

- Effectiveness.

- Productivity.
- Safety.
- Satisfaction.

Some of the more important items from this list are defined in the glossary, others can be found in ISO Standard ISO/IEC 9126-1 (2001).

It is obvious that the ISO definition is purely directed towards the goals of the system or service provider. In particular, it does not give any insight into the perceptive dimensions of quality which might be important for the user when interacting with a service. Nevertheless, four constituents of the internal/external quality and the quality of use are particularly related to the user: Effectiveness, efficiency, usability, and user satisfaction. A fifth term requires some explanation, namely naturalness. These items will be briefly addressed in the following paragraphs.

Both *effectiveness* and *efficiency* are related to the performance in achieving the task goal (operational functions) a service has been built for. For example, ETSI Technical Report ETR 095 (1993), p. 11, defines performance as the "degree of task achievement by utilizing a telecommunication service", and states that "performance can be described by effectiveness and efficiency". Effectiveness is an absolute index which describes how well the goal was reached, with respect to the accuracy and completeness of the goals, see e.g. ETSI Technical Report ETR 095 (1993), p. 13:

> "Effectiveness: The accuracy and completeness with which specified users can achieve specified goals in particular environments."

Measures of effectiveness which are reported in literature are e.g. task success or the kappa metrics. Efficiency, on the other hand, is a relative measure of goal achievement in relation to the resources used (ETSI Technical Report ETR 095, 1993, p. 13);

> "Efficiency: The resources expended in relation to the accuracy and completeness of goals achieved."

Commonly used metrics are e.g. the dialogue duration or the number of turns uttered by the system or the user.

Efficiency seems to be a measure which is directly related to the ability of reaching the task goals when resources are limited. It is however not sufficient in characterizing the interaction, as it was already discussed by Walker (1994):

> "A common assumption in work on collaborative problem solving is that interaction should be efficient. When language is the mode of interaction, the measure of efficiency has been, in the main, the number of utterances required to complete the dialogue [...]. One problem with this efficiency measure is that it ignores the cognitive effort required by resource limited agents in collaborative problem solving. Another problem is that

> an utterance-based efficiency measure shows no sensitivity to the required quality and robustness of the problem solution."

Walker states that human agents engage in problem-solving dialogue also in an "inefficient" manner. Namely, they make use of informally redundant utterances (IRUs) containing facts that are already mutually believed. On a corpus of 55 problem-solving HHI dialogues, Walker (1993) found 12% of such IRUs, and concluded that these utterances serve an important cognitive function. In a different experiment (Walker, 1994), she showed that strategies which are inefficient under assumptions of perfect reasoners with unlimited attention, are efficient with resource-limited agents. Different tasks require different cognitive demands, and thus place different requirements on the agent's collaborative behavior. Especially tasks which require a high level of belief coordination benefit from communicative strategies that include redundancy, as well as fault-intolerant tasks for which the redundancy rehearses the effects of critical actions.

Efficiency and cognitive demand are criteria characterizing a system with which a user is able to achieve his or her task goals. *Usability*, however, is generally defined in a much broader sense, and describes the capability of a system or service to be understood, learned and used by specified users under specified conditions. The following definition of usability, which is not identical to the ones given by ISO or ETSI, will be used here (Möller, 2000, p. 198):

> "Usability: Suitability of a system or service to fulfill the user's requirements. Includes effectiveness and efficiency of the system and results in user satisfaction."

In contrast to other usability definitions, user satisfaction or pleasantness are not prerequisites, but consequences of usability. *User satisfaction* is an indicator of the system's perceived usefulness and usability from the intended user group. It includes whether the user gets the information he/she wants, is comfortable with the system, and gets the information within an acceptable elapsed time (Maier et al., 1997). Dybkjær and Bernsen (2000) propose a mandatory list of 15 usability evaluation criteria for SDS evaluation:

- Modality appropriateness.
- Input recognition adequacy.
- Naturalness of user speech relative to the task(s), including coverage of user vocabulary and grammar.
- Output voice quality.
- Output phrasing adequacy.
- Feedback adequacy.
- Adequacy of dialogue initiative relative to the task(s).

- Naturalness of the dialogue structure relative to the task(s).
- Sufficiency of task and domain coverage.
- Sufficiency of the system's reasoning capabilities.
- Sufficiency of interaction guidance (information about system capabilities, limitations and operations).
- Error handling adequacy.
- Sufficiency of adaptation to user differences.
- Number of interaction problems.
- User satisfaction.

One term in this list needs further clarification, namely the notion of *naturalness*. An interaction with an SDS can be described as natural if it closely resembles the interaction between humans (Boyce and Gorin, 1996). By making an interaction natural, users can bring into the dialogue what they already know about language and conversation from everyday interaction with humans. A possible definition of naturalness can for example be based on the Turing Test, which is a long-debated and still applied methodology to determine to what extent it is possible to convince a user that an automated system is human-like. Following Turing's argumentation, if an interrogator cannot distinguish a system from a human interlocutor by questioning, then it would be unreasonable not to call the computer intelligent (Turing, 1950). A prize (Loebner prize) is annually awarded to the most human natural language processing (NLP) system in this respect. Currently awarded systems usually rely on written language input.

In view of the number of criteria which influence the user's perception of a system or service, the question arises whether it is possible to organize these criteria in a way which makes them distinguishable and relatable for quality analysis and synthesis. The following section describes a proposal for such a taxonomy which has recently been published (Möller, 2002a). It will be used throughout this book in order to distinguish quality aspects, to classify quality elements and features, and to develop quality prediction models.

2.3.1 QoS Taxonomy

In contrast to the definitions of QoS from a system developer's point of view, the following taxonomy puts the focus on the user. The overall picture is presented in Figure 2.9. It illustrates broad categories (white boxes) which can be sub-divided into aspects (gray boxes), and their relationships (arrows). As the user is the decision point for each quality aspect, user factors have to be seen in a distributed way over the whole picture. This fact has tentatively been illustrated by the gray cans on the upper side of the taxonomy, but will not be addressed further here. The remaining categories are discussed in the following.

Walker et al. (1997) identified three factors which carry an influence on the performance of SDSs, and which therefore are thought to contribute to its quality perceived by the user: Agent factors (mainly related to the dialogue and the system itself), task factors (related to how the SDS captures the task it has been developed for) and environmental factors (e.g. factors related to the acoustic environment and to the transmission channel). Because the taxonomy refers to the service as a whole, a fourth point is added here, namely contextual factors such as costs, type of access, or the availability. All four types of factors subsume quality elements which can be expected to carry an influence on the quality perceived by the user. The corresponding quality features are summarized into aspects and categories in the following lower part of the picture.

The agent factors carry an influence on three quality categories. On the speech level, input and output quality will have a major influence. Quality features for speech output have been largely investigated in the literature, and include intelligibility, naturalness, or listening-effort. They will depend on the whole system set-up, and on the situation and task the user is confronted with. Quality features related to the speech input from the user (and thus to the system's recognition and understanding capabilities) are far less obvious. They are, in addition, much more difficult to investigate, because the user only receives an indirect feedback on the system's capabilities, namely from the system reactions which are influenced by the dialogue as a whole. Both speech input and output may be largely impacted by environmental factors.

On the language and dialogue level, cooperativity has been identified as a key requirement for high-quality services. The classification of cooperativity into aspects, which was proposed by Bernsen et al. and which is related to Grice's maxims (see Section 2.2.3), is mainly adopted here, with one exception: The partner asymmetry aspect is captured under a separate category called dialogue symmetry, together with the aspects of initiative and interaction control. Dialogue cooperativity will thus cover the aspects informativeness, truth and evidence, relevance, manner, the user's background knowledge, and meta-communication handling strategies.

As it has been defined above, efficiency describes the effort and resources expended in relation to the accuracy and completeness with which users can reach specified goals. It is proposed to differentiate three types of efficiency. Communication efficiency relates to the efficiency of the dialogic interaction, and includes – besides the aspects of speed and conciseness – also the smoothness of the dialogue (which is sometimes called "dialogue quality"). Note that this is a significant difference to other notions of efficiency which only address

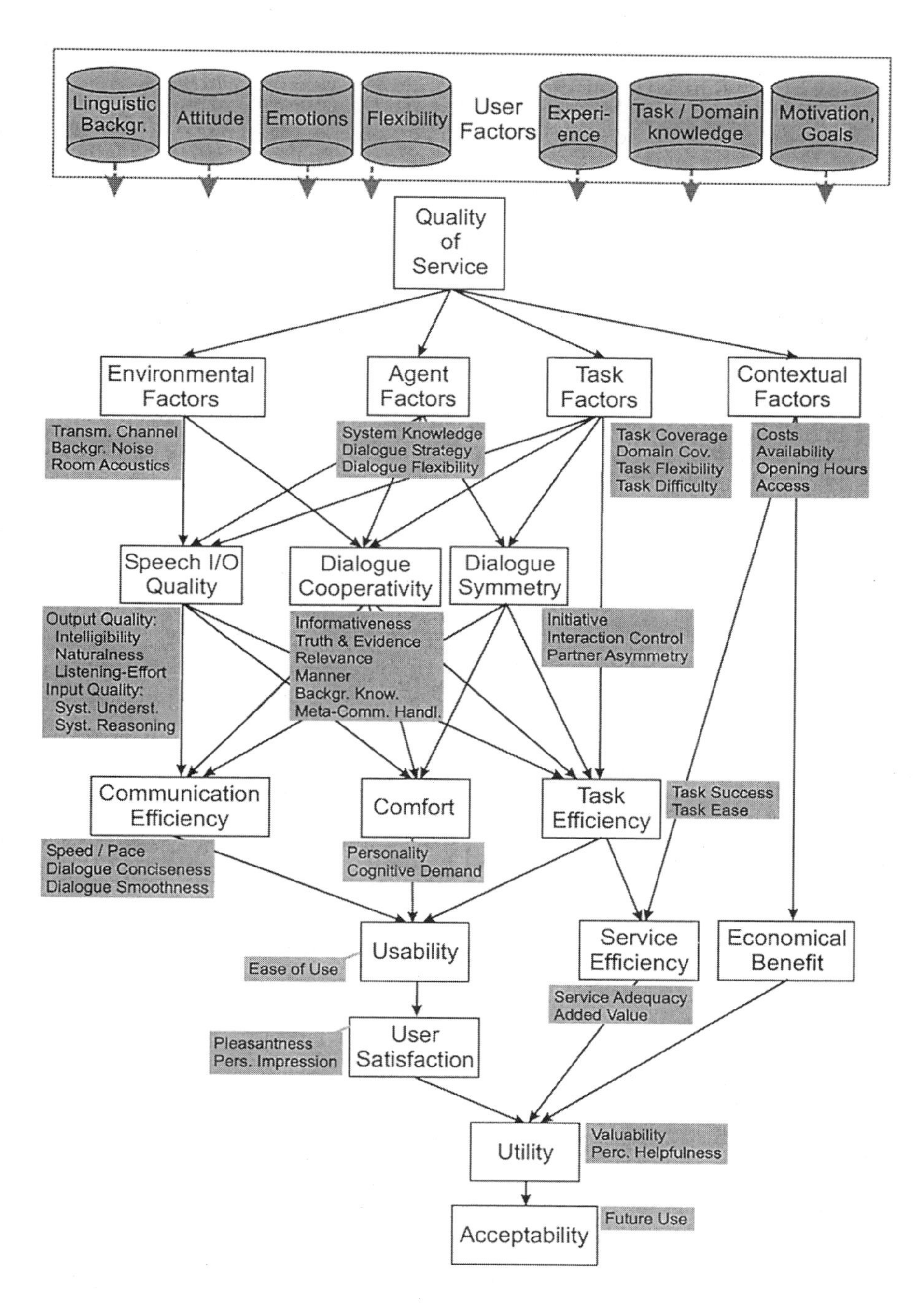

Figure 2.9. Taxonomy of QoS aspects for task-orientated HMI.

the efforts and resources, but not the accuracy and completeness (effectiveness) of the goals to be reached. Task efficiency is related to the success of the system in accomplishing the task; it covers task success as well as task ease[8]. Service efficiency is the adequacy of the service as a whole for the purpose defined by the user. It also includes the 'added value' which is contributed to the service, e.g. in comparison to other means of information (comparable interfaces or human operators).

In addition to efficiency aspects, other aspects exist which relate to the agent itself, as well as its perception by the user during the dialogue. These aspects will be subsumed in the following under the category 'comfort', although other terms might exist which better describe the according perceptions of the user. Comfort covers the agent's 'social personality' (perceived friendliness, politeness, etc.), as well as the cognitive demand required from the user.

Usability has been defined as the suitability of a system or service to fulfill the user's requirements. It considers mainly the ease of using the system and may result in user satisfaction. It does, however, not cover service efficiency or economical benefit, which carry an influence on the utility (usability in relation to the financial costs and to other contextual factors) of the service. Walker et al. also state that "user satisfaction ratings [...] have frequently been used in the literature as an external indicator of the usability of an agent" (Walker et al., 1998b, p. 320). As in Kamm and Walker (1997), it is assumed that user satisfaction is predictive of other system designer objectives, e.g. the willingness to use or pay for a service. Acceptability, which is commonly defined on this more or less 'economic' level, can therefore be seen in a relationship to usability and utility. It is a multidimensional property of a service, describing how readily a customer will use the service. The acceptability of a service (AoS) can be represented as the ratio of potential users to the quantity of the target user group, see the definitions on AoS adopted by EURESCOM Project P.807 Deliverable 1 (1998).

The taxonomy shows that a large number of aspects contribute to what can be called communication efficiency, usability, or user satisfaction. Several interrelations (and a certain degree of inevitable overlap) exist between the categories and aspects, and these interrelations are marked by arrows. They will become more apparent when taking a closer look to the underlying quality features which can be associated with each aspect. This association will be presented in the following section.

[8]Task efficiency has to be distinguished from communication efficiency, because aspects like speed and conciseness are not necessarily targeted by all types of dialogue systems. For example, dialogue systems which enable a domain conversation without resolving a pre-specified task (such as education, training or entertainment systems) may aim for long or otherwise "inefficient" interactions to resolve the underlying task, see e.g. Bernsen (2004) or Traum et al. (2004).

2.3.2 Quality Features

Quality features are recognized and designated characteristics of the service which are relevant to its quality (Jekosch, 2000). They can only be measured by asking users in realistic scenarios, in a subjective way. Several studies with this aim have been reported in the literature, and an overview is given in Section 3.8.6. The author analyzed twelve investigations and extracted quality features from the questions the users were asked (as far as they have been reported). These quality features were grouped according to the QoS taxonomy. Additional features are related to dialogue cooperativity and have been extracted from the definition of design guidelines given in Section 2.2.3. The features do not solely refer to telephone services, but will be valid for a broader class of systems and services based on SDSs.

In Tables 2.2 and 2.3, an overview is given of the quality features which result from this analysis. The list is not necessarily complete, in the sense that it will contain *all* quality features which are relevant for the success/non-success of a service. Nevertheless, the tables illustrate the relationship between quality aspects and user percepts. They can be used for two different purposes: On the one hand, they help to identify the aspects of quality which are addressed by a given user judgment on a specific question. On the other hand, they can be used to systematically set up evaluation schemes for individual quality aspects. Such an evaluation scheme will generally encompass both subjective user judgments and interaction parameters logged during the interaction. Both purposes will be addressed in later sections of this book.

2.3.3 Validation and Discussion

The new taxonomy is useful for classifying quality features as well as instrumentally or expert-derived measures (interaction parameters) which are related to service quality, usability, and acceptability. A validation against experimental (empirical) data on a large number of systems and services still has to be performed. First steps in this direction will be presented in Chapter 6, for a prototypical information service accessible over the phone. Nevertheless, there are a number of facts reported in literature which suggest that the taxonomy captures both the definitions of quality aspects given above as well as intuitive assumptions of individual authors.

First of all, in his review of both subjective evaluations as well as dialogue- or system-related measures, the author didn't encounter quality features or interaction parameters which would not be covered by the taxonomy. Example categorizations will be given in Sections 3.8.5 and 3.8.6. As stated above, the

Table 2.2. Dialogue-related quality features.

Category	*Aspect*	*Quality features*
Dialogue cooperativity	Informativeness	– Accuracy / specificity of information – Completeness of information – Clarity of information – Conciseness of information – System feedback adequacy
	Truth and evidence	– Credibility of information – Consistency of information – Reliability of information – Perceived system reasoning
	Relevance	– System feedback adequacy – Perceived system understanding – Perceived system reasoning – Naturalness of interaction
	Manner	– Clarity / non-ambiguity of expression – Consistency of expression – Conciseness of expression – Transparency of interaction – Order of interaction – Respect of natural information packages
	Background knowledge	– Congruence with user's task/domain knowledge – Congruence with user experience – Suitability of user adaptation – Inference adequacy – Interaction guidance
	Meta-comm. handling	– Repair handling adequacy – Clarification handling adequacy – Help capability – Repetition capability
Dialogue symmetry	Initiative	– Interaction guidance – Naturalness of interaction
	Interaction control	– Flexibility of interaction – Perceived control capability – Barge-in capability – Cancel capability
	Partner asymmetry	– Transparency of interaction – Transparency of task / domain coverage – Interaction guidance – Naturalness of interaction – Respect of natural information packages
Speech I/O quality	Speech output quality	– Intelligibility – Naturalness of speech – Listening-effort required from the user
	Speech input quality	– Perceived system understanding – Perceived system reasoning

Table 2.3. Communication-, task- and service-related quality features.

Category	*Aspect*	*Quality features*
Communication efficiency	Speed	– Perceived interaction pace – Perceived response time
	Conciseness	– Perceived interaction length – Perceived interaction duration
	Smoothness	– Perceived system errors – Repair handling adequacy – Clarification handling adequacy – Transparency of interaction – Congruence with user experience
Comfort	Agent personality	– Politeness – Friendliness – Naturalness of behavior
	Cognitive demand	– Ease of communication – Concentration required from the user – Stress / fluster
Task efficiency	Task success	– Adequacy of task / domain coverage – Validity of task results – Precision of task results – Reliability of task results
	Task ease	– Perceived helpfulness – Task guidance – Transparency of task / domain coverage
Service efficiency	Service adequacy	– Access adequacy – Availability – Modality adequacy – Task adequacy – Perceived service functionality – Perceived usefulness
	Added value	– Service improvement – Comparable interface
Usability	Ease of use	– Service operability – Service understandability – Service learnability
User satisfaction		– Pleasantness – Personal impression
Utility		– Valuability – Perceived helpfulness
Acceptability		– Future use

separation of environmental, agent and task factors was motivated by Walker et al. (1997). The same categories appear in the characterization of spoken dialogue systems given by Fraser (1997), plus an additional user factor, which obviously is nested in the quality aspects due to the fact that it is the user who decides on quality. The context factor is also recognized by Dybkjær and Bernsen (2000). Dialogue cooperativity is a category which is based on a relatively sophisticated theoretical, as well as empirical, background. It has proven useful especially in the system design and set-up phase, and first results in evaluation have also been reported (Bernsen et al., 1998; Niculescu, 2002). The dialogue symmetry category captures the remaining partner asymmetry aspect, and has been designed separately to additionally cover initiative and interaction control aspects. To the author's knowledge, no similar category has been reported yet[9]. The relationship between efficiency, usability and user satisfaction has already been discussed in Section 2.3.

Recognized quality prediction models for SDS-based services (see Section 6.3) make use of a similar definition of user satisfaction. In the PARADISE framework (Walker et al., 1997), user satisfaction is composed of maximal task success and minimal dialogue costs – thus a type of efficiency in the way it was defined here. This concept is still congruent with the proposed taxonomy. On the other hand, the separation into "efficiency measures" and "quality measures" used in the PARADISE model does not seem to be fine-graded enough. It is proposed that the taxonomy be used to classify different measures beforehand, according to the quality aspect they refer to. Based on the categories, a multi-level prediction model can be defined, first summarizing similar measures (belonging to the same category) into intermediate indices, and then combining the contributions of different indices into an estimation of user satisfaction. An example for such a hierarchical model will be presented in Section 6.3.3.

2.4 System Specification, Design and Evaluation

In order to design systems which deliver a high quality to their users, quality has to be a criterium in all phases of system specification, design, and evaluation. The scenario which is depicted in Figure 2.2 shows two main systems which will carry an influence on the quality which is experienced by the human user: the transmission network including its user interfaces, and the spoken dialogue system. If one of the systems does not show an adequate performance, the user will generally not be able to differentiate whether the overall quality is affected by the transmission channel or by the machine agent. Thus, an integrated view on both systems – i.e. an end-to-end consideration including the whole communication channel – is useful. It becomes even more important

[9]Recently, the importance of different types of symmetry in a domain conversation has been recognized by Bernsen et al. (2004).

because many transmission experts are not familiar with the requirements of speech technology, and many speech technology experts do not know which transmission impairments are to be expected for their systems in the near future.

Transmission network planning has been a topic of traditional telephony for a number of years. Its strategy has recently changed due to the diversified and liberalized market. An overview of the methods which are currently applied when a new network or network branch is installed will be given in Section 2.4.1. It will particularly address quality considerations which are made during the planning phase, and illustrate the major quality prediction approaches used by network operators. These quality prediction models will be used in Chapters 4 and 5 in order to estimate the transmission channel impact in the HMI scenario.

For the design of SDSs, planning strategies are less well developed. As an SDS is largely software-reliant, general software engineering life-cycle models can be used for its design. On the basis of experiences from the Danish SDS for flight ticket reservation, Bernsen et al. (1998) propose a general model, consisting of seven steps:

- *Survey*: Provides a reasonable and reliable estimate of the project feasibility.

- *Requirement specification*: Identifies strategic goals, system goals and constraints, and resource constraints.

- *Evaluation specification*: Establishes a set of criteria to be used in the final evaluation of the system, which refer to system goals and constraints.

- *Design specification*: Describes how to build a system which will satisfy the requirement specification and evaluation criteria, in a sufficiently detailed way to enable system implementation.

- *Analysis and design*: Set up the design specification in order to implement the system.

- *System set-up*: Either directly (implement-test-revise) or via simulation (simulate-evaluate-revise), see Section 2.4.3.

- *Acceptance tests*: For example controlled user tests, field tests, or final acceptance tests according to the pre-defined evaluation criteria.

All these phases will be briefly addressed in Sections 2.4.2 to 2.4.4. They should be preceded by an intensive analysis of user requirements in order to guarantee that the system does not miss the needs of its potential users.

Several successful design and implementation examples are described in the literature. They cover most of the steps to a varying degree, and usually involve a simulation of the final system. The most detailed description is given for the Danish dialogue system for flight ticket reservation by Bernsen et al.

(1998). It follows the design procedure indicated above in a principled way, and is a very valuable source of information for SDS developers. Another design example is the Spanish dialogue system for train information, see San-Segundo et al. (2001a,b). For this system, a 5-step procedure is used, including database analysis and design by intuition, design by observation of human-human interactions, simulation, iterative improvement, and evaluation.

2.4.1 Transmission Network Planning

The strategy for planning speech transmission networks has lived a fundamental change during the last decade. Historically, the main transmission impairments caused by analogue technology were loss and noise, and according transmission plans (for the PSTN) and loss plans (for PBXs) were agreed upon by the telecom administrations and by private companies. The interconnection between public and private networks was strictly defined. For each transmission parameter, permitted ranges and limits were given, based on the assumption that the parameter under consideration would cause the most important degradation in the specific connection. Examples of such definitions can be found in the respective ITU-T Recommendations, namely for loudness ratings which are described below (ITU-T Rec. G.111, 1993; ITU-T Rec. G.121, 1993), delay (ITU-T Rec. G.114, 2003), talker echo (ITU-T Rec. G.131, 1996), or listener echo (ITU-T Rec. G.126, 1993).

The situation completely changed with the de-regulation and liberalization of transmission networks, resulting in the set-up of new global private networks and the use of modern transmission technologies. Due to digital transmission and switching, parameters which have been important for analogue networks (frequency shape, circuit noise, crosstalk, loss and level variation over time, etc.) are less important, and other parameters (echo, delay, signal-processing-originated impairments, acoustic characteristics of the terminals) become dominant in such networks. Generally, several impairments are present simultaneously, and they interact with each other. Thus, the definition of strict constraints for individual parameters is no longer applicable, as it would lead to too pessimistic (worst-case) estimations, and finally prevent the use of modern technologies.

Instead, transmission planning in modern networks should be based on an end-to-end consideration of the whole transmission channel, from the mouth of the talker (or the speech generation output of the SDS) to the ear of the listener (or the ASR input). For narrow-band and handset-terminated networks, this approach is described in detail in ITU-T Rec. G.108 (1999). In principle, the approach involves the following steps:

- Determination of specific requirements and network features, e.g. type and structure of the network, routing, interconnection, and far-end termination.

- Determination of potential configurations, including the most critical configuration. This step generally requires a lot of experience and planning practice.

- Determination of transmission parameters on the basis of a reference configuration, for all private networks, public networks, tie trunks, and leased lines.

- Estimation of the expected end-to-end quality with the help of a quality prediction model: For HHI scenarios using the E-model, for ASR e.g. using a modified version described in Section 4.5.

- Analysis of the results and assignment of speech communication quality categories.

The main planning task is the collection of necessary information about the various network components in the configuration under consideration, and their contribution to the transmission impairments of the overall connection. This task is facilitated by a so-called 'reference connection' which is depicted in Figure 2.10. It collects all parameters which are relevant for the estimation of the overall quality of a prototypical connection, using a network planning model (e.g. the E-model, see Section 2.4.1.1). A list of the currently used parameters and of the default values and permitted ranges is given in Table 2.4.

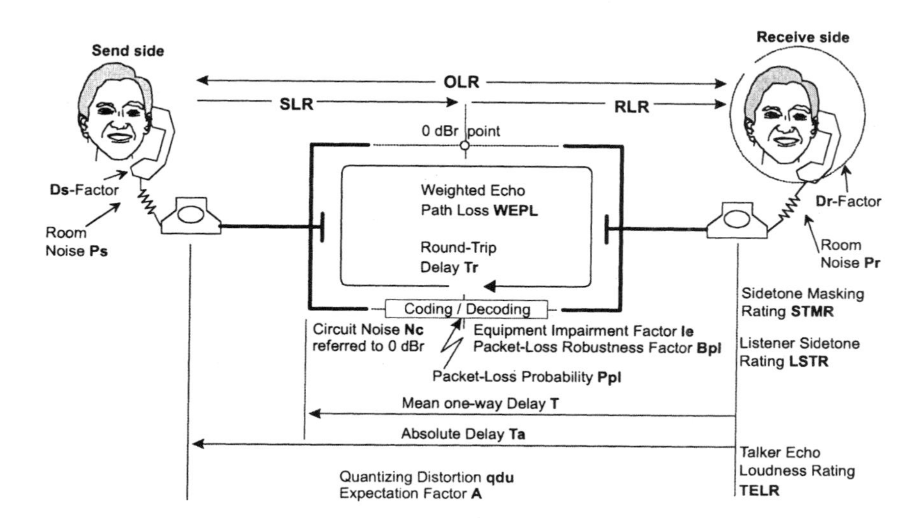

Figure 2.10. Reference connection for network planning purposes, see ITU-T Rec. G.107 (2003).

Table 2.4. Parameters of the reference connection, default values and permitted ranges, according to ITU-T Rec. G.107 (2003)[1].

Parameter	*Abbr.*	*Unit*	*Default value*	*Permitted range*
Send loudness rating	SLR	dB	8	0...18
Receive loudness rating	RLR	dB	2	-5...+14
Sidetone masking rating	$STMR$	dB	15	10...20
Listener sidetone rating	$LSTR$	dB	18	13...23
D-value of handset telephone, send side	Ds		3	-3...+3
D-value of handset telephone, receive side	Dr		3	-3...+3
Talker echo loudness rating	$TELR$	dB	65	5...65
Weighted echo path loss	$WEPL$	dB	110	5...110
Mean one-way delay of the echo path	T	ms	0	0...500
Round-trip delay in a four-wire loop	Tr	dB	0	0...1000
Pure delay of the connection	Ta	ms	0	0...500
Number of quantizing distortion units	qdu		1	1...14
Equipment impairment factor	Ie		0	0...40
Packet loss robustness factor	Bpl		1	1...40
Random packet loss probability	Ppl	%	0	0...20
Circuit noise referred to the 0 dBr-point	Nc	dBm0p	-70	-80...-40
Noise floor the receive side	$Nfor$	dBmp	-64	
Room noise at the send side	Ps	dB(A)	35	35...85
Room noise at the receive side	Pr	dB(A)	35	35...85
Expectation factor	A		0	0...20

[1] The units dBmp and dBm0p are commonly used in telephony. They describe the absolute power level (dBm) a signal has at a virtual 0 dB reference point in the network, behind the SLR' filter in Figure 4.1. The index p defines that the power level has to be weighted psophometrically, e.g. using a psophometer defined in ITU-T Rec. P.53 (1994).

The reference connection describes the whole transmission path, from the mouth of the talker (or from the speech synthesis system) to the ear of the listener (or to the speech recognizer), in a parametric way. Parameters of this description are one-dimensional planning values. They can be obtained by measuring frequency responses of the transmission paths, or noise power spectra, and by using defined frequency-weighting algorithms.

The characteristics of the transmission paths (main speech transmission path, echo path, sidetone path) are expressed in terms of so-called 'loudness ratings', which reflect the perceived loudness of the path, and the corresponding mean delays. For a definition of loudness ratings see ITU-T Rec. P.79 (1999). Noises are described by their psophometrically weighted power levels (for circuit noise and noise floor, see ITU-T Rec. P.53, 1994) or by standard A-weighted power levels (for ambient noise). Waveform codecs and effects of A/D-D/A conversions are expressed in terms of a signal-to-quantizing-noise ratio. The effects of non-linear codecs operating at medium to low bit-rates cannot easily be described by an instrumentally measurable parameter. For the purpose of network planning, they are covered by another scalar value, the so-called equipment impairment factor Ie. It describes the additional amount of degradation which is introduced by the coder-decoder pair in comparison to other impairments. The exact description of all the planning values can be found in ITU-T Rec. G.107 (2003) and in the corresponding ITU-T P-Series Recommendations.

2.4.1.1 Quality Prediction Models for HHI over the Phone

The aim of quality prediction models is to establish relationships between signals or parameters describing the network or system on the one hand, and perceptive quality features on the other hand. The solution may be considered successful when a high correlation between auditory test results and input signals or parameters is reached. However, a high correlation does not necessarily mean that valid quality predictions can be obtained for scenarios which have not been covered in the test databases used for model definition. It is important not to extrapolate beyond the verified range of validity. Otherwise, the predicted quality will be useless or may mislead the service provider.

According to a classification of prediction models proposed by Möller and Raake (2002), five criteria have to be taken into account:

- The application scope of the model (network planning, optimization, monitoring).
- The considered network components or configurations (mouth-to-ear, codecs only, etc.).
- The predicted quality features (integral quality, intelligibility, listening-effort, conversational aspects, etc.).
- The model's input parameters.
- The amount of psychoacoustically or perceptively motivated contents vs. empirically motivated knowledge.

These criteria will be briefly discussed in the following section.

Prediction models can serve different purposes, depending on the phase of planning, set-up, operation and optimization, which the network under consideration finds itself in. In the planning phase, it is possible to predict the quality that the user of the connection may encounter, provided that all or most of the planning parameters shown in Figure 2.10 are known beforehand. In such a case, comparative calculations of different set-ups can help to find weaknesses during network set-up, or to decide whether to apply an additional piece of equipment (e.g. a codec, echo canceller). When parts of the network equipment are available, it is also possible to perform measurements with artificial or typical speech signals, e.g. for equipment optimization. Signal-based measures (see below) are a good means for this purpose. On the other hand, it is also possible to measure specific signals, speech/noise levels, or other parameters (e.g. delay times) in operating networks. From these measurements, quality can be estimated for the specific network configuration, and potential weaknesses and network malfunctions can be identified.

Depending on which parts of the network are already available, different types of input parameters can be used for predicting quality. These parameters may refer to one or several network elements or to the whole transmission channel, mouth-to-ear. As a result, the predictions based on the parameters/signals will only be valid for this or a comparable type of equipment, with identical degradations.

Most models aim at predicting integral quality in terms of a mean opinion score (MOS, i.e. the mean judgment on a 5-point absolute category rating scale) or a transmission rating (R, see below). The predicted quality index may express user opinion (which has been collected in specific situations, e.g. in a laboratory listening-only test environment), or they can be associated with quality of service classes defined for network design, see ITU-T Rec. G.109 (1999). Quality indices will only be valid for the specified range of impairments which were part of the tests for which the prediction model has been optimized. For example, if time-variant degradations have not been included in the tests, the model cannot be expected to provide reasonable predictions for such impairments. Still, the associated quality features may become dominant in specific network configurations. In this case, the validity of the predicted results is questionable.

The mapping between input parameters and output indices can be performed by incorporating different degrees of psychoacoustic knowledge: Either a pure curve-fitting to the results of auditory tests is performed, or certain known characteristics of the human auditory perception are taken into account (e.g. masking, loudness perception, etc.) and are modelled. Usually, both ways are chosen, and the models contain some perceptively motivated parts and some purely empirically determined parts (strictly speaking, however, all psychoacoustically motivated knowledge is also empirically determined).

According to these classification criteria, three major types of models can be distinguished:

- Signal-based comparative measures.
- Network planning models.
- Monitoring models.

Signal-based comparative measures predict one-way voice transmission quality for single network components. They have mainly been developed for predicting the effects of waveform and non-waveform codecs on transmitted speech, but new versions also partly cover the effects of background noise and transmission errors. Such comparative measures are a powerful tool when new codecs have to be tested and compared to others whose performances are known, and when auditory tests are too time-consuming or expensive. For example, several versions of a newly developed codec can be evaluated comparatively, in order to find out the optimum parameter setting, and to gain an impression of the achievable quality level in comparison to other codecs on the market.

Quality predictions are performed on the basis of the input and output signals from a part of the transmission chain (Hauenstein, 1997; Vary et al., 1998). The two signals are first pre-processed to equalize them for loudness and delay differences, which – unless in extreme cases – do not result in an audible degradation. Sometimes, a correction for linear frequency distortions and a voice activity detection are included as well. The pre-processed signals are then transformed into an internal representation. This internal representation is commonly chosen according to perceptive considerations. Hauenstein (1997) discusses a couple of different psychoacoustically motivated loudness representations. Experience shows that it is not necessary to use very elaborate internal representations in order to achieve accurate quality predictions. Nevertheless, in comparison to the other model classes, the amount of psychoacoustically motivated contents is considerable.

Between the two internal representations, a perceptively weighted distance or similarity measure can be computed. Alternatively, the transmitted signal can be separated into the source signal component and a disturbance component. In this case, instead of measuring a distance or similarity, an annoyance measure is computed. The chosen measure is computed for every segment of the input and output speech signals. Then, a temporal mean value is calculated over all the segments, sometimes taking into account threshold effects or time-scale memory effects. This mean distance, similarity, or annoyance is finally transformed into an estimation of an integral quality judgment, e.g. the MOS value, using an empirically determined curve (e.g. an *arctan* function). Curve fitting is performed on the basis of training and adaptation data, i.e. speech samples recorded in real-live networks as well as laboratory-produced speech data, accompanied by MOS values from auditory listening-only tests. Thus, the models' quality predictions reflect the integral quality judgment in a listening-only test situation.

Several signal-based comparative measures have been proven to enable accurate quality predictions within the application range they have been developed for. The ITU-T currently recommends the PESQ model (ITU-T Rec. P.862, 2001; Rix et al., 2002; Beerends et al., 2002), which incorporates to a certain extent the effects of background noise and time-variant transmission impairments. Other model examples are the Telekom Objective Speech Quality Assessment (TOSQA) method which equilibrates some of the linear distortions of the transmission channel (Berger, 1998), the measures proposed by Hauenstein (1997), or by Hansen (1998) and Hansen and Kollmeier (2000). New developments are underway in order to include the effect of the user interface, making use of acoustic measurements of the input and the output signal, and in order to include conversational aspects (delay and echo). More details on new signal-based quality measures can be found in the given literature and in recent ITU-T Contributions to Study Group 12.

In contrast to signal-based measures, *network planning models* naturally refer to the whole transmission channel from the mouth of the talker to the ear of the listener. They also take conversational aspects (namely the effects of pure delay, sidetone and echo) into account. The models cover most of the traditional analogue and digital transmission equipment, handset telephones, low bit-rate codecs, and partly also the effects of transmission errors. Input parameters are scalar values or frequency responses which can be measured instrumentally off-line, or for which realistic planning values are available. For example, the model recommended by the ITU-T, the E-model (ITU-T Rec. G.107, 2003), uses the parameters depicted in Figure 2.10 as an input. When realistic planning values are available, it is possible to predict quality in the network planning phase, before the network has been fully set up.

Network planning models aim at predicting overall quality, including conversational aspects, and to a certain extent contextual factors, e.g. effects of user expectation towards different types of services, see Möller (2002b, 2000) and Möller and Riedel (1999)[10]. In view of the multitude of perceptually diverse quality dimensions, a reduction or integration of dimensions to a one-dimensional index has to be performed by the model. Thus, the output of the model cannot be seen as an exact description of the underlying quality dimensions, but rather as a rough estimation of the overall quality impact. In the planning process of networks and services, this information is often sufficient for the designers.

The model which is most widely used in telephony is the so-called E-model. It has been developed by a group of experts within the European Telecommunications Standards Institute (ETSI) and is now the only model recommended by both ITU-T and ETSI (ETSI Technical Report ETR 250, 1996; ITU-T Rec. G.107, 2003). Its origins are mainly based on work performed by Johannesson (1997), and on older network planning models developed by the former telephone monopolists (ITU-T Suppl. 3 to P-Series Rec., 1993). The model

[10]The effect of user expectation on auditory quality judgments has also been quantified for other judgment tasks, e.g. for sound quality assessment (Västfjäll, 2003).

has later been modified to better predict the effects of ambient noise, quantizing distortion, and time-variant impairments like lost frames or packets. The current model version is described in detail in ITU-T Rec. G.107 (2003).

The idea underlying the E-model is to transform the effects of individual impairments (e.g. those caused by noise, echo, delay, etc.) first to an intermediate 'transmission rating scale'. During this transformation, instrumentally measurable parameters of the transmission path are transformed into the respective amount of degradation they provoke, called 'impairment factors'. Three types of impairment factors, reflecting three types of degradations, are calculated:

- All types of degradations which occur simultaneously to the speech signal, e.g. a too loud connection, quantizing noise, or a non-optimum sidetone, are expressed by the simultaneous impairment factor Is.
- All degradations occurring delayed to the speech signals, e.g. the effects of pure delay (in a conversation) or of listener and talker echo, are expressed by the delayed impairment factor Id.
- All degradations resulting from low bit-rate codecs, partly also under transmission error conditions, are expressed by the effective equipment impairment factor $Ie{,}eff$. $Ie{,}eff$ takes the equipment impairment factors for the error-free case, Ie, into account.

These types of degradations do not necessarily reflect the quality dimensions which can be obtained in a multidimensional auditory scaling experiment. In fact, such dimensions have been identified as "intelligibility" or "overall clarity", "naturalness" or "fidelity", loudness, color of sound, or the distinction between background and signal distortions (McGee, 1964; McDermott, 1969; Bappert and Blauert, 1994). Instead, the impairment factors of the E-model have been chosen for practical reasons, to distinguish between parameters which can easily be measured and handled in the network planning process.

The different impairment factors are subtracted from the highest possible transmission rating level Ro which is determined by the overall signal-to-noise ratio of the connection. This ratio is calculated assuming a standard active speech level of -26 dB below the overload point of the digital system, cf. the definition of the active speech level in ITU-T Rec. P.56 (1993), and taking the SLR and RLR loudness ratings, the circuit noise Nc and $Nfor$, as well as the ambient room noise into account. An allowance for the transmission rating level is made to reflect the differences in user expectation towards networks differing from the standard wireline one (e.g. cordless or mobile phones), expressed by a so-called 'advantage of access' factor A. For a discussion of this factor see Möller (2000). In result, the overall transmission rating factor R of the connection can be calculated as

$$R = Ro - Is - Id - Ie{,}eff + A \tag{2.4.1}$$

Table 2.5. Relationship between the transmission rating factor R and categories of speech transmission quality according to ITU-T Rec. G.109 (1999).

Transmission rating range	*Speech transmission quality category*
$100 \leq R \leq 90$	best
$90 < R \leq 80$	high
$80 < R \leq 70$	medium
$70 < R \leq 60$	low
$60 < R \leq 50$	poor
$50 < R$	not recommended

This transmission rating factor is the principal output of the E-model. It reflects the overall quality level of the connection which is described by the input parameters discussed in the last section. For normal parameter settings $R \in [0;100]$. R can be transformed to an estimation of a mean user judgment on a 5-point ACR quality scale defined in ITU-T Rec. P.800 (1996), using the fixed S-shaped relationship

$$\mathrm{MOS} = \begin{cases} 1 & for \quad R < 0 \\ 1 + 0.035 \cdot R + R(R-60)(100-R) \cdot 7 \cdot 10^{-6} & for \quad 0 \leq R \leq 100 \\ 4.5 & for \quad R > 100 \end{cases} \tag{2.4.2}$$

Both the transmission rating factor R and the estimated mean opinion score MOS give an indication of the overall quality of the connection. They can be related to network planning quality classes defined in ITU-T Rec. G.109 (1999), see Table 2.5. For the network planner, not only the overall R value is important, but also the single contributions (Ro, Is, Id and $Ie{,}eff$), because they provide an indication on the sources of the quality degradations and potential reduction solutions (e.g. by introducing an echo canceller). Other formulae exist for relating R to the percentage of users rating a connection good or better ($\%GoB$) or poor or worse ($\%PoW$).

The exact formulae for calculating Ro, Is, Id, and $Ie{,}eff$ are given in ITU-T Rec. G.107 (2003). For Ie and A, fixed values are defined in ITU-T Appendix I to Rec. G.113 (2002) and ITU-T Rec. G.107 (2003). Another example of a network planning model is the SUBMOD model developed by British Telecom (ITU-T Suppl. 3 to P-Series Rec., 1993), which is based on ideas from Richards (1973).

If the network has already been set up, it is possible to obtain realistic measurements of major parts of the network equipment. The measurements can be performed either off-line (intrusively, when the equipment is put out of network operation), or on-line in operating networks (non-intrusive measurement). In operating networks, however, it might be difficult to access the user interfaces; therefore, standard values are taken for this part of the transmission chain. The measured input parameters or signals can be used as an input to the signal-based or network planning models (so-called *monitoring models*). In this way, it becomes possible to monitor quality for the specific network under consideration. Different models and model combinations can be envisaged, and details can be found in the literature (Möller and Raake, 2002; ITU-T Rec. P.562, 2004; Ludwig, 2003).

From the principles used by the models, the quality aspects which may be predicted become obvious. Current signal-based measures predict only one-way voice transmission quality for specific parts of the transmission channel that they have been optimized for. These predictions usually reach a high accuracy because adequate input parameters are available. In contrast to this, network planning models like the E-model base their predictions on simplified and perhaps imprecisely estimated planning values. In addition to one-way voice transmission quality, they cover conversational aspects and to a certain extent the effects caused by the service and its context of use. All models which have been described in this section address HHI over the phone. Investigations on how they may be used in HMI for predicting ASR performance are described in Chapter 4, and for synthesized speech in Chapter 5.

2.4.2 SDS Specification

The specification phase of an SDS may be of crucial importance for the success of a service. An appropriate specification will give an indication of the scale of the whole task, increases the modularity of a system, allows early problem spotting, and is particularly suited to check the functionality of the system to be set up. The specification should be initialized by a survey of user requirements: Who are the potential users, and where, why and how will they use the service?

Before starting with an exact specification of a service and the underlying system, the target functionality has to be clarified. Several authors point out that system functionality may be a very critical issue for the success of a service. For example, Lamel et al. (1998b) reported that the prototype users of their French ARISE system for train information did not differentiate between the service functionality (operative functions) and the system responses which may be critically determined by the technical functions. In the case that the system informs the user about its limitations, the system response may be appropriate under the given constraints, but completely dissatisfying for the user. Thus, systems which are well-designed from a technological and from an interaction point of view may be unusable because of a restricted functionality.

In order to design systems and services which are usable, human factor issues should be taken into account early in the specification phase (Dybkjær and Bernsen, 2000). The specification should cover all aspects which potentially influence the system usability, including its ease of use, its capability to perform a natural, flexible and robust dialogue with the user, a sufficient task domain coverage, and contextual factors in the deployment of the SDS (e.g. service improvement or economical benefit). The following information needs to be specified:

- Application domain and task. Although developers are seeking application-independent systems, there are a number of principle design decisions which are dependent on the specific application under consideration. Within a domain, different tasks may require completely differing solutions, e.g. an information task may be insensible to security requirements whereas the corresponding reservation may require the communication of a credit card number and thus may be inappropriate for the speech modality. The application will also determine the linguistic aspects of the interaction (vocabulary, syntax, etc.).

- User and task requirements. They may be determined from recordings of human services if the corresponding situation exists, or via interviews in case of new tasks which have no prior history in HHI.

- Intended user group.

- Contextual factors. They may be amongst the most important factors influencing user's satisfaction with SDSs, and include service improvement (longer opening hours, introduction of new functionalities, avoid queues, etc.) and economical benefits (e.g. users pay less for an SDS service than for a human one), see Dybkjær and Bernsen (2000).

- Common knowledge which will have to be shared between the human user and the SDS. This knowledge will arise from the application domain and task, and will have to be specified in terms of an initial vocabulary and language model, the required speech understanding capability, and the speech output capability.

- Common knowledge which will have to be shared between the SDS and the underlying application, and the corresponding interface (e.g. SQL).

- Knowledge to be included in the user model, cf. the discussion of user models in Section 2.1.3.4.

- Principle dialogue strategies to be used in the interaction, and potential description solutions (e.g. finite state machines, dialogue grammar, flowcharts).
- Hardware and software platform, i.e. the computing environment including communication protocols, application system interfaces, etc.

These general specification topics partly overlap with the characterization of individual system components for system analysis and evaluation. They form a prerequisite to the system design and implementation phase. The evaluation specification will be discussed in Section 3.1, together with the assessment and evaluation methods.

2.4.3 SDS Design

On a basis of the specification, system designers have the task to describe how to build the service. This description has to be made in a sufficiently detailed way in order to permit system implementation. System designers may consult end users as well as domain or industrial experts for support (Atwell et al., 2000).

Such a consultation may be established in a principled way, as was done in the European REWARD (REal World Application of Robust Dialogue) project, see e.g Failenschmid (1998). This project aimed to provide domain specialists with a more active role in the design process of SDSs. Graphical dialogue design tools and SDS engines were provided to the domain experts which had little or no knowledge of speech technology, and only technical assistance was given to them by speech technologists. The design decisions taken by the domain experts were taken in a way which addressed as directly as possible the users' expectations, while the technical experts concentrated on the possibility to achieve a function or task in a technically sophisticated way.

From the system designer's point of view, three design approaches and two combinations can be distinguished (Fraser, 1997, p. 571-594):

- *Design by intuition*: Starting from the specification, the task is analyzed in detail in order to establish parameters and routes for task accomplishment. The routes are specified in linguistic terms by introspection, and are based on expert intuition. Such a methodology is mostly suited for system-initiative dialogues and structured tasks, with a limited use of vocabulary and language. Because of the large space of possibilities, intuitions about user performance are generally unreliable, and intuitions on HMI are sparse anyway. Design by intuition can be facilitated by structured task analysis and design representations, as well as by usability criteria checklists, as will be described below.
- *Design by observation of HHI*: This methodology avoids the limitations of intuition by giving data evidence. It helps to build domain and task

understanding, to create initial vocabularies, language models, and dialogue descriptions. It gives information about the user goals, the items needed to satisfy the goals, and the strategies and information used during negotiation (San-Segundo et al., 2001a,b). The main problem of design by observation is that an extrapolation is performed from HHI to HMI. Such an extrapolation may be critical even for narrow tasks, because of the described differences between HHI and HMI, see Section 2.2. In particular, some aspects which are important in HMI cannot be observed in HHI, e.g. the initial setting of user expectations by the greeting, input confirmation and re-prompt, or the connection to a human operator in case of system failure.

- *Design by simulation*: The most popular method is the Wizard-of-Oz (WoZ) technique. The name is based on Baum's novel, where the "great and terrible" wizard turns out to be no more than a mechanical device operated by a man hiding behind a curtain (Baum, 1900). The technique is sometimes also called PNAMBIC (Pay No Attention to the Man BehInd the Curtain). In a WoZ simulation, a human wizard plays the role of the computer. The wizard takes spoken input, processes it in some principled way, and generates spoken system responses. The degree to which components are simulated can vary, and commonly so-called 'bionic wizards' (half human, half machine) are used. WoZ simulations can be largely facilitated by the use of rapid prototyping tools, see below. The use of WoZ simulations in the system evaluation phase is addressed in Section 3.8.

- *Iterative WoZ methodology*: This iterative methodology makes use of WoZ simulations in a principled way. In the pre-experimental phase, the application domain is analyzed in order to define the domain knowledge (database), subject scenarios, and a first experimental set-up for the simulation (location, hardware/software, subjects). In the first experimental phase, a WoZ simulation is performed in which very few constraints are put on the wizard, e.g. only some limitations of what the wizard is allowed to say. The data collected in this simulation and in the pre-experimental phase are used to develop initial linguistic resources (vocabulary, grammar, language model) and a dialogue model. In subsequent phases, the WoZ simulation is repeated, however putting more restrictions on what the wizard is allowed to understand and to say, and how to behave. Potentially, a bionic wizard is used in later simulation steps. This procedure is repeated until a fully automated system is available. The methodology is expected to provide a stable set-up after three to four iterations (Fraser and Gilbert, 1991b; Bernsen et al., 1998).

- *System-in-the-loop*: The idea of this methodology is to collect data with an existing system, in order to enhance the vocabulary, the language models, etc. The use of a real system generally provides good and realistic data, but only for the domain captured by the current system, and perhaps for small steps beyond. A main difficulty is that the methodology requires a fully working system.

Usually, a combination of approaches is used when a new system is set up. Designers start from the specification and their intuition, which should be described in a formalized way in order to be useful in the system design phase. On the basis of the intuitions and of observations from HHI, a cycle of WoZ simulations is carried out. During the WoZ cycles, more and more components of the final system are used, until a fully working system is obtained. This system is then enhanced during a system-in-the-loop paradigm.

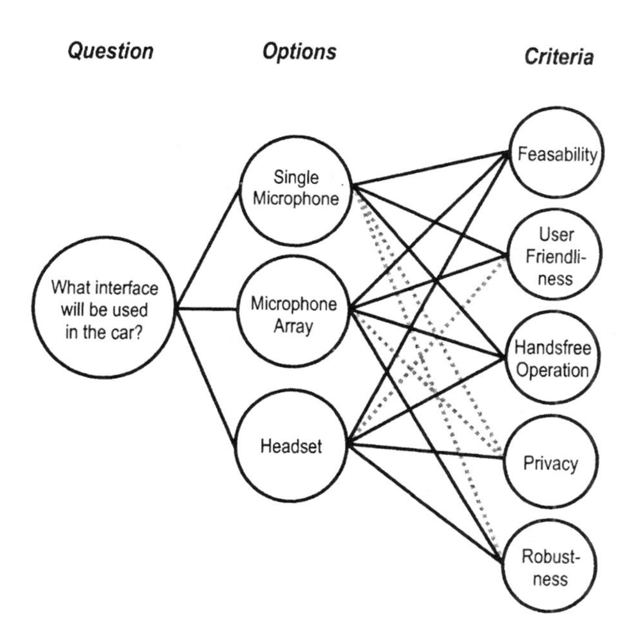

Figure 2.11. Example for a design decision addressed with the QOC (Questions-Options-Criteria) method, see de Ruyter and Hoonhout (2002). Criteria are positively (black solid lines) or negatively (gray dashed lines) met by choosing one of the options.

Design based on intuition can largely be facilitated by presenting the space of design decisions in a systemized way, because the quality elements of an SDS are less well-defined than those of a transmission channel. A systemized representation illustrates the interdependence of design constraints, and helps to identify contradicting goals and requirements. An example for such a representation is the Design Space Development and Design Rationale (DSD/DR), see Bernsen et al. (1998). In this approach, the requirements are represented in a frame which also captures the designer commitments at a certain point in the decision process. A DR frame represents the reasoning about a certain design problem, capturing the options, trade-offs, and reasons why a particular solution was chosen.

An alternative way is the so-called Questions-Options-Criteria (QOC) rationale (MacLean et al., 1991; Bellotti et al., 1991). In this rationale, the design space is characterized by questions identifying key design issues, options providing possible answers to the questions, and criteria for assessing and comparing the options. All possible options (answers) to a question are assessed positively or negatively (or via +/- scaling), each by a number of criteria. An example is given in Figure 2.11, taken from the European IST project INSPIRE (INfotainment management with SPeech Interaction via REmote microphones and telephone interfaces), see de Ruyter and Hoonhout (2002). Questions have to be posed in a way that they provide an adequate context and structure to the design space (Bellotti et al., 1991). The methodology assists with early design reasoning as well as the later comprehension and propagation of the resulting design decisions.

Apart from formalized representations of design decisions, general design guidelines and "checklists" are a commonly agreed basis for usability engineering, see e.g. the guidelines proposed by ETSI for telephone user interfaces (ETSI Technical Report ETR 051, 1992; ETSI Technical Report ETR 147, 1994). For SDS design, Dybkjær and Bernsen (2000) defined a number of "best practice" guidelines, including the following:

- Good speech recognition capability: The user should be confident that the system successfully receives what he/she says.

- Good speech understanding capability: Speaking to an SDS should be as easy and natural as possible.

- Good output voice quality: The system's voice should be clear and intelligible, not be distorted or noisy, show a natural intonation and prosody, an appropriate speaking rate, be pleasant to listen to, and require no extra listening-effort.

- Adequate output phrasing: The system should have a cooperative way of expression and provide correct and relevant speech output with sufficient information content. The output should be clear and unambiguous, in a familiar language.

- Adequate feedback about processes and about information: The user should notice what the system is doing, what information has been understood by the system, and which actions have been taken. The amount and style of feedback should be adapted to the user and the dialogue situation, and depends on the risk and costs involved with the task.

- Adequate initiative control, domain coverage and reasoning capabilities: The system should make the user understand which tasks it is able to carry out, how they are structured, addressed, and accessed.

- Sufficient interaction guidance: Clear cues for turn-taking and barge-in should be supported, help mechanisms should be provided, and a distinction between system experts/novices and task experts/novices should be made.

- Adequate error handling: Errors can be handled via meta-communication for repair or clarification, initiated either by the system or by the user.

Different (but partly overlapping) guidelines have been set up by Suhm (2003), on the basis of a taxonomy of speech interface limitations.

Additional guidelines specifically address the system's output speech. System prompts are critical because people often judge a system mainly by the quality of the speech output, and not by its recognition capability (Souvignier et al., 2000). Fraser (1997), p. 592, collects the following prompt design guidelines:

- Be as brief and simple as possible.

- Use a consistent linguistic style.

- Finish each prompt with an explicit question.

- Allow barge-in.

- Use a single speaker for each function.

- Use a prompt voice which gives a friendly personality to the system.

- Remember that instructions presented at the beginning of the dialogue are not always remembered by the user.

- In case of re-prompting, provide additional information and guidance.

- Do not pose as a human as long as the system cannot understand as well as a human (Basson et al., 1996).

Even when prompts are designed according to these guidelines, the system may still be pretty boring in the eyes (ears) of its users. Aspects like the metaphor, i.e. the transfer of meaning due to similarities in the external form or function, and the impression and feeling which is created have to be supported by the speech output. Speech output can be amended by other audio output, e.g. auditory signs ("earcons") or landmarks, in order to reach this goal.

System prompts will have an important effect on the user's behavior, and may stimulate users to model the system's language (Zoltan-Ford, 1991; Basson et al., 1996). In order to prevent dialogues from having too rigid a style due to specific system prompts, adaptive systems may be able to "zoom in" to more specific questions (alternatives questions, yes/no questions) or to "zoom out" to more general ones (open questions), depending on the success or failure of system questions (Veldhuijzen van Zanten, 1999). The selection of the right system prompts also depends on the intended user group: Whereas naïve users often prefer directed prompts, open prompts may be a better solution for users which are familiar with the system (Williams et al., 2003; Witt and Williams, 2003).

Respect of design guidelines will help to minimize the risks which are inherent in intuitive design approaches. However, they do not guarantee that all relevant design issues are adequately addressed. In particular, they do not provide any help in the event of conflicting guidelines, because no weighting of the individual items can be given.

Design by simulation is a very useful way to close the gaps which intuition may leave. A discussion about important factors of WoZ experiments will be given in conjunction with the assessment and evaluation methods in Section 3.8. Results which have been obtained in a WoZ simulation are often very useful and justify the effort required to set up the simulation environment. They are however limited to a *simulated* system which should not be confounded with a working system in a real application situation. The step between a WoZ simulation and a working system is manifested in all the environmental, agent, task, and contextual factors, and it should not be underestimated. Polifroni et al. (1998) observed for their JUPITER weather information service that the ASR error rates for the first system working in a real-world environment tripled in comparison to the performance in a WoZ simulation. Within a year, both word and sentence error rates could be reduced again by a factor of three. During the installation of new systems, it thus has to be carefully considered how to treat ASR errors in early system development stages. Apart from leaving the system unchanged, it is possible to try to detect and ignore these errors by using a different rejection threshold than the one of the optimized system (Rosset et al., 1999), or using a different confirmation strategy.

Design decision-taking and testing can be largely facilitated by rapid prototyping tools. A number of such tools are described and compared in DISC Deliverable D2.7a (1999) and by McTear (2002). They include tools which enable the description of the dialogue management component, and others integrating different system components (ASR, TTS, etc.) to a running prototype. The most well-known examples are:

- A suite of markup languages covering dialog, speech synthesis, speech recognition, call control, and other aspects of interactive voice response applications defined by the W3C Voice Browser Working Group[11]. The most prominent part is the Voice eXtensible Markup Language VoiceXML for creating mixed-initiative dialog systems with ASR/DTMF input and synthesized speech output. Additional parts are the Speech Synthesis Markup Language, the Speech Recognition Grammar Specification, and the Call Control XML.

- The Rapid Application Developer (RAD) provided together with the Speech Toolkit by the Oregon Graduate Institute (now OHSU, Hillsboro, USA-Oregon), see Sutton et al. (1996, 1998). It consists of a graphical editor for implementing finite state machines which is amended by several modules for information input and output (ASR, TTS, animated head, etc.). With the help of extension modules to RAD it is also possible to implement more flexible dialogue control models (McTear et al., 2000). This tool has been used for setting up the restaurant information system described in Section 6.

- DDLTool, a graphical editor which supports the representation of dialogue management software in the Dialogue Description Language DDL, see Bernsen et al. (1998). DDL consists of three layers with different levels of abstraction: A graphical layer for overall dialogue structure (based on the specification and description language SDL), a frame layer for defining the slot filling, and a textual layer for declarations, assignments, computational expressions, events, etc. DDLTool is part of the Generic Dialogue System platform developed at CPK, DK-Aalborg (Baekgaard, 1995, 1996), and has been used in the Sunstar and in the Danish flight reservation projects.

- SpeechBuilder developed at MIT, see Glass and Weinstein (2001). It allows mixed-initiative dialogue systems to be developed on the basis of a database, semantic concepts, and example sentences to be defined by the developer. SpeechBuilder automatically configures ASR, speech understanding, language generation, and discourse components. It makes use of all major components of the GALAXY system (Seneff, 1998).

- The dialogue environment TESADIS for speech interfaces to databases, in which the system designer can specify the application task and parameters needed from the user in a purely declarative way, see Feldes et al. (1998). Linguistic knowledge is extracted automatically from templates to be provided to the design environment. The environment is connected to an

[11]See `http://www.w3.org/voice`.

interpretation module (ASR and speech understanding), a generation module (including TTS), a data manager, a dialogue manager, and a telephone interface.

- Several proprietary solutions, including the Philips SpeechMania© system with a dialogue creation and management tool based on the dialogue description language HDDL (Aust and Oerder, 1995), the Natural Language Speech Assistant (NLSA) from Unisys, the Nuance Voice Platform™, and the Vocalis SpeechWare©.

- Voice application management systems which enable easy service design and support life-cycle management of SDS-based services, e.g. VoiceObjects©. Such systems are able to drive different speech platforms (phone server, ASR and TTS) and application back-ends by dynamically generating markup code (e.g. VoiceXML).

Most of these tools have reportedly been used both for system design as well as for assessment and evaluation.

2.4.4 System Assessment and Evaluation

System assessment and evaluation plays an important role for system developers, operators, and users. For system developers, it allows progress of a single system to be monitored, and it can facilitate comparisons across systems. For system operators and users, it shows the potential advantages a user will derive from using the system, and the level of training which is required to use the system effectively (Sikorski and Allen, 1997). Independently of this, it guides research to the areas where improvements are necessary (Hirschman and Thompson, 1997).

Apparently, the motivation for evaluation often differs between developers, users and evaluation funders (Hirschman, 1998):

- Developers want technology-centered evaluation methods, e.g. diagnostic evaluation for a system-in-the-loop.

- Users want user-centered evaluation, with real users in realistic environments.

- Funders want to demonstrate that their funding has advanced the field, and the utility of an emerging technology (e.g. by embedding the technology into an application).

Although these needs are different, they do not need to be contradictory. In particular, a close relation should be kept between technology evaluation and usage evaluation. Good technology is necessary, but not sufficient for successful system development.

Until now, there is no universally agreed-upon distinction between the terms 'assessment' and 'evaluation'. They are usually assigned to a specific task and motivation of evaluation. Most authors differentiate between three or four terms (Jekosch, 2000; Fraser, 1997; Hirschman and Thompson, 1997):

- *Evaluation* of existing systems for a given purpose: According to Jekosch (2000), p. 109, the term evaluation is used for the "determination of the fitness of a system for a purpose – will it do what is required, how well, at what costs, etc. Typically for a prospective user, may be comparative or not, may require considerable work to identify user's needs". In the terminology of Hirschman and Thompson (1997) this is called "adequacy evaluation".

- *Assessment* of system (component) performance: According to Jekosch (2000), the term assessment is used to describe the "measurement of system performance with respect to one or more criteria. Typically used to compare like with like, whether two alternative implementations of a technology, or successive generations of the same implementation". Hirschman and Thompson (1997) use the term "performance evaluation" for this purpose.

- *Diagnosis* of system (component) performance: This term captures the "production of a system performance profile with respect to some taxonomisation of the space of possible inputs. Typically used by system developers, but sometimes offered to end-users as well" (Jekosch, 2000). This is sometimes called "diagnostic evaluation" (Hirschman and Thompson, 1997).

- *Prediction* of future behavior of a system in a given environment: In some cases, this is called "predictive evaluation" (ISO Technical Report ISO/TR 19358, 2002). The author does not consider this as a specific type of assessment or evaluation; instead, prediction is based on the outcome of assessment or evaluation experiments, and can be seen as an application of the obtained results for system development and improvement.

These motivations are not mutually exclusive, and consequently assessment, evaluation and diagnosis are not orthogonal.

Unfortunately, the terminological differentiation between evaluation and assessment is not universal. Other authors use it to differentiate between "black box" and "glass box" methods, e.g. Pallett and Fourcin (1997). These terms relate to the transparency of the system during the assessment or evaluation process. In a glass box situation, the internal characteristics of a system are known and accessible during the evaluation process. This allows system behavior to be analyzed in a diagnostic way from the perspective of the system designer. A black box approach assumes that the internal characteristics of the system under consideration are invisible to the evaluator, and the system can only be described by its input and output behavior. In between these two extremities, some authors locate a white box (internal characteristics are known from a specification) or a gray box (parts of the internal characteristics are known, others are unknown) situation.

Several authors differentiate between "objective evaluation" and "subjective evaluation", e.g. Bernsen et al. (1998), Minker (1998), or ISO Technical Report ISO/TR 19358 (2002). In this terminology, "subjective evaluation" describes approaches in which human test subjects are directly involved during the measurement (ISO Technical Report ISO/TR 19358, 2002), e.g. for reporting quality judgments they made of the system (Bernsen et al., 1998). In contrast to this, "objective evaluation" refers to approaches in which humans are not directly involved in the measurement (e.g. tests carried out with prerecorded speech), or in which instrumentally measurable parameters related to some aspect of system performance are collected (Bernsen et al., 1998). This differentiation is not only ill-defined (what does "directly" mean?), but it is partly wrong because human subjects are *always* involved in determining the performance of a spoken language interface. The degree of involvement may vary, e.g. from recording natural utterances in HHI and linking it off-line to a spoken dialogue system (Rothkrantz et al., 1997), or constructing a second system which interacts with the system under test in a similar way as a human user (Araki and Doshita, 1997), to a human interaction with a system under laboratory or real-life conditions. In each situation, measures of performance can be obtained either instrumentally, from human expert evaluators, or from human test subjects, and relations between them can be established with the help of quality prediction models.

In the following chapters, the differentiation will therefore be made between (subjective) *quality judgments*, (instrumentally or expert-derived) *interaction parameters*, and (instrumental) *quality predictions*. The situation in which quality judgments or interaction parameters are collected is a different issue, and it definitely has an influence on the results obtained.

Following this terminology, assessment and evaluation methods can be categorized according to the following criteria:

- *Motivation* for assessment/evaluation:
  - Evaluation of the fitness of an existing system for a given purpose.
  - Assessment of system (component) performance.
  - Diagnostic profile of system performance.
- *Object* of the assessment/evaluation: Individual component vs. overall system. This choice also depends on the degree of system integration and availability, and a WoZ simulation might be evaluated instead of a real system during early stages of system development.
- *Environment* for assessment/evaluation:

- Laboratory: Enables repeatable experiments under controlled conditions, with only the desired variable(s) changed between interactions. However, a laboratory environment is unrealistic and leads to a different user motivation, and the user population which can be covered with reasonable effort is limited.
- Field: The field situation guarantees realistic scenarios, user motivations, and acoustic environments. Experiments are generally not repeatable, and the environmental and situative conditions vary between the interactions.

- *System transparency*: Glass box vs. black box.
  - Glass box: Assessment of the performance of one or several system components, potentially including its contribution to overall system performance. Requires access to internal components at some key points of the system.
  - Black box: Assumes that the internal characteristics and components of the system are invisible to the evaluator. Only the input-output relation of the system is considered, without regarding the specific mechanisms linking input to output.

- *Type of measurement method*: Instrumental or expert-based measurement of system and interaction parameters, vs. quality judgments obtained from human users.

- *Reference*: Qualitative assessment and evaluation describing the "absolute" values of instrumentally measurable or perceived system characteristics, vs. quantitative assessment and evaluation with respect to a measurable reference or benchmark.

- *Nature of functions* to be evaluated: Intrinsic criteria related to the system's objective, vs. extrinsic criteria related to the function of the system in its environmental use, see Sparck-Jones and Gallier (1996), p. 19. The choice of criteria is partly determined by the environment in which the assessment/evaluation takes place.

Other criteria exist which are useful from a methodological point of view in order to discriminate and describe quality measurements, e.g. the ones included in the general projection model for speech quality measurements from Jekosch (2000), p. 112. They will be disregarded here because they are rarely used in the assessment and evaluation of spoken dialogue systems. On the basis of the listed criteria, assessment and evaluation methods can be chosen or have to be designed. An overview of such methods will be given in Chapter 3.

2.5 Summary

Spoken dialogue systems enabling task-orientated human-machine interaction over the phone offer a relatively new type of service to their users. Because of the inexperience of most users, and because of the the fact that the agent at the far end is a machine and not a human, interactions with spoken dialogue systems follow rules which are different from the ones of a human-to-human telephone interaction. Nevertheless, a comparable operator-based service will form one reference against which the quality of SDS-based services is judged, and with which they have to compete in order to be successful and acceptable to their users.

The quality of the interaction with a spoken dialogue system will depend on the characteristics of the system itself, as well as on the characteristics of the transmission channel and the environment the user is situated in. The physical and algorithmic characteristics of these quality elements have been addressed in Section 2.1. They can be classified with the help of an interactive speech theory developed by Bernsen et al. (1998), showing the interaction loop via a speech, language, control and context layer. In this interaction loop, the user behavior differs from the one in a normal human-to-human interaction situation. Acknowledging that the capabilities of the system are limited, the user adapts to this fact by producing language and speech with different (often simplified) characteristics, and by adjusting its initiative. Thus, in spite of the limitations, a successful dialogue and task achievement can be reached, because both interaction participants try to behave cooperatively.

Cooperativity is a key requirement for a successful interaction. This fact is captured by a set of guidelines which support successful system development, and which are based on Grice's maxims of cooperativity in human communication. Apart from cooperativity, other dimensions are important for reaching a high interaction quality for the user. In the definition adopted here, quality can be seen as the result of a judgment and a perception process, in which the user compares the perceived characteristics of the services with his/her desires or expectations. Thus, quality can only be measured subjectively, by introspection. The quality features perceived by the user are influenced by the physical and algorithmic characteristics of the quality elements, but not in the sense of a one-to-one relationship, because both are separated by a complex perception process.

Influencing factors on quality result from the machine agent, from the talking and listening environment, from the task to be carried out, and from the context of use. These factors are in a complex relationship to different notions of quality (performance, effectiveness, efficiency, usability, user satisfaction, utility and acceptability), as it is described by a new taxonomy for the quality of SDS-based services which is given in Section 2.3.1. The taxonomy can be helpful for system developers in three different ways: (1) Quality elements of the SDS and the transmission network can be identified; (2) Quality features perceived by the user can be described, together with adequate (subjective) assessment methods; and (3) Prediction models can be developed to estimate quality from instrumentally or expert-derived interaction parameters during the system design phase.

In order to design systems which deliver a high quality to their users, quality has to be a criterium in all phases of system specification, design, and evaluation. In particular, both the characteristics of the transmission channel as well as the ones of the SDS have to be addressed. This integrated view on the whole interaction scenario is useful because many transmission experts are not familiar with the requirements of speech technology, and many speech technology experts do not know which transmission impairments are to be expected for their systems in the near future. It also corresponds to the user's point of view (end-to-end consideration). Commonly used specification and design practices were discussed in Section 2.4. For transmission networks, these practices are already well defined, and appropriate quality prediction models allow quality estimations to be obtained in early planning stages. The situation is different for spoken dialogue systems, where iterative design principles based on intuition, simulation and running systems have to be used. Such an approach intensifies the need for adequate assessment and evaluation methods. The respective methods will be discussed in detail in Chapter 3, and they will be applied to exemplary speech recognizers (Chapter 4), speech synthesizers (Chapter 5), and to whole services based on SDSs (Chapter 6).

Chapter 3

ASSESSMENT AND EVALUATION METHODS

In parallel to the improvements made in speech and language technology during the past 20 years, the need for assessment and evaluation methods is steadily increasing. A number of campaigns for assessing the performance of speech recognizers and the intelligibility of synthesized speech have already been launched at the end of the 1980s and the beginning of the 1990s. In the US, comparative assessment of speech recognition and language understanding was mainly organized under the DARPA program. In Europe, the activities were of a less permanent nature, and included the SAM projects (Multi-Lingual Speech Input/Output Assessment, Methodology and Standardization; ESPRIT Projects 2589 and 6819), the EAGLES initiative (Expert Advisory Group on Language Engineering Standards, see Gibbon et al., 1997), the Francophone Aupelf-Uref speech and language evaluation actions (Mariani, 1998), and the Sqale (Steeneken and van Leeuwen, 1995; Young et al., 1997a), Class (Jacquemin et al., 2000), and DISC projects (Bernsen and Dybkjær, 1997). Most of the early campaigns addressed the performance of individual speech technology components, because fully working systems were only sparsely available. The focus has changed in the last few years, and several programs have now been extended towards whole dialogue system evaluation, e.g. the DARPA Communicator program (Levin et al., 2000; Walker et al., 2002b) or the activities in the EU IST program (Mariani and Lamel, 1998).

Assessment on a component level may turn out to have very limited practical value. The whole system is more than a sum of its composing parts, because the performance of one system component heavily depends on its input – which is at the same time the output of another system component. Thus, it is rarely possible to meaningfully compare isolated system components by indicating metrics which have been collected in a glass box approach. The interdependence of system components plays a significant role, and this aspect can only be captured by additionally testing the whole system in a black box way. For example, it is important to know in how far a good dialogue manager can compensate for a poor speech understanding performance, or whether a poor dialogue manager can squander the achievements of good speech understanding (Fraser, 1995). Such questions address the overall quality of an SDS, and they are still far from being answered. Assessment and evaluation should yield information on the system component and on the overall system level, because the description of system components alone may be misleading for capturing the quality of the overall system.

A full description of the quality aspects of an SDS can only be obtained by using a combination of assessment and evaluation methods. On the one hand, these methods should be able to collect information about the performance of individual system components, and about the performance of the whole system. Interaction parameters which were defined in Section 2.3 are an adequate means for describing different aspects of system (component) performance. On the other hand, the methods should capture as far as possible the quality perceptions of the user. The latter aim can only be reached in an interaction experiment by directly asking the user. Both performance-related and quality-related information may be collected in a single experiment, but require different methods to be applied in the experimental set-up. The combination of subjective judgments and system performance metrics allows significant problems in system operation to be identified and resolved which otherwise would remain undetected, e.g. wrong system parameter settings, vocabulary deficiencies, voice activity detection problems, etc. (Kamm et al., 1997a).

Because the running of experiments with human test subjects is generally expensive and time-consuming, attempts have been made to automatize evaluation. Several authors propose to replace the human part in the interaction by another system, leading to machine-machine interaction which takes into account the interrelation of system components and the system's interactive ability as a whole. On a language level, Walker (1994) reports on experiments with two simulated agents carrying out a room design task. Agents are modelled with a scalable attention/working memory, and their communicative strategies can be selected according to a desired interaction style. In this way, the effect of task, communication strategy, and of "cognitive demand" can be investigated. A comparison is drawn to a corpus of recorded HHI dialogues, but no verification of the methodology is reported. Similar experiments have been described by Carletta (1992) for the Edinburgh Map Task, with agents which can be parametrized according to their communicative and error recovery strategies.

For a speech-based system, Araki and Doshita (1997) and López-Cózar et al. (2003) propose a system-to-system evaluation. Araki and Doshita (1997) install a mediator program between the dialogue system and the simulated user. It introduces random noise into the communication channel, for simulating speech recognition errors. The aim of the method is to measure the system's robustness against ASR errors, and its ability to repair or manage such misrecognized sentences by a robust linguistic processor, or by the dialogue management strategy. System performance is measured by the task achievement rate (ability of problem solving) and by the average number of turns needed for task completion (conciseness of the dialogue), for a given recognition error rate which can be adjusted via the noise setting of the mediator program. López-Cózar et al. (2003) propose a rule-based "user simulator" which feeds the dialogue system under test. It generates user prompts from a corpus of utterances previously collected in HHI, and re-recorded by a number of speakers. Automatized evaluation starting from HHI test corpora is also used in the Simcall testbed, for the evaluation of an automatic call center application (Rothkrantz et al., 1997). The testbed makes use of a corpus of human-human dialogues and is thus restricted to the recognition and linguistic processing of expressions occurring in this corpus, including speaker dependency and environmental factors.

Although providing some detailed information on the interrelation of system components, such an automatic evaluation is very restricted in principle, namely for the following reasons:

- An automated system is, by definition, unable to evaluate dimensions of quality as they would be perceived by a user. There are no indications that the automated evaluation output correlates with human quality perception, and – if so – for which systems, tasks or situations this might be the case.
- An SDS can be optimized for a good performance in an automatized evaluation without respecting the rules of HHI – in extreme cases without using naturally spoken language at all. However, users will expect that these rules are respected by the machine agent.
- The results which can be obtained with automatized evaluation are strongly dependent on the models which are inherently used for describing the task, the system, the user, and the dialogue.

As a consequence, the interaction between the system and its human users can be assumed as the only valid source of information for describing a large set of system and service quality aspects.

It has become obvious that the validity of the obtained results is a critical requirement for the assessment and evaluation of speech technology systems. Both assessment and evaluation can be seen as measurement processes, and consequently the methods and methodologies used have to fulfill the following fundamental requirements which are generally expected from measurements:

- *Validity*: The method should be able to measure what it is intended to measure.
- *Reliability*: The method should be able to provide stable results across repeated administrations of the same measurement.

- *Sensitivity*: The method should be able to measure small variations in what it is intended to measure.
- *Objectivity*: The method should reach inter-individual agreement on the measurement results.
- *Robustness*: The method should be able to provide results independent from variables that are extraneous to the construct being measured.

The fulfillment of these requirements has to be checked in each assessment or evaluation process. They may not only be violated when new assessment methods have been developed. Also well-established methods are often misapplied or misinterpreted, because the aim they have been developed for is not completely clear to the evaluator.

In order to avoid such misuse, the target and the circumstances of an assessment or evaluation experiment should be made explicit, and they should be documented. In the DISC project, a template has been developed for this purpose (Bernsen and Dybkjær, 2000). Based on this template and on the classification of methods given in Section 2.4.4, the following criteria can be defined:

- Motivation of assessment/evaluation (e.g. a detailed analysis of the system's recovery mechanisms, or the estimated satisfaction of future users).
- Object of assessment/evaluation (e.g. the speech recognizer, the dialogue manager, or the whole system).
- Environment for assessment/evaluation (e.g. in a controlled laboratory experiment or in a field test).
- Type of measurement methods (e.g. via an instrumental measurement of interaction parameters, or via open or closed quality judgments obtained from the users).
- Symptoms to look for (e.g. user clarification questions or ASR rejections).
- Life cycle phase in which the assessment/evaluation takes place (e.g. for a simulation, a prototype version, or for a fully working system).
- Accessibility of the system and its components (e.g. in a glass box or in a black box approach).
- Reference used for the measurements (e.g. qualitative measures of absolute system performance, or quantitative values with respect to a measurable reference or benchmark).
- Support tools which are available for the assessment/evaluation.

These criteria form a basic set of documentation which should be provided with assessment or evaluation experiments. The documentation may be implemented in terms of an item list as given here, or via a detailed experimental description as it will be done in Chapters 4 to 6.

It is the aim of this chapter to discuss assessment and evaluation methods for single SDS components as well as for whole systems and services with respect to these criteria. The starting point is the definition of factors influencing the quality of telephone services based on SDSs, as they are included in the QoS taxonomy of Section 2.3.1. They characterize the system in its environmental, task and contextual setting, and include all system components. Common to most types of performance assessment are the notion of reference (Section 3.2) and the collection of data (Section 3.3) which will be addressed in separate sections. Then, assessment methods for individual components of SDSs will be discussed, namely for ASR (Section 3.4), for speech and natural language understanding (Section 3.5), for speaker recognition (Section 3.6), and for speech output (Section 3.7). The final Section 3.8 deals with the assessment and evaluation of entire spoken dialogue systems, including the dialogue management component.

3.1 Characterization

Following the taxonomy of QoS aspects given in Section 2.3.1, five types of factors characterize the interaction situations addressed in this book: Agent factors, task factors, user factors, environmental factors, and contextual factors. They are partly defined in the system specification phase (Section 2.4.2), and partly result from decisions taken during the system design and implementation phases. These factors will carry an influence on the performance of the system (components) and on the quality perceived by the user. Thus, they should be taken into account when selecting or designing an assessment or evaluation experiment.

3.1.1 Agent Factors

The system as an interaction agent can be characterized in a technical way, namely by defining the characteristics of the individual system components and their interconnection in a pipelined or hub architecture, or by specifying the agent's operational functions. The most important agent functions to be captured are the speech recognition capability, the natural language understanding capability, the dialogue management capability, the response generation capability, and the speech output capability. The natural language understanding and the response generation components are closely linked to the neighbouring components, namely the dialogue manager on one side, and the speech recognizer or the speech synthesizer on the other. Thus, the interfaces to these components have to be precisely described. For multimodal agents, the characterization has to be extended with respect to the number of different media used for input/output, the processing time per medium, the way in which the media are used (in parallel, combined, alternate, etc.), and the input and output modalities provided by each medium.

3.1.1.1 ASR Characterization

From a functional point of view, ASR systems can be classified according to the following parameters (see van Leeuwen and Steeneken, 1997):

- Vocabulary size, e.g. small, medium, or large vocabulary speech recognizers.
- Vocabulary complexity, e.g. with respect to the confusability of words.
- Speech type, e.g. isolated words, connected words, continuous speech, spontaneous speech including discontinuities such as coughs, hesitations, interruptions, restarts, etc.
- Language: Mono-lingual or multi-lingual recognizers, language dependency of recognition results, language portability.
- Speaker dependency, e.g. speaker-dependent, speaker-independent or speaker-adaptive recognizers.
- Type and complexity of grammar. The complexity of a grammar can be determined in terms of its perplexity, which is a measure of how well a word sequence can be predicted by the language model.
- Training method, e.g. multiple training of explicitly uttered isolated words, or embedded training on strings of words of which the starting and ending points are not defined.

On the other hand, ASR components can be described in terms of general technical characteristics which may be implemented differently in individual systems (Lamel et al., 2000b). The following technical characteristics have partly been used in the DISC project:

- Signal capture: Sampling frequency, signal bandwidth, quantization, windowing.
- Feature analysis, e.g. mel-scaled cepstral coefficients, energy, and first or second order derivatives.
- Fundamental speech units, e.g. phone models or word models, modelling of silence or other non-speech sounds.

- Lexicon: Number of entries for each word, with one or several pronunciations; generated either from dictionaries or from grapheme-to-phoneme converters; additional entries for filler words and noises; expected coverage of the vocabulary with respect to the target vocabulary.
- Acoustic model: Type of model, e.g. MLP networks or HMMs; training data and parameters; post-processing of the model.
- Language model: Type of model, e.g. a statistical N-gram back-off language model, or a context-free grammar; training material, e.g. a large general-purpose training corpus or data collected in a WoZ experiment; individual word modelling or classes for specific categories (e.g. dates or names); dialogue-state-independent or dialogue-state-dependent models.
- Type of decoder, e.g. HMM-based.
- Use of prosodic information.

3.1.1.2 Speaker Recognition Characterization

Like ASR systems, speaker recognition systems can be characterized from a functional and a technical point of view. The functional description includes the following items:

- Task typology: The two main areas are speaker verification and speaker identification, and additional related tasks include speaker matching, speaker labelling, speaker alignment, or speaker change detection (Bimbot and Chollet, 1997).
- Text-dependency, e.g. text-dependent, text-independent, or text-prompted.
- Training method.

The technical characterization is slightly different from the one for ASR systems. The reader is referred to Furui (1996, 2001a) for a detailed discussion.

3.1.1.3 Language Understanding Characterization

The following characteristics are important for the language understanding capability of the system:

- Semantic description of the task, e.g. via slots.
- Syntactic-semantic analysis: General parsing capability, e.g. full parsing or robust partial parsing; number and complexity of allowed syntax, e.g. the number of alternatives available at a given level.
- Contextual analysis: Number and complexity of rules.

- Interaction with ASR and dialogue management modules: Type and amount of input and output information (single hypotheses, ranked lists, etc.), dependency of syntactic-semantic and contextual interpretation on the dialogue state.

3.1.1.4 Dialogue Manager Characterization

The approach taken for dialogue management can be defined from a technical point of view, e.g. a dialogue grammar, a plan-based approach, or a collaborative approach (see Section 2.1.3.4). The most important characteristics of the dialogue manager are the type and amount of knowledge implemented in the manager, the distribution of initiative between the system and the user, and the system's meta-communication strategies:

- Dialogue manager knowledge: Dialogue history model (information that has been exchanged in the dialogue so far), task and domain models (scenario, plans, goals and subgoals, objects and their characteristics), world knowledge model, conversational model, and user model.
- Initiative: System-initiative, mixed-initiative, or user-initiative.
- Confirmation strategy: Explicit confirmation, implicit confirmation, "echo" confirmation, summarizing confirmation.
- Repair, clarification and recovery strategies.
- Dialogue manager adaptivity: Constitutive managers that have to learn new notions in their normal operation, or adaptive managers which might include a dynamic user model and might be able to learn the user's communicative strategies. Bernsen (2003) differentiates between design-time adaptation, user-driven customization, and online adaptation towards the user.

In addition to the interaction with the user, the interaction with the application system has to be defined, including the interface (language), and potential control mechanisms for handling the dynamics of the application system.

3.1.1.5 Speech Generation Characterization

Speech generation includes the potential generation of a textual response, and the translation into spoken language. Most systems use one of three types of speech generation: Pre-recorded speech, template sentences, or text-to-speech. The following characteristics have to be defined:

- Interaction with the dialogue manager: Type and amount of input information provided by the dialogue manager, e.g. orthographic or annotated text, focus or prosodic information, etc.

- Response generation: Strategy (e.g. formal grammar or simple templates), flexibility (pre-defined vocabulary or open vocabulary), type and amount of information to be included in each utterance, form of the message (syntax, choice of words).
- System voice: Number of voices, gender, professionalism, training, prosodic quality, recording conditions, adaptivity.
- Language: Mono-lingual or multi-lingual synthesizers, language identification capability, language portability.
- Type of speech generation: Pre-recorded messages, template sentences, text-to-speech, concept-to-speech.
- Text-to-speech characteristics: Strategy (e.g. model-based or corpus-based), text pre-processing capabilities, model parameters, unit corpus characteristics (types and length of units, coverage of the target vocabulary, etc.), concatenation and/or selection algorithms, prosody generation strategies (fundamental frequency, duration, intensity), etc.
- Contextual characteristics: Speaking style, speaking rate, contextual adaptivity.

3.1.2 Task Factors

The task which can be carried out by the user is a determining factor of the interaction. Common tasks include information access and retrieval (e.g. train, flight or weather information), negotiation (e.g. hotel booking, date negotiation), or conversation monitoring (e.g. a speech-to-speech translation system). The task can be characterized with respect to the type of task, task domain, task complexity, task frequency, task consequences, and portability:

- Task type: Bernsen et al. (1998) differentiate between
  - well-structured tasks, having a stereotypical structure that prescribes which piece of information must be exchanged, and often also in which natural order, and
  - ill-structured tasks, containing a large number of optional sub-tasks whose nature and order are difficult to predict,

  as well as between

  - homogenous and
  - heterogenous, which means *inherently* a combination of several different tasks which are different by their actual nature (e.g. ordering plus information plus device control).

- Task domain: Richness, scalability, number of users that are familiar with the domain, usefulness for the domain, generalizability, etc.
- Task complexity: Number of covered scenarios, maximum number of subgoals, number of subtasks which can be achieved in parallel, minimum number of exchanges necessary to solve the problem, expected complexity of syntax/vocabulary, etc. Dybkjær et al. (1996) show that increased task complexity requires a more sophisticated dialogue, and more and better technologies (ASR, vocabulary, syntactic/semantic analysis, handling of discourse phenomena, task and domain models, user models, speech generation).
- Task frequency, i.e. the frequency with which users can be expected to use the system for the given task. Systems for call routing or flight information (so-called "walk-up-and-use systems") will be used with relatively low frequency, so that potential users cannot be expected to have knowledge about the system, nor to show learning effects (remember behavior from previous calls) or to accept training.
- Task consequences, e.g. security issues.
- Task portability.

3.1.3 User Factors

In most cases, the characterization of the user is limited to a broad categorization with respect to his/her task, domain and world background, because an exact description of factors important for an individual user (attitude, motivation, emotions, flexibility) cannot be achieved. The following characteristics are often given in assessment and evaluation protocols:

- Number of users.
- Age and gender: They are expected to carry an influence on the fundamental frequency and the speech spectrum, but also on the dialogue interaction.
- Level of experience: Novice vs. expert, occasional user vs. regular user, trained user vs. untrained user.
- Level of expertise in the application domain: Professional users vs. private users.
- Explicit motivation for using the service.
- Physical status, vocal effort, speaking rate, etc.
- Native language, accent, dialect, etc.

For specialized applications, it might be necessary to be more explicit in specifying experience and expertise, e.g. with respect to the knowledge of task goals, the ability to develop strategies to optimize task performance, and the ability to use the devices necessary to perform the task, cf. the characterization of users for a battlefield observation simulator given by Life et al. (1988).

3.1.4 Environmental Factors

The environment contains the entire physical context of the interaction. A full characterization will generally be impossible, and only the factors which affect the speech signal should be described, namely:

- Type and proximity/position of microphones.
- Transmission channel: For telephone channels, the description can be performed on different levels, e.g. in terms of the transmission, switching and terminal equipment used in the connection, or in terms of the parameters of a reference configuration, see Section 2.4.1.
- Room acoustic situation: Includes reverberation, coloration, ambient noise levels and spectra, concurrent speakers, etc.

3.1.5 Contextual factors

These are non-physical factors characterizing the context of use of the service under consideration. Typical factors include:

- Facility of access: Availability of telephone numbers, links to and from other services, etc.
- Service availability: Opening hours, potential restrictions of access.
- Costs: Fixed and time-dependent costs of the interaction, specific account conditions, etc.
- Services with similar functionality: Have to be compared with respect to all other contextual factors.

3.2 Reference

The term 'performance' has been defined as the ability of a system to provide the function it has been designed for. Consequently, a simple approach to measure performance is to compare the function provided by the system with a reference describing the target function. Indeed, the comparison to a specified reference forms the basis for measuring the performance of a variety of speech and language technology components (Simpson and Fraser, 1993):

- ASR: Performance is measured by comparing a recognized string to an expert's transcription of what has actually been said.

- Text understanding: Performance is expressed by the ability to fill in slots of a reference frame, which is constructed on the basis of expert judgments.

- Speech understanding: Performance is expressed by the ability to construct the same database query from an utterance as an expert does.

- Synthesized speech: Intelligibility is often evaluated by comparing transcriptions of what the user understood with the input string of the speech synthesizer.

Because of the automation possibility, the early evaluations carried out in the DARPA program were heavily relying on the comparison to a reference. In the air travel information domain (ATIS), speech understanding capability was calculated from a common answer specification protocol which compares the system's database result to a canonical database answer (Bates et al., 1990). The protocol had to be very elaborated in order to reach agreement across different systems and test sites.

Unfortunately, the comparison to a reference becomes very restricted for an interaction spanning over several utterances, because dialogues are generally complex (i.e. they can accomplish multiple tasks which require multiple metrics) and dynamic. In order to evaluate complete dialogues, the context and interaction history have to be taken into account. First proposals in this direction have been made by Hirschman et al. (1990). They evaluated reference answers for specific tasks, each task consisting of an utterance and an encoding of the dialogue state (canonical context). The canonical context consisted of the query-answer pairs seen by the subject up to the point of the specific utterance. As long as the same input utterances and canonical context were used, the approach allowed to compare different systems and took the dynamic dialogue characteristics into account.

A different example of reference-based performance measures for dialogue assessment is the dialogue breadth test described by Bates and Ayuso (1991). For each canonical context, a large set of different user utterances was defined, namely 150 different dialogue starters consisting of 15 first queries with each 10 associated second queries. These starters were used as an input to the ATIS speech understanding and database access components at each participating site, and 100 test results (database answers) were generated for each starter. The answers were compared to reference answers, and the results (simple correct/wrong decision) were scored automatically. Such a dialogue breadth test evaluates the ability of a speech understanding system to handle context-sensitive information for a multitude of potential user input utterances. It does not tend to produce a reference dialogue, and focusses on local dialogue structures. However, it is unrealistic to extend such an approach to global dialogue structures.

Despite the described approaches, the principle of comparing system output to a reference is problematic *per se* when evaluating complete dialogues. This comparison implies that a system performs well when it produces a correct answer, which is very doubtful because some answers may be much more efficient for reaching the task goals than others. Dialogues are very complex structures, and often contain ambiguous queries which are not interpretable without making extensive use of the dialogue history (Bates and Ayuso, 1991). A user utterance is generally based on what was said before, and the system has to decide whether constraints made in earlier utterances are still valid for the current dialogue context. When the system is unable to answer a query, that might have a severe impact on the course of the dialogue, and on the meaningfulness of user utterances. An incorrect interpretation of user utterances by the system will cause a divergence in the subsequent dialogue tree, and thus make the reference unusable for the rest of the dialogue.

Instead of defining references for the individual parts of the dialogue system, Paek (2001) proposes to take the human (wizard) performance as the "gold standard", or alternatively – in the case that it performs better – the performance of a GUI or touch-tone interface providing the same functionality. By varying the system parameter that best matches the desired performance with respect to a specific performance metric (e.g. task success) in a WoZ experiment, and holding all other input and output between system and the user constant, a comparison can be drawn between wizard and system performance, and a "gold impurity graph" (i.e. the deviation from the "gold standard" as a function of the varied component) can be established. In this way, it would be possible to extrapolate to performance ranges which cannot be reached with state-of-the-art technology. The approach is however very problematic for several reasons:

- Humans behave differently in HHI than in HMI. Thus, the strategy that a human wizard will take when he/she is confronted with system modules which have a limited performance (e.g. with the output of a moderately performing speech recognizer) will differ considerably from the strategy a system can take under the same circumstances. Even when HHI is regarded as the benchmark for the whole HMI scenario, it cannot be taken as a reference for the individual steps which have to be performed in order to reach the task goal.

- Paek suggests to substantiate performance claims across task domains by using the deviation from the "gold standard" as an index. This is highly questionable, because a new domain will require new dialogue strategies, both from the system and from the wizard taken as the reference.

- Paek claims that the distance of the "gold standard" to the absolute upper bound of performance can be used to measure the "intellectual complexity" of the task. This implies that task complexity is low when humans perform well on a task. In view of the specialized human reasoning capabilities this hypothesis seems to be inadequate.

In conclusion, the notion of a reference for the determination of system performance only seems to be appropriate for a limited set of well-defined functions. For the global performance of dialogue management components such a simple approach is impractical and inappropriate. In a different sense, however, reference plays an important role for the quality features perceived by the user. The user judges quality in relation to an internal reference which describes the desired or expected composition of the unit to be judged.

3.3 Data Collection

In order to design SDSs which are able to cope with a variety of user utterances and dialogue situations, large databases have to be collected. The collection often takes place at different sites, in order to distribute the burden of recording and annotation, and to provide a more diverse pool of data than could be obtained at one single site. Examples of multi-site data collection include the MADCOW group in the DARPA program (Hirschman et al., 1993), the POLYCOST telephone speech database for speaker recognition, developed under the European COST 250 action (Hennebert et al., 2000), as well as the SpeechDat projects (e.g. Elenius (1999) for the Swedish part of the SpeechDat database). Several groups try to make speech and language corpora publicly available via distribution centers like LDC and ELRA/ELDA. Such institutions distribute diverse corpora which have been collected at different sites, sometimes in parallel for different languages, and under specific acoustic/channel characteristics and recording situations (in-car environment, telephone speech, etc.).

According to the purpose they are used for, three types of corpora have to be distinguished:

- Training data: Typically large corpora which are used for the training of acoustic and language models.
- Development data: They are used to tune the system's performance to the characteristics of the final application.
- Test data: Are exclusively used for testing purposes, and should be used sparsely in order to avoid unreliable results due to overtraining of systems.

The type and amount of training data will largely affect the performance of existing algorithms, and it limits the development of new ones. Developmental data are very important in order to make a laboratory system fit for real-world application. Test data serve the assessment and evaluation of system components, and will largely affect their development, because enhancements generally concentrate on the weak points of the system which are mainly found in the evaluation phases.

The mentioned types of corpora differ in several characteristics. From a language point of view, the corpus size, the number of dialogues, utterances and words, the total duration, the average length of a dialogue, the average length of an utterance, or the average number of words per utterance or dialogue are important characteristics. The speech is largely affected by the speaker characteristics (language, accents, voice characteristics, speech idiosyncracies, speech style, professionalism, training, fatigue, etc.), the recording set-up and environment, and the recording supervision and instructions given to the speakers. Both speech and language are influenced by the interaction scenario, namely whether a speaker is instructed to read or repeat pre-defined utterances, or whether the interaction takes place between two human speakers, a human speaker and a wizard simulating a system, or a human and a real system (Hirschman et al., 1993). In addition, the particular task, the domain, and the underlying application system play an important role. As already shown in Section 2.2.1, the characteristics of the system or simulation output also carry an influence on the user's speech and language.

In the early stages of system development, system components are often trained on general-purpose databases. These databases are large and cover a variety of speaker and language characteristics, but they are not well suited for the application domain of the future system. In order to tune the system components to the speech, transmission and language characteristics which are expected in the application situation, additional data is collected in WoZ experiments. The WoZ approach is necessary as long as no fully working system is available, but it encounters the problem that the data may be different from the ones collected in HMI situations. For example, speaker adaptation to low recognition performance (hyper-articulation, etc.) is often not covered in such data (Shriberg et al., 1992). After a first system version has been set up, it is possible to collect more realistic data in a system-in-the-loop paradigm (Baggia et al., 1994).

It is important to test an application on an *independent* set of utterances and dialogues, which the system is likely to encounter later on. This ensures that developers handle the phenomena in a way which allows generalization, and with a priority which correlates to the proportion to their likelihood of occurrence (Price, 1990). For a translation system, Gates et al. (1997) report that – at the end of a development/evaluation cycle, a retest on the data used for development typically results in about 90% of correct translations, with considerably lower translation performance on unseen data. Thus, testing on data which has been used for system development will not help to discover system inadequacies and to improve the system.

There are nevertheless several opportunities for linking the efforts of data collection and evaluation. Because the data which has been collected during the system evaluation phase shows many of the characteristics which can be regarded as typical for the later application scenario, it will be a valuable source of information for similar tasks and applications. On the other hand, evaluation efforts are sometimes linked to the prior availability of corpora. This was the case in the Parseval evaluation (Hirschman, 1998), where the availability of the Penn Treebank opened up new research capabilities in parsing and semantic structure analysis.

3.4 Speech Recognition Assessment

Although the task of speech recognizer assessment seems to be very simple – namely to compare the hypothesized output of the recognizer to a reference output produced by an expert – there are a number of different approaches for ASR performance measurement. Pallett and Fourcin (1997) noted the following ones:

- Application-orientated assessment, based on the use of general-purpose databases which are collected under representative conditions.
- Assessment with the help of calibrated databases which are designed to represent a broad spectrum of operational and environmental conditions.
- Use of reference methods in order to achieve cross-site comparability, based on the use of reference recognizers, reference speech databases, or referring to human recognition.
- Diagnostic methods, based on the use of specific vocabularies or specifically designed word sequences.
- Techniques using artificial test signals to achieve control over the experimental design and/or language independence.
- Predictive methods using system parameters and/or speech knowledge as an input.

These approaches mainly differ with respect to the experimental assessment conditions. In fact, there are a number of factors which carry an influence on the outcome of recognizer performance tests (Pallett, 1985). They are related to the type (isolated words, connected words, continuous speech, spontaneous speech) and the linguistic characteristics of the input speech (vocabulary size, fraction of out-of-vocabulary words, sentence perplexity, phonetic similarity between words), the speakers (age, gender, dialect, speech idiosyncracies, variability, motivation, fatigue, workload, experience), the task or domain for which the recognizer is designed (vocabulary and grammar), and to the environment (microphone, acoustic environment, bandwidth, transmission impairments, signal level).

When recognizers are to be compared on a large scale, all listed influencing factors have to be described and controlled. This was the case for benchmark tests which have been implemented in the DARPA spoken language program since 1987. The tests are based on specified databases (e.g. the Wall Street Journal corpus or the ATIS corpus), and data processing and result analysis have been carefully defined in advance. The data is divided into independent sets of common training and development data supplied to all participants, and evaluation data which is only available to the assessment organizer. An overview of results obtained in this way can be found in the proceedings of the DARPA Speech and Natural Language Workshops (Pallett et al., 1993; Young, 1997; Pallett, 1998). Newer activities concentrate on the evaluation of conversational speech over the telephone, the so-called HUB-5 evaluations, see the NIST HUB-5 Evaluation Plan (2000, 2001). Databases for this assessment consist of a set of conversations or parts of conversations which have been collected over the telephone. In Europe, an effort has been made for a multilingual benchmark test in the frame of the Squale project (Steeneken and van Leeuwen, 1995; Young et al., 1997a), involving four different languages.

Continuous speech recognizers generally provide a word string hypothesis as an output. In order to judge whether the string correctly represents what has been said, a reference transcription has to be provided. These transcriptions are available for the common databases indicated above, or they have to be produced when testing is done on a specific new set of data. For each utterance, hypothesized and reference string are first aligned on a word level, e.g. using a Dynamic Programming (DP) matching algorithm (Picone et al., 1990, 1986) which assigns different penalties to deleted, inserted or substituted words. Such a simple alignment tends to underestimate the errors at high error rates, and therefore improved alignment schemes have been proposed, using a distance measure which depends on the phonological distance between words (Young, 1997).

On the basis of the alignment, the number of correctly determined words c_w, of substitutions s_w, of insertions i_w, and of deletions d_w is counted. These counts can be related to the total number of words in the reference W, and result in two alternative measures of recognition performance, the word error rate WER and the word accuracy WA:

$$WER = \frac{s_w + i_w + d_w}{W} \quad (\%) \qquad (3.4.1)$$

$$WA = 1 - \frac{s_w + i_w + d_w}{W} \quad (\%) \tag{3.4.2}$$
$$= \frac{c_w - i_w}{c_w + s_w + d_w} \quad (\%)$$
$$= 1 - WER$$

Complementary performance measures can be defined on the sentence level. Designating the number of substituted sentences s_s, inserted sentences i_s, deleted sentences d_s, and the total number of sentences S, a sentence error rate SER and a sentence accuracy SA can be calculated as follows:

$$SER = \frac{s_s + i_s + d_s}{S} \quad (\%) \tag{3.4.3}$$

$$SA = 1 - \frac{s_s + i_s + d_s}{S} \quad (\%) \tag{3.4.4}$$
$$= 1 - SER$$

In general, SA is lower than WA, because a single misrecognized word in a sentence impacts the SA measure. It may however become higher than word accuracy, especially when many single-word sentences are correctly recognized.

Isolated word recognizers should provide an output hypothesis for each input word or utterance. Input and output words can be directly compared, and the performance measures can be defined as follows:

$$WER_{iso} = \frac{s_w + d_w}{W} \quad (\%) \tag{3.4.5}$$

$$WA_{iso} = 1 - \frac{s_w + d_w}{W} \quad (\%) \tag{3.4.6}$$
$$= 1 - WER_{iso}$$

Instead of the insertions, the number of false alarms in a time period can be counted, see van Leeuwen and Steeneken (1997). An additional performance measure for both continuous and isolated word recognizers is the response time, defined as the average time it takes to output the recognized word string after the input has been uttered.

In conjunction with these measures, different types of significance tests are recommended in order to determine whether a difference in accuracy or error rate is statistically significant. For sentence errors, the McNemar (MN) test is recommended by DARPA, and for word error rates the Matched-Pair-Sentence-Segment-Word-Error test (MAPSSWE), see Pallett et al. (1990b). The McNemar test requires errors made by an algorithm to be independent events, which can be regarded as being satisfied only for sentences and isolated words. The MAPSSWE test assumes that the output of the algorithm can be divided into segments (e.g. phrases or sentences) in which the errors are independent of errors made in other segments (Gillick and Cox, 1989). In recent years, also the Signed-Pair test and the Wilcoxon signed rank (WSR) test have been proposed for comparing word accuracy and word error rates. Signed-Pair and WSR are standard non-parametric tests to determine whether two pairs of samples are from the same distribution. These tests are relevant to the word error rates found for individual speakers, and they are particularly sensitive to the number of speakers in the test set.

The information which is provided by WER and SER is however not always sufficient to describe and compare CSR performance (Strik et al., 2000). Especially SER seems to be problematic, because it penalizes an entire sentence independently of the number of errors in the sentence. Strik et al. (2001) propose two additional measures on the sentence level, namely the number of errors per sentence, NES, and the word error per sentence, WES, defined as follows:

$$NES(k) = s_w(k) + d_w(k) + i_w(k) \quad (3.4.7)$$

$$WES(k) = \frac{NES(k)}{w(k)} \quad (3.4.8)$$

with $s_w(k)$, $i_w(k)$ and $d_w(k)$ being the number of substituted, inserted and deleted words in sentence k, and $w(k)$ the total number of words in sentence k. WER is, in principle, a weighted average of $WES(k)$, with weighting factors $w(k)$. WER is related to the average $\overline{NES}$ by a constant factor, namely

$$\overline{NES} = \frac{\sum_{k=1}^{S} NES(k)}{S} = \frac{WER \cdot W}{S} \quad (3.4.9)$$

$$\overline{WES} = \frac{\sum_{k=1}^{S} WES(k)}{S} \quad (3.4.10)$$

thus

$$WER = \sum_{k=1}^{S} \frac{NES(k)}{W} = \sum_{k=1}^{S} \frac{w(k) \cdot WES(k)}{W} \quad (3.4.11)$$

W being the total number of words in the test set. The advantage of these measures is that they provide more information on the degree of recognition accuracy at the sentence level than just right or wrong. NES and WES should be investigated with either the T-test or the WSR test to determine statistical significance (Strik et al., 2001).

Unfortunately, speech recognizers are not always confronted with "legal" words which are contained in the vocabulary. Following Kamm et al. (1997a), user input can be broadly classified into speech and non-speech, speech being either in-vocabulary (only legal words and phrases), embedded (legal words and phrases embedded into extraneous speech), multiple keywords, or out-of-vocabulary (OOV). In an analysis of speech data from a voice dialling service according to these categories, Narayanan et al. (1998) found about 56.6% in-vocabulary words, 11.5% OOV words, and 32% no speech. On the basis of this classification, additional performance measures can be calculated (Kamm et al., 1997a):

- HC_{U1}: The total number of in-vocabulary correctly recognized utterances, divided by the total number of in-vocabulary utterances; indicates the performance on in-vocabulary speech in a user-centric way, because correctly rejected utterances are regarded as an error.

- HC_{S1}: The total number of in-vocabulary correctly recognized or correctly rejected utterances, divided by the total number of in-vocabulary utterances; indicates the performance on in-vocabulary speech in a system-centric way, in contrast to HC_{U1}.

- HC_{U2}: The total number of utterances correctly recognized, divided by the total number of utterances with foreground speech; indicates the performance on all user input, in a user-centric way.

- HC_{S2}: The total number of in-vocabulary, embedded and related utterances correctly recognized or correctly rejected, divided by the total number of utterances with foreground speech; the system-centric analogon to HC_{U2}.

These measures may give a more analytic view of the reasons for recognition errors from the user's utterances.

For a diagnostic assessment of speech recognizers, it is possible to generate confusion matrices, either on the word or on the phonemic level, see e.g. Steeneken and van Velden (1989a), van Leeuwen and Steeneken (1997), and van Leeuwen and de Louwere (1999). Such confusion matrices point at weaknesses of the specific recognizers, and may help to improve them in a very efficient way. Word confusability may also be modelled and predicted, and several proposals have been made in this respect, e.g. by Moore (1985), Roe and Riley (1994), Lindberg (1994) and Trancoso et al. (1995). The models can be used e.g. to simulate the recognition performance in a WoZ test set-up – even beyond the capabilities of current state-of-the-art recognizers, and to select adequate vocabularies and grammars. Lindberg (1994) describes the development of a prediction model for the effects of vocabulary and SNR, ignoring supra-segmental and contextual effects, however, which have been shown to affect phoneme similarities used as an input to the model (Bronkhorst et al., 1993).

In order to characterize the performance of a speech recognizer independently of the speech database, several authors have proposed to use human recognition performance as a reference, e.g. Moore (1985) or Cox et al. (1997). The idea is to describe ASR performance in terms of a controlled degradation of the human recognition performance in order to reach the same level as the recognizer. When noise is used as the controlled degradation, the signal-to-noise ratio at which the human performance is equal to the recognizer one's is called Human Equivalent Noise Ratio, HENR (Moore, 1985). Cox et al. (1997) experimented with several types of degradations (additive white noise, additive speech amplitude-modulated noise, de-tuned AM receiver distortion, speech spectrum inversion, noise added to an LPC error-signal or LPC coefficients, and time-frequency-modulation TFM). They propose to use TFM, because no prominent threshold effects were found for this type of degradation, and because an approximately linear decrease of the human recognition performance as a function of the modulation depth was found. The variation due to different levels of TFM was much greater than that due to different listening subjects.

Steeneken and van Velden (1989a,b) propose to diagnostically assess recognizers by manipulating human speech (so-called RAMOS method). They selected a specific CVC test vocabulary and manipulated the formant spectrum, the fundamental frequency and the temporal distribution via an LPC analysis and re-synthesis. Peckham et al. (1990) describe a predictive method for recognizer sensitivity analysis (RSA) and performance prediction under real-life conditions. It is based on the assumption that speech variability affecting ASR performance can be described by a small number of parameters, and that the performance of the ASR system under consideration can be predicted by a particular combination of these parameters. The parameters relate mostly to speaker variability (voice quality, speaking consistency) and to vocabulary effects.

The effects of language and of the environment have been investigated with the help of different reference recognizer implementations and specific noise databases, for example:

- The COST 249 SpeechDat multilingual reference recognizer for multilingual recognition research (Johansen et al., 2000). It is a phoneme recognizer with a language-independent training procedure, relying on the HTK toolkit (Young et al., 1997b) and the SpeechDat(II) database (Höge et al., 1999).

- Standards for distributed speech recognition developed by the ETSI STQ-AURORA DSR Working Group (Hirsch and Pearce, 2000; ETSI Standard ES 201 108, 2000). In distributed speech recognition, the feature analysis is carried out in the telephone user interface, and the recognition at a central location in the network. This facilitates a secure transmission of the extracted features over impaired mobile channels. So far, AURORA has defined a clean and noisy speech database (based on TIDigits), front-ends for feature extraction, and an HMM recognition back-end which is based on the HTK toolkit (Young et al., 2000). Recognizers set up in this way can be used to obtain comparable recognition results for speaker-independent recognition of connected words in the presence of additive noise and convolutional distortions. An example in this respect is given in Chapter 4.

When new recognizers and environmental conditions are investigated, such references may provide a certain degree of comparability. They should however not be misunderstood as universal references, because the purpose they have been developed for is specific and limited.

The question arises as to which level of ASR performance is necessary in order to reach a sufficient level of user satisfaction in actual application scenarios. Chang (2000) estimates that for most subscriber applications a minimum required WA would be around 90% or better. He emphasizes the need for testing these figures in a realistic application scenario, taking into account task complexity, speaker factors, and transmission conditions. In the French prototype RailTel evaluation, recognition errors occurred in 34.8% of the queries, but the scenario failure rate was only 28% (Lamel et al., 1997). 80% of the task failures were due to recognition and understanding errors, 14% due to dialogue management errors, and 6% due to information retrieval errors. The WER in this case was 18%.

The figures show that WA or WER may not be a good indicator of the usefulness of the recognizer in an SDS application. In practice, it is not always clear what counts as a word (treatment of word boundaries in the recognizer), and inflections often do not affect the meaning representation, but the WER. Different words of a user utterance have different information contents, and by robust partial parsing the meaning of a sentence can often be understood without a complete recognition of each single word (Hutchinson, 2001). Part of these shortcomings can be circumvented by counting a WER only on information-carrying words, see e.g. Möller and Bourlard (2002). When the speech recognizer is coupled to a language understanding module, a better measure of the whole recognition-plus-understanding performance is the frequency of returning the correct meaning by the understanding component, e.g. in terms of correct attribute-value pairs. Such measures will be discussed in the next section.

3.5 Speech and Natural Language Understanding Assessment

Whereas for dictation tasks word accuracy will be an appropriate measure of system performance, many other tasks do not rely on a word-by-word recognition, rather than on the extraction of semantic units from the user input. Such an extraction may be effective even if the WER is high. It is therefore necessary to describe the ability of the language understanding component to extract the semantic content of a user utterance which is important for the subsequent dialogue. The term 'understanding' is commonly used for this purpose in the literature, and it will be used this way in the following section.

Sematic concept extraction can be defined by taking speech or natural language as an input. 'Speech understanding' is defined here synonymously to 'spoken language understanding': The input is speech, the outputs are semantic concepts. It is closely related to 'natural language understanding' which takes transcriptions or texts as an input. Natural language understanding can be expressed in terms of tasks such as message understanding, text understanding, or information extraction, and application domains such as database query or machine translation. Understanding can be reached either by a collaborative interaction between system and user, making use of questions and confirmation (the approach which is of interest for SDSs), or by extracting information without any feedback or interaction (e.g. text summarization). The focus of the following discussion will be on interactive speech and language understanding, giving some references to other domains of language understanding where appropriate. For a more detailed discussion of the activities undertaken in natural language processing areas the reader is referred to Sparck-Jones and Gallier (1996).

Assessment of speech and natural language understanding seems to be more difficult than speech recognition assessment (Price, 1990), because

- the phenomena of interest occur less frequently, i.e. the number of syntactic/semantic phenomena is lower than the number of phones/words,
- semantic is far more domain-dependent than phonetics or phonology, and
- because of the lack of agreement of what constitutes a 'correct' semantic analysis.

In fact, the output of an ASR system is clearly defined as a character string containing the words that were actually spoken, and it is relatively easy to define the reference answer and compare it to the output of a particular ASR system. Each of these steps – specifying an output format, determining the reference output, and comparing reference answer to the system output – is a difficult task for speech and language understanding systems (Bates et al., 1990). The principle of a corpus-based technology assessment can however be applied in both cases.

Speech understanding has been assessed first by monitoring the ability of systems to generate appropriate database queries from spoken input questions (Bates et al., 1990; Pallett et al., 1990a; Price, 1990). The performance was thus related to a template-filling task which can be scored on measures of precision and completeness. As speech is taken as the input, the results show the influence of the speech recognition, language understanding and the database components of a system. The ATIS speech understanding evaluation campaigns organized under the DARPA program between 1990 and 1994 are a good example for this approach (Price, 1990). Test data was first collected in WoZ scenarios, transcribed according to pre-defined conventions, and classified into one of three classes: Class A utterances which can be interpreted without additional context, class D utterances which require a prior context setting, and class X utterances which cannot be evaluated in terms of a reference database answer (Bates et al., 1990). Initial evaluation was carried out uniquely on the class A utterances, and consisted in a comparison to two canonical reference answers which were generated by a wizard: A minimal canonical answer which includes all mandatory tuples, and a maximum answer that can also include supplementary tuples, and which is generated algorithmically from the minimal answer. A correct answer must show all tuples of the minimum answer and not more tuples than contained in the maximum answer.

For the comparison between system answer and reference answer, interpretation principles turned out to be necessary (Bates et al., 1990). The answers were made in a common answer specification format (Ramshaw and Boisen, 1990; Hirschman et al., 1993; Polifroni et al., 1998) which permitted automatic scoring with a comparator program. Each answer was counted as correct ($AN{:}CO$), incorrect ($AN{:}IC$), or failed, i.e. no answer ($AN{:}FA$). Two global system understanding scores were calculated from these figures, namely the DARPA score $DARPA_s$ and the weighted error $DARPA_e$ (Polifroni et al., 1992; Goodine et al., 1992), relating these figures to the number of user questions per dialogue, # USER QUESTIONS:

$$DARPA_s = \frac{AN{:}CO - AN{:}IC}{\#\ \text{USER QUESTIONS}} \quad (3.5.1)$$

$$DARPA_e = \frac{AN{:}FA + 2 \cdot AN{:}IC}{\#\ \text{USER QUESTIONS}} \quad (3.5.2)$$

The weighting of failed and incorrect system answers is chosen somehow arbitrarily. Experiments from Hirschman and Pao (1993) showed that subjects were able to detect system errors before making their next query in about 90% of the cases, so that only a part of the system errors caused subjects to loose time in recovery. This lead to a reduced weighting factor for the incorrect system answers of about 1.25. The measures defined by DARPA did not take into account the number of partially correct system answers, $AN{:}PA$, e.g. when a system provides an uncomplete list of available options to the user. For the experiments described in Chapter 6, this was taken into account by modifying the weighted error to

$$DARPA_{me} = \frac{AN{:}FA + 2 \cdot (AN{:}IC + AN{:}PA)}{\#\ \text{USER QUESTIONS}} \quad (3.5.3)$$

The weighting of two for the partially answered questions has also been chosen arbitrarily, taking the expected effect of errors in the particular application into account. Other weightings might be appropriate for different tasks.

The DARPA ATIS evaluations coincided with the beginning of formal evaluation of written text, as was addressed in the Message Understanding Conference series. An overview of the results achieved in that framework is given by Hirschman (1998) and Sundheim (2001). The evaluation also made use of a reference answer key prepared by a human analyst, and system answers were classified in terms of correct or incorrect items. Different scoring functions were calculated from these basic counts, but mostly precision (number of slots filled correctly divided by the number of fills the system attempted) and recall (number of slots filled correctly divided by the number of possible correct fills taken from the human-prepared reference) were used. The Message Understanding Conference series was followed by a Document Understanding Conference series concentrating on text summarization[1]. Approaches similar to the ATIS one have also been made for natural language database queries, see Jarke et al. (1985) and Sparck-Jones and Gallier (1996).

Although the evaluation scheme followed by DARPA was highly elaborated and automatized, it had several inherent limitations. Adhering to the principle of reference answers, it was not designed for giving information on interaction characteristics, or on the usefulness of system answers. In a dialogue, users do not always direct complete queries to the system, but provide single pieces of information which have to be interpreted in the light of the dialogue history (class D utterances according to the classification given above, which have never been fully addressed by ATIS). For an SDS-based service, better performance measures may be obtained by analyzing the semantic concepts which have been extracted from each individual utterance with the help of the parser. Thus, alternative simple measures of speech and language understanding are the number of user utterances which have been parsed correctly (*PA*:*CO*), partially correctly (*PA*:*PA*), or which failed the parsing process (*PA*:*FA*). The semantic concepts are mainly expressed by so-called attribute-value matrices, AVMs. In such a matrix, each piece of information which can and has to be understood by the system in order to fulfill the task (e.g. the departure city or date for a railway information service) is associated with its respective value (e.g. from Torino or on Monday 21st). Such a pair is called an attribute-value pair, AVP.

On the basis of reference AVP keys determined by an expert, similar metrics as the ones used for ASR performance characterization can be defined. Designating s_{AVP}, i_{AVP} and d_{AVP} the number of substituted, inserted and deleted

[1]See `http://duc.nist.gov/`.

AVPs in an utterance, respectively, and N_{AVP} the total number of AVPs in the utterance, measures of the semantic concept accuracy similar to the word accuracy and word error rate are defined as follows:

$$IC = 1 - \frac{s_{AVP} + i_{AVP} + d_{AVP}}{N_{AVP}} \quad (\%) \tag{3.5.4}$$

$$CER = \frac{s_{AVP} + i_{AVP} + d_{AVP}}{N_{AVP}} \quad (\%) \tag{3.5.5}$$

IC is called the information content or concept accuracy (Gerbino et al., 1993; Simpson and Fraser, 1993; Boros et al., 1996; Billi et al., 1996), and CER the concept error rate, sometimes keyword error rate. Incorrectly extracted concepts may be labelled with respect to the expected source of the problem, namely the ASR or the language understanding component (Lamel et al., 1997). When all concepts in a specific utterance are regarded as an ensemble, an understanding accuracy UA can be defined as

$$UA = \frac{PA:CO}{\#\ \text{USER TURNS}} \quad (\%) \tag{3.5.6}$$

This entity is similar to the SA for describing ASR performance. A comparison between SA and UA which is documented for the JUPITER weather information system (Zue et al., 2000) shows that many utterances that contain recognition errors can nevertheless be correctly understood.

Apart from the utterance-level speech understanding metrics, Glass et al. (2000) define two understanding metrics on the dialogue level. These metrics quantify how efficiently a user can provide new information to a system (query density, QD), and how efficiently the system can absorb information from the user (concept efficiency, CE). QD measures the mean number of new concepts introduced by the user:

$$QD = \frac{1}{N_d} \sum_{i=1}^{N_d} \frac{N_u(i)}{N_q(i)} \tag{3.5.7}$$

where N_d is the number of dialogues, $N_q(i)$ is the total number of user queries in the ith dialogue, and $N_u(i)$ is the number of unique concepts newly understood by the system in the ith dialogue. CE quantifies the average number of turns which is necessary for each concept to be understood by the system:

$$CE = \frac{1}{N_d} \sum_{i=1}^{N_d} \frac{N_u(i)}{N_c(i)} \tag{3.5.8}$$

where $N_c(i)$ is the total number of concepts in the ith dialogue. Here, a concept is counted whenever it was uttered by the user and not yet understood by the system.

A set of 13 dialogue-level speech understanding metrics has been defined by Higashinaka et al. (2003). It consists of five traditional metrics for the current dialogue turn (s_{AVP}, i_{AVP}, d_{AVP}, IC and CER), the same five metrics calculated only for slots which have been updated during the current turn, as well as of three metrics calculated on the filled slots only. The authors used this set for predicting dialogue duration via regression models similar to the one of the PARADISE framework, see Section 6.3.1. The concept accuracy on updated slots turned out to be a major predictor of dialogue duration.

All mentioned approaches regard the language understanding component in principle as a black box, and only assess its output on the basis of the extracted concepts. The "Challenge" campaign initiated by the French CNRS tries to give more generic (open to other domains) and more diagnostic (describing the sources of errors) information on the sources of errors in speech understanding (Antoine et al., 2002). Other approaches in this direction are the Declaration-Control-Reference method and variants of it, see Antoine et al. (2000) and Kurdi and Ahafhaf (2002). A general overview of assessment methods for parsers can be found in Black (1997) or Carroll et al. (1998).

3.6 Speaker Recognition Assessment

Speaker recognition systems are integrated into a part of SDS applications in which security is a relevant issue. This might be the case for bank account information and transactions, domestic control, or other transaction services operated over the phone. The systems will not be addressed in full detail here. Instead, the reader is referred to a complete discussion given by Bimbot and Chollet (1997) from where most of the following information has been taken.

The performance of speaker recognition systems is influenced by a number of factors which are mainly related to the characteristics of the speech signal used as an input. Consequently, these factors should be catered for in the assessment process:

- Input speech quality: Talking environment, speech acquisition system, speech transmission channel, input speech level.

- Temporal drift: Variation of voice characteristics over time, training effects.

- Speech quantity and variety for training and testing.

- Speaker population size and typology.

- Speaker objectives (motivation) and other human factors: Cooperative registered users, uncooperative registered users, acquainted impostors with knowledge of a genuine user, unacquainted impostors, casual users and impostors. The general motivation and the task motivation are additional factors, e.g. with respect to the potential benefits of a successful recognition or an imposture.

The test protocol has to be designed in order to take the relevant factors into account and model them in the test speaker population. For example, Bimbot and Chollet (1997) recommend two distinct imposture experiments, one with same-gender imposture attempts, and one with cross-gender impostures. Also for speaker recognition, standardized databases and a set of baseline experiments have been developed (Hennebert et al., 2000; Melin and Lindberg, 1996), e.g. in the COST 250 action "Speaker Recognition in Telephony". In this way the comparability between systems and test sites can be considerably increased.

Two main types of speaker recognition have to be distinguished: Speaker verification (SV), whose task is to decide whether or not an unlabelled voice belongs to a specific known reference speaker, and speaker identification (SI), whose task is to classify an unlabelled voice token as belonging to one of a set of reference speakers. SI may be performed on an open or closed set of speakers. In practical applications, an open-set SI requires an additional outcome of rejection, in the case that the applicant speaker is an impostor. In SV, an acceptance decision is required when the applicant speaker is the genuine speaker, or a rejection when the applicant speaker is an impostor.

Depending on the recognition task, three types of error may occur:

- Misclassification γ, when a registered speaker is mistaken for another registered speaker (open-set and closed-set SI).
- False acceptance β, when an impostor is accepted as the speaker he claimed he was (SV and open-set SI).
- False rejection α, when a genuine speaker is rejected (SV and open-set SI)

All these errors can be quantified in terms of a performance metrics. They can be specified for each speaker i (denoted γ_i, $\beta_{\bar{i}}$ and α_i), or average values can be determined as a mean for all speakers (denoted $\overline{\gamma}$, $\overline{\beta}$ and $\overline{\alpha}$), or for the speech sample test set (denoted γ, β and α). Sometimes, it is also useful to specify gender-balanced values when the test population is composed of males and females.

Starting from this metrics, different types of performance characteristics may be specified. For closed-set SI, the three misclassification rates (γ_i, $\overline{\gamma}$, γ) are defined, from which the latter two are identical for an identical number or utterances per speaker. Additional information provides a mistrust rate for the probability that the speaker is not really the classified one, and a confidence rank as a similarity measure. For SV, the false rejection rates (α_i, $\overline{\alpha}$, α), the false acceptance rates against a speaker i ($\beta_{\bar{i}}$, $\overline{\beta}$, β), and the imposture rates from an impostor j ($\tilde{\beta}_j$, $\overline{\tilde{\beta}}$, $\tilde{\beta}$) can be given, where $\overline{\beta} = \overline{\tilde{\beta}}$ and $\beta = \tilde{\beta}$. The distribution between false acceptance and false rejection will normally depend on a threshold which can be set in different ways (speaker-dependent or speaker-independent, a priori or a posteriori), according to the application requirements. In order to show the influence of the threshold setting, an equal error rate EER can be calculated for which $\alpha = \beta = \epsilon$, or a Receiver Operating Characteristic (ROC) can be determined showing the false acceptance rate as a function of the false rejection rate, $\beta = f(\alpha)$. In open-set SI, all three types of errors can occur. Misclassification can be either subsumed under the false acceptance cases, resulting in the same set of measures as in SV, or it can be be kept distinct, leading to three different measures α, β and γ. In this case, the ROC curve has to be plotted in a three-dimensional space $\phi(\alpha, \beta, \gamma) = 0$.

Apart from the error rates, other types of information may be useful for describing the system performance. Oglesby (1995) used a set of five attributes which form a "performance profile" for a SV system under test. Apart from the mentioned measures, he included the memory requirements for the speaker models, the level of difficulty inherent in the speaker population, and the quality and the quantity of available speech material. The author argues that such a performance profile gives a much completer picture of a SV system than a simple EER does. From a user's point of view, the performance measures may be accompanied by the number of successful access cases from the user, and the average time for a successful access. These figures can then be set into a relationship with the user's satisfaction ratings with respect to enrolment, usability, and security or expected chance for fraud.

As for speech recognition, there are no clearly defined levels of required performance for a speaker recognizer. Such a level will surely depend on the specific application area and on the task which may be carried out. Requirements may differ between the service provider's and the user's point of view. Caminero et al. (2002) report on evaluation results from users of a speaker verification system (SAFE). Although the majority of the users judged the "perceived security" of the system between 'high' and 'very high', the opinion of whether the system could be used in practical application scenarios differed considerably: 90% judged that the system might be used for domestic control, 85% for access to bank account information, 80% for transaction services over the phone, and only 60% for access to bank account operations.

3.7 Speech Output Assessment

Although the assessment and evaluation of synthesized speech has been a topic of research for more than a decade, there is still no universal agreement (and a lot of misuse) on appropriate test methods for specific purposes. In fact, the speech output component can be assessed and evaluated on different levels such as segmental intelligibility, naturalness, pleasantness, comprehension tasks, or global acceptance in a real-life application. The choice of an appropriate test method depends on the aim of the assessment/evaluation (system characterization, comparative evaluation for a given purpose, etc.); thus, no universal test method can be defined which would provide optimum results in all cases. Nearly all tests which are applied to assess speech output quality rely on the acoustic signal, and involve human test subjects. This section will therefore address these types of tests. It will principally concentrate on the generation of the acoustic signal, be it synthesized or naturally produced, and disregard the response generation which can finally only be evaluated in the context of the whole dialogue system, see Section 3.8. A more detailed overview of assessment and evaluation methods can be found e.g. in van Bezooijen and van Heuven (1997) or in Francis and Nusbaum (1999).

A thoughtless application of standardized test methods may lead to wrong assumptions about whether the quality level provided by current state-of-the-art speech synthesis is sufficient for SDS-based services. Whereas most text-to-speech systems often yield a high segmental intelligibility in laboratory tests (in some cases comparable to that of natural speech), application-typical evaluations often show a poorer performance (Silverman et al., 1990; Balestri et al., 1992). Potential reasons for this discrepancy are the test material and task, the interference between segmental and supra-segmental structures in the speech material, the more complex information to be conveyed in a real-life situation, the cognitive load on the user, or the mutual influence of acoustic and prosodic components of a TTS system.

In order to properly apply the right test method for a specific assessment or evaluation task, it is useful to systemize the space of available and potential test methodologies and methods. Van Bezooijen and van Heuven (1997) present some steps towards a taxonomy for this purpose. They distinguish the following test parameters:

- Glass box vs. black box tests.
- Laboratory vs. field tests.
- Linguistic vs. acoustic aspects to be tested.
- Use of humans vs. automated measurement objects.
- Judgment testing vs. functional testing.
- Global vs. analytical assessment and evaluation.

A more ambitious systematization is proposed by Jekosch (2000, 2001). She identifies functional parameters of speech quality measurements:

- Objective of the measurement: Includes function and conditions of the measurement.

- Measurand: Includes feature aspects (form, function, content) and function of the measurand.

- Measuring object: Speech and language material and its generation.

- Measurement procedure: Method of measurement or investigative process, and principle of the measurement.

The parameters can be used to define and develop appropriate test methods for a given purpose. Jekosch and Pols (1994) propose the following steps in order to reach this goal: (1) Analysis of the application conditions, resulting in a feature profile of the application scenario; (2) Definition of the best possible test matching this feature profile; (3) Comparison of what is desired and what is available in terms of standardized assessment tests; (4) Adaptation of available test(s) or development of one's own test.

In most cases, the speech output component of an SDS is assessed in an auditory listening-only test with human subjects. Test subjects should be chosen to be representative of the later user group, unless specific diagnostic judgments are to be collected. In general, untrained test subjects can be expected to provide more valid judgments of 'absolute' quality levels than trained ones or experts, in the sense that they are not influenced by their knowledge of the system under investigation. On the other hand, experts or trained test subjects can produce more analytical results. The experience of test subjects is a critical factor when synthesized speech is to be assessed, because intelligibility may increase with the exposure of non-expert listeners to synthetic speech (van Bezooijen and van Heuven, 1997).

Subjects are asked to judge test stimuli which are presented to them in a listening-only situation. The stimuli differ with respect to their length and complexity, the linguistic level (words, sentences, paragraphs), the stimulus set (open or closed), the meaningfulness (meaningful, meaningless, mixed), and the representativeness (phonetically or phonemically balanced or unbalanced distribution, representative for the application or not). The subjects' task is either to identify or verify the stimuli in an open or closed set of answers, or to judge different aspects of what they perceived. Judgment may be done in a completely unguided way, or via interviews or questionnaires, often using judgment scales. It refers to an individual stimulus or to a comparison between two or more stimuli. In either case, the judgment will reflect some type of implicit or explicit reference.

The question of reference is an important one for the quality assessment and evaluation of synthesized speech. In contrast to references for speech recognition or speech understanding, it refers however to the perception of the user. When no explicit references are given to the user, he/she will make use of his/her internal references in the judgment. Explicit references can be either topline references, baseline references, or scalable references. Such references can be chosen on a segmental (e.g. high-quality or coded speech as a topline, or concatenations of co-articulatory neutral phones as a baseline), prosodic (natural prosody as a topline, and original durations and flat melody as a baseline), voice characteristic (target speaker as a topline for a personalized speech output), or on an overall quality level, see van Bezooijen and van Heuven (1997).

A scalable reference which is often used for the evaluation of transmitted speech in telephony is calibrated signal-correlated noise generated with the help of a modulated noise reference unit, MNRU (ITU-T Rec. P.810, 1996). Because it is perceptively not similar to the degradations of current speech synthesizers, the use of an MNRU often leads to reference conditions outside the range of systems to be assessed (Salza et al., 1996; Klaus et al., 1997). Time-and-frequency warping (TFW) has been developed as an alternative, producing a controlled "wow and flutter" effect by speeding up and slowing down the speech signal (Johnston, 1997). It is however still perceptively different from the one produced by modern corpus-based synthesizers.

The experimental design has to be chosen to equilibrate between test conditions, speech material, and voices, e.g. using a Graeco Latin Square or a Balanced Block design (Cochran and Cox, 1992). The length of individual test sessions should be limited to a maximum which the test subjects can tolerate without fatigue. Speech samples should be played back with a high-quality test management equipment in order not to introduce additional degradations to the ones under investigation (e.g. the ones stemming from the synthesized speech samples, and potential transmission degradations, see Chapter 5). They should be calibrated to a common level, e.g. -26dB below the overload point of the digital system which is the recommended level for narrow-band telephony. On the acoustic side, this level should correspond to a listening level of 79 dB SPL.

The listening set-up should reflect the situation which will be encountered in the later real-life application. For a telephone-based dialogue service, handset or hands-free terminals should be used as listening user interfaces. Because of the variety of different telephone handsets available, an 'ideal' handset with a frequency response calibrated to the one of an intermediate reference system, IRS (ITU-T Rec. P.48, 1988), is commonly used. Test results are finally analyzed by means of an analysis of variance (ANOVA) to test the significance of the experiment factors, and to find confidence intervals for the individual mean values. More general information on the test set-up and administration can be found in ITU-T Rec. P.800 (1996) or in Arden (1997).

When the speech output module as a whole is to be evaluated in ins functional context, black box test methods using judgment scales are commonly applied. Different aspects of global quality such as intelligibility, naturalness, comprehensibility, listening-effort, or cognitive load should nevertheless be taken into account. The principle of functional testing will be discussed in more detail in Section 5.1. The method which is currently recommended by the ITU-T is a standard listening-only test, with stimuli which are representative for SDS-based telephone services, see ITU-T Rec. P.85 (1994). In addition to the judgment task, test subjects have to answer content-related questions so that their focus of attention remains on a content level during the test. It is recommended that the following set of five-point category scales[2] is given to the subjects in two separate questionnaires (type Q and I):

- Acceptance: Do you think that this voice could be used for such an information service by telephone? Yes; no. (Q and I)

- Overall impression: How do you rate the quality of the sound of what you have just heard? Excellent; good; fair; poor; bad. (Q and I)

- Listening effort: How would you describe the effort you were required to make in order to understand the message? Complete relaxation possible, no effort required; attention necessary, no appreciable effort required; moderate effort required; effort required; no meaning understood with any feasible effort. (I)

- Comprehension problems: Did you find certain words hard to understand? Never; rarely; occasionally; often; all of the time. (I)

- Articulation: Were the sounds distinguishable? Yes, very clear; yes, clear enough; fairly clear; no, not very clear; no, not at all. (I)

- Pronunciation: Did you notice any anomalies in pronunciation? No; yes, but not annoying; yes, slightly annoying; yes, annoying; yes, very annoying. (Q)

- Speaking rate: The average speed or delivery was: Much faster than preferred; faster than preferred; preferred; slower than preferred; much slower than preferred. (Q)

[2] A brief discussion on scaling is given in Section 3.8.6.

- Voice pleasantness: How would you describe the voice? Very pleasant; pleasant; fair; unpleasant; very unpleasant. (Q)

An example for a functional test based on this principle is described in Chapter 5. Other approaches include judgments on naturalness and intelligibility, e.g. the SAM overall quality test (van Bezooijen and van Heuven, 1997).

In order to obtain analytic information on the individual components of a speech synthesizer, a number of specific glass box tests have been developed. They refer to linguistic aspects like text pre-processing, grapheme-to-phoneme conversion, word stress, morphological decomposition, syntactic parsing, and sentence stress, as well as to acoustic aspects like segmental quality at the word or sentence level, prosodic aspects, and voice characteristics. For a discussion of the most important methods see van Bezooijen and van Heuven (1997) and van Bezooijen and Pols (1990). On the segmental level, examples include the diagnostic rhyme test (DRT) and the modified rhyme test (MRT), the SAM Standard Segmental Test, the CLuster IDentification test (CLID), the Bellcore test, and tests with semantically unpredictable sentences (SUS). Prosodic evaluation can be done either on a formal or on a functional level, and using different presentation methods and scales (paired comparison or single stimulus, category judgment or magnitude estimation). Mariniak and Mersdorf (1994) and Sonntag and Portele (1997) describe methods for assessing the prosody of synthetic speech without interference from the segmental level, using test stimuli that convey only intensity, fundamental frequency, and temporal structure (e.g. re-iterant intonation by Mersdorf (2001), or artificial voice signals, sinusoidal waveforms, sawtooth signals, etc.). Other tests concentrate on the prosodic function, e.g. in terms of illocutionary acts (SAM Prosodic Function Test), see van Bezooijen and van Heuven (1997).

A specific acoustic aspect is the voice of the machine agent. Voice characteristics are the mean pitch level, mean loudness, mean tempo, harshness, creak, whisper, tongue body orientation, dialect, accent, etc. They help the listener to make an idea of the speakers mood, personality, physical size, gender, age, regional background, socio-economic status, health, and identity. This information is not consciously used by the listener, but helps him to infer information, and may have practical consequences as to the listener's attitude towards the machine agent, and to his/her interpretation of the agent's message. A general aspect of the voice which is often assessed is voice pleasantness, e.g. using the approach in ITU-T Rec. P.85 (1994). More diagnostic assessment of voice characteristics is mainly restricted to the judgment of natural speech, see van Bezooijen and van Heuven (1997). However, these authors state that the effect of voice characteristics on the overall quality of services is still rather unclear.

Several comparative studies between different evaluation methods have been reported in the literature. Kraft and Portele (1995) compared five German synthesis systems using a cluster identification test for segmental intelligibility, a paired-comparison test for addressing general acceptance of the sentence level, and a category rating test on the paragraph level. The authors conclude that each test yielded results in its own right, and that a comprehensive assessment of speech synthesis systems demands cross-tests in order to relate individual quality aspects to each other. Salza et al. (1996) used a single stimulus rating according to ITU-T Rec. P.85 (1994) (but without comprehension questions) and a paired comparison technique. They found good agreement between the two methods in terms of overall quality. The most important aspects used by the subjects to differentiate between systems were global impression, voice, articulation and pronunciation.

3.8 SDS Assessment and Evaluation

At the beginning of this chapter it was stated that the assessment or system components, in the way it was described in the previous sections, is not sufficient for addressing the overall quality of an SDS-based service. Analytical measures of system performance are a valuable source of information in describing how the individual parts of the system fulfill their task. They may however sometimes miss the relevant contributors to the overall performance of the system, and to the quality perceived by the user. For example, erroneous speech recognition or speech understanding may be compensated for by the discourse processing component, without affecting the overall system quality. For this reason, interaction experiments with real or test users are indispensable when the quality of an SDS and of a telecommunication service relying on it are to be determined.

In laboratory experiments, both types of information can be obtained in parallel: During the dialogue of a user with the system under test, interaction parameters can be collected. These parameters can partly be measured instrumentally, from log files which are produced by the dialogue system. Other parameters can only be determined with the help of experts who annotate a completed dialogue with respect to certain characteristics (e.g. task fulfillment, contextual appropriateness of system utterances, etc.). After each interaction, test subjects are given a questionnaire, or they are interviewed in order to collect judgments on the perceived quality features.

In a field test situation with real users, instrumentally logged interaction parameters are often the unique source of information for the service provider in order to monitor the quality of the system. The amount of data which can be collected with an operating service may however become very large. In this case, it is important to define a core set of metrics which describe system performance, and to have tools at hand which automatize a large part of the data analysis process. The task of the human evaluator is then to interpret this data, and to estimate the effect of the collected performance measures on the quality which would be perceived by a (prototypical) user. Provided that both types of information are available, relationships between interaction parameters and subjective judgments can be established. An example for such a complete evaluation is given in Chapter 6.

In the following subsections, the principle set-up and the parameters of evaluation experiments with entire spoken dialogue systems are described. The experiments can either be carried out with fully working systems, or with the help of a wizard simulating missing parts of the system, or the system as a whole. In order to obtain valid results, the (simulated) system, the test users, and the experimental task have to fulfil several requirements, see Sections 3.8.1 to 3.8.3. The interactions are logged and annotated by a human expert (Section 3.8.4), so that interaction parameters can be calculated. Staring from a literature survey, the author collected a large set of such interaction parameters. They are presented in Section 3.8.5 and discussed with respect to the QoS taxonomy. The same taxonomy can be used to classify the quality judgements obtained from the users, see Section 3.8.6. Finally, a short overview of evaluation methods addressing the usability of systems and services is given (Section 3.8.7). The section concludes with a list of references to assessment and evaluation examples documented in the recent literature.

3.8.1 Experimental Set-Up

In order to carry out interaction experiments with human (test) users, a set-up providing the full functionality of the system has to be implemented. The exact nature of the set-up will depend on the availability of system components, and thus on the system development phase. If system components have not yet been implemented, or if an implementation would be unfeasible (e.g. due to the lack of data) or uneconomic, simulation of the respective components or of the system as a whole is required.

The simulation of the interactive system by a human being, i.e. the Wizard-of-Oz (WoZ) simulation, is a well-accepted technique in the system development phase. At the same time, it serves as a tool for evaluation of the system-in-the-loop, or of the bionic system (half system, half wizard). The idea is to simulate the system taking spoken language as an input, process it in some *principled* way, and generate spoken language responses to the user. In order to provide a realistic telephone service situation, speech input and output should be provided to the users via a simulated or real telephone connection, using a standard user interface. Detailed descriptions of the set-up of WoZ experiments can be found in Fraser and Gilbert (1991b), Bernsen et al. (1998), Andernach et al. (1993), and Dahlbäck et al. (1993).

The interaction between the human user and the wizard can be characterized by a number of variables which are either under the control of the experimenter (control variables), accessible and measurable by the experimenter (response variables), or confounding factors where the experimenter has no interest in or no control over. Fraser and Gilbert (1991b) identified the following three major types of variables:

- Subject variables: Recognition by the subject (acoustic recognition, lexical recognition), production by the subject (accent, voice quality, dialect, verbosity, politeness), subject's knowledge (domain expertise, system expertise, prior information about the system), etc.
- Wizard variables: Recognition (acoustic, lexical, syntactic and pragmatic phenomena), production (voice quality, intonation, syntax, response time), dialogue model, system capabilities, training, etc.
- Communication channel variables: General speech input/output characteristics (transmission channel, user interface), filter variables (e.g. deliberately introduced recognition errors, de-humanized voice), duplex capability or barge-in, etc.

Some of these variables will be control variables of the experiment, e.g. those related to the dialogue model or to the speech input and output capability of the simulated system. Confounding factors can be catered for by careful experimental design procedures, namely by a complete or partially complete within-subject design.

WoZ simulations can be used advantageously in cases where the human capacities are superior to those of computers, as it is currently the case for speech understanding or speech output. Because the system can be evaluated before it has been fully set up, the performance of certain system components can be simulated to a degree which is beyond the current state-of-the-art. Thus, an extrapolation to technologies which will be available in the future becomes possible (Jack et al., 1992). WoZ simulation allows testing of feasibility, coverage, and adequacy prior to implementation, in a very economic way. High degrees of novelty and complex interaction models may be easier to simulate in WoZ than to implement in an implement-test-revise approach. However, the latter is likely to gain ground as standard software and prototyping tools emerge, and in industrial settings where platforms are largely available. WoZ is nevertheless worthwhile if the application is at high risk, and the costs to re-build the system are sufficiently high (Bernsen et al., 1998).

A main characteristic of a WoZ simulation is that the test subjects do not realize that the system they are interacting with is simulated. Evidence given by Fraser and Gilbert (1991b) and Dahlbäck et al. (1993) shows that this goal can be reached in nearly 100% of all cases if the simulation is carefully designed. The most important aspect for the illusion of the subject is the speech input and output capability of the system. Several authors emphasize that the illusion of a dialogue with a computer should be supported by voice distortion, e.g.

Fraser and Gilbert (1991a) and Amalberti et al. (1993). However, Dybkjær et al. (1993) report that no significant effect of voice disguise could be observed in their experiments, probably because other system parameters had already caused the same effect (e.g. system directedness).

WoZ simulations should provide a realistic simulation of the system's functionality. Therefore, an exact description of the system functionality and of the system behavior is needed before the WoZ simulation can be set up. It is important that the wizard adheres to this description, and ignores any superior knowledge and skills which he/she has compared to the system to be tested. This requires a significant amount of training and support for the wizard. Because a human would intuitively use its superior skills, the work of the wizard should be automatized as far as possible. A number of tools have been developed for this purpose. They usually consist in a representation of the interaction model, e.g. in terms of a visual graph (Bernsen et al., 1998) or of a rapid prototyping software tool (Dudda, 2001; Skowronek, 2002), filters for the system input and output channel (e.g. structured audio playback, voice disguise, and recognition simulators), and other support tools like interaction logging (audio, text, video) and domain support (e.g. timetables). The following tools can be seen as typical examples:

- The JIM (Just sIMulation) software for the initiation of contact to the test subjects via telephone, the delivery of dialogue prompts according to the dialogue state which is specified by a finite-state network, the registering of keystrokes from the wizard as result of the user utterances, the on-line generation of recognition errors, and the logging of supplementary data such as timing, statistics, etc. (Jack et al., 1992; Foster et al., 1993).

- The ARNE simulation environment consisting of a response editor with canned texts and templates, a database query editor, the ability to access various background systems, and an interaction log with time stamps (Dahlbäck et al., 1993).

- A JAVA-based GUI for flexible response generation and speech output to the user, based on synthesized or pre-recorded speech (Rehmann, 1999).

- A CSLU-based WoZ workbench for simulating a restaurant information system, see Dudda (2001) and Skowronek (2002). The workbench consists of an automatic finite-state-model for implementing the dialogue manager (including variable confirmation strategies), a recognition simulation tool (see Section 6.1.2), a flexible speech output generation from pre-recorded or synthesized speech, and a number of wizard support tools for administering the experimental set-up and data analysis. The workbench will be described in more detail in Section 6.1, and it was used in all experiments of Chapter 6.

With the help of WoZ simulations, it is easily possible to set up parametrizable versions of a system. The CSLU-based WoZ workbench and the JIM simulation allow speech input performance to be set in a controlled way, making use of the wizard's transcription of the user utterance and a defined error generation protocol. The CSLU workbench is also able to generate different types of speech output (pre-recorded and synthesized) for different parts of the dialogue. Different confirmation strategies can be applied, in a fully or semi-automatic way. Smith and Gordon (1997) report on studies where the initiative of the system is parametrizable. Such parametrizable simulations are very efficient tools for system enhancement, because they help to identify those elements of a system which most critically affect quality.

3.8.2 Test Subjects

The general rule for psychoacoustic experiments is that the choice of test subjects should be guided by the purpose of the test. For example, analytic assessment of specific system characteristics will only be possible for trained test subjects who are experts of the system under consideration. However, this group will not be able to judge overall aspects of system quality in a way which would not be influenced by their knowledge of the system. Valid overall quality judgments can only be expected from test subjects which match as close as possible the group of future service users.

An overview of user factors has already been given in Section 3.1.3. Some of these factors are responsible for the acoustic and linguistic characteristics of the speech produced by the user, namely age and gender, physical status, speaking rate, vocal effort, native language, dialect, or accent. Because these factors may be very critical for the speech recognition and understanding performance, test subjects with significantly different characteristics will not be able to use the system in a comparable way. Thus, quality judgments obtained from a user group differing in the acoustic and language characteristics might not reflect the quality which can be expected for the target user group. User groups are however variable and ill-defined. A service which is open to the general public will sooner or later be confronted with a large range of different users. Testing with specified users outside the target user group will therefore provide a measure of system robustness with respect to the user characteristics.

A second group of user factors is related to the experience and expertise with the system, the task, and the domain. Several investigations show that user experience affects a large range of speech and dialogue characteristics. Delogu et al. (1993) report that users have the tendency to solve more problems per call when they get used to the system, and that the interaction gets shorter. Kamm et al. (1997a) showed that the number of in-vocabulary utterances increased when the users became familiar with the system. At the same time, the task completion rate increased. In the MASK kiosk evaluation (Lamel et al., 1998a,

2002), system familiarity lead to a reduced number of user inputs and help messages, and to a reduced transaction time. Also in this case the task success rate increased. Shriberg et al. (1992) report higher recognition accuracy with increasing system familiarity (specifically for talkers with low initial recognition performance), probably due to a lower perplexity of the words produced by the users, and to a lower number of OOV words. For two subsequent dialogues carried out with a home banking system, Larsen (2004) reports a reduction in dialogue duration by 10 to 15%, a significant reduction of task failure, and a significant increase in the number of user initiatives between the two dialogues.

Kamm et al. (1998) compared the task performance and quality judgments of novice users without prior training, novice users who were given a four-minute tutorial, as well as expert users familiar with the system. It turned out that user experience with the system had an impact on both task performance (perceived and instrumental measures of task completion) and user satisfaction with the system. Novice users who were given a tutorial performed almost at the expert level, and their satisfaction was higher than for non-tutorial novices. Although task performance of the non-tutorial novices increased within three dialogues, the corresponding satisfaction scores did not reach the level of tutorial novices. Most of the dialogue cost measures were significantly higher for the non-tutorial novices than for both other groups.

Users seem to develop specific interaction patterns when they get familiar with a system. Sturm et al. (2002a) suppose that such a pattern is a perceived optimal balance between the effort each individual user has to put into the interaction, and the efficiency (defined as the time for task completion) with which the interaction takes place. In their evaluation of a multimodal train timetable information service, they found that nearly all users developed stable patterns with the system, but that the patterns were not identical for all users. Thus, even after training sessions the system still has to cope with different interaction approaches from the individual users. Cookson (1988) observed that the interaction pattern may depend on the recognition accuracy which can be reached for certain users. In her evaluation of the VODIS system, male and female users developed a different behavior, i.e. they used different words for the same command, because the overall recognition rates differed significantly between these two user groups.

The interaction pattern a user develops may also reflect his or her beliefs of the machine agent. Souvignier et al. (2000) point out that the user may have a "cognitive model" of the system which reflects what is regarded as the current system belief. Such a model is partly determined by the utterances given to the system, and partly by the utterances coming from the system. The user generally assumes that his/her utterances are well understood by the system. In case of misunderstandings, the user gets confused, and dialogue flow problems are likely to occur. Another source of divergence between the user's cognitive model and the system's beliefs is that the system has access to secondary information sources such as an application database. The user may be surprised if confronted with information which he/she didn't provide. To avoid this problem, it is important that the system beliefs are made transparent to the user. Thus, a compromise has to be found between system verbosity, reliability, and dialogue duration. This compromise may also depend on the system and task/domain expertise of the user.

3.8.3 Experimental Task

A user factor which cannot be described easily is the motivation for using a service. Because of the lack of a real motivation, laboratory tests often make use of experimental tasks which the subjects have to carry out. The experimental task provides an explicit goal, but this goal should not be confused with a goal which a user would like to reach in a real-life situation. Because of this discrepancy, valid user judgments on system usefulness and acceptability cannot easily be obtained in a laboratory test set-up.

In a laboratory test, the experimental task is defined by a scenario description. A scenario describes a particular task which the subject has to perform through interaction with the system, e.g. to collect information about a specific train connection, or to search for a specific restaurant (Bernsen et al., 1998). Using a pre-defined scenario gives maximum control over the task carried out by the user, while at the same time covering a wide range of possible situations (and possible problems) in the interaction. Scenarios can be designed on purpose for testing specific system functionalities (so-called development scenarios), or for covering a wide range of potential interaction situations which is desirable for evaluation. Thus, development scenarios are usually different from evaluation scenarios.

Scenarios help to find different weaknesses in a dialogue, and thereby to increase the usability and acceptability of the final system. They define user goals in terms of the task and the sub-domain addressed in a dialogue, and are a pre-requisite to determine whether the user achieved his/her goal. Without a pre-defined scenario it will be extremely difficult to compare results obtained in different dialogues, because the user requests will differ and may fall outside the system domain knowledge. If the influence of the task is a factor which has to be investigated in the experiment, the experimenter needs to ensure that all users execute the same tasks. This can only be reached by pre-defined scenarios.

Unfortunately, pre-defined scenarios may have some negative effects on the user's behavior. Although they do not provide a real-life goal for the test subjects, scenarios prime the users on how to interact with the system. Written scenarios may invite the test subjects to imitate the language given in the scenario, leading to read-aloud instead of spontaneous speech. Walker et al. (1998a) showed that the choice of scenarios influenced the solution strategies which were most effective for resolving the task. In particular, it seemed that scenarios defined in a table format primed the users not to take the initiative, and gave the impression that the user's role would be restricted to providing values for the items listed in the table (Walker et al., 2002a). Lamel et al. (1997) report that test subjects carrying out pre-defined scenarios are not particularly concerned about the response of the system, as they do not really need the information. As a result, task success did not seem to have an influence on the usability judgments of the test subjects. Goodine et al. (1992) report that many test subjects did not read the instructions carefully, and ignored or misinterpreted key restrictions in the scenarios. Sturm et al. (1999) observed that subjects were more willing to accept incorrect information than can be expected in real-life situations, because they do not really need the provided information, and sometimes they do not even notice that they were given the wrong information. The same fact was observed by Niculescu (2002). Sanderman et al. (1998) reported problems in using scenarios for eliciting complex negotiations, because subjects often did not respect the described constraints, either because they did not pay attention to or did not understand what was requested.

The priming effect on the user's language can be reduced with the help of graphical scenario descriptions. Graphical scenarios have successfully been used by Dudda (2001), Dybkjær et al. (1995) and Bernsen et al. (1998), and examples can be found in Appendix D.2. Bernsen et al. (1998) and Dybkjær et al. (1995) report on comparative experiments with written and graphical scenarios. They show that the massive priming effect of written scenarios could be nearly completely avoided by a graphical representation, but that the diversity of linguistic items (total number of words, number of OOV words) was similar in both cases. Apparently, language diversity cannot be increased with graphical scenario representations, and still has to be assured by collecting utterances from a sufficiently high number of different users, e.g. in a field test situation. Another attempt to reduce priming was made in the DARPA Communicator program, presenting recorded speech descriptions of the tasks to the test subjects and advising them to take own notes (Walker et al., 2002a). In this way, it is hoped that the involved comprehension and memory processes would leave the subjects with an encoding of the meaning of the task description, but not with a representation of the surface form. An empirical proof of this assumption, however, has not yet been given.

3.8.4 Dialogue Analysis and Annotation

In the system development and operation phases, it is very useful for evaluation experts to analyze a corpus of recorded dialogues by means of log files, and to investigate system and user behavior at specific points in the dialogue. Tracing of recorded dialogues helps to identify and localize interaction problems very efficiently, and to find principled solutions which will also enhance the system behavior in other dialogue situations. At the same time, it is possible to annotate the dialogue in order to extract quantitative information which can be used to describe system performance on different levels. Both aspects will be briefly addressed in the following section.

Dialogue analysis should be performed in a formalized way in order to efficiently identify and classify interaction problems. Bernsen et al. (1998) describe such a formalized analysis which is based on the cooperativity guidelines presented in Section 2.2.3. Each interaction problem is marked and labelled with the expected source of the problem: Either a dialogue design error, or a "user error". Assuming that each design error can be seen as a case of non-cooperative system behavior, the violated guideline can be identified, and a cure in terms of a change of the interaction model can be proposed. A "user error" is defined as "a case in which a user does not behave in accordance with the full normative model of the dialogue". The normative model consists of explicit designer instructions provided to the user via the scenario, explicit system instructions to the user, explicit system utterances in the course of the dialogue, and implicit system instructions. The following types of "user errors" are distinguished:

- E1: Misunderstanding of scenario. This error can only occur in controlled laboratory tests.

- E2: Ignoring clear system feedback. May be reduced by encouraging attentive listening.

- E3: Responding to a question different from the clear system question, either (a) by a straight wrong response, or (b) by an indirect user response which would be acceptable in HHI, but which cannot be handled due to system's lack of inference capabilities.

- E4: Change through comments. This error would be acceptable in HHI, and results from the system's limited understanding or interaction capabilities.

- E5: Asking questions. Once again, this is acceptable in HHI and requires better mixed-initiative capabilities of the system.

- E6: Answering several questions at a time, either (a) due to natural "information packages", e.g. date and time, or (b) to naturally occurring slips of tongue.

- E7: Thinking aloud.

- E8: Straight non-cooperativity from the user.

An analysis carried out on interactions with the Danish flight inquiry system showed the E3b, E4, E5 and E6a are not really user errors, because they may

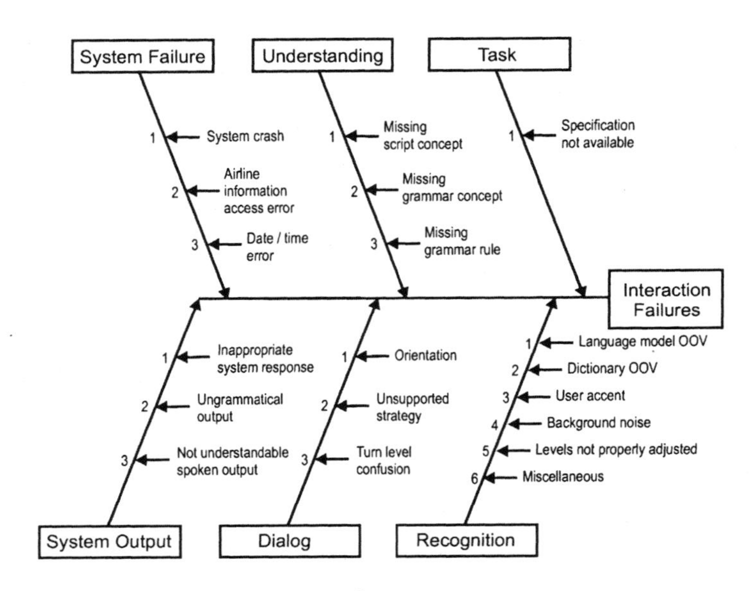

Figure 3.1. Categorization scheme for causes of interaction failure in the Communicator system (Constantinides and Rudnicky, 1999).

have been caused by cognitive overload, and thus indicate a system design problem. They may be reduced by changing the interaction model.

A different proposal to classify interaction problems was made by Constantinides and Rudnicky (1999), grounded on the analysis of safety-critical systems. The aim of their analysis scheme is to identify the source of interaction problems in terms of the responsible system component. A system expert or external evaluator traces a recorded dialogue with the help of information sources like audio files, log files with the decoded and parsed utterances, or database information. The expert then characterizes interaction failures (e.g. no task success, system does not pay attention to user action, sessions terminated prematurely, expression of confusion or frustration by the user, inappropriate user output generated by the system) according to the items of a "fishbone" diagram, and briefly describes how the conversation ended. Fishbone categories were chosen to visually organize causes-and-effects in a particular system, see Figure 3.1. They are described by typifying examples and questions which help to localize each interaction problem in the right category. Bengler (2000) proposes a different, less elaborated error taxonomy for classifying errors in driving situations.

In order to quantify the interaction behavior of the system and of the user, and to calculate interaction parameters, it is necessary to annotate dialogue transcriptions. Dialogues can be annotated on different levels, e.g. in terms of transactions, conversational games, or moves (Carletta et al., 1997). When annotation is carried out on an utterance level, it is difficult to explicitly cover system feedback and mixed-initiative. Annotation on a dialogue level may however miss important information on the utterance level. Most annotation schemes differ with respect to the details of the target categories, and consequently with respect to the extent to which inter-expert agreement can be reached. In general, annotation of low-level linguistic phenomena is relatively straightforward, since agreement on the choices of units can often be reached. On the other hand, higher level annotation depends on the choice of the underlying linguistic theories which are often not universally accepted (Flammia and Zue, 1995). Thus, high level annotation is usually less reliable. One approach to dealing with this problem is to provide a set of minimal theory-neutral annotations, as has been used in the Penn Treebank (Marcus et al., 1993). Another way is to annotate a dialogue simultaneously on several levels of abstraction, see e.g. Heeman et al. (2002).

The reliability of classification tasks performed by experts or naïve coders has been addressed by Carletta and his colleagues (Carletta, 1996; Carletta et al., 1997). Different types of reliability have to be distinguished: Test-retest reliability (stability), tested by asking a single coder to code the same data several times; inter-coder reliability (reproducibility), tested by training several coders and comparing their results; and accuracy, which requires coders to code in the same way as a known defined standard. Carletta (1996) proposes the κ coefficient of agreement in order to measure the pairwise agreement of coders performing category judgment tasks, as was defined by Siegel and Castellan (1988). κ is corrected for the expected chance agreement, and defined as follows:

$$\kappa = \frac{P(A) - P(E)}{1 - P(E)} \tag{3.8.1}$$

where $P(A)$ is the proportion of times that the coders agree, and $P(E)$ the proportion of times that they are expected to agree by chance. When there is no other agreement than that which would be expected by chance, κ is zero, and for total agreement $\kappa = 1$. For dialogue annotation tasks, $\kappa > 0.8$ can be seen as a good reliability, whereas for $0.67 < \kappa < 0.8$ only tentative conclusions should be drawn[3]. κ can also be used as a metric for task success, based on the agreement between AVPs for the actual dialogue and the reference AVPs. Different measures of task success will be discussed in Section 3.8.5.

Dialogue annotation can be largely facilitated and made more reliable with the help of software tools. Such tools support the annotation expert by a graphical representation of the allowed categories, or by giving the possibility to

[3] κ may also become negative when $P(A) < P(E)$.

listen to user and system turns, showing ASR and language understanding output, or the application database content (Polifroni et al., 1998). The EU DiET program (Diagnostic and Evaluation Tools for Natural Language Applications) developed a comprehensive environment for the construction, annotation and maintenance of structured reference data, including tools for the glass box evaluation of natural language applications (Netter et al., 1998). Other examples include "Nota Bene" from MIT (Flammia and Zue, 1995), the MATE workbench (Klein et al., 1998), or DialogueView for annotation on different abstraction levels (Heeman et al., 2002).

Several annotation schemes have been developed for collecting information which can directly be used in the system evaluation phase. Walker et al. (2001) describe the DATE dialogue act tagger (Dialogue Act Tagging for Evaluation) which is used in the DARPA Communicator program: DATE classifies each system utterance according to three orthogonal dimensions: A speech act dimension (capturing the communicative goal), a conversational dimension (about task, about communication, about situation/frame), and a task-subtask dimension which is domain-dependent (e.g. departure city or ground hotel reservation). Using the DATE tool, utterances can be identified and labelled automatically by comparison to a database of hand-labelled templates. Depending on the databases used for training and testing, as well as on the interaction situation through which the data has been collected (HHI or HMI), automatic tagging performance ranges between 49 and 99% (Hastie et al., 2002a; Prasad and Walker, 2002). DATE tags have been used as input parameters to the PARADISE quality prediction framework, see Section 6.3.1.3. It has to be emphasized that the tagging only refers to the system utterances, which can be expected to be more homogenous than user utterances.

Devillers et al. (2002) describe an annotation scheme which tries to capture dialogue progression and user emotions. User emotions are annotated by experts from the audio log files. Dialogue progression is presented on two axes: An axe P presenting the "good" progression of the dialogue, and an axe A representing the "accidents" between the system and the user. Dialogues are annotated by incrementally assigning values of +1 to either the P or A axis for each turn (resulting in an overall number of turns $A + P$). The authors determine a residual error which represents the difference between a perfect (without misunderstandings or errors) and the real dialogue. The residual error is incremented when A is incremented, and decremented when P is incremented. Dialogue progress annotation was used to predict dialogue "smoothness", which is expected to be positively correlated to P, and negatively to A and to the residual error.

Evaluation annotation tools are most useful if they are able to automatically extract interaction parameters from the annotated dialogues. Such interaction parameters are expected to be related to user quality perceptions, and to give an impression of the overall quality of the system or service. For the experimental evaluations described in Chapter 6, a Tcl/Tk-based annotation tool has been developed by Skowronek (2002). It is designed to extract most of the known interaction parameters from log files of laboratory interactions with the restaurant information system BoRIS. The tool facilitates the annotation by an expert, in that it gives a relatively precise definition of each annotation task, see Appendix D.3. Following these definitions, the expert has to perform the following steps on each dialogue:

- Definition of the scenario AVM (has to be performed only once for each scenario).
- Literal transcription of the user utterances. In the case of simulated ASR, the wizard's transcriptions during the interaction are taken as the initial transcriptions, and the annotation task is limited to the correction of typing errors.
- Marking of user barge-in attempts.
- Definition of the modified AVM. The initial scenario AVM has to be modified in case of user inattention, or because the systems did not find an appropriate solution and asked for modifications.
- Tagging of task success, based on an automatic proposal calculated from the AVMs.
- Tagging of contextual appropriateness for each system utterance (cf. next section).
- Tagging of system and user correction turns.
- Tagging of cancel attempts from the user.
- Tagging of user help requests.
- Tagging of user questions, and whether these questions have been correctly, incorrectly or partially correctly answered or not.
- Tagging of AVPs extracted by the system from each user utterance, with respect to correct identifications, substitutions, deletions or insertions.

After the final annotation step, the tool automatically calculates a list of interaction parameters and writes them to an evaluation log file. Nearly all known interaction parameters which were applicable to the system under consideration could be extracted, see Section 6.1.3. A similar XML-based tool has been developed by Charfuelán et al. (2002), however with a more limited range of interaction parameters. This tool also allows annotated dialogues to be traced in retrospective, in order to collect diagnostic information on interaction failures.

3.8.5 Interaction Parameters

It has been pointed out that user judgments are the only way to investigate quality percepts. They are, however, time-consuming and expensive to collect. For the developers of SDSs, it is therefore interesting to identify parameters describing specific aspects of the interaction. Interaction parameters may be instrumentally measurable, or they can be extracted from log files with the help of expert annotations, cf. the discussion in the previous section. Although they provide useful information on the perceived quality of the service, there is no general relationship between interaction parameters and specific quality features. Word accuracy, which is a common measure to describe the performance of a speech recognizer, can be taken as an example. The designer can tune the ASR system to increase the word accuracy, but it cannot be determined beforehand how this will affect perceived system understanding, system usability, or user satisfaction.

Interaction parameters can be collected during and after user interactions with the system under consideration. They refer to the characteristics of the system, of the user, and of the interaction between both. Usually, these influences cannot be separated, because the user behavior is strongly influenced by the one of the system. Nevertheless, it is possible to decide whether a specific parameter mainly describes the behavior of the system or that of the user (elicited by the system), and some glass box measures clearly refer to system (component) capabilities. Interaction parameters can be calculated on a word, sentence or utterance, or on a dialogue level. In case of word and utterance level parameters, average values are often calculated for each dialogue. Parameters may be collected in WoZ scenarios instead of real user-system interactions, but one has to be aware of the limitations of a human wizard, e.g. with respect to the response delay. Thus, it has to be ensured that the parameters reflect the behavior of the system to be set up, and not the limitations of the human wizard. Parameters collected in a WoZ scenario may however be of value for judging the experimental set-up and the system development: For example, the number of ad-hoc generated system responses in a bionic wizard experiment gives an indication of the coverage of interaction situations by the available dialogue model (Bernsen et al., 1998).

SDSs are of such high complexity that a description of system behavior and a comparison between systems needs to be based on a multitude of different parameters (Simpson and Fraser, 1993). In this way, evaluation results can be expected to better capture different quality dimensions. In the following, a review of parameters which have been used in assessment and evaluation experiments during the past ten years is presented. These parameters can broadly be labelled as follows:

- Dialogue- and communication-related parameters.

- Meta-communication-related parameters.
- Cooperativity-related parameters.
- Task-related parameters.
- Speech-input-related parameters.

A complete list of parameters is given in Tables A.1 to A.16 of Appendix A, including a definition, the potential values they can take, the required transparency of the system (black box or glass box), the type of measurement required to determine the parameter (instrumental or expert-based), the interaction level they refer to (word, utterance or dialogue level), and whether they primarily address the behavior of the system or that of the user. The parameters will be briefly discussed in this section.

Parameters which refer to the overall dialogue and to the communication of information give a very rough indication of how the interaction takes place, without specifying the communicative function of the individual turns in detail. Parameters belonging to this group are duration parameters (overall dialogue duration, duration of system and user turns, system and user response delay), and word- and turn-related parameters (average number of system and user turns, average number of system and user words, words per system and per user turn, number of system and user questions). Two additional parameters have to be noted: The query density gives an indication of how efficiently a user can provide new information to a system, and the concept efficiency describes how efficiently the system can absorb this information from the user. These parameters have already been defined in Section 3.5. They will be grouped under the more general communication category here, because they result from the system's interaction capabilities as a whole, and not purely from the language understanding capabilities. All measures are of global character and refer to the dialogue as a whole, although they are partly calculated on an utterance level. Global parameters are sometimes problematic, because the individual differences in cognitive skill may be large in relation to the system-originated differences, and because subjects might learn strategies for task solution which have a significant impact on global parameters.

The second group of parameters refers to the system's meta-communication capabilities. These parameters quantify the number of system and user utterances which are part of meta-communication, i.e. the communication about communication. Meta-communication is an important issue in HMI because of the limited understanding and reasoning capabilities of the machine agent. Most of the parameters are calculated as the absolute number of utterances in a dialogue which relate to a specific interaction problem, and are then averaged over a set of dialogues. They include the number of help requests from the user, of time-out prompts from the system, of system rejections of user utterances in the case that no semantic content could be extracted from a user utterance (ASR rejections), of diagnostic system error messages, of barge-in attempts from the user, and of user attempts to cancel a previous action. The ability of the system (and of the user) to recover from interaction problems is described in an explicit way by the correction rate, namely the percentage of all (system or user) turns which are primarily concerned with rectifying an interaction problem, and in an implicit way by the IR measure, which quantifies the capacity of the system to regain utterances which have partially failed to be recognized or understood. In contrast to the global measures, all meta-communication-related parameters describe the function of system and user utterances in the communication process.

Cooperativity has been identified as a key aspect of successful HMI. Unfortunately, it is difficult to quantify whether a system behaves cooperatively or not. Several of the dialogue- and meta-communication-related parameters somehow relate to system cooperativity, but they do not attempt to quantify this aspect. A direct measure of cooperativity is the contextual appropriateness parameter CA, first introduced by Simpson and Fraser (1993). Each system utterance has to be judged by experts as to whether it violates one or more of Grice's maxims for cooperativity, see Section 2.2.3. The utterances are classified into the categories of appropriate (not violating Grice's maxims), inappropriate (violating one or more maxim), appropriate/innappropriate (the experts cannot reach agreement in their classification), incomprehensible (the content of the utterance cannot be discerned in the dialogue context), or total failure (no linguistic response from the system). It has to be noted that the classification is not always straightforward, and that interpretation principles may be necessary. Appendix D.3 gives some interpretation principles for the restaurant information system used in the experiments of Chapter 6. Other schemes for classifying appropriateness have been suggested, e.g. by Traum et al. (2004) for a multi-character virtual reality training simulation.

Current state-of-the-art systems enable task-orientated interactions between system and user, and task success is a key issue for the usefulness of a service. Task success may best be determined in a laboratory situation where explicit tasks are given to the test subjects, see Section 3.8.3. However, realistic measures of task success have to take into account potential deviations from the scenario by the user, either because he/she didn't pay attention to the instructions given in the test, or because of his/her inattentiveness to the system utterances, or because the task was unresolvable and had to be modified in the course of the dialogue. Modification of the experimental task is considered in most definitions of task success which are reported in the topic literature. Success may be reached by simply providing the right answer to the constraints set in the instructions, by constraint relaxation from the system or from the user (or both), or by spotting that no answer exists for the defined task. Task failure may be tentatively attributed to the system's or to the user's behavior, the latter however being influenced by the system (cf. the discussion on user errors in Section 3.8.4). Other simple descriptions of task success disregard the possibility of scenario deviations and take a binary decision on the existence and correctness of a task solution reported by the user (Goodine et al., 1992).

A slightly more elaborate approach to determine task success is the κ coefficient which has already been introduced to describe the reliability of coding schemes, see Formula 3.8.1. The κ coefficient for task success is based on the individual AVPs which describe the semantic content of the scenario and the solution reported by the user, and is corrected for the expected chance agreement (Walker et al., 1997). A confusion matrix $M(i, j)$ can be set up for the attributes in the key and in the reported solution. Then, the agreement between key and solution $P(A)$ and the chance agreement $P(E)$ can be calculated from this matrix, see Table A.9. $M(i, j)$ can be calculated for individual dialogues, or for a set of dialogues which belong to a specific system or system configuration.

The task success measures described so far rely on the availability of a simple task coding scheme, namely in terms of an AVM. However, some tasks cannot be characterized as easily, e.g. TV program information (Beringer et al., 2002b). In this case, more elaborated approaches to task success are needed, approaches which usually depend on the type of task under consideration. Proposals have also been made to measure task solution quality. For example, a train connection can be rated with respect to the journey time, the fare, or the number of changes required (the distance not being of primary importance for the user). By their nature, such solution quality measures are heavily dependent on the task itself.

A number of measures related to speech recognition and speech understanding have already been discussed in Sections 3.4 and 3.5. For speech recognition, the most important are WA and WER on the word level, and SA and SER on the utterance (sentence) level. Additional measures include NES and WES, as well as the HC metrics. For speech understanding, two common approaches have to be differentiated. The first one is based on the classification of system answers to user questions into categories of correctly answered, partially correctly answered, incorrectly answered, or failed answers. DARPA measures can be calculated from these categories. The second way is to classify the system's parsing capabilities, either in terms of correctly parsed utterances, or of correctly identified AVPs. On the basis of the identified AVPs global measures such as IC, CER and UA can be calculated.

The majority of interaction parameters listed in the tables describe the behavior of the system, which is obvious because it is the system and service quality which is of interest. In addition to these, user-related parameters can be defined. They are specific to the test user group, but may nevertheless be closely related to quality features perceived by the user. Delogu et al. (1993) indicate several parameters which are related to the performance of the user in accomplishing the task (task comprehension, number of completed tasks, number of times the user ignores greeting formalities, etc.), and to the user's flexibility (number of user recovery attempts, number of successful recoveries from the user). Sutton et al. (1995) propose a classification scheme for user answers to a spoken census questionnaire, and distinguish between adequate user answers, inadequate answers, qualified answers expressing uncertainty, requests for clarification from the user, interruptions, and refusals. Hirschman et al. (1993) classify user responses as to whether they provide new information, repeat previously given information, or rephrase it. Such a classification captures the strategies which users apply to recover from misunderstandings, and helps the system developer to choose optimal recovery strategies as well.

The mentioned interaction parameters are related to different quality aspects which can be identified by means of the QoS taxonomy described in Section 2.3.1. In Tables 3.1 and 3.2, a tentative classification has been performed which is based on the definition of the respective parameters, as well as on common sense. Details of this classification may be disputed, because some parameters relate to several categories of the taxonomy. The proposed classification will be used as a basis for a thorough analysis of empirical data collected with the mentioned restaurant information system BoRIS, see Section 6.2.

Interestingly, a number of parameters can be found which relate to the lower level categories, with the exception of the speech output quality category. In fact, only very few approaches which instrumentally address speech output quality have been made. Instrumental measures related to speech intelligibility are defined e.g. in IEC Standard 60268-16 (1998), but they have not been designed to describe the intelligibility of synthesized speech in a telephone environment. Chu and Peng (2001) propose a concatenation cost measure which can be calculated from the input text and the speech database of a concatenative TTS system, and which shows high correlations to MOS scores obtained in an auditory experiment. The measure is however specific to the TTS system and its concatenation corpus, and it is questionable in how far general predictors of overall quality – or of naturalness, as claimed by Chu and Peng – can be constructed on the basis of concatenation cost measures. Ehrette et al. (2003) try to predict mean user judgments on different aspects of a naturally produced system voice with the help of instrumentally extracted parameters describing prosody (global and dynamic behavior), speech spectrum, waveform, and articulation. Although the prediction accuracy is relatively good, the number of input parameters needed for a successful prediction is very high compared to the number of predicted user judgments. So far, the model has only been tested on a single system utterance pronounced by 20 different speakers.

Table 3.1. Classification of interaction parameters in the QoS taxonomy (1).

Category	*Aspect*	*Interaction parameter*
Dialogue cooperativity		– CA: Contextual appropriateness
	Informativeness	– # USER QUESTIONS – # HELP REQUESTS
	Truth and evidence	– $AN{:}CO$, $AN{:}IC$, $AN{:}PA$, $AN{:}FA$ – $DARPA_s$, $DARPA_e$, $DARPA_{me}$
	Relevance	– # BARGE-INS
	Manner	– # SYSTEM TURNS – # SYSTEM WORDS – $WPST$: Words per system turn
	Background knowledge	– # HELP REQUESTS – # CANCEL ATTEMPTS – # BARGE-INS – # TIME-OUT PROMPTS
	Meta-comm. handling	– # SYSTEM ERROR MESSAGES – # HELP REQUESTS – # CANCEL ATTEMPTS – SCT, SCR: System correction – IR: Implicit recovery
Dialogue symmetry	Initiative	– # SYSTEM WORDS, # USER WORDS – # SYSTEM TURNS, # USER TURNS – $WPST$, $WPUT$: Words per system / user turn – # SYSTEM QUESTIONS, # USER QUESTIONS – SCT, UCT, SCR, UCR: Correction rates
	Interaction control	– # BARGE-INS – # HELP REQUESTS – # CANCEL ATTEMPTS – UCT, UCR: User correction – # TIME-OUT PROMPTS
	Partner asymmetry	– # BARGE-INS – # TIME-OUT PROMPTS
Speech I/O quality	Speech output quality	
	Speech input quality	– WA, WER, SA, SER – NES, WES – HC_{U1}, HC_{U2}, HC_{S1}, HC_{S2} – # ASR REJECTIONS – IC, CER: Information content, content error rate – UA: Understanding accuracy – # SYSTEM ERROR MESSAGES – $PA{:}CO$, $PA{:}PA$, $PA{:}FA$: Parsed user utterances – QD, CE

Table 3.2. Classification of interaction parameters in the QoS taxonomy (2).

Category	*Aspect*	*Interaction parameter*
Communic. efficiency	Speed	– TD: Turn duration (STD, UTD) – RD: Response delay (SRD, URD) – # TIME-OUT PROMPTS – # BARGE-INS
	Conciseness	– DD: Dialogue duration – # TURNS (# SYSTEM TURNS, # USER TURNS) – # WORDS (# SYSTEM WORDS, # USER WORDS) – $WPST$, $WPUT$: Words per system/user turn – QD, CE
	Smoothness	– # SYSTEM ERROR MESSAGES – SCT, UCT, SCR, UCR: Correction rates – # CANCEL ATTEMPTS – # HELP REQUESTS – # ASR REJECTIONS – # BARGE-INS – # TIME-OUT PROMPTS
Comfort	Agent personality	
	Cognitive demand	– # TIME-OUT PROMPTS – URD: User response delay
Task efficiency	Task success	– TS: Task success – κ: Kappa coefficient – Task solution – Solution correctness – Solution quality
	Task ease	– # HELP REQUESTS
Service efficiency	Service adequacy	
	Added value	
Usability	Ease of use	
User satisfaction		
Utility		
Acceptability		

For the higher levels of the QoS taxonomy (agent personality, service efficiency, usability, user satisfaction, utility and acceptability), no interaction parameters can be identified which would "naturally" relate to these aspects. Relationships may however turn out when analyzing empirical data for a specific system. The missing classes may indicate a fundamental impossibility to predict complex aspects of interaction quality on the basis of interaction parameters. A deeper analysis of prediction approaches will be presented in Section 6.3.

An interpretation of interaction parameters may be based on experimental findings which are, however, often specific to the considered system or service. An an example, an increased number of time-out prompts may indicate that the user does not know what to say at specific points in a dialogue, or that he/she is confused about system actions (Walker et al., 1998a). Increasing barge-in attempts may simply reflect that the user learned that it is possible to interrupt the system. In contrast, a reduced number may equally indicate that the user does not know what to say to the system. Lengthy user utterances may result from a large amount of initiative attributed to the user. Because this may be problematic for the speech recognition and understanding components of the system, it may be desirable to reduce the user utterance length by transferring initiative back to the system. In general, a decrease of meta-communication-related parameter values (especially of user-initiated meta-communication) can be expected to increase system robustness, dialogue smoothness, and communication efficiency (Bernsen et al., 1998).

3.8.6 Quality Judgments

In order to obtain information about quality features perceived by the user, subjective judgments have to be collected. Two different principles can be applied in the collection: Either to identify the relevant quality features in a more or less unguided way, or to quantify pre-determined aspects of quality as responses to closed questions or judgment scaling tasks. Both ways have their advantages and inconveniences: Open inquiries help to find quality dimensions which would otherwise remain undetected (Pirker et al., 1999), and to identify the aspects of quality which are most relevant from the user's point of view. In this way, the interpretation of closed quantitative judgments can be facilitated. Closed questions or scaling tasks facilitate comparison between subjects and experiments, and give an exact means to quantify user perceptions. They can be carried out relatively easily, and untrained subjects often prefer this method of judgment.

Many researchers adhere to the advantages of closed judgment tasks and collect user quality judgments on a set of closed scales which are labelled according to the aspect to be judged. The scaling will yield valid and reliable results when two main requirements are satisfied: The items to be judged have

	strongly disagree	disagree	neither agree nor disagree	agree	strongly agree
The system was easy to understand.	☐	☐	☐	☐	☐

Figure 3.2. Judgment on a statement in a way which was proposed by Likert (1932).

Quality of the speech:

excellent	good	fair	poor	bad
5	4	3	2	1

Figure 3.3. Graphical representation of a 5-point ACR quality judgment scale (ITU-T Rec. P.800, 1996).

to be chosen adequately and meaningfully, and the scaling measurement has to follow well-established rules. Scaling methods are described in detail in the psychometrics literature, e.g. by Guilford (1954) or by Borg and Staufenbiel (1993). For collecting quality judgments on SDS-based services, most authors use absolute category rating (ACR) scales. An ACR scale consists of a number of discrete categories one of which has to be chosen by the test subject. The categories are displayed visually and may be labelled with attributes for each category, or for the extreme (left-most and right-most) categories only. One possibility is to formulate a statement (e.g. "The system was easy to use."), and then perform the rating on the five categories which are depicted in Figure 3.2. Such a method is based on proposals made by Likert (1932). Numbers are attributed to the categories, depending on whether the statement is positive (from 0 for "strongly disagree" to 4 for "strongly agree") or negative (from 4 for "strongly disagree" to 0 for "strongly agree"), and ratings are summed up for all subjects. Another possibility is to define self-explaining labels for each category, as it is proposed e.g. by the ITU-T for ACR tests. The most well-known scale of this type is the 5-point ACR quality scale defined in ITU-T Rec. P.800 (1996), see Figure 3.3. The mean value over all ratings on this scale is called the mean opinion score, MOS.

Although the MOS scale is commonly used for speech quality judgments, it has a number of disadvantages which result from the choice of available answer options and labels. A discussion can be found in Möller (2000), pp. 68-72. In order to overcome some of the disadvantages, a continuous rating scale, which was first proposed by Bodden and Jekosch (1996), has been used in the experiments of Chapter 6, see Figure 3.4. This scale tries to minimize saturation effects occurring at the scale extremities, supports equal-width categories graphically, and incites the test subjects to make their judgments as fine-graded

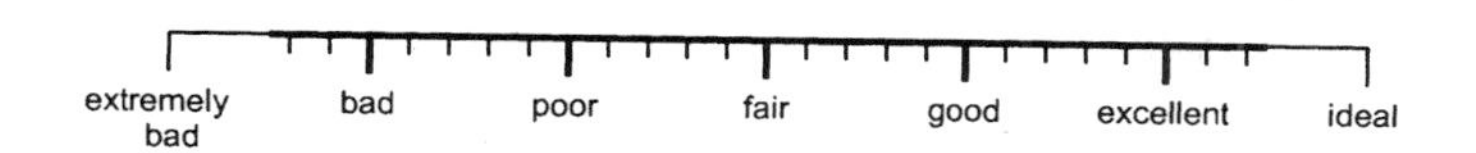

Figure 3.4. Continuous rating scale according to Bodden and Jekosch (1996).

as possible. The scale has been used in two ways: Either for rating attitudes towards a statement, in a similar way as proposed by Likert (but providing only labels for the end categories), or by labelling each main scale tick with an appropriate self-explaining label.

The judgments which are obtained on ACR quality scales or on Likert-type scales can be analyzed by means of barcharts or cumulative distributions. Although the distributions are not necessarily gaussian, it is often recommended to calculate arithmetic mean values (and not medians) over all ratings obtained with a specified system configuration (ITU-T Rec. P.800, 1996; ITU-T Handbook on Telephonometry, 1992; van Bezooijen and van Heuven, 1997). For the mean values, confidence limits are evaluated and significance tests performed by conventional analysis of variance (ANOVA). The assumptions underlying an analysis of variance (gaussian distribution and homogeneity of variances) are not always satisfied; still, this method seems to be robust enough to provide reasonable results also in case of departures from the statistically ideal conditions. In case of a statistically significant effect of one of the variates (system configuration and/or voice, test subject, scenario, order of conditions in the experiment, test session, etc.), a post-hoc test can be used to perform pairwise comparisons among the means and to determine the sources of differences. The Tukey Honestly Significant Difference (HSD) test is recommended for this purpose (Tukey, 1977; ITU-T Rec. P.800, 1996). From a statistical point of view, however, the results should be summarized in terms of a median or mode, and non-parametric tests like the one according to Kruskal and Wallis have to be used for the analysis.

Well-constructed scales will not provide valid information when the quality feature to be judged upon is ill-defined, or when it is not appropriately chosen for the object to be judged. In the literature, a number of questionnaires is reported, each containing different judgment items and scales. An exemplary choice will be briefly described in the following paragraphs. In a laboratory set-up, the questionnaires can be given to the test subjects directly after performing an interaction (potentially reflecting the impression after this interaction), or after a number of interactions (providing some integration over the past experiences). In a field test, the compilation of the questionnaires cannot be strictly controlled, and the judgment usually refers to a number of interactions carried out in a broadly defined time period. It may occur that negative experiences are more prominent and have a stronger influence in the time integration process than positive ones (Duncanson, 1969).

Jack et al. (1992) used two types of questionnaires in the WoZ field test of an automated telephone service for exchanging credit card information: A spoken interview which was carried out with the test subjects directly after the interaction, and a postal questionnaire which was sent back to the evaluator. The telephone interview contained mostly questions for which related instrumentally measurable values were known from log files, e.g.

- how many mistakes the subjects thought they would have made,
- how many mistakes the subjects thought the service made during the interaction,
- how long the subjects thought it took to read a credit card number,

as well as an overall judgment of the service quality. The postal questionnaire consisted of 22 statements to be judged on 7-point Likert-type response scales, which were however not indicated by the authors.

Kamm et al. (1997a) evaluated a voice server developed at AT&T for call control and messaging (Annie) with the help of two surveys, one given after a 5-day interaction period, and one given three months after the initial experiments. The test subjects were asked to judge the following criteria:

- Perceived system failures.
- Perceived success in performing specific tasks offered by the system.
- Perception of the system terminating an interaction unexpectedly.
- Perceived recognition performance.
- Perceived system speed.
- Ease to interrupt system prompts.
- Ease to cancel previous commands.
- Ease of use.

In later evaluations of Annie (Kamm et al., 1998), of the email reader ELVIS (Walker et al., 1998a), and of the train timetable information system TOOT (Litman et al., 1998; Litman and Pan, 1999), a standard 8-to-10-item questionnaire was used at AT&T:

- Did you complete the task? (task completion)
- Was the system easy to understand in this conversation? (TTS performance)

- In this conversation, did the system understand what you said? (ASR performance)
- In this conversation, was it easy to find the schedule you wanted? (task ease)
- Was the pace of interaction with the system appropriate in this conversation? (interaction pace)
- In this conversation, did you know what you could say at each point of the dialogue? (user expertise)
- How often was the system sluggish and slow to reply to you in this conversation? (system response)
- Did the system work the way you expected it to in this conversation? (expected behavior)
- In this conversation, how did the system's voice interface compare to a touch-tone interface? (comparable interface)
- From your current experience with using our system, do you think you'd use this regularly to access train schedules when you are away from your desk? (future use)

Responses usually ranged over five pre-defined values, except for the task completion judgment where the alternatives "yes", "no" and "maybe" were defined. The scores were summed up to calculate a so-called "cumulative satisfaction score" for each dialogue. However, calculation of such an average value attributes an equal weighting to each quality attribute – a hypothesis which is highly questionable with respect to the internal weighting of each attribute by the user.

In a field trial evaluation of the Italian RailTel train timetable information system, a larger number of questions was given to the participating subjects (Billi and Lamel, 1997; Lamel et al., 1997). The questions were grouped into the five categories attitude (A), ease of use (EU), efficiency (E), reliability (R), and user friendliness (UF), and included the following aspects:

- Ease of use. (EU)
- Confusion provoked by the system. (EU)
- Friendliness. (UF)
- Complexity. (EU)
- Future use. (A)
- Reliability. (R)

- Control ability. (UF)
- Concentration required. (EU)
- Efficiency. (E)
- Fluster provoked by the system. (UF)
- Speed. (UF)
- Stress. (A)
- Preference for a human service. (A)
- Complicated to use. (EU)
- Enjoyable to use. (A)
- Need for improvement. (E)
- Politeness. (UF)
- Information obtained from the system. (R)
- Speed compared to a human service. (E)
- Perceived system understanding. (R)

In addition, information on the subjects' travel habits and their computer experience were collected. A slightly different questionnaire was used for the follow-up system ARISE, see Baggia et al. (1998, 2000). The Dutch version of the ARISE system was evaluated on twelve Likert-type scales addressing the subjects' feelings during the interaction, and nine 5-point ACR scales addressing individual aspects of the system. This experiment is described by Sanderman et al. (1998).

Gerbino et al. (1993) report on experiments for evaluating an email reader service developed at CSELT, for the Italian language. Although the test was conducted with naïve users, relatively analytic judgments were requested:

- Was your approach to the system natural?
- Did the system understand, on average, your request?
- When your sentences were not understood, how did you overcome the problem? (repeat, rephrase, eliminate pauses and hesitations, speak louder, change the microphone handle)
- Judge some system features: Is the syntactic coverage enough? Does the dialogue management allow you to control the development of interaction?

> Does the system interpret your requests according to your expectations? In the case of understanding errors, are the recovery strategies useful and complete? Is the use of isolated words for confirmation too restrictive? Are the system answers brief and intelligible?

It remains questionable whether non-expert users are able to give such diagnostic information on individual aspects and components of a system. Nevertheless, the authors did not report any problems which would have been expressed by the test subjects.

For the evaluation of the Danish flight information system, a questionnaire with a number of 5-point ACR scales was used (Bernsen et al., 1998). Ratings on the following attributes/statements were obtained from the subjects: Acceptable, satisfactory, fast, efficient, flexible, kind, stimulating, simple, predictable, reliable, desirable, useful now, useful in the future, made few errors, output quality, ability to use free language, correction was easy, task performance was easy, user preferred the system over a human operator. Dintruff et al. (1985) evaluated the quality of voice input and output for an early speech-based prototype in the office environment. He used twenty 10-point scales which were labelled with antonym adjectives (so-called semantic differential), in the way that has been proposed by Osgood et al. (1957). The questionnaire contained ten "feeling items" (uncomfortable/ comfortable, passive/ active, tense/ relaxed, angry/ friendly, sad/ happy, depersonalized/ individualized, bored/ interested, weak/ strong, inhibited/ spontaneous, dissatisfied/ satisfied) and ten "attitude items" (unfavorable/ favorable, hard to use/ easy to use, unreliable/ reliable, slow/ fast, useless/ useful, rigid/ flexible, inefficient/ efficient, worthless/ valuable, inaccurate/ accurate, inappropriate/ appropriate). For each group a mean "overall feeling measure" and an "overall attitude measure" were calculated. Casali et al. (1990) used 13 bipolar adjective scales to assess user satisfaction in a simulated system for data entry in a computerized database: An acceptable/ unacceptable scale, and 12 other bipolar scales (fast/ slow, accurate/ inaccurate, consistent/ inconsistent, pleasing/ irritating, dependable/ undependable, natural/ unnatural, complete/ incomplete, comfortable/ uncomfortable, friendly/ unfriendly, facilitating/ distracting, simple/ complicated, useless/ useful). Judgments on these scales were combined to form a so-called "acceptability index". Love et al. (1994) designed a questionnaire for automated telephone services which included 18 general service usability attributes, and four attributes which were specific to speech-based telephone services. The attributes were judged on 7-point Likert-type response scales.

The question arises as to which attributes are really appropriate for the judgment task, distinguishable to the test subjects, and potentially independent. Optimally, the evaluator should choose a reduced set of attributes which are easily and commonly interpretable by the subjects, and which provide a maximum amount of information about all relevant quality features of the multidimensional perceptual quality space. The choice of appropriate attributes can be the topic of a limited pre-test to the final experiment, e.g. in the way it was performed by Dudda (2001). Section 6.1.4 gives an example of how the questionnaires for the evaluation of the restaurant information system BoRIS were constructed.

Hone and Graham (2001, 2000) report on a systematic approach to develop a scientifically well-founded questionnaire for the subjective assessment of speech system interfaces (mainly speech input systems), called SASSI. Their aim is to avoid the two main weaknesses of investigations like those cited, namely (1) that the measurement techniques have not been satisfactorily validated against alternative subjective or instrumental measures, and (2) there are no reports on the reliability of the measurement techniques. On the basis of a large pool of attitude statements, an iterative design-application-analysis-redesign procedure was started. The authors extracted a questionnaire with 50 statements which had to be judged on 7-point Likert-type scales. The application of the questionnaire in order to identify the underlying quality dimensions will be discussed in Section 6.2.2. In that section, a number of empirical investigations will be analyzed in order to identify quality dimensions which are commonly observed for SDS-based services. A similarly systematic questionnaire development has only be reported for the questionnaire used by Edinburgh University and British Telecom in the "Intelligent Dialogue Project" (Larsen, 2003).

Even without analyzing empirical data, it is possible to classify the quality aspects and features which are addressed by the mentioned investigations. Once again, the QoS taxonomy described in Section 2.3.1 will be used for this purpose. Tables 3.3 to 3.5 list the quality judgment items or quality features which are addressed in the cited references with respect to the categories of the taxonomy. It can be seen that, up to the user satisfaction level, all quality aspects are addressed by individual items of the questionnaires. On the utility and acceptability level, price and service aspects become important which cannot easily be addressed in a laboratory test situation. In particular the price is an aspect which is often neglected when evaluating SDS-based services. The relationship between speech transmission quality and the price of a telephone service has been tentatively investigated in Möller (2000), pp. 141-145, but the results will not easily be transferable to HMI services. Judgments on "acceptability" obtained in a laboratory setting should also be treated with care, as they might not reflect the actual user behavior when a service is available on the market.

The classification in Tables 3.3 to 3.5 serves two main purposes. On the one hand, quality judgments which have been collected in an experiment can be judged as to which aspect of service quality they reflect. This is important because the individual aspects will contribute with a specific weight to

Table 3.3. Classification of user quality judgments in the QoS taxonomy (1). K97: Kamm et al. (1997a); W98a: Walker et al. (1998a), Litman et al. (1998); L97: Billi and Lamel (1997), Lamel et al. (1997); G93: Gerbino et al. (1993); B98: Bernsen et al. (1998), Dybkjær and Dybkjær (1993); D85: Dintruff et al. (1985); C90: Casali et al. (1990); H01: Hone and Graham (2001), Hone and Graham (2000); L94: Love et al. (1994).

Category	*Aspect*	*Quality judgment item*
Dialogue cooperativity	Informativeness	– Accuracy of information (D85, C90, H01) – Completeness of information (C90)
	Truth and evidence	– System reliability (L97, B98, D85, H01) – Consistency of information (C90)
	Relevance	– Frequency of system errors (B98, H01) – Perceived system understanding (W98a, L97, G93) – Naturalness of system behavior (C90)
	Manner	– Predictability of system behavior (B98) – Naturalness of system behavior (C90)
	Background knowledge	– Predictability of system behavior (B98, H01) – User knew what to do/say at each point in the dialogue (W98a, L94, H01)
	Meta-comm. handling	– Ease to correct mistakes (B98) – Frequency of system errors (B98) – Frequency of system failures (K97) – Frequency of unexpected termination from the system (K97) – Ease to cancel actions (K97)
Dialogue symmetry	Initiative	– Naturalness of system behavior (C90) – Perceived naturalness of user's own behavior (G93) – Perceived user activity (D85)
	Interaction control	– System flexibility (B98, D85, H01) – Ease to interrupt system prompts (K97) – Ease to cancel actions (K97) – Ease to correct mistakes (B98) – Degree of control over the service (L97, L94, H01) – Perceived spontaneity of the user's behavior (D85)
	Partner asymmetry	– Naturalness of system behavior (C90) – Perceived naturalness of user's own behavior (G93) – User knew what to do/say at each point in the dialogue (W98a, L94, H01)
Speech I/O quality	Speech output quality	– Speech intelligibility (W98a) – Voice clarity (L94) – Concentration required (L97, L94)
	Speech input quality	– Perceived recognition performance (K97) – Perceived system understanding (W98a, L97, G93)

Table 3.4. Classification of user quality judgments in the QoS taxonomy (2). K97: Kamm et al. (1997a); W98a: Walker et al. (1998a), Litman et al. (1998); L97: Billi and Lamel (1997), Lamel et al. (1997); G93: Gerbino et al. (1993); B98: Bernsen et al. (1998), Dybkjær and Dybkjær (1993); D85: Dintruff et al. (1985); C90: Casali et al. (1990); H01: Hone and Graham (2001), Hone and Graham (2000); L94: Love et al. (1994); P92: Price et al. (1992).

Category	*Aspect*	*Quality judgment item*
Communic. efficiency	Speed	– System speed (K97, B98, D85, H01, C90, L94) – Interaction pace (W98a) – System response sluggishness (W98a)
	Conciseness	– Perceived system efficiency (L97, H01) – Perceived service efficiency (L94)
	Smoothness	– Confusion provoked by the system (L97, L94) – System complexity (L97, B98, C90, L94) – Ease to correct mistakes (B98) – Predictability of system behavior (B98)
Comfort	Agent personality	– System friendliness (L97, C90, L94, H01) – System politeness (L97, B98) – Naturalness of system behavior (C90) – System individuality/personalization (D85)
	Cognitive demand	– Perceived comfort (D85, C90) – Concentration required (L97, L94, H01) – Fluster caused by the system (L97, L94) – Stress caused by the system (L97, L94) – Tension caused by the system (D85, H01) – Stimulation experienced from the system (B98)
Task efficiency	Task success	– Perceived task success (K97, L97) – Perceived task completion (W98a) – Perceived service reliability (L94) – Perceived reliability of task results (B98) – Information accuracy (D85, C90) – Completeness of task results (C90) – Frequency of system failures (K97) – Frequency of unexp. termination from the sys. (K97) – Perceived service efficiency (L94)
	Task ease	– Perceived task ease (W98a, B98) – Perceived task complexity
Service efficiency	Service adequacy	– System value (D85) – System appropriateness (D85) – Perceived system usefulness (B98, D85, C90, H01)
	Added value	– Preference for other interfaces (W98a, P92) – Preference for human operator (L97, B98, L94) – System desirability (B98)

Table 3.5. Classification of user quality judgments in the QoS taxonomy (3). K97: Kamm et al. (1997a); W98a: Walker et al. (1998a), Litman et al. (1998); L97: Billi and Lamel (1997), Lamel et al. (1997); B98: Bernsen et al. (1998), Dybkjær and Dybkjær (1993); D85: Dintruff et al. (1985); C90: Casali et al. (1990); H01: Hone and Graham (2001), Hone and Graham (2000); L94: Love et al. (1994); P92: Price et al. (1992).

Category	*Aspect*	*Quality judgment item*
Usability	Ease of use	– Ease to use the system/service (K97, D85, L94, H01)
		– System habitability (H01)
		– Ease to learn how to use the system (P92, H01)
User satisfaction		– Perceived satisfaction with the system (B98, D85)
		– Degree of enjoyment (L97, L94, H01)
		– User happiness (D85)
		– System likability (H01)
		– Degree of frustration (L94, H01)
		– Degree of irritation (C90, H01)
		– Need for improvement (L97, L94)
Utility		
Acceptability		– Perceived system acceptability (B98, C90)
		– Willingness to use the service in the future (W98a)

global quality aspects like usability or user satisfaction. Relationships can be established between the judgments obtained on different levels, and analyses with this goal will be presented in Chapter 6. On the other hand, it becomes possible to construct questionnaires which address specific quality aspects in a detailed way, and neglect other aspects which might be of no use to the evaluator. Such an optimized questionnaire will be more efficient in providing the relevant information.

3.8.7 Usability Evaluation

Apart from addressing individual aspects of usability by the mentioned questionnaires, dedicated usability evaluation methods are available. Usability can either be evaluated with real users performing specific tests, or by usability inspection methods with the help of evaluation experts. Both methods are complementary to each other, in that usability inspection methods may be able to detect usability problems which remain overlooked by user testing, and vice-versa (Nielsen and Mack, 1994). In fact, a large degree of non-overlap between the two has been observed. Thus, usability design should combine empirical tests and usability inspections.

Usability inspection methods have been developed in the human factors community since the end of the 1980s. They aim at finding usability problems in an existing user interface design, potentially rating the severity of problems, making recommendations on how to fix the problems, and hereby improving the usability of the system. Such methods allow the knowledge and experience of user interface designers to be easily applied in optimizing new systems. An important part of usability inspection consists of counting and classifying usability problems which are observed in HMI, and some of the methods which have already been presented in Section 3.8.4 (e.g. the error classification from Bernsen et al., 1998, or the fishbone diagram proposed by Constantinides and Rudnicky, 1999) can be seen as usability inspection approaches. Usability inspection should however not only be efficient in detecting problems, but also in weighting them according to their severity (there is no use in resolving unimportant problems), and especially in suggesting design changes and improvements. Because many inspection methods rely on the design specification rather than on the design implementation, they may be applied relatively early in the system design process.

Nielsen and Mack (1994) distinguish the following eight types of usability inspection methods:

- *Heuristic evaluation*: This informal method involves usability specialists who judge whether a dialogue element conforms to established usability principles, the so-called heuristics. Nielsen (1994) gives examples of such heuristic principles which may have to be extended in order to reflect the characteristics of spoken language interaction (Klemmert et al., 2001; Suhm, 2003).

- *Guideline reviews*: Inspections where an interface is checked for conformance with a comprehensive list of usability guidelines. Because the overall number of guidelines may be very high, this approach requires a high degree of expertise.

- *Pluralistic walkthroughs*: Meetings where users, developers and human factors experts step together through a scenario, discussing usability issues associated with dialogue elements which are involved in each scenario step.

- *Consistency inspections*: An interface is inspected by several designers representing multiple design aspects, and then rated as to whether it is consistent with all design issues.

- *Standards inspections*: An expert investigates a specific interface for compliance with a defined standard.

- *Cognitive walkthroughs*: Simulate a user's problem-solving process at each step in the HMI, and check whether the user's goals and action memory can be assumed to lead to the next correct action. Are typically cast in the form of questions about the relationship between task goals attributed to the user, and the interface actions needed to accomplish them.
- *Formal usability inspections*: A formalized method involving a usability inspection team. Each team member has a particular task in the inspection process, e.g. as a moderator, design owner, or inspector. Meetings are organized to prepare and carry out the inspection, and to analyze its results.
- *Feature inspections*: Focusses on the operational functions of the user interface, and whether the provided functions meet the requirements of the intended end users.

Most of these methods are discussed in detail in Nielsen and Mack (1994). The choice of the right method depends on the objectives of the evaluation, the availability of guidelines, the experience of the evaluator, and time and money constraints. An example for using heuristic evaluation for designing a speech-controlled internet browser is described by Klemmert et al. (2001), and for telephone-based spoken dialogue systems by Suhm (2003).

Usability evaluation with controlled user experiments is the second alternative. Such tests can be carried out either in an "objective, non-intrusive" or in a "subjective, intrusive" way (Gleiss, 1992). Non-intrusive methods try to capture the behavior of the human user in a natural and undisturbed way, e.g. by observation with audio-visual equipment, or by logging with a recording device. Intrusive methods require an active involvement of the users, e.g. by responding to questionnaires or interviews (cf. the last section), by group discussions, or in a self-descriptive way, i.e. requiring a verbal protocol which reflects the user's thoughts or opinions during or after the interaction. These methods are described in more detail in ETSI Technical Report ETR 095 (1993).

3.8.8 SDS Evaluation Examples

The following incomplete list gives references to evaluation activities which have been performed during the past ten years or so. Most investigations combine different types of assessment and evaluation, e.g. the extraction of interaction parameters and the collection of user quality judgments. They refer to individual system components or to the system as a whole. Although the evaluation methodology was often not in the focus of the investigations, they may serve as illustrations for the methods and methodologies described in this chapter.

- DARPA ATIS flight information system at MIT and SRI, see Goodine et al. (1992) and Polifroni et al. (1992): Scenario-guided evaluation of two system configurations (mainly with respect to speech understanding), extraction of interaction parameters and user questionnaire judgments. Joint MIT/SRI experiments in the DARPA framework, see Price et al. (1992): User questionnaire and interaction parameter extraction.

- SUNDIAL system for information services over the phone, see Simpson and Fraser (1993) or Peckham and Fraser (1994): First example of a combined black box and glass box evaluation.
- CSELT email server developed as a part of the SUNDIAL project, see Gerbino et al. (1993): Scenario-guided evaluation with a user questionnaire and expert assistance; during the dialogues, experts observed the test subjects, took notes of the difficulties the users encountered, and gave advice in the case of unrecoverable situations; logging of user-system interactions and subject-expert interactions, collection of interaction parameters.
- Italian train information system partly based on SUNDIAL, see Danieli and Gerbino (1995): Scenario-guided evaluation and definition of a large number of interaction parameters.
- Danish home-banking application (part of the European OVID project), see Jack and Lefèvre (1997) and Larsen (1999): Interaction parameter extraction and user quality judgments; multidimensional analysis of user judgments.
- RailTel and ARISE train information systems, see Billi and Lamel (1997), Lamel et al. (1997), Bennacef et al. (1996), Baggia et al. (1998), and Baggia et al. (2000): Field trials carried out in three countries, with a pre-defined scenario; questionnaire with usability-related questions and open answer possibilities; interaction parameter collection.
- Comparative evaluation of two Italian rail information inquiry systems (RailTel and Dialogos), see Billi et al. (1996): Laboratory and field tests with pre-defined scenarios and user questionnaires; extraction of interaction parameters; in one pilot study, an expert psychologist interviewed the subjects to extract quality dimensions; at the end of the test, a second interview with the psychologist took place, and a final questionnaire was constructed as the outcome of the experiment.
- Annie messaging server from AT&T, see Kamm et al. (1997a): Scenario-guided evaluation and limited field test; logging of interactions, collection of interaction parameters and of user quality judgments; first approach to establish relationships between user ratings and system performance measures.
- TOOT train information server, ELVIS email server, and Annie messaging server from AT&T, see Litman et al. (1998), Walker et al. (1998a), Walker

et al. (1998b), and Kamm et al. (1998): Scenario-guided comparison between different system configurations and different user groups; extraction of interaction parameters, user quality judgments, and application of the PARADISE modelling approach.

- Danish dialogue system for flight information, see Dybkjær et al. (1996) and Bernsen et al. (1998): WoZ-based experiments, consisting of a diagnostic evaluation during the design process, a performance evaluation between the WoZ steps, and an adequacy evaluation via questionnaire; experiments with the implemented system, namely a black box evaluation resulting in design specifications, a controlled scenario-guided user test, a diagnostic evaluation on the user test corpus, and a performance evaluation.

- Austrian cinema ticket reservation system, see Pirker et al. (1999) and Loderer (1998): WoZ-based evaluation of different system configurations by extracting interaction parameters, collecting users' quality judgments, and observing the subjects' behavior.

- BoRIS restaurant information system, see Dudda (2001), Niculescu (2002) and Skowronek (2002): Scenario-guided evaluation with different system configurations; interaction parameter extraction, and quality judgments via interview and questionnaire; analysis of relationships between interaction parameters and quality judgments, and prediction modelling approaches.

- German VerbMobil translation system, see e.g. Jekosch et al. (1997), Jost (1997), Krause (1997), Malenke et al. (2000), and Steffens and Paulus (2000): WoZ-based evaluation to explore system acceptance and patterns of usage; phase I evaluation: regular competitive evaluation of the different speech recognizers, and a final end-to-end translation evaluation; phase II: dialogue end-to-end evaluation with selected trained users resolving pre-defined scenarios, turn end-to-end evaluation, and module evaluation; speech synthesis component assessment with respect to segmental intelligibility (cluster identification tests) and global "acceptability" (evaluation in a conversation game scenario with user questionnaire and interview).

- DARPA Communicator program for the development of advanced spoken dialogue systems, see Levin et al. (2000), Walker et al. (2002b), and Walker et al. (2002a): Comparative evaluation on different systems; a scenario-guided within-subject design (i.e. all subjects had to solve multiple problems with all systems) was chosen in 2000; in the 2001 evaluation, frequent travellers were instructed to solve their own travel problems (however with some fixed tasks to allow for comparison).

- Multimodal MASK kiosk for rail enquiries, see Lamel et al. (1998a, 2002): User trials for assessing the performance of the prototype kiosk, plus additional trials to compare different I/O modalities, and to compare to a standard ticket machine; interaction logging, interaction parameter extraction, and user quality judgment collection via a questionnaire.

An recent overview of evaluation activities has also been compiled by Dybkjær et al. (2004), indicating additional projects as well as web addresses.

3.9 Summary

Two points of view can be taken to determine the quality of services which are based on SDSs. The service provider has a look at the system, including its individual components (ASR, language understanding, dialogue management, and speech output), and analyzes the performance of the components and of the system as a whole with respect to certain pre-defined functions. The user has a wholistic view on the service. He or she judges the overall perception of the system characteristics in the course of the interaction, with respect to his/her internal references. This second point of view should be seen as a reference for quality, because it is finally the user who decides whether he/she likes to use the service, and thus on the service acceptability. It is the user, too, who puts the weighting on different dimensions of quality: Whether a certain system characteristic is very important, less important, or not remarkable at all, depends on the user's perceptions of quality. Thus, a user-orientated view on quality is *always* necessary when the quality of interfaces to the human user has to be determined.

Although the user may take the final decision, he/she is not always the right measuring organ to determine all quality aspects. Because of the wholistic view taken, naïve users cannot be expected to provide analytic information as to whether individual system components need improvement, and how this will affect the overall quality of the system. Technology assessment and evaluation thus has to go hand in hand with user-orientated evaluation. Technological assessment and evaluation of individual system components will indicate weak points in a system, show the sources of malfunctions, and finally lead to better systems which can be expected to provide a higher quality to their users. A good evaluation practice always encompasses both a thorough analysis of individual system functions and of the system components in their functional context, and a detailed view on the system from a representative group of users. This chapter presented methods for both types of assessment and evaluation.

On the speech input side, the task of the system components is relatively clear and well-defined. The system shall be able to recognize what the user said, and extract the meaning, i.e. interpret the contents in the light of the interaction history. The recognition process usually ends with the provision of a written word (sequence) hypothesis, or of a number of hypotheses. The correctness of the hypothesis can be expressed by different measures on the word or utterance level. The meaning of an utterance is far more difficult to capture, as the early approaches in the DARPA framework showed. For simple information and transaction tasks which are commonly addressed by current state-of-the-art SDSs, an appropriate task and meaning representation is an attribute-value matrix (AVM) consisting of slots (attribute-value pairs) which contain individual pieces of information. The correctness of language understanding can then be expressed by the correctness of the AVM. It should however be noted that this simple way of meaning extraction has its limits: Ill-structured and heterogenous tasks like the search for a TV film and subsequent VCR recording will require more sophisticated task and meaning representations.

On the speech output side, the user is the only measurement organ who can decide on quality aspects. The evaluator's task here is to identify and quantify the relevant quality dimensions which are responsible for the overall impression of the speech output in its functional context. Whereas a number of specific test methods are available to quantify aspects like intelligibility on a segmental or word level, the results obtained this way will not be very informative as to how the output speech will be accepted by the user (they will however provide useful information for the synthesis developer). Listening-only or interaction tests carried out in a representative context will be a better mean to quantify the quality which is apparent to the user. An example for such an evaluation will be presented in Chapter 5.

Dialogue management and response generation components are the most difficult parts of the system to address individually. By design they have a strong functional relationship with all other system components, and this interrelation has to be taken into account in the evaluation process. Their performance with respect to explicit or implicit operational functions can tentatively be captured in terms of interaction parameters. Interaction parameters refer to the communication of information and the overall course of the dialogue, to the systems cooperativity, to its meta-communication capabilities, to the task to be carried out, and to the system's speech input capabilities. They can be determined on a word, utterance, or dialogue level, either instrumentally or with the help of experts analyzing and annotating dialogue log files. The annotation is an important step in the system evaluation phase, as it helps to identify interaction problems and system malfunctions very effectively. Software tools support the evaluator in this task.

Interaction parameters can be determined from test interactions of human users with the system under development. As long as the system has not been fully set up, Wizard-of-Oz (WoZ) simulations may be used, in which a human experimenter replaces missing parts of the system by behaving in a principled way. WoZ simulations are a commonly-used and well-accepted tool in the system development and evaluation phase. They may be cheaper when different system behavior options are to be compared, and allow technical capabilities to be extrapolated beyond the current state-of-the-art, e.g. with respect to recognition performance or speech output. In order to provide valid results, WoZ simulations have to fulfil several requirements so that the user gets the impression of interacting with a real system in a real situation. These requirements have been discussed with respect to the simulation set-up, the user group, and the experimental task. As soon as fully working systems become available, interactions may be logged in a field test setting, providing a more realistic pool of evaluation data.

In parallel to the interaction parameters, users have to judge different aspects of perceived quality. The judgments are usually collected on closed rating or scaling tasks, although open answer and comment possibilities may provide additional and important information as well. A number of judgment items which can be addressed in such questionnaires has been extracted from the literature. They reflect quality features which are the perceptive dimensions in the user's mind. The items have been categorized with respect to the quality aspect they address, using the QoS taxonomy which was developed in Chapter 2. This taxonomy also helped to classify a large number of interaction parameters. On the basis of these classifications, structured evaluation schemes can be developed for specific quality aspects in an efficient, target-orientated way. The classifications form the basis for detailed interaction experiments with a prototypical restaurant information system which are described in Chapter 6. These experiments will highlight the interrelationships between interaction parameters and quality judgments, and open new ideas for predicting quality in early stages of system development.

Chapter 4

SPEECH RECOGNITION PERFORMANCE OVER THE PHONE

It has been stated above that the performance of the speech recognition component in a spoken dialogue system is influenced by several factors which are mostly related to the environment (transmission channel, ambient noise, room acoustics), but partly also to the agent and the task to be carried out. This situation is illustrated in the QoS taxonomy given in Figure 2.9. Wyard (1993) tried to quantify the relative influence of different factors related to the environment and to the talker on the performance of commercial isolated-digit recognizers. He found that for three of the four tested recognizers, the influence ranking was

1 Telephone line characteristics

2 Talker characteristics

3 Acoustic properties of the telephone handset

4 Ambient noise

and for the fourth recognizer, the ranking between the second and the third position was inverted. Thus, the telephone transmission channel seems to exercise a dominant effect of the performance of ASR used in a telephonic environment.

As a consequence, it has to be assumed that the overall quality, usability and acceptability of a telephone-based spoken dialogue system is largely affected by the transmission channel. The channel plays an important role on the input and the output sides of the system's user interface. On the one hand, it limits the performance of state-of-the-art speech recognition and speaker identification systems. In this way it influences the subsequent language processing stages, such as speech understanding and dialogue management. On the other hand, speech output – be it synthetically or naturally produced – is degraded on its way back to the human user. For the recognition side (this chapter), the decrease in ASR performance can be measured via a transcription-based parameter like word error rate, recognition accuracy, etc. This parameter will be relevant for other aspects of system quality like user satisfaction or acceptability. Investigations in this respect are reported in Chapter 6. For the speech output side, human users have to rate the quality in a scenario which is typical for the application under consideration. Their quality judgments will be addressed in Chapter 5.

Modern transmission channels introduce a large range of perceptively diverse types of impairment. Some of these impairments have already been investigated in the past with respect to their impact on speech recognizer performance. An exemplary choice of relevant literature is given in Section 4.1. However, most investigations mainly address individual aspects of the transmission channel, with the aim of making a particular speech recognizer robust against the degradation under consideration. The new experimental investigations reported in the remainder of this chapter will paint a more global picture, taking a parametric description of the whole transmission channel, mouth-to-ear, as an input. For such a 'general' transmission channel, the performance of different speech recognizers will be addressed in an analytical way. The results will have a generic character, because the description of the transmission channel is generic, and because different types of recognizers with different target vocabularies, languages, training material etc. have been used. Parts of the experiments have already been published elsewhere by the author (Möller and Bourlard, 2002, 2000; Möller and Kavallieratou, 2002).

In principle, ASR robustness can be achieved in four different ways: (1) By an appropriate extraction of robust features in the front-end, (2) by a transformation of noisy speech features to clean speech features, (3) by adaptation of the references towards the current environment, or (4) by including noisy and/or distorted references in the training database. All four ways (and combinations of them) have been used in the past to increase ASR robustness towards environmental factors. For example, robust front-ends for specific transmission channels have been proposed (Mokbel et al., 1998; Gallardo-Antolín et al., 2001), spectral subtraction and channel equalization techniques have been used (Hermansky et al., 1991; Mokbel et al., 1993, 1995), MAP or Bayesian adaptation of HMM parameters have been employed (Lee et al., 1991; Mokbel et al., 1997; Hirsch, 2001), and acoustic models have been trained with impaired speech data (Puel and André-Obrecht, 1997). Overviews of different approaches are given by Mokbel et al. (1997, 1998) and Viikki (2001). Robust HMM architectures have also been described, e.g. for impairments which will be encountered in GSM cellular networks (interruptions and impulsive noise) by Karray et al. (1998). Making use of these approaches, the recognition performance for the addressed type of degradation (namely the one which was taken into account in the development of the system) can be improved.

Flexible systems, however, should be able to cope with a variety of transmission channels. Unfortunately, it cannot be guaranteed that a recognizer will perform similarly well for new types of degradations, or for combinations of them. Wyard (1993) argues that the joint effects of different impairments have to be taken into account if systems are to be developed successfully. Using databases recorded in real networks, corpus-based approaches include combined degradations in the training material. However, they do not provide any control over the type and amount of impairments, and they require large databases to be recorded in different networks and under different environmental (ambient noise) conditions. These databases are expensive and time-consuming to set up (e.g. Chang (2000); Das et al. (1999); Höge et al. (1997), for the SpeechDat project; Hennebert et al. (2000), for speaker recognition). The results obtained by other authors, however, suggest that it may be possible to reach a similar recognition performance when training on high-quality speech recordings compared to task-dependent database training, provided that adequate normalization techniques are used, cf. the results from Neumeyer et al. (1994).

Apart from the environmental factors, the target vocabulary will have a strong influence on the overall performance of the recognizer. Analytic investigations in this respect are useful for the designers of speech recognition systems, but only to a limited extent for the designers of spoken dialogue system applications. For example, Steeneken and van Velden (1989a) developed the RAMOS (Recognition Assessment by Manipulation Of Speech) method to characterize environmental impact on ASR performance with a specifically designed CVC test vocabulary. They found that noise with relatively strong high frequency components (white noise) affected recognizer performance on initial consonants more than low-frequency noise. Frequency shifts of the first formant had only a minor impact on the ASR performance for initial consonants and vowels, but significantly reduced human recognition performance. Steeneken and Varga (1993) compared three different recognizer vocabularies with respect to their discriminative power for other (environmental) factors. They found that the most complex CVC vocabulary – although being the smallest – had the best discriminating capabilities with respect to other factors like noise.

Such results are unfortunately very difficult to interpret with respect to the impact on overall system quality aspects in realistic application scenarios. Thus, the focus of this chapter will be on overall recognition performance – in terms of correctly identified items – for the target application, and not on aspects linked to a specific recognizer. The reported experiments will also be limited to the speech recognition component, and will thus not allow direct conclusions to be drawn for natural language understanding and subsequent language processing components. Relationships between speech recognition and natural language understanding parameters are shown in Chapter 6, and they will be discussed with respect to their influence on overall system quality aspects.

It is the aim of this and of the following chapter to investigate the overall performance of speech technology devices (speech recognition or speech synthesis) in relation to transmission impairments. In order to be as flexible and economical as possible, the approach starts at an early stage of system development. A simulation model is set up which generates all the relevant transmission channel degradations in a controlled way. The modelled degradations are those which are encountered in traditional and modern telecommunication networks. Several degradations can be implemented simultaneously; thus, it becomes possible to address the combined effects of different types of impairments (e.g. codecs operating on noisy speech) in a realistic way. Due to its real-time capability, the simulation model can be operated just as well in a one-way (transmission) or in a bi-directional (conversation) mode.

In contrast to proposals made by Tarcisio et al. (1999) or Giuliani et al. (1999), no detailed filtering technique, which would necessitate the measurement of impulse responses in real-life networks, is applied. Input parameters to the proposed simulation model are planning values, which are commonly used in the planning process of telecommunication networks, see Section 2.4.1. Based on these planning values, the ASR impact can easily be investigated before the respective network has been set up. In this way, it is not only possible to adapt the ASR system to a class of transmission channels. In addition, telecommunication network planners obtain diagnostic information on the impact of specific characteristics of their networks, and may use this opportunity to select suitable components.

Telecommunication networks are normally not designed for the requirements of speech technology devices. As a consequence, it cannot be guaranteed that ASR will perform well under all realistic transmission channel conditions. In order to illustrate this fact, the quality predictions provided by network planning models for human-to-human interaction will be compared to ASR performance presenting one aspect of human-machine interaction quality. This possibility is only provided by using the simulation model, because identical transmission conditions can be guaranteed. The investigations will form a basis for quality network planning for ASR, and show limitations and future extension possibilities of network planning models like the E-model with respect to human-machine interaction. Taking the E-model as an example, this will be illustrated by proposing modifications to the model algorithm in order to reach a good agreement between model predictions and observed ASR performance. In addition, for codec-type degradations signal-based comparative measures will be investigated with respect to their potential for predicting ASR performance.

After an overview of the literature on the impact of individual impairments of ASR performance (Section 4.1), the architecture and implementation of the simulation system are described in Section 4.2. In the experiments reported in this chapter, the simulation system has been applied to the assessment of the impact of modern telephone channels without time-variant distortions on ASR performance. Two prototype ASR systems, which are part of a telephone-based information server, and a third standardized ASR system able to recognize connected digits have been used for this purpose. These systems are briefly described in Section 4.3. The results are given in Section 4.4, and they are compared to the quality degradation which can be expected for human-to-human interaction, using the E-model and the PESQ and TOSQA models. The E-model algorithm is subsequently modified in order to reach a better agreement between quality predictions and observed ASR performance, see Section 4.5. A discussion and an outlook on potential extensions of the simulation model, as well as on transmission quality aspects of HHI and HMI, conclude this chapter.

Further applications of the simulation technique may be considered. One consists of a controlled degradation of large databases of clean speech, which can be used for model training and adaptation. In this way, it becomes possible to multiply the amount of available data with respect to network characteristics which are expected to be representative for the later application situation. The described on-line tool can also be applied to the evaluation of transmission channel effects on synthesized speech (Chapter 5), and on the whole spoken dialogue system, in realistic conversation scenarios. For example, the influence of the ASR performance degradation on subsequent stages of speech understanding and dialogue management can be investigated and modelled. First steps in this direction are reported in Chapter 6.

4.1 Impact of Individual Transmission Impairments on ASR Performance

Several impairments caused by the transmission channel limit the performance of speech recognition and speaker verification systems. Moreno and Stern (1994) investigate the main sources of degradation resulting from analogue telephone transmission, using a commercially available analogue telephone line simulator which introduces noise and linear distortions (no non-linear influences). They found that the bandwidth limitation (300-3400 Hz) and linear distortions of the channel are only responsible for about 1/3 of the performance degradation. This decrease was mainly due to the non-flat frequency response, whereas phase distortions had almost no influence on the ASR performance rates. For the rest of the degradations, C-message weighted noise, impulsive noise, and interference tones at 180 Hz impaired ASR performance most. Except for high levels of C-message weighted noise, the degradations could be reduced by cepstral compensation techniques.

A number of studies address the effects of speech codecs on ASR performance. Lilly and Paliwal (1996) compared codecs operating at bit-rates between 40 and 4.8 kbit/s with respect to their influence on two different speech recognizers: an isolated word recognizer with whole-word models, and a phoneme-based continuous speech recognizer. These recognizers made use of LPC-derived cepstral coefficients or MFCCs in the feature-extraction stage. The authors found that MFCCs were less affected by coding than LPC-derived cepstral coefficients. In general, ASR performance degraded further when the bit-rate was lower, but, for example, the LD-CELP codec (G.728) operating at 16 kbit/s yielded better performance than the ADPCM codec (G.726) operating at 24 kbit/s for the whole-word models[1]. Tandeming of high bit-rate ADPCM coders only had a minor influence on the ASR performance, but with low bit-rate codecs the tandem degradation was stronger.

A similar general tendency was found in the experiments from Euler and Zinke (1994). They investigate the codec influence on the performance of an isolated word speech recognizer and on a speaker verification system. For the three codecs in the test (LD-CELP (G.728) at 16 kbit/s, RPE-LTP (GSM-FR) at 13.2 kbit/s, and a 4.8 kbit/s CELP codec) an increasing error rate with decreasing bit-rate was observed. The introduction of delta coefficients in the feature vectors makes the recognizer more robust and reduces the word error rate by approx. 50%. As would be expected, the mismatch between training and test conditions is responsible for a large part of the performance degradation, and training with coded speech can significantly improve the robustness. Nevertheless, some degradation from the low bit-rate coding remains even when training on coded speech. In contrast to this finding, Hirsch (2002) observed best overall recognition performance when training on speech data without additional coding. He tested the influence of GSM and AMR codecs operating at different bit-rates in a standardized AURORA set-up, with two different feature extraction algorithms. Recognition performance was considerably different between the two feature extraction front-ends, whereas the different AMR codec implementations (bit-rates between 4.75 and 12.2 kbit/s) lead to comparable performance results.

The performance degradation for coded speech seems to be partly due to insufficient modelling of vocal tract information in the coder. Turunen and Vlaj (2001) investigated several different codecs using diverse speech modelling approaches. They conclude that ASR performance remains fairly good if the vocal tract is modelled sufficiently well. Such a modelling can be reached by LPC parameters or by other appropriate representations. The modelling of the residual signal and the quantization do not seem to be very important. An additional degradation due to codec tandeming was almost linearly related to the number of consecutively connected codecs.

[1]For a discussion of commonly used coding algorithms see Vary et al. (1998) or O'Shaughnessy (2000). It has to be noted that most investigations focus on speech codecs alone, without any channel coding. This is also the case for the experiments described in this and in the following chapter. Channel coding, however, becomes important when the effect of transmission errors is addressed.

For a speaker verification system developed as a part of the European IST project SAFE (Secure-Access Front End), Caminero et al. (2002) evaluated the performance of the ASR component and of the whole speaker verification module under combined noise and codec degradations. Codecs included GSM-FR, GSM-HR, AMR, G.723.1 and G.729, all operating on speech signals degraded by different levels of artificially added stationary noise. Their results show that the speech recognizer sensitivity to additive noise strongly depends on the codec. In particular, the GSM-FR and the G.723.1 codec significantly degrade recognition performance when operating on speech signals with low SNR. Only a part of this effect can be ruled out by training on coded speech. For the whole speaker verification system, equal error rates (EER) varied between 5.5% for standard PSTN transmission to 8.8% for G.723.1 transmission. Increased EERs were mainly observed for the GSM-FR, G.723.1 and G.729 codecs. Hardt et al. (1998) found increased EERs when confronting their text-dependent speaker verification system with data coded with the GSM-HR codec. When cascading ADPCM (G.726) codecs, the error rates reached higher values than just the sum of the respective single encoding errors. It was shown that EERs heavily depend on the choice of feature coefficients, and that optimum features have to be chosen for each transmission condition anew.

The cited investigations take only stationary impairments into account. Modern networks, however, will show time-variant channel characteristics due to frames or packets which get lost during the connection, or due to jitter buffer overflow. The reduced ASR performance caused by such time-variant impairments will strongly depend on the exact way they are generated. Influencing factors will be, for example, the number of coded speech frames per packet, the type of speech coder, the frequency and distribution of lost frames or packets, and potential error concealment techniques. Unfortunately, there are still no agreed-upon parameters which could characterize the related impairments, nor any standardized models for generating such impairments artificially. Thus, most of the results which have been published recently on this topic have only limited generic informative value. Interesting examples include the investigations from Gallardo-Antolín and co-workers for GSM and IP-based speech transmission (Gallardo-Antolín et al., 2001; Peláez-Moreno et al., 2001), from Metze et al. (2001) for VoIP connections, and from Kiss (2000) on distributed speech recognition over mobile radio channels.

4.2 Transmission Channel Simulation

In order to quantify the effects of different degradations caused by the transmission channel, it is important to have full control over all parts of the channel. This cannot be attained in real-life networks, where the configuration of each individual connection depends on the traffic load in the network and on other operator-specific constraints. However, the set-up of commercial equipment in a laboratory is expensive and restricts the number and type of connections which can be tested.

For these reasons, a flexible simulation system has been developed and implemented at IKA. It generates most of the degradations which occur in traditional (analogue/digital) and mobile telephone networks in a controlled way. The simulation has been designed on the basis of the reference connection which was discussed in Section 2.4.1. This reference connection description has been transformed into a signal flow-graph, see Figure 4.1. It was originally developed for investigating the transmission channel impact on natural speech (Möller, 2000), and includes all relevant impairments affecting the conversational situation. The whole structure is implemented on DSP hardware which can be wired and programmed via software (Tucker Davis Technologies, System 2). In this way, the system is flexible enough to change between different parameter settings quickly, while the use of hardware allows the simulation to run in real or close-to-real time.

The triangles represent programmable filters (which can be used as attenuators as well), the rectangles delay lines (for T, Ta and Tr), codecs, or the channel bandpass filter. Low bit-rate codecs have been implemented on another DSP hardware (ASPI Elf board), and they can be cascaded asynchronously up to three times, going back to the signal representation. Different types of user interfaces can be connected to the simulation system in a four-wire mode, in order to avoid reflections which may occur with standard analogue telephone sets using two wires.

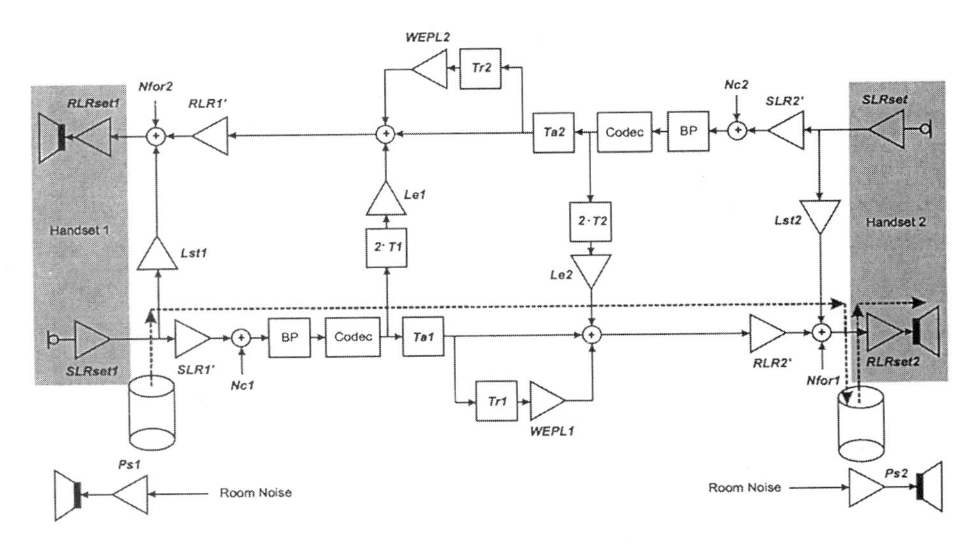

Figure 4.1. Telephone line simulation system used in the experiment. Indices 1 and 2 indicate the direction of the transmission.

The simulation allows the following types of impairments to be generated:

- Attenuation and frequency distortion of the main transmission path, expressed in terms of loudness ratings, namely the send loudness rating, SLR, and the receive loudness rating, RLR). Both loudness ratings contain a fixed part reflecting the electro-acoustic sensitivities of the user interface ($SLRset$ and $RLRset$), and a variable part which can be adjusted (SLR' and RLR'). The characteristics of the handset used in the experiment were first measured with an artificial head (head and torso simulator), see ITU-T Rec. P.64 (1999) for a description of the measurement method. They were then adjusted via SLR' and RLR' to a desired frequency shape which is defined by the ITU-T, a so-called modified intermediate reference system, IRS (ITU-T Rec. P.48, 1988; ITU-T Rec. P.830, 1996). In the case of high-quality microphone recordings or of synthesized speech (cf. the next chapter), the IRS characteristic can be directly implemented using the SLR' filter.

- Continuous white circuit noise, representing all the potentially distributed noise sources, both on the channel (Nc, narrow-band because it is filtered with the BP filter) and at the receive side ($Nfor$, wideband restricted by the electro-acoustic coupling at the receiver handset).

- Transmission channel bandwidth impact: BP with 300-3400 Hz according to ITU-T Rec. G.712 (2001), i.e. the standard narrow-band telephone bandwidth, or a wideband characteristic 50-7000 Hz according to ITU-T Rec. G.722 (1988). For the reported experiment, only the narrow-band filter was used.

- Impact of low bit-rate speech codecs: Several codecs standardized by the ITU-T, as well as a North American cellular codec were implemented. They include logarithmic PCM at 64 kbit/s (ITU-T Rec. G.711, 1988), ADPCM at 32 kbit/s (ITU-T Rec. G.726, 1990), a low-delay CELP coder at 16 kbit/s (ITU-T Rec. G.728, 1992), a conjugate-structure algebraic CELP coder (ITU-T Rec. G.729, 1996), and a vector sum excited linear predictive coder (IS-54). A description of the coding principles can be found e.g. in Vary et al. (1998).

- Quantizing noise resulting from waveform codecs (e.g. PCM) or from D/A-A/D conversions was implemented using a modulated noise reference unit, MNRU (ITU-T Rec. P.810, 1996), at the position of the codec. The corresponding degradation is expressed in terms of the signal-to-quantizing-noise ratio Q.

- Ambient room noise of A-weighted power level Ps at the send side, and Pr at the receive side ($Pr = Ps2$).

- Pure overall delay Ta.
- Talker echo with one-way delay T and attenuation Le. The corresponding loudness rating $TELR$ of the talker echo path can be calculated by $TELR = SLR + RLR + Le$.
- Listener echo with round-trip delay Tr and an attenuation with respect to the direct speech (corresponding loudness rating $WEPL$ of the closed echo loop).
- Sidetone with attenuation Lst (loudness rating for direct speech: $STMR = SLRset + RLRset + Lst - 1$; loudness rating for ambient noise: $LSTR = STMR + Ds$, Ds reflecting a weighted difference between the handset sensitivity for direct sound and for diffuse sound).

More details on the simulation model and on the individual parameters can be found in Möller (2000) and in ETSI Technical Report ETR 250 (1996).

Comparing Figures 4.1 and 2.10, it can be seen that all the relevant transmission paths and all the stationary impairments in the planning structure are covered by the simulation model. There is a small difference to real-life networks in the simulation of the echo path: Whereas the talker echo normally originates from a reflection at the far end and passes through two codecs, the simulation only takes one codec into account. This allowance was made to avoid instability, which can otherwise result from a closed loop formed by the two echo paths.

The simulation is integrated in a test environment which consists of two test cabinets (e.g. for recording or carrying out conversational tests) and a control room. Background noise can be inserted in both test cabinets, so that realistic ambient noise scenarios can be set up. This means that the speaking style variation due to ambient noise (Lombard reflex) as well as due to bad transmission channels is guaranteed to be realistic. In the experiment reported in this chapter, the simulation was used in a one-way transmission mode, replacing the second handset interface with a speech recognizer. For the experiments in Chapter 5, the speech input terminal has been replaced by a harddisk playing back the digitally pre-recorded or synthesized speech samples. Finally, in the experiments of Chapter 6, the simulation is used in the full conversational mode. Depending on the task, simplified solutions can easily be deduced from the full structure of Figure 4.1, and can be implemented either using standard filter structures (as was done in the reported experiments) or specifically measured ones.

When recording speech samples at the left bin of Figure 4.1, it is important to implement the sidetone path ($Lst1$), and in case of noticeable echo also the talker echo path ($Le1$), because the feedback they provide (of speech and background noise at the send side) might influence the speaking style – an effect which cannot be neglected in ASR. Also the influence of ambient noise should be catered for by performing recordings in a noisy environment. Matassoni et al. (2001) performed a comparison of ASR performance between a system trained with speech recorded under real driving conditions (SpeechDatCar and VODIS II projects), and a second system trained with speech with artificially added noise. They illustrated a considerable advantage for a system trained on real-life data instead of artificially added noise data.

In general, the use of simulation equipment has to be validated before confidence can be laid into the results. The simulation system described here has been verified with respect to the signal transmission characteristics (frequency responses of the transmission paths, signal and noise levels, transfer characteristics of codecs), as well as with respect to the quality judgments obtained in listening-only and conversational experiments. Details on the verification are described in Raake and Möller (1999). In view of the large number of input parameters to the simulation system, such a verification can unfortunately never be exhaustive in the sense that all potential parameter combinations could be verified.

Nevertheless, the use of simulation systems for this type of experiments can also be disputed. For example, Wyard (1993) states that it has to be guaranteed that the simulated data gives the same results as real-life data does *for the same experimental purpose*, and that a validation for another purpose (e.g. for transmission quality experiments) is not enough. In their evaluation of wireline and cellular transmission channel impacts on a large-vocabulary recognizer, Rao et al. (2000) found that bandwidth limitation and coding only explained about half of the degradation which they observed for a real-life cellular channel. They argue that the codec operating on noisy speech as well as spectral or temporal signal distortions may be responsible for the additional amount of degradation. This is in line with Wyard's argumentation, namely that a recognizer might be very sensitive to slight differences between real and simulated data which are not important in other contexts, and interaction effects may occur which are not captured by the simulation. The latter argument is addressed in the proposed system by a relatively complete simulation of all impairments which are taken into account in network planning. The former argument could unfortunately not be verified or falsified here, due to a lack of recognition data from real-life networks. Such data will be uniquely available to network operators or service providers. However, both types of validations carried out so far did not point at specific factors which might limit the use of the described simulation technique.

4.3 Recognizer and Test Set-Up

The simulation model will now be used to assess the impact of several types of telephone degradation on the performance of speech recognizers. Three different recognizers are chosen for this purpose. Two of them are part of a spoken dialogue system which provides information on restaurants in the city of Martigny, Switzerland (Swiss-French version) or Bochum, Germany (German version). This spoken dialogue system is integrated into a server which enables voice and internet access, and which has been implemented under the Swiss CTI-funded project InfoVOX. The whole system will be described in more detail in Chapter 6. The third recognizer is a more-or-less standardized HMM recognizer which has been defined in the framework of the ETSI AURORA project for distributed ASR in car environments (Hirsch and Pearce, 2000). It has been built using the HTK toolkit and performs connected digit recognition for English. Training and test data for this system are available through ELRA (AURORA 1.0 database), whereas the German and the Swiss-French recognizer have been tested on specific speech data which stem from Wizard-of-Oz experiments in the restaurant information domain.

The Swiss-French system is a large-vocabulary continuous speech recognizer for the Swiss-French language. It makes use of a hybrid HMM/ANN architecture (Bourlard and Morgan, 1998). ANN weights as well as HMM phone models and phone prior probabilities have been trained on the Swiss-French PolyPhone database (Chollet et al., 1996), using 4,293 prompted information service calls (2,407 female, 1,886 male speakers) collected over the Swiss telephone network. The recognizer's dictionary was built from 255 initial Wizard-of-Oz dialogue transcriptions on the restaurant information task. These dialogues were carried out at IDIAP, Martigny, and EPFL, Lausanne, in the frame of the InfoVOX project. The same transcriptions were used to set up 2-gram and 3-gram language models. Log-RASTA feature coefficients (Hermansky and Morgan, 1994) were used for the acoustic model, consisting of 12 MFCC coefficients, 12 derivatives, and the energy and energy derivatives. A 10th order LPC analysis and 17 critical band filters were used for the MFCC calculation.

The German system is a partly commercially available small-vocabulary HMM recognizer for command and control applications. It can recognize connected words in a keyword-spotting mode. Acoustic models have been trained on speech recorded in a low-noise office environment and band-limited to 4 kHz. The dictionary has been adapted from the respective Swiss-French version, and contains 395 German words of the restaurant domain, including proper place names (which have been transcribed manually). Due to commercial reasons, no detailed information on the architecture and on the acoustic features and models of the recognizer is available to the author. As it is not the aim to investigate the features of the specific recognizer, this fact is tolerable for the given purpose.

The AURORA recognizer has been set up using the HTK software package version 3.0, see Young et al. (2000). Its task is the recognition of connected digit strings in English. Training and recognition parameters of this system have been defined in such a way as to compare recognition results when applying different feature extraction schemes, see the description given by Hirsch and Pearce (2000). The training material consists of the TIDigits database (Leonard and Doddington, 1991) to which different types of noise have been added in a defined way. Digits are modelled as whole-word HMMs with 16 states per word, simple left-to-right models without skips between states, and 3 Gaussian mixtures per state. Feature vectors consist of 12 cepstral coefficients and the logarithmic frame energy, plus their first and second order derivatives.

It has to be noted that the particular recognizers are not of primary interest here. Two of them (Swiss-French and German) reflect typical solutions which are commonly used in spoken dialogue systems. This means that the outcome of the described experiments may be representative for similar application scenarios. Whereas a reasonable estimation of the relative performance in relation to the amount of transmission channel degradation can be obtained, the absolute performance of these two recognizers is not yet competitive. This is due to the fact that the whole system is still in the prototype stage and has not been optimized for the specific application scenario. The third recognizer (AURORA) has been chosen to provide comparative data to other investigations. It is not a typical example for the application under consideration.

Because the German and the Swiss-French system are still in the prototype stage, test data is relatively restricted. This is not a severe limitation, as only the relative performance degradation is interesting here, and not the absolute numbers. The Swiss-French system was tested with 150 test utterances which were collected from 10 speakers (6m, 4f) in a quiet library environment ($Ps \approx 35$ dB(A)). 15 utterances that were comparable in dialogue structure (though not identical) to the WoZ transcriptions were solicited from each subject. Each contained at least two keyword specifiers, which are used in the speech understanding module of the dialogue system. Speakers were asked to read the utterances aloud in a natural way. The German system was tested using recordings of 10 speakers (5m, 5f) which were made in a low-noise test cabinet ($Ps \approx 35$ dB(A)). Each speaker was asked to read the 395 German keywords of the recognizer's vocabulary in a natural way. All of them were part of the restaurant task context and were being used in the speech understanding module. In both cases recordings were made via a traditionally shaped wireline telephone handset. Training and test material for the AURORA system consisted of part of the AURORA 1.0 database which is available through ELRA. This system has been trained in two different settings: The first set consisted of the clean speech files only (indicated 'clean'), and the second of a mixture of clean and noisy speech files, where different types of noise have been added artificially to the speech signals (so-called multi-condition training), see Hirsch and Pearce (2000).

Table 4.1. Parameter settings for the recognition experiment. An 'X' indicates that the circuit condition has been included in the according test. S: Swiss-French recognizer; G: German recognizer; E: English AURORA recognizer. All other transmission parameters were adjusted to the default values given in Table 2.4.

No.	*Nc (dBm0p)*	*Nfor (dBmp)*	*Codec / MNRU*	*Note*	*Test condition*					*E-model*	
					S, all w.	*S, keyw.*	*G*	*E, clean*	*E, multi*	*R*	*MOS*
0	-	-	-	no transmission	x	x	x	x	x	100	4.5
1	-100	-100	-	low noise, no codec	x	x	x	x	x	100	4.5
2	-100	-100	G.711	low noise	x	x	x	x	x	100	4.5
3	-70	-100	G.711	low noise			x			100	4.5
4	-60	-100	G.711	moderate nb. noise			x			91.3	4.37
5	-50	-100	G.711	moderate nb. noise			x			76.7	3.89
6	-40	-100	G.711	high nb. noise			x			61.8	3.19
7	-30	-100	G.711	high nb. noise			x			46.9	2.41
8	-70	-70	G.711	low noise			x			99.2	4.49
9	-70	-64	G.711	default connection	x	x	x	x	x	93.2	4.41
10	-70	-60	G.711	moderate wb. noise			x			88.1	4.29
11	-70	-50	G.711	moderate wb. noise			x			73.7	3.77
12	-70	-40	G.711	high wb. noise			x			58.8	3.04
13	-70	-30	G.711	high wb. noise			x			43.9	2.26
14	-100	-64	G.711	low nb. noise	x	x		x	x	94.1	4.43
15	-60	-64	G.711	moderate nb. noise	x	x		x	x	88.3	4.29
16	-55	-64	G.711	moderate nb. noise	x	x		x	x	82.9	4.13
17	-50	-64	G.711	moderate nb. noise	x	x		x	x	76.3	3.88
18	-40	-64	G.711	high nb. noise	x	x		x	x	61.8	3.19
19	-30	-64	G.711	high nb. noise	x	x		x	x	46.8	2.41
20	-70	-64	G.726	ADPCM coding	x	x	x	x	x	86.2	4.23
21	-70	-64	G.728	LD-CELP coding	x	x	x	x	x	86.2	4.23
22	-70	-64	G.729	CS-ACELP coding	x	x	x	x	x	83.2	4.14
23	-70	-64	IS-54	VSELP coding	x	x	x	x	x	73.2	3.74
24	-70	-64	G.726*G.726	ADPCM tandem	x	x	x	x	x	79.2	3.99
25	-70	-64	IS-54*IS-54	VSELP tandem	x	x	x	x	x	53.2	2.74
26	-70	-64	G.729*IS-54	mixed tandem	x	x	x	x	x	63.2	3.26
27	-70	-64	MNRU, Q=30 dB	low sign.-corr. noise	x	x	x	x	x	90.0	4.34
28	-70	-64	MNRU, Q=20 dB	moderate sign.-corr. noise	x	x	x	x	x	67.0	3.45
29	-70	-64	MNRU, Q=15 dB	moderate sign.-corr. noise	x	x	x	x	x	48.0	2.47
30	-70	-64	MNRU, Q=10 dB	moderate sign.-corr. noise	x	x	x	x	x	33.4	1.75
31	-70	-64	MNRU, Q=5 dB	high sign.-corr. noise	x	x	x	x	x	23.7	1.37
32	-70	-64	MNRU, Q=0 dB	high sign.-corr. noise	x	x	x	x	x	19.0	1.22
33	-100	-100	IS-54	VSELP, low noise			x			89.4	4.33
34	-70	-100	IS-54	VSELP, low noise			x			83.9	4.16
35	-60	-100	IS-54	VSELP, moderate noise			x			71.3	3.66
36	-50	-100	IS-54	VSELP, moderate noise			x			56.7	2.93
37	-40	-100	IS-54	VSELP, high noise			x			41.8	2.15
38	-30	-100	IS-54	VSELP, high noise			x			26.9	1.48
39	-55	-64	IS-54	VSELP, moderate noise	x	x		x	x	62.9	3.25
40	-40	-64	IS-54	VSELP, high noise	x	x		x	x	41.8	2.15

The test utterances were digitally recorded and then transmitted through the simulation model, cf. the dashed line in Figure 4.1. At the output of the simulator, the degraded utterances were collected and then processed by the recognizer. All in all, 40 different settings of the simulation model were tested. The exact parameter settings are given in Table 4.1, which indicates only the parameters differing from the default setting. The connections include different levels of narrow-band or wideband circuit noise (No. 2-19), several codecs operating at bit-rates between 32 and 8 kbit/s (No. 20-26), quantizing noise modelled by means of a modulated noise reference unit at the position of the codec (No. 27-32), as well as combinations of non-linear codec distortions and circuit noise (No. 33-40). The other parameters of the simulation model, which are not addressed in the specific configuration, were set to their default values as defined in ITU-T Rec. G.107 (2003), see Table 2.4.

It has to be mentioned that the tested impairments solely reflect the listening-only situation, and for the sake of comparison, did not include background noise. In realistic dialogue scenarios, however, conversational impairments can be tested as well. For the the ASR component, it can be assumed that talker echo on telephone connections will be a major problem when barge-in capability is provided. In such a case, adequate echo cancelling strategies have to be implemented. The performance of the ASR component will then depend on the echo cancelling strategy, as well as on the rejection threshold the recognizer has been adjusted to.

4.4 Recognition Results

In this section, the viewpoint of a transmission network planner is taken, who has to guarantee that the transmission system performs well for both human-to-human and human-machine interaction. A prerequisite for the former is an adequate speech quality, for the latter a good ASR performance. Thus, the degradation in recognition performance due to the transmission channel will be investigated and compared to the quality degradation which can be expected in a human-to-human communication. This is a comparison between two unequal partners, which nevertheless have some similar underlying principles.

Speech quality has been defined as the result of a perception and assessment process, in which the assessing subject establishes a relation between the perceived characteristics of the speech signal on the one hand, and the desired or expected characteristics on the other (Jekosch, 2000). Thus, speech quality is a subjective entity, and is not completely determined by the acoustic signal reaching the listener's ear. Intelligibility, i.e. the ability to recognize what is said, forms just one dimension of speech quality. It also has to be measured subjectively, using auditory experiments. The performance of a speech recognizer, in contrast, can be measured instrumentally, with the help of expert transcriptions of the user's speech. As for speech quality, it also depends on the 'background knowledge', which is mainly included in the acoustic and language models of the recognizer.

From a system designer's point of view, comparing the unequal partners seems to be justifiable. Both are prerequisites for reasonable communication or interaction quality. Whereas speech quality is a direct, subjective quality measure judged by a human perceiving subject, recognizer performance is only *one* interaction parameter which will be relevant for the overall quality of the human-machine interaction. For the planner of transmission systems, it is important that good speech quality as well as good recognition performance are provided by the system, because speech transmission channels are increasingly being used with both, human and ASR back-ends.

On the other hand, if the underlying recognition mechanisms are to be investigated, the human and the machine ability to identify speech items should be compared. Some authors argue that such a comparison may be pointless in principle, because (a) the performance measures are normally different (word accuracy for ASR, relative speed and accuracy of processing under varying conditions for human speech recognition), and (b) the vocabulary size and the amount of 'training material' is different in both cases. Lippmann (1997) illustrated that machine ASR accuracy still lags about one order of magnitude behind that of humans. Moore and Cutler (2001) conclude that even the increase in training material will not bridge that gap, but a change in the recognition approach is needed, which better exploits the information available in the existing data. Thus, a more thorough understanding of the mechanisms underlying human speech recognition may lead to more structured models for ASR in the future. Unfortunately, identified links are often not obvious to implement.

System designers make use of the E-model to predict quality for the network configuration which is depicted in Figure 2.10. As this structure is implemented in the simulation model, it is possible to obtain speech communication quality estimates for all the tested transmission channels, based on the settings of the planning values which are used as an input to the simulation model. Alternatively, signal-based comparative measures can be used to obtain quality estimates for specific parts of the transmission channel, using signals which have been collected at the input and the output side of the part under consideration as an input. It has to be noted that both R and MOS values obtained from the models are only predictions, and do not necessarily correspond to user judgments in real conversation scenarios. Nevertheless, the validity of quality predictions has been tested extensively (Möller, 2000; Möller and Raake, 2002), and found to be in relatively good agreement with auditory test data for most of the tested impairments.

The object of the investigation will be the recognizer performance, presented in relation to the amount of transmission channel degradation introduced by the simulation, e.g. the noise level, type of codec, etc. Recognizer performance is first calculated with the help of aligned transcriptions in terms of the percentage of correctly identified words $\frac{c_W}{W}$, and the corresponding error rates (substitutions, insertions and deletions; $\frac{c_w}{W} = 1 - \frac{s_w}{W} - \frac{d_w}{W}$), which are not reproduced here. The alignment is performed according to the NIST evaluation scheme, using the SCLITE software (NIST Speech Recognition Scoring Toolkit, 2001). For the Swiss-French continuous speech recognizer, the performance is evaluated twice, both for all the words in the vocabulary and for just the keywords which are used in the speech understanding module. The German recognizer carries out a keyword-spotting, so the evaluation is performed uniquely on keywords. The AURORA recognizer is always evaluated with respect to the complete connected digit string.

4.4.1 Normalization

Because the object of the experiment is the relative recognizer performance with respect to the performance without transmission degradation ($topline$), an adjustment to a normalized performance range [$perf_{min}$;$perf_{max}$] is carried out. A linear transformation is used for this purpose:

$$\text{adjusted recognition rate} = \frac{c_w/W}{topline} \cdot (perf_{max} - perf_{min}) + perf_{min} \qquad (4.4.1)$$

All recognition scores are normalized to a range which can be compared to the quality index predicted by the E-model. The normalization also helps to draw comparisons between the recognizers. As it was described in Section 2.4.1, the E-model predicts speech quality in terms of a transmission rating factor R [0;100], which can be transformed via the non-linear relationship of Formula 2.4.2 into estimations of mean users' quality judgments on the 5-point ACR quality scale, the mean opinion scores MOS [1;4.5]. Because the relationship is non-linear, it is worth investigating both prediction outputs of the E-model. For R, the recognition rate has to be adjusted to a range of $perf_{min} = 0$ to $perf_{max} = 100$, for MOS to a range of $perf_{min} = 1$ to $perf_{max} = 4.5$. Based on the R values, classes of speech transmission quality are defined in ITU-T Rec. G.109 (1999), see Table 2.5. They indicate how the calculated R values have to be interpreted in the case of HHI.

The $topline$ parameter is defined here as the recognition rate for the input speech material without any telephone channel transmission, collected at the left bin in Figure 4.1. The according values for each recognizer are indicated in Table 4.2. They reached 98.8% (clean training) and 98.6% (multi-condition training) for the AURORA recognizer, and 68.1% for the German recognizer. For the Swiss-French continuous recognizer, the $topline$ values were 57.4% for all words in the vocabulary, and 69.5% for the keywords only which are used in the speech understanding module. Obviously, the recognizers differ in their absolute performance because the applications they have been built for are different. This fact is tolerable, as the interest here is the relative degradation of recognition performance as a function of the physically measurable channel characteristics. As only prototype versions of both recognizers were available

Table 4.2. Absolute recognition rates for condition No. 0 (topline performance), No. 1 (only bandwidth limitation and IRS filtering), No. 2 (same as 1 but with log. PCM coding), and for condition No. 9 (default parameter values according to Table 2.4, including noise). S: Swiss-French recognizer; G: German recognizer; E: English AURORA recognizer.

Condition	*S, all w.*	*S, keyw.*	*G*	*E, clean*	*E, multi*
0	57.4	69.5	68.1	98.8	98.6
1	45.1	57.0	67.8	96.6	96.0
2	45.9	57.8	66.6	91.9	95.3
9	47.2	59.5	64.1	94.9	92.2

at the time the experiments were carried out, the relatively low performance due to the mismatch between training and testing was foreseen.

Because the default channel performance is sometimes significantly lower than the *topline* performance, the normalized recognition performance curves do not necessarily reach the highest possible level (100 or 4.5). This fact can be clearly observed for the Swiss-French recognizer, where recognition performance drops by about 10% for the default channel, see Table 4.2. The strict bandwidth limitation applied in the current simulation model (G.712 filter) and the IRS filtering seem to be responsible for this decrease, cf. the comparison between conditions No. 0 and No. 1 in the table. This recognizer has been trained on a telephone database with very diverse transmission channels and probably diverse bandwidth limitations; thus, the strict G.712 bandpass filter seems to cause a mismatch between training and testing conditions. On the other hand, the default noise levels and the G.711 log. PCM coding do not cause a degradation in performance for this recognizer (in fact, they even show a slight increase), because noise was well represented in the database. For the German recognizer, the degradation between clean (condition No. 0) and default (condition No. 9) channel characteristics seems to be mainly due to the default noise levels, and for the AURORA recognizer the sources of the degradation are both bandwidth limitation and noise.

In the following figures, the E-model speech quality predictions in terms of R and MOS are compared to the normalized recognition performance, for all three recognizers used in the test. The test results have been separated for the different types of transmission impairments and are depicted in Figures 4.2 to 4.10. In each figure (except 4.7), the left diagram shows a comparison between the transmission rating R and the normalized recognition performance [0;100], the right diagram between MOS and the corresponding normalized performance [1;4.5]. Higher values indicate better performances for both R

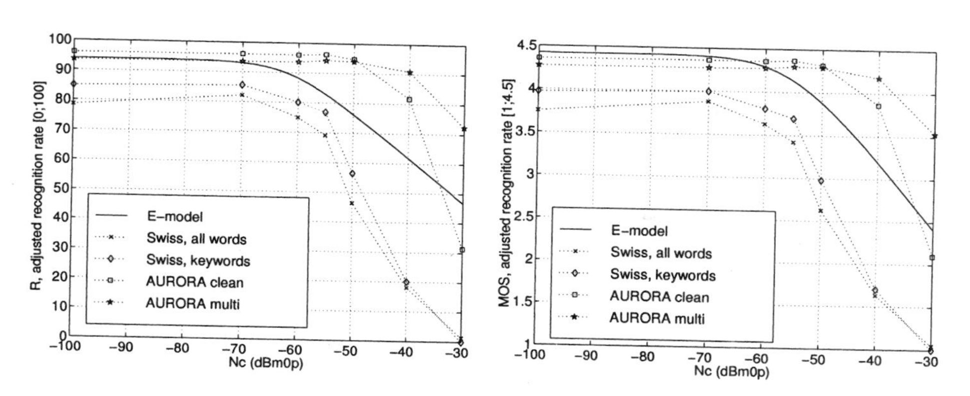

Figure 4.2. Comparison of adjusted recognition rates and E-model prediction, Swiss-French and AURORA recognizers. Variable parameter: Nc. $Nfor = -64$ dBm0p.

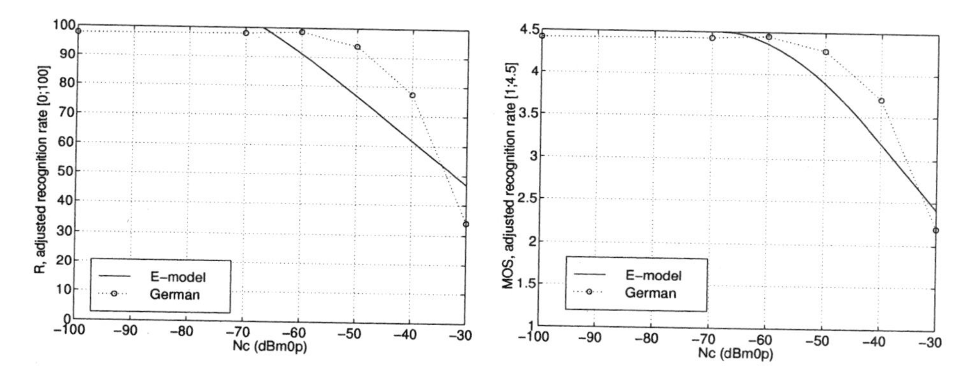

Figure 4.3. Comparison of adjusted recognition rates and E-model prediction, German recognizer. Variable parameter: Nc. $Nfor = -100$ dBmp.

and MOS. The discussion here can only show general tendencies in terms of the shape of the corresponding performance curves; a deeper analysis is required to define 'acceptable' limits for recognition performance (which will depend on the system the recognizer is used in) and for speech quality (e.g. on the basis of Table 2.5).

4.4.2 Impact of Circuit Noise

Figures 4.2 and 4.3 show the degradations due to narrow-band (300-3400 Hz) circuit noise Nc. Because two different settings of the noise floor $Nfor$ were used, the predictions from the E-model differ slightly between the Swiss-French and AURORA recognizer conditions on the one hand, and the German recognizer conditions on the other. For the German and the AURORA recognizers, a considerable decrease in recognition performance can be observed starting at noise levels of around Nc = -50...-40 dBm0p. Assuming an active speech level of -19dBm on the line (ETSI Technical Report ETR 250, 1996, p. 67), this corresponds to an SNR of 21...31 dB. As would have been expected, training on noisy speech (multi-condition training) makes the AURORA recognizer more robust against additive noise. The performance deterioration of the Swiss-French system occurs at lower Nc levels than the one of the German and the AURORA system. All in all, the performance degradation of the German recognizer and the AURORA recognizer trained on clean speech is very similar; the overall performance of the Swiss-French system (evaluated both for all words in the vocabulary and for the keywords only) is much lower, as this system seems to be much more affected by the strict bandwidth limitation (see discussion above).

In comparison to the E-model predictions, the recognition performance decrease is in all cases (with exception of the AURORA recognizer with multi-condition training) much steeper than the R and MOS decrease. Thus, a kind of threshold effect can be observed for all recognizers. The exact position of the threshold seems to be specific for each recognizer, and (as the comparison between clean and multi-condition training shows) also depends on the training material. The agreement between adjusted recognition rates and E-model predictions is slightly better on the MOS than on the R scale, but both curves are very similar. For the Swiss-French system, the performance curves for all words and for the keywords only are mainly parallel, except for very high noise levels where they coincide. For all recognizers, the optimum performance is not reached at the lowest noise level, but for $Nc \sim$ -70...-60 dBm0p. This is due to the training material, which was probably recorded at similar noise levels.

When wideband noise of level $Nfor$ is added instead of channel-filtered noise, the agreement between recognition performance degradation and predicted speech quality degradation is relatively good, see Figure 4.4. The decrease in performance occurs at nearly the same noise level as was predicted by the E-model, though it is much steeper for high noise levels. Once again, the MOS predictions are closer to the adjusted recognition rates than the transmission rating R.

4.4.3 Impact of Signal-Correlated Noise

Figure 4.5 shows the effect of signal-correlated noise which has been generated by a modulated noise reference unit (MNRU) at the position of the codec. The abscissa parameter is the signal-to-quantizing-noise ratio Q. Compared to the Swiss-French recognizer, the German and the AURORA recognizers are slightly more robust, in that the recognition performance decrease occurs at lower SNR values. The shape of the recognition performance curves for the

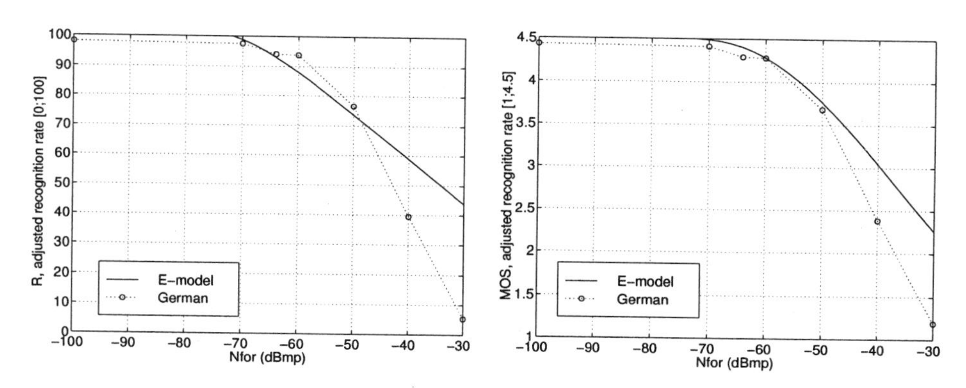

Figure 4.4. Comparison of adjusted recognition rates and E-model prediction, German recognizer. Variable parameter: $Nfor$. $Nc = -70$ dBmp.

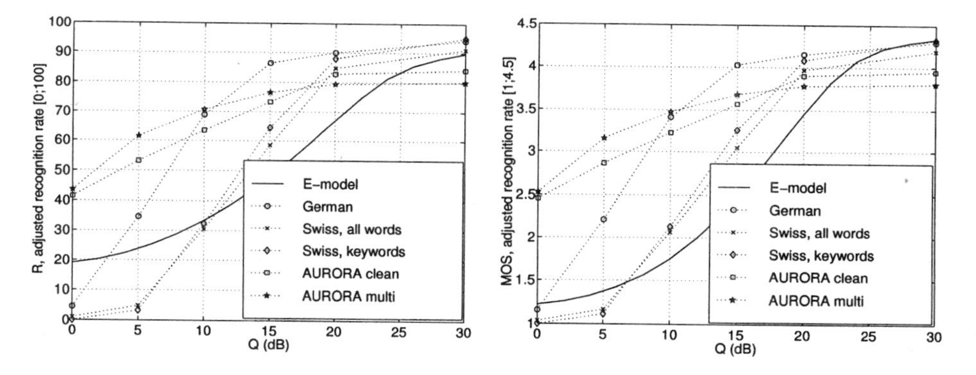

Figure 4.5. Comparison of adjusted recognition rates and E-model prediction. Variable parameter: signal-to-quantizing-noise ratio Q.

German and the Swiss-French recognizers is close to the E-model prediction for MOS, but the decrease occurs at lower SNR values. For the AURORA recognizer, the shape of the curve is much flatter. Although this recognizer does not reach the optimum performance level even for high signal-to-quantizing-noise ratios Q, it is particularly robust against high levels of quantizing noise (more than 40% of the optimum value even for $Q = 0$ dB). With multi-condition training, this recognizer becomes even more robust, at the cost of a slight performance decrease for the high signal-to-quantizing-noise ratios. As a general tendency, human-to-human communication seems to be more critical to signal-correlated noise degradations than ASR performance.

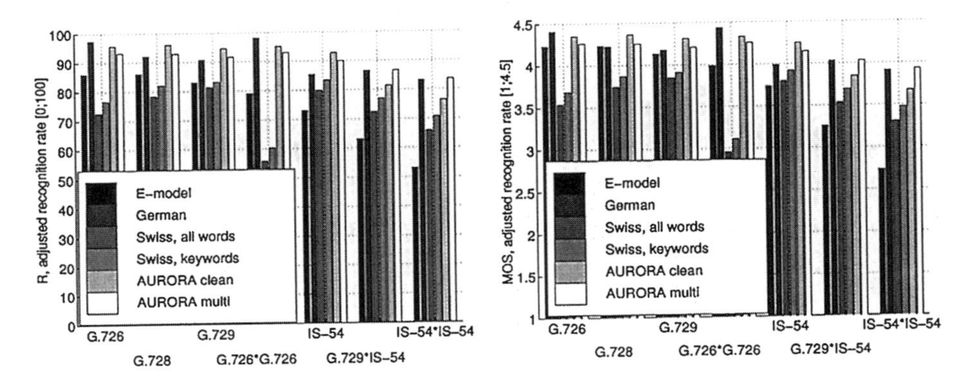

Figure 4.6. Comparison of adjusted recognition rates and E-model prediction. Variable parameter: codec.

4.4.4 Impact of Low Bit-Rate Coding

Non-linear codecs are commonly used in modern telephone networks. They introduce different types of impairment which are often neither comparable to correlated or uncorrelated noise, nor to linear distortions. In Figure 4.6, recognition performance degradation and E-model predictions are compared for the following codecs: ADPCM coding at 32 kbit/s (G.726), low-delay CELP coding at 16 kbit/s (G.728), conjugate-structure algebraic CELP at 8 kbit/s (G.729), vector sum excited linear predictive coding at 7.95 kbit/s, as is used in the first generation North-American TDMA cellular system (IS-54), as well as tandems of these codecs. The bars are depicted in decreasing predicted quality order.

It can be seen that there is no close agreement between estimated speech quality and recognition performance, neither for MOS nor for R predictions. The Swiss-French recognizer seems to be particularly sensitive to ADPCM (G.726) coding. This type of degradation is similar to the signal-correlated noise produced by the MNRU (Figure 4.5), where the same tendency has been observed. The German recognizer, on the other hand, is particularly insensitive to this codec, resulting in high recognition performance for the ADPCM codec in single as well as tandem operation. This recognizer also seems to be quite insensitive to codec tandeming in general, whereas the Swiss-French recognizer's performance deteriorates considerably. Except for codec tandems, the AURORA recognizer is very insensitive to the effects of low bit-rate codecs. This finding is independent from the type of training material (clean or multi-condition training). In the case of tandems, the decrease in performance observed for this recognizer is still very moderate; it is even more robust in the case of multi-condition training. All in all, the significant decrease in recogni-

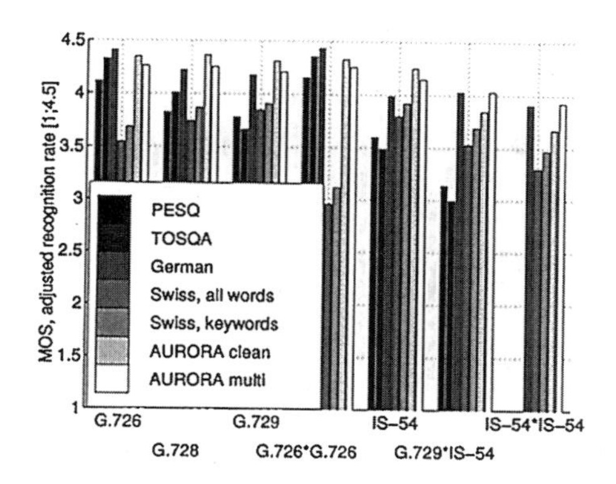

Figure 4.7. Comparison of adjusted recognition rates and PESQ and TOSQA model predictions. Variable parameter: codec.

tion performance predicted by the E-model for low bit-rate codecs could not be observed for the recognizers included in this test.

Codec impairments are also predicted by signal-based comparative measures like PESQ or TOSQA. These measures estimate an MOS value for the codec under consideration, based on the input and output signals. In principle, it is thus possible to estimate the degradation introduced by the codec with the help of the recorded input and output signals. However, in real-life planning situations no speech samples from the input user interface will be available and – apart from limited test corpora – also no signals from the output side. In the system set-up phase this is a fundamental problem, and it is a fundamental difference to the network planning models which rely on planning values only. As a consequence, a slightly different approach is taken here: Reference speech material will be used as an input to the signal-based comparative measures instead of the material which is taken as an input to the recognizer. Such reference material is available in ITU-T Suppl. 23 to P-Series Rec. (1998) and consists of speech samples (connected short sentences) which have been recorded in three different languages under controlled laboratory conditions. This material is also recommended for the instrumental derivation of equipment impairment factors Ie used in the E-model, see ITU-T Rec. P.834 (2002).

The speech files have been prepared as recommended by the ITU-T (ITU-T Suppl. 23 to P-Series Rec., 1998) and processed through reference implementations of the codecs given in conditions No. 20 to 26 (with exception of the IS-54*IS-54 tandem which was not available at the time the experiment was carried out). The individual results have been published by the author in ITU-T Delayed Contribution D.29 (2001). They have subsequently been normalized in a way similar to Formula 4.4.1, taking the maximum value which is predicted for the G.711 log. PCM codec (PESQ MOS = 4.27, TOSQA MOS = 4.19) as the *topline* value. In this way, the PESQ and TOSQA predictions are adjusted to the range predicted by the E-model [1;4.5]. The predictions can now be compared to the adjusted recognition rates, see Figure 4.7.

It turns out that both models predict very similar MOS values. For the G.726 ADPCM codec (single and tandem operation), the predictions are close to the adjusted recognition rates of the German and the AURORA recognizer. The performance of the Swiss-French recognizer is significantly inferior. For the G.728 and G.729 codecs, the predictions are more pessimistic, in the range of the lowest adjusted recognition rates observed in the experiment. The IS-54 and the G.729*IS-54 codec tandems are predicted far more pessimistically than is indicated by the recognition rates. In particular for the tandem a particularly low quality index is predicted. The same tendency was already observed for the E-model. The IS-54*IS-54 tandem has not been included in the test conditions of the signal-based comparative measures. Overall, it seems these measures do not provide better predictions of ASR performance than network planning models like the E-model do. As for the E-model, it has however to be noted that they have not been developed for that purpose.

Lilly and Paliwal (1996) found their recognizers to be insensitive to tandeming at high (32kbit/s) bit-rates, but more sensitive to tandeming at low bit-rates; this is just the opposite of what is observed for the Swiss-French system, whereas it is comparable to the behavior of the AURORA system. Apart from the ADPCM and the IS-54 codecs, the rank order between codecs predicted by the E-model is roughly maintained. ADPCM coding seems to be a problem for the Swiss-French recognizer, whereas the IS-54 codec is better tolerated by all recognizers than it would have been expected from the E-model predictions. As a general tendency, the overall amount of degradation in recognition performance is smaller than it is predicted by the E-model for speech quality. This may be a consequence of using robust features which are expected to be relatively insensitive to a convolution-type degradation.

4.4.5 Impact of Combined Impairments

In Figures 4.8 to 4.10, the effect of the IS-54 codec operating on noisy speech signals is investigated as an example for combinations of different impairments. For speech quality in HHI, the E-model predicts additivity of such impairments on the transmission rating scale. As can be seen from the figures, the MOS curves are also nearly parallel.

The behavior is different from that shown by two of the recognizers used in this experiment. Both the German and the Swiss-French recognizer show an intersection of the curves with and without codec. Whereas the transmission without codec yields higher recognition performance for low noise levels, recognition over high-noise channels is better when the IS-54 codec is included.

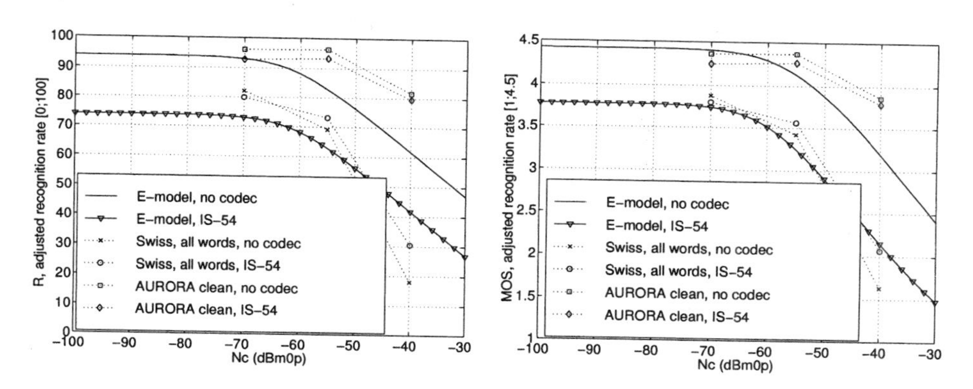

Figure 4.8. Comparison of adjusted recognition rates and E-model prediction, Swiss-French and AURORA (clean training) recognizers. Variable parameters: Nc and codec.

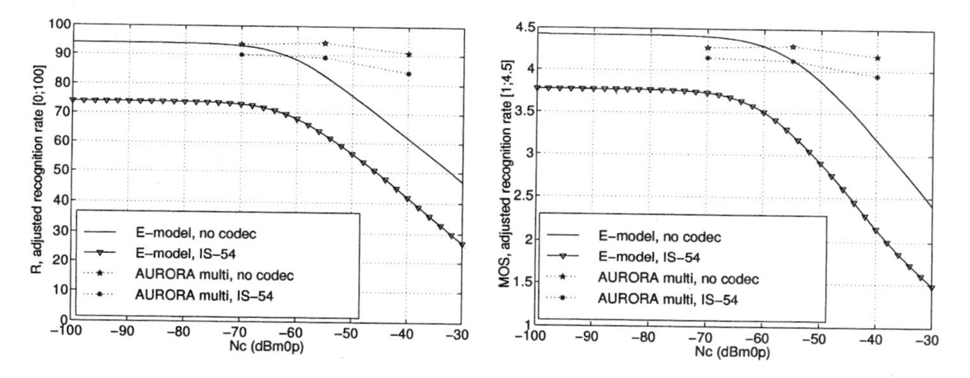

Figure 4.9. Comparison of adjusted recognition rates and E-model prediction, AURORA (multi-condition training) recognizer. Variable parameters: Nc and codec.

Apparently, this codec seems to suppress some of the noise which significantly affects recognition performance of the German and the Swiss-French recognizer. No explanation can be given for the surprisingly high German recognition rate at $Nc = 50$ dBm0p, when combined with the IS-54 codec. Neither the corresponding connection without codec nor the Swiss-French system show such high rates.

The AURORA recognizer, on the other hand, was found to be particularly robust to uncorrelated narrow-band noise. As a consequence, the codec does not seem to have a 'filtering' or 'masking' effect for the high noise levels. Instead, the curves for transmission channels with and without the IS-54 codec are mainly parallel, both in the clean and in the multi-condition training versions.

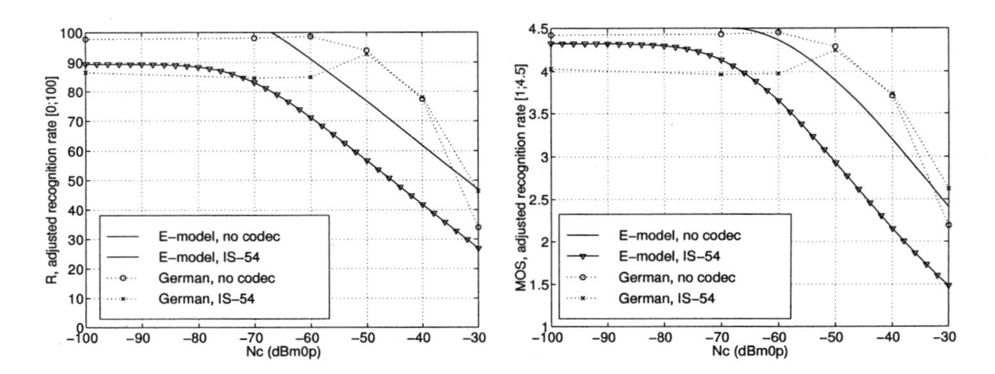

Figure 4.10. Comparison of adjusted recognition rates and E-model prediction, German recognizer. Variable parameters: Nc and codec.

In contrast to the E-model predictions, the offset between the curves is, however, very small. This can be explained by the very small influence of the IS-54 codec which has already been observed in Figure 4.6.

For speech quality, the E-model assumes additivity of different types of impairments on the transmission rating scale R. The presented results indicate that this additivity property might not be satisfied with respect to recognition performance in some cases. However, only one particular combination has been tested so far. It would be interesting to investigate more combinations of different impairments following the analytic method described in this chapter. Before such results are available, a final conclusion on the applicability of the E-model for the prediction of ASR performance degradation will not be possible.

4.5 E-Model Modification for ASR Performance Prediction

The quality prediction models discussed so far, and in particular the E-model, have been developed for estimating the effects of transmission impairments on speech quality in human-to-human communication over the phone. Thus, it cannot be expected that the predictions are also a good indicator for ASR performance. Whereas the overall agreement is not bad, there are obviously large differences between E-model predictions and observed recognition rates for some impairments. These differences can be minimized by an appropriate modification of the model.

Such a modification will be discussed in this section. It addresses the E-model predictions for uncorrelated narrow-band and wideband circuit noise, as well as for quantizing noise. The model will not be modified with respect to the predictions for low bit-rate codecs, because these predictions are very simple in nature. More precisely, coding effects on error-free channels are covered by a single, independent equipment impairment factor Ie for which tabulated values are given; thus, no real modelling is performed in the E-model for this type of impairment.

Uncorrelated noise is captured in the basic signal-to-noise-ratio Ro of the E-model. This ratio is defined by

$$Ro = 15 - 1.5\,(SLR + No) \tag{4.5.1}$$

where No (dBm0p) is the total noise level on the line. It is calculated by the power addition of the four different noise sources:

$$No = 10 \cdot \log_{10}(10^{Nc/10} + 10^{Nos/10} + 10^{Nor/10} + 10^{Nfo/10}) \tag{4.5.2}$$

Nc is the sum of all circuit noise powers, referred to the 0 dBr point. Nos (dBm0p) is the equivalent circuit noise at the 0 dBr point, caused by room noise at the send side of level Ps:

$$Nos = Ps - SLR - Ds - 100 + 0.008\,(Ps - OLR - Ds - 14)^2 \tag{4.5.3}$$

In the same way, an equivalent circuit noise for the room noise Pr at the receive side is calculated:

$$Nor = RLR - 121 + Pre + 0.008\,(Pre - 35)^2 \tag{4.5.4}$$

where the term Pre (dBm0p) represents Pr modified by the listener sidetone:

$$Pre = Pr + 10\log_{10}\left[1 + 10^{(10-LSTR)/10}\right] \tag{4.5.5}$$

The noise floor, $Nfor = -64$ dBmp, is referred to the 0 dBr point

$$Nfo = Nfor + RLR \tag{4.5.6}$$

Power addition of these four noise sources allows Ro to be calculated via Formula 4.5.1.

For the prediction of recognition performance, Formula 4.5.1 is modified in a way which emphasizes the threshold for high noise levels which was observed in the experiment:

$$Ro = 0 - 2.2\,(SLR + No) \tag{4.5.7}$$

In addition, the effect of the noise floor is amended by a parameter Nro (in dBm0p) which covers the particular sensitivity of each recognizer towards noise:

$$Nfo = Nfor + RLR + Nro \tag{4.5.8}$$

The rest of the formulae for calculating Ro (4.5.2 to 4.5.5) remains unchanged.

With respect to quantizing noise, the E-model makes use of so-called quantizing distortion units qdu as the input parameter. One qdu represents the quantizing noise which is introduced by a logarithmic PCM coding-decoding process as it is defined in ITU-T Rec. G.711 (1988). This unit is related to the signal-to-quantizing-noise ratio Q via an empirically determined formula (Coleman et al., 1988; South and Usai, 1992):

$$Q = 37 - 15 \log_{10} (qdu) \tag{4.5.9}$$

The signal-to-quantizing-noise ratio Q is subsequently transformed into an equivalent continuous circuit noise G, as it was determined by Richards (1973):

$$G = 1.07 + 0.258\, Q + 0.0602\, Q^2 \tag{4.5.10}$$

From G, the impairment factor Iq is calculated in the following way:

$$Iq = 15 \log_{10} (1 + 10^Y + 10^Z) \tag{4.5.11}$$

where

$$Y = \frac{Ro - 100}{15} + \frac{46}{8.4} - \frac{G}{9} \tag{4.5.12}$$

and

$$Z = \frac{46}{30} - \frac{G}{40} \tag{4.5.13}$$

This impairment factor is an additive part of the simultaneous impairment factor Is. The exact formulae are given in ITU-T Rec. G.107 (2003).

This part of the model has proven unsuccessful for predicting the impact of signal-correlated noise on ASR performance, see Figure 4.5. It has thus been modified in order to reflect the high recognizer-specific robustness towards signal-correlated noise. For this aim, Q is replaced by

$$Qx = Q + Qo \tag{4.5.14}$$

in Formula 4.5.10, Qo being a recognizer-specific robustness factor with respect to signal-correlated noise. In this way, G can be calculated, and subsequently Iq which is now defined by

$$Iq = 45 \log_{10} (1 + 10^Y + 10^Z) + Iqo \tag{4.5.15}$$

with Iqo being a recognizer-specific constant, Y being defined by Formula 4.5.12, and Z by

$$Z = \frac{46}{15} - \frac{G}{40} \tag{4.5.16}$$

These modifications contain three robustness parameters (Nro, Qo and Iqo) which are specific for each recognizer. The values which are given in Table 4.3

Table 4.3. Recognizer-specific parameters of the modified E-model for predicting ASR performance.

Parameter	*Swiss-French*	*German*	*AURORA*
Nro (dBm0p)	0	10	13
Qo (dB)	26	28	33
Iqo	33	0	0

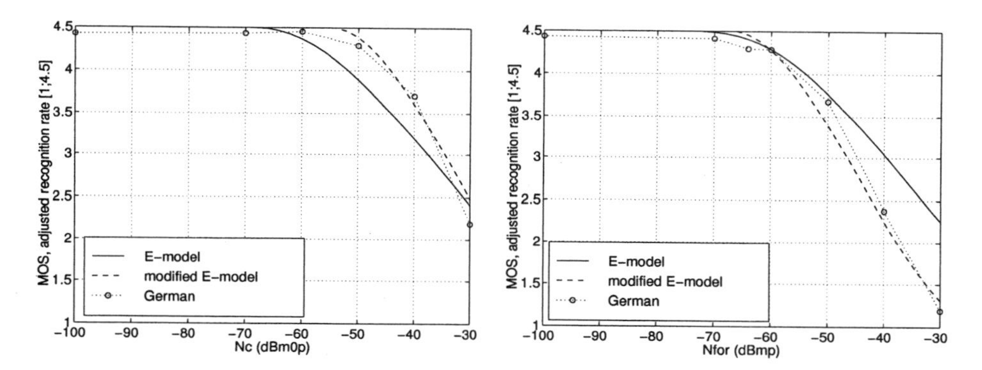

Figure 4.11. Comparison of adjusted recognition rates and E-model vs. modified E-model predictions, German recognizer. Variable parameters: Nc (left) and $Nfor$ (right).

have been derived in order to obtain a relatively good match with the experimental results.

This modified version of the E-model provides estimations which better fit with the normalized recognition performance results observed in the experiments. In Figures 4.11 and 4.12, the MOS predictions of the conventional and of the modified E-model are compared to the adjusted recognition rate curves for narrow-band and wideband circuit noise. Only the MOS predictions are given here, because, in general, they showed a slightly better agreement with the recognition rates than the R values, see Figures 4.2 to 4.10.

Figure 4.11 shows the predictions for the German recognizer and narrow-band circuit noise Nc (left), or wideband circuit noise $Nfor$ (right), respectively. In both cases, the modifications of the E-model lead to a much better fit of the observed recognition rates. In particular, the higher factor (2.2 instead of 1.5) linking the overall noise level No to Ro leads to a steeper decrease of the curve for rising noise levels, which is better in agreement with the behavior of this recognizer. Similarly, the model modification leads to a better prediction of the ASR performance reached by the Swiss-French and the AURORA

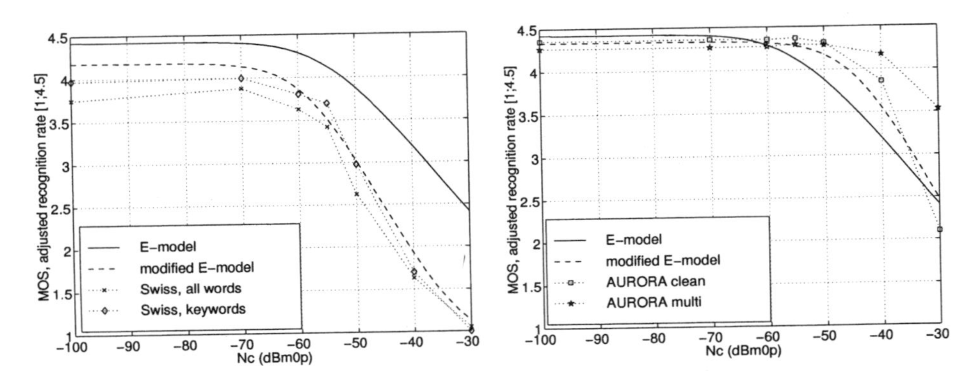

Figure 4.12. Comparison of adjusted recognition rates and E-model vs. modified E-model predictions, Swiss-French (left) and AURORA (right) recognizers. Variable parameter: Nc.

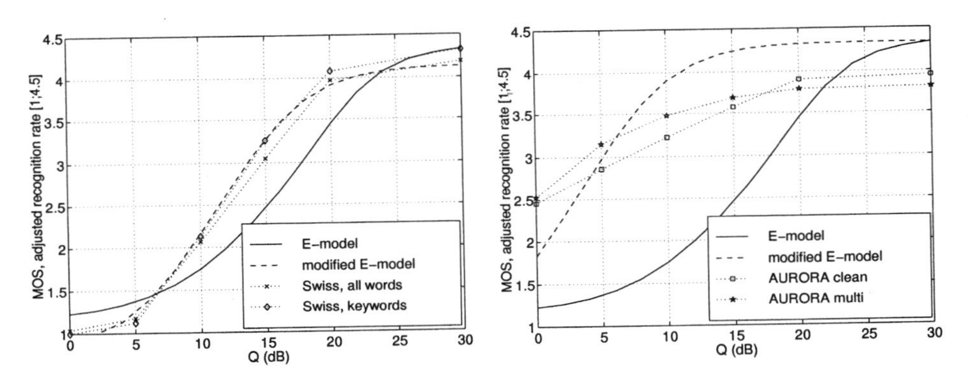

Figure 4.13. Comparison of adjusted recognition rates and E-model vs. modified E-model predictions, Swiss-French (left) and AURORA (right) recognizers. Variable parameter: signal-to-quantizing-noise ratio Q.

recognizers, see Figure 4.12. Whereas for the Swiss-French recognizer both performance results (evaluation over all words and over the keywords only) are covered, the robustness parameters of the AURORA recognizer have been optimized for the version trained on clean speech. For the multi-condition training, the model parameters given in Table 4.3 would have to be adjusted again, leading to different values.

For the effects of signal-correlated noise, the modification of the E-model leads to curves which fit the performance results of the Swiss-French and the German recognizer relatively well, see Figures 4.13 and 4.14. For the AURORA recognizer, however, the relatively flat shape of the performance curve contradicts a good fit. Because the curve is in principle S-shaped, no optimized

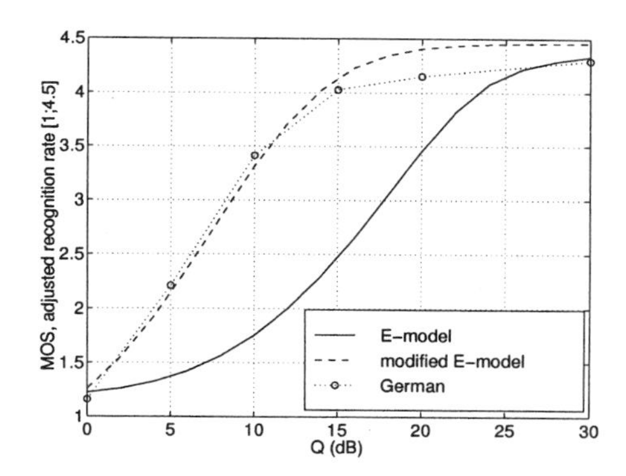

Figure 4.14. Comparison of adjusted recognition rates and E-model vs. modified E-model predictions, German recognizer. Variable parameter: signal-to-quantizing-noise ratio Q.

prediction can be reached without modifying additional parameters of the E-model algorithm. The parameter settings chosen here (see Table 4.3) lead to a too optimistic prediction for the higher signal-to-quantizing-noise ratios ($Q \geq 5$ dB) and a too pessimistic prediction for lower values of Q.

In principle, the proposed modification of the E-model algorithm shows that network planning models which have been developed for predicting the effects of transmission impairments on speech quality in human-to-human communication scenarios can be optimized for predicting the effects on recognizer performance. The modifications were chosen in a way to keep the original E-model algorithm as unchanged as possible. Three additional parameters had to be introduced which describe the robustness of each specific recognizer, observed to be different in the experiment. For predicting the effects of low bit-rate codecs, the equipment impairment factor values Ie used by the E-model would have to be modified as well. The results show that – with one exception – a relatively good agreement with the observed recognition rates can be reached for all recognizers in the experiment.

4.6 Conclusions from the Experiment

The comparison between recognition performance and E-model predictions for speech quality reveals similarities, but also differences between the two entities. The findings have to be interpreted separately for the transmission channel conditions and for the recognizers used in the experiments.

The (normalized) amount of recognition performance degradation due to noise is similar to that predicted by the E-model for most recognizers. With respect to narrow-band and wideband uncorrelated noise, the E-model predictions are in the middle of the range of results covered by the recognizers. The agreement is slightly better on the MOS scale than on the transmission rating scale. However, for all these noises, the ASR performance decrease is steeper than the predicted quality decrease from the E-model. This might be an indication of a threshold effect occurring in the recognizer: Recognition performance is acceptable up to a specific threshold of noise and drops quickly when the noise level exceeds this threshold. The threshold is dependent on the recognizer and on the training material. The exact level of this threshold for a particular recognizer setting has to be defined in terms of the recognition performance which is required for a specific application. Different values for such a minimum requirement have been provided by system developers.

For signal-correlated noise, two of the recognizers (the German and the Swiss-French one) show a behavior which is similar to the predictions of the E-model. However, the decrease of the recognition rate occurs at lower SNR values, indicating that recognizers are more "robust" against this type of degradation than humans are. It has to be noted that this robustness comparison refers to different aspects, namely the recognizer's ability to identify words vs. the human quality perception. The AURORA recognizer is relatively insensitive to high levels of signal-correlated noise, but its overall performance is already affected by high signal-to-quantizing-noise ratios.

The correlation between predicted speech quality degradation and recognition performance degradation is less clear when low bit-rate codecs are considered. This may indicate that the E-model puts emphasis on quality dimensions like naturalness or sound quality, which are perhaps not so important for good recognition performance. More experimental data is needed to justify this hypothesis. The signal-based comparative measures PESQ and TOSQA do not provide better predictions of ASR performance for this type of impairment. Whereas the German and the AURORA recognizers seem to be relatively insensitive to codec-produced distortions, the Swiss-French system is particularly sensitive to ADPCM coding. On the other hand, the IS-54 VSELP coder does not affect recognition performance very strongly, but is expected to have a considerable impact on human speech quality.

The combination of IS-54 coding and circuit noise has been tested as an example for combined impairments. The resulting recognition performance curves do not agree well with the E-model predictions for two of the recognizers. In particular, some "masking" between the two degradations seems to be present (the noise degradation is masked by the subsequent codec for higher noise levels), resulting in an intersection of the performance curves which cannot be observed for the E-model prediction curves. If this difference in behavior can be reproduced for other combinations of impairments, the whole principle underlying the E-model might be difficult to apply to predicting recognition performance. However, doubt has already been cast on this principle by auditory experiments combining background noise and codec distortions (Möller, 2000), or packet loss and noise or echo (Raake, 2003, 2004).

So far, the influence of the transmission channel on three specific speech recognizers was tested, with different languages and test/training material. For uncorrelated noise (narrow-band and wideband) all three recognizers behave similarly, whereas fundamental differences can be observed for signal-correlated noise. The German and the AURORA recognizers seem to be more robust, in the sense that a deterioration in performance occurs at a higher noise level, or lower SNR. The behavior of the recognizers is also different for speech coded at low bit-rates. Here, one recognizer was particularly affected by the ADPCM coding, whereas the other two were insensitive to this degradation. It has to be emphasized that the reported experiments do not permit a general quality comparison to be drawn between the three recognizers. In fact, this was not the aim of the presented study. Instead, the figures have a relative significance with respect to the impairment-free case. The recognizers have been assessed in a near-application scenario, but they can in no way be considered to be optimized systems.

The reported experiment is of course limited with respect to the number and types of degradations which could be investigated. Therefore, further experiments have to be carried out, and extensions to the simulation model are planned. One characteristic to be investigated is wideband (50-7000 Hz) speech transmission which will become more common for IP-based networks. Another topic is time-variant channel characteristics, like random bit errors, bursty error patterns, or lost frames. This type of impairment is common in mobile as well as IP-based networks. The described simulation system has been updated recently in order to include such channel characteristics, cf. the descriptions by Rehmann et al. (2002) and Rehmann (2001). However, approaches to model the effects of time-variant impairments, without using auditory tests, are still very limited. Tests have shown that the speech material and the time distribution of errors in the speech sample have an influence on the quality perceived by humans – a fact which might similarly play a role in assessing recognizer performance.

Another characteristic which has to be modelled more extensively is the user interface. Apart from standard handset telephones, modern networks will be operated from hands-free and headset terminals. Such terminals have a different sensitivity to the received speech material (e.g. because of room acoustic properties) as well as to ambient background noise, when compared to handsets. The effect of background noise is taken into account by the described simulation model when databases are produced. Variations in speaking style (e.g. Lombard reflex) will be reflected in the speech material, as long as the recordings are made with realistic user interfaces.

4.7 Summary

In this chapter, the transmission channel impact on ASR performance is investigated in an analytical way, using three different recognizers: Two prototypical recognizers which are common for telephone-based information servers, and a standardized set-up developed under the AURORA framework for distributed ASR. In order to gain control over the transmission channel, a simulation model has been developed. It implements all types of stationary impairments which can be found in traditional analogue as well as digital telephone networks, including narrow-band and wideband uncorrelated noise, signal-correlated noise, linear frequency distortions, and non-linear codec distortions. ASR performance degradation is compared to the degradation in speech quality between humans, as is predicted by a network planning model (the E-model) and two signal-based comparative measures (PESQ and TOSQA).

It turns out that some interesting differences exist between the ASR system performance and speech quality in human-to-human communication. Such a difference is to be expected, because prediction models like the E-model have been purely optimized for HHI. They predict speech communication quality in a two-way interaction scenario between humans, taking partly also user-interface- and service-related aspects into account. ASR performance, on the other hand, is an interaction parameter which can be related to speech input quality, cf. the QoS taxonomy defined in Section 2.3.1. The quantitative contribution of recognition accuracy to global quality aspects of spoken dialogue systems like usability, user satisfaction, or acceptability will be discussed in Chapter 6.

The results for uncorrelated and signal-correlated noise give a relatively constant level of recognition performance for high signal-to-noise ratios. This drops considerably at a certain noise level which is typical for the recognizer and its training database. For uncorrelated noise, the decrease is steeper than the decrease in speech quality predicted by the E-model. For signal-correlated noise, the decrease is similar to the one predicted by the E-model for two of the recognizers. Non-linear distortions originating from low bit-rate codecs are generally better tolerated by the recognizers than in a human-to-human communication scenario. However, this rule is not without exception, as the results for the ADPCM codec and the Swiss-French recognizer show.

By modifying the model algorithm, it is possible to enhance the E-model prediction accuracy for ASR performance. This has been demonstrated for the effects of uncorrelated and signal-correlated noise. The modifications introduce three additional model parameters which are related to the robustness of each recognizer. With the help of these parameters, it becomes possible to reach a high level of prediction accuracy for all recognizers in the experiment, and for most of the impairments. Additional modifications with respect to the predictions of low bit-rate codecs have not been proposed, as the E-model algorithm describes the effects in a very simplified way, using a one-dimensional impairment factor for which tabulated values are given in ITU-T Appendix I to Rec. G.113 (2002). A modification of the model at this point would signify to simply modify the tabulated data.

The results have some implications, both for the developers of speech technology devices (speech recognition, speech detection, speaker recognition, dialogue management), as well as for the planners of speech transmission networks. Speech recognizers may show weaknesses for certain types of transmission degradations, which are either typical for recognizers in general, or specific to a particular recognizer. The simulation model presented in this section helps to identify these weaknesses and subsequently to enhance recognizer performance. For example, specific training material can be produced for optimizing acoustic models. The recognition results for the two AURORA recognizer settings emphasize the need for training material which has characteristics similar to the application's later scenario's. Such training material can be produced very efficiently using the presented simulation model, as it allows available clean speech databases to be multiplied.

Speech technology aspects should also be considered by transmission planning experts. In particular, codec and combined degradations show that telephone networks which are planned according to the needs of human-to-human communication do not necessarily satisfy the requirements of modern speech technology devices. Thus, what is tolerable according to the E-model (which is the only recommended planning tool for mouth-to-ear quality planning) is not always tolerated by speech recognizers. Modifications to the model algorithms help to bridge this gap, but they will make use of additional parameters which have to be determined beforehand for the speech technology equipment under consideration. Fortunately, in most cases the speech recognizers are more robust against degradations than is expected for humans in a normal communication scenario. Only in two of the experimental conditions (No. 20 and No. 24) was a remarkable decrease in performance observed for the Swiss-French recognizer where the corresponding E-model predictions were far less pessimistic. One could argue that this is only a problem for speech technology developers. However, telecommunication networks are not static, but evolve very quickly due to a changing technical and economic background. As a consequence, speech technology which has been adapted to specific, current transmission equipment is not necessarily robust towards new types of speech processing devices (e.g. new codecs). The current standardization processes for new codecs only includes auditory speech quality tests, but no tests with speech recognizers.

Apart from assessing the transmission channel impact on speech recognition, other applications of the simulation system become obvious. For example, synthesized speech can be assessed under realistic transmission channel situations, as will be done in the following chapter. Nevertheless, speech input and speech output quality should not be seen in isolation. It is also interesting to investigate the effect of the degraded recognition on the subsequent dialogue flow, e.g. by installing the simulation model between the speech recognizer and the test users' interface. Although in Chapter 6 a slightly different approach has been taken (namely to use a recognition simulator with an adjustable recognition rate), the results obtained in this chapter form the link between environmental factors and recognition performance. The latter is an input parameter to the experiments described in Chapter 6.

Both speech recognizers and speech synthesizers may be adapted to the current channel characteristics. The characteristics may be determined online, e.g. using in-service, non-intrusive measurement devices specified in ITU-T Rec. P.561 (2002), and can then be mapped onto parameters which are identical to the ones used for the simulation model. For details on the mapping see ITU-T Rec. P.562 (2004). By simply comparing the parameters describing the network characteristics, adequate acoustic models can be chosen for the speech recognizer, or Lombard speech can be generated by an adaptive speech synthesizer. In this way, the speech input and output modules of a spoken dialogue system can efficiently be optimized towards environmental factors. These factors influence global aspects of the quality of the provided services.

Chapter 5

QUALITY OF SYNTHESIZED SPEECH OVER THE PHONE

In recent years, synthesized speech has reached a quality level which allows it to be integrated into many real-life applications. In particular, text-to-speech (TTS) can fruitfully be used in systems enabling the interaction with an information database or a transaction server, e.g. via the telephone network. There are several reasons which make it attractive in this respect (Fettke, 2001):

- Synthesized speech allows services to be built with quickly changing content; examples include news, sports, stock market information, or weather forecasts.
- Synthesized speech is indispensable for services requiring open text input, e.g. email readers.
- Time and money for studio recordings can be saved when using synthesized speech.

As a result, more and more services make use of synthesized speech in specific dialogue situations, or for the whole dialogue. Because the apparent voice of the interaction partner is a prominent constituent of the whole system, its quality has to satisfy the user's demands and requirements to a high degree.

In this chapter, the quality of the synthesized speech in the application scenario of a telephone-based spoken dialogue system is addressed. In such a scenario, overall quality as well as its perceptive dimensions cannot be regarded in a 'neutralized' way. Rather, the application scenario defines context- and environment-specific factors which have to be taken into account during the evaluation. In order to illustrate such an evaluation, a case study has been performed, taking the restaurant information system used in the last chapter as the application example. The context-specific factors are covered by a specific test design, reflecting the later context of use of the whole system. The requirements for such a test design are addressed in Section 5.1 which condenses a broader discussion given in Möller et al. (2001). The environmental factors are the conditions of the experiment which have been altered. Parts of this study are described in a separate article (Möller, 2004), and new modelling examples with signal-based comparative measures have been added.

The case study is described in the main body of this chapter, in a way which focuses on the transmission channel influences, and not on a specific synthesis system. For that reason, two prototypical speech synthesizers are taken as examples. After a discussion of the requirements for assessment and evaluation experiments in the application scenario addressed here, a short overview of relevant literature is given in Sections 5.2 and 5.3. The main experiment is described in Section 5.4. The obtained results not only illustrate the transmission channel influences in an analytic way, but are also compared to the estimations of quality prediction models which have been originally designed for HHI over the phone, cf. Section 2.4.1.1. In this way, it is shown how far it is possible to estimate the quality of a future system component operating over a telephone transmission channel with defined characteristics, taking a basic quality index and the transmission channel parameters as an input.

5.1 Functional Testing of Synthesized Speech

In the assessment of speech, it is very useful to differentiate between the acoustic form (surface structure), and the content or meaning the listener assigns to it (deep structure). The content has a strong influence on the perception of speech, as well as on the perceived quality of the auditory event. Nevertheless, in a natural communication situation the recipient does normally not differentiate between surface structure and deep structure. Exceptions to this rule can be encountered when the form is so bad (e.g. due to transmission impairments or to background noise) that the listener is unable to work out the intended meaning. For natural speech, the form is a natural fact, and it normally does not catch the listener's attention. In contrast to natural speech, the form of synthesized speech is artificial and sounds strange so that the listener tends to focus on form characteristics first. After some habituation, functional aspects increase in importance again (Delogu et al., 1997).

This difference in focus should be reflected in the evaluation of synthesized speech. The quality of speech, be it synthesized or naturally produced, can thus only be determined taking the whole interaction situation into account. It follows that the designer of the speech output component has to specify the system functionality with regard to an average user's communication requirements. There is a notable difference in perceived quality between applications where the synthesized speech enables a new system functionality and others where it only replaces recordings of naturally produced speech in existing systems.

It becomes clear that there are strong links between the synthesis component and the whole application system it is used for. As a consequence, the decision of whether synthesized speech is to be preferred to naturally produced utterances or not, or which one of several available synthesis components is to be selected, can only be taken in view of the specific speech and application system function. The quality of the speech device thus has to be determined in use. If attention is given to the functional aspect in the assessment procedure, more than just an indication of the speech output quality can be obtained. Also the contributions of the speech output component to system usability, user satisfaction, and acceptability can be identified.

In the past, the quality assessment of synthesized speech often served the development and improvement of the synthesis itself. For this purpose, analytic tests as well as multidimensional analyses of synthesized speech samples were mainly used as assessment tools. In most of the tests, synthesized speech was regarded in a 'neutralized' way, independent of a specific application scenario (Salza et al., 1996; Klaus et al., 1997). Several investigations have been performed to analyze the performance of different components of speech synthesis systems, e.g. in Benoît et al. (1991), Kraft and Portele (1995), van Bezooijen and Pols (1990), and van Santen (1993). This work has been accompanied by research on different aspects of the assessment method itself, e.g. scaling (Pavlovic et al., 1990; Delogu et al., 1991; Bappert and Blauert, 1994), test material (Silverman et al., 1990), listener dimensions (Howard-Jones, 1992), cognition (Delogu et al., 1997), cognitive load (Silverman et al., 1990), etc.

Because the speech output component has to be seen in conjunction with other modules of the SDS, assessment methods will be used here which differ from the cited ones. Ideally, field tests will allow valid assessment results regarding the application scenario to be obtained. Laboratory tests carried out on the isolated speech synthesis component can generally only predict the success of the speech output module in the final application to a limited extent. However, field tests have severe disadvantages as well. Lack of control over environmental conditions of the test situation is a problem in field tests, and they are generally far more costly. In the developmental phase, the future system is often not yet available at the time its features have to be assessed. As a consequence, simulations have to be used which can be set up more easily in laboratory conditions. Nevertheless, it is essential that tests performed in laboratories are designed to cover predominant aspects of the application system behavior in use.

The user group and their typical attitudes towards new technologies and experience with comparable systems also cause alterations in quality perception (Möller, 2002b). Unfortunately, systems can generally not be designed in order to satisfy individual users' needs. It is therefore necessary to investigate and specify the requirements of potential user groups before staring the evaluation. The task of the system designer is then to specify an average user profile, and to select system elements which best satisfy the defined needs. Generally, users attribute the quality they have perceived to the system as a whole, and not specifically to the (synthesized) speech. With regard to quality assessment it follows that a clear distinction has to be made between speech and system quality assessment. In this chapter, the focus lies on the quality of synthesized speech in a specific application scenario, and in particular on the influence of the context and environmental factors. How this quality influences the overall usability, user satisfaction, and acceptability of the system as a whole is addressed in the following chapter.

In summary, the assessment of functional quality aspects of speech requires an analytic description of the environmental and contextual factors, as well as an expectation profile of the potential user group towards the application (user factors). An appropriate test methodology can then either be chosen from existing ones, or it has to be designed. The development of assessment methods must be guided by the following questions:

1 Which quality features are of primary interest (test objective)? The focus of the assessment should be put on these features.

2 Which type of assessment (laboratory tests, user surveys, field tests) is the most appropriate for the assessment task? Compromises will sometimes have to be made because of time and budget restrictions.

3 Are all quality features with potential relevance for the user included in the assessment situation? The test situation and set-up should reflect the future application as far as possible.

4 Is the test material representative of the application?

5 Is the group of subjects representative of the user group?

6 Is the assessment task comparable to the future application in its focus of attention as well as in its difficulty?

7 Are appropriate scaling methods or other evaluation forms (questionnaires, interview questions) available to assess all relevant quality features? Can additional information be obtained by means of non-formalized evaluation?

8 Can test results later be analyzed and interpreted in a way that answers the posed questions and problems? Statistical standards have to be satisfied in the analysis of test results.

9 Can test results be generalized to other situations and application scenarios? Under which circumstances, and with what degree of validity?

Standardized test methods satisfy some but not all of these requirements in a specific assessment task. They carry the advantage of having frequently been used in the past, and thus they can be run relatively quickly under fixed and controlled conditions at different locations. The test design is mainly determined by criteria (1), (2), (7) and (8). A severe disadvantage, however, is the fact that the requirements of a specific application are mainly disregarded, and that the assessment methods often do not carry the right focus to yield valid information with respect to the test objective. Speech is regarded in a 'neutralized' way, omitting many of its functional and content-related aspects. If nevertheless pre-elaborated tests are carried out, points (1), (3), (4), (6) and (9) should be critically checked in order to guarantee valid assessment results for the application system of interest. Functional assessment of speech quality should be seen as an addition to standardized methods, in that it takes the content or meaning of speech in use into account. An example for such a test is given in Section 5.4. Functional tests are not informal. They have to satisfy the same general criteria as standardized tests in order to provide reliable results with respect to psychophysical and psychoacoustic criteria as well as statistical criteria.

5.2 Intelligibility and Quality in Good and Telephonic Conditions

The quality which can be attributed to the transmitted speech is a multidimensional entity. It results from a perception and a judgment process, cf. the definition of speech quality given by Jekosch (2000), and includes perceptual dimensions such as articulation and intelligibility, naturalness, speech sound quality, noisiness, or listening-effort. A number of investigations are reported in the literature which analyze these perceptual dimensions and appropriate assessment methods are available for quantifying selected dimensions, both with naturally produced as well as with synthesized speech. For the latter, an overview was given in van Bezooijen and van Heuven (1997). The set of quality dimensions which is relevant for a specific service will depend on the speech material, on the characteristics of the transmission channel and on the talking and listening environment. Unfortunately, only few investigations take this fact into account in an analytic way.

Several studies address the intelligibility of synthesized speech, see e.g. Delogu et al. (1995), Spiegel et al. (1990), and Jekosch (2000). They reveal that fundamental perceptual differences exist between synthesized and naturally produced speech. One main difference seems to be that synthesized speech is poorer from a phonetic point of view, i.e. that there are fewer variations and fewer redundancies. This lack of phonetic variation results in intelligibility differences when both types of speech are transmitted over telephone channels (Salza et al., 1996; Delogu et al., 1995). Spiegel et al. (1990) investigated the intelligibility of synthesized and naturally produced speech which was transmitted over telephone channels, generated by noise addition and bandwidth restriction. They used the "Bellcore" test corpus for assessing the intelligibility of one or more consonants in initial and final word position, and found out that the intelligibility of natural speech is particularly affected for voiceless fricatives, due to the combination of bandwidth restriction and noise. Several consonant clusters were identified which were particularly unintelligible for synthesized speech, but not for natural speech (e.g. stop consonants in combination with liquids, glides or nasals). Nearly all common confusions observed for the natural speech samples in their investigation could be explained by the bandwidth limitation and the noise introduced by the simulated telephone transmission.

Significant intelligibility differences were also found for wideband, high-quality transmission conditions compared to telephonic ones, the latter being generated by restricting bandwidth and adding noise. Delogu et al. (1995) evaluated the segmental intelligibility of speech transmitted over high-quality and telephone channels, using an open response test. A difference of around 10 to 21% was observed between the high-quality and the telephonic channel conditions for intelligibility of synthetic speech, whereas the difference for natural speech was only around 5%. Balestri et al. (1992) report on a comparative (natural vs. synthesized) speech intelligibility test for a reverse telephone directory application. Although their test was designed to be very application-specific (using representative text material, a specific prosodic structure, a real-life listener task, etc.), the results give some general indications about intelligibility differences in high-quality wideband (headphone listening) and telephonic (handset listening) conditions. All the experimental factors such as synthetic speech, bandwidth reduction, listening condition and log. PCM coding introduced perceptual degradations which also affected intelligibility. In particular, the intelligibility of natural speech was less affected by the transmission channel restriction and the handset listening (97.8% to 96.5%) than the synthesized speech (94.0% to 88.1%). The authors explain this finding with a higher cognitive demand which synthesized speech puts on its listeners. The telephone environment had only little effects on the initial and final consonants of natural speech, but on all (initial, medial and final) consonant positions for synthesized speech. Because intelligibility is a key dimension of overall quality, it has to be assumed that telephone channels will also exercise a different overall quality impact on synthesized compared to naturally produced speech.

5.3 Test- and System-Related Influences

A number of investigations comparing different TTS systems in a specific application scenario are reported in literature. Apart from the value they have for the application under consideration, the results also identify test-dependent factors influencing the overall quality or individual quality dimensions. Arden (1997) describes the comparative evaluation of the BT Laureate TTS system with two commercially available systems. His results show that the listening level is an important experimental factor, and that different TTS systems perform best at different listening levels. The subjective judgments for longer text paragraphs were lower than for individual sentences, indicating that the length of the speech stimuli has an effect on the quality judgments. As was already reported for natural speech, a 5-point ACR listening-effort scale is more sensitive than a comparable overall quality scale, in terms that the judgment range used by the test subjects is larger.

In ITU-T Contribution COM 12-176 (1987), Gleiss describes two experiments assessing the quality of the Infovox formant synthesizer implemented in a weather forecast service which is operated over the phone. In a field test, quality aspects of synthesized speech like pronunciation, stress, speed, intelligibility, distinctness, comprehensibility, naturalness, and pleasantness were judged by the test subjects on different scales. They were also interviewed with respect to their ratings on overall quality and acceptability. In an accompanying laboratory test, natural and synthesized speech were the subject of a paired comparison with an additional acceptability evaluation. A comparison of the results obtained in both tests showed that the ranking of the overall quality and of the naturalness and pleasantness dimensions was the same in both tests, although obtained with different scaling methods. On the other hand, intelligibility was rated more critically in the field test than in the laboratory test – probably because in the laboratory test the text material was known to the subjects beforehand. Both paired comparison and single stimulus ratings lead to comparable results, and interestingly the laboratory and the field test yielded about the same "acceptability" level. In the laboratory test, naturally produced speech taken as a reference did not distort the results for the synthesized speech samples.

In several application scenarios, it is not necessary to use synthesized speech throughout the whole dialogue. For example, a system for providing train timetable information can deliver the fixed system prompts by naturally recorded speech, and the changing locations by synthesized speech. McInnes et al. (1999) describe an experiment evaluating the use of mixed natural/TTS prompts for the speech output of a flight confirmation service. In their experiment, the test subjects had to extract specific information from the system messages, and to judge six statements on Likert scales. The authors found that synthesized speech was rated worse than naturally produced one in all cases, when used exclusively (against all natural speakers) as well as in mixed prompts. The longer the parts of the TTS, the worse the rating was. For the natural prompts, professional voices were preferred to amateur ones, and the quality difference between a professional and an amateur speaker was larger than that between two professional speakers. For the mixed prompts, the authors observed a preference for using one single voice (naturally produced and synthesized parts from the same speaker) instead of mixed voices from different speakers. TTS intelligibility was particularly low for the unfamiliar or unexpected pieces of information.

5.4 Transmission Channel Influences

For the designer of transmission channels as well as of spoken dialogue systems, analytic investigations of perceptual speech quality dimensions are not always easy to interpret. Many designers prefer to have a one-dimensional quality index at hand for describing and optimizing the overall quality of the planned service, and finally for optimizing its acceptability (i.e. the number of actual users related to the target user group size). This overall quality index will depend in some way on the transmission channel characteristics. When the impact of the channel can be quantified on the basis of parameters which are known beforehand, it becomes possible to plan networks according to quality considerations. This is important both for the designers of networks as well as for the designers of SDSs, because the networks have to enable the transmission of naturally produced as well as synthesized speech.

The experiment described beneath analyzes the impact of a general telephone transmission channel on the quality of synthesized speech. A typical application scenario is taken as an example, namely a restaurant information system using synthesized speech. In particular, two questions will be addressed:

- Is the impact of the transmission channel on the quality of synthesized speech different from the impact on naturally produced speech?

- How far can prediction models describing the quality impact on naturally produced speech (HHI) be used for predicting the effects on synthesized speech (HMI)?

No comparison will be made to a wideband, high quality headphone situation, because the latter is not a reference for the application scenario. Instead, the results will be interpreted in a relative way, in comparison to the results obtained for the undisturbed channel (only bandwidth restriction and frequency distortion from the handset).

The speech synthesizer itself is not the focus of the reported investigations, but only the degradation which is caused by the transmission channel. Two typical synthesizers have been taken as examples, which do not necessarily reflect the latest developments in speech synthesis, and which have not been optimized for the application scenario. In particular, both synthesizers concatenate short segments (mainly diphones) via an overlap-add algorithm, irrespective of the prosodic match of the units. Unit-selection synthesis systems have not been chosen, because it is expected that such systems deliver an overall quality which is highly dependent on the size and optimization of the unit inventory, and which varies in the 'long term' (over longer parts of phrases) with the number and positions of the concatenations. As investigations by Nebbia et al. (1998) and others show, the overall quality and intelligibility for unit-selection synthesis can be expected to reach a higher absolute level. Thus, differences which are observed in the described experiment may become smaller for unit-selection synthesizers. Formant synthesizers, on the other hand, have not been chosen, because they are only rarely used in application systems.

In order to obtain controlled transmission channel conditions, the simulation model described in the last chapter has been used. It generates impairments which are typical for traditional as well as modern (mobile or IP-based) networks in a controlled and flexible way. The focus of the reported experiment is on the channel degradations, and not particularly on the user interfaces, which have been chosen to represent standard handset telephones. The acoustic effects of other user interfaces will be the topic of follow-up experiments which still have to be performed.

The input parameters of the simulation system characterize the channel from mouth to ear. They are subsequently used to estimate the overall quality which can be expected for human-to-human communication over a telephone connection with the given characteristics, using a network planning model. Alternatively, system prompts at the input and the output of the transmission channel simulation are used as an input to the signal-based comparative measures. A comparison between the model predictions and the auditory test results will show whether it is possible to estimate the telephone transmission channel impact on synthesized speech with the methods typically used in telephony.

The transmission channel simulation and typical quality prediction models have already been described in Sections 4.2 and 2.4.1.1. Section 5.4.1 discusses the source speech material used for the auditory experiment, its processing through the transmission channel simulation, as well as the test design and rating procedure. The experimental results are analyzed in Section 5.4.2, separately for each type of degradation. An outlook in Section 5.4.3 discusses the experimental results with respect to the two questions raised above, and identifies future work to be carried out with additional synthesis systems and other transmission channel characteristics.

5.4.1 Experimental Set-Up

It is the aim of the proposed experiment to analyze the effects of different transmission channels. Therefore, it was decided to include a large number of channel settings in the test stimuli. As a consequence, it was not possible to test the synthesized speech in a realistic conversation (interaction) scenario. A real interaction scenario would have allowed conversational degradations to be addressed, but would have required far longer test sessions, leading to subject fatigue. As a compromise, a listening-only test scenario was chosen.

5.4.1.1 Speech Material

The scenario addressed in the experiment is the server for information about the restaurants in the town of Bochum which was briefly described in the last chapter (a more detailed description of the dialogue capabilities will follow in Chapter 6). It can be accessed via the telephone network. A prototype version of the system exists for the German language and is based on a Swiss-French prototype (Möller and Bourlard, 2002). It possesses ASR and rough speech understanding capabilities, a finite-state dialogue model, database access, and speech output. The responses generated by the system are either fixed messages (e.g. welcome message, standard meta-communication questions) or responses (questions or statements) which provide specific task information to the user. They address the type of food served in a specific restaurant (e.g. Greek, Italian, pizza), the location of the restaurant (town center, near to the railway station, Bochum-Querenburg, etc.), its opening hours, and the price level. Such task-related responses are particularly suitable for being produced by synthesized speech, because they contain variable pieces of information which may have to be updated when the database is changed. Consequently, these sentences have been selected for the experiment.

In order to generate sentences of standard length and complexity which nevertheless provide new information to the listener, five sentence templates have been defined. Each template contains two variable pieces of information which are inserted at different positions in the sentence. The templates and the variable parts are given in Appendix B. On the basis of the sentences, speech files have been generated by two synthesis systems (male voices), and recorded from two natural speakers (1 male, 1 female). The speakers have been selected to represent different voice characteristics, and were recorded in an anechoic environment; they were not professional speakers.

Synthesis system 1 is a concatenative synthesizer developed at IKA. It consists of a symbolic pre-processing unit (SyRUB), see Böhm (1993), which is linked to the synthesizer IKAphon (Köster, 2003). F_0 and phone length modelling is performed as described by Böhm (1993). The inventory consists of variable length units which are concatenated as described by Kraft (1997). They have been recorded by a professional male speaker and are available in a linear PCM (16 bit) coding format. Synthesis system 2 is a multilingual concatenative LPC diphone synthesizer which is commercially available. It makes use of vector-quantized LPC and a parametrized glottal waveform for synthesis. Details on this synthesis can be found in Sproat (1997). It has to be noted that the speech material has not been specifically optimized after generation – a fact which has to be taken into account when interpreting the experimental results. In particular, pronunciation errors or inadequate prosody have not been corrected. Because one of the synthesizers is a commercial system which is not fully available to the author, a potential correction would have to be carried out on the signal level, e.g. using an overlap-add technique. An effect which is probably linked to an inadequate prosody will be shown in Section 5.4.2.4.

Because the listening level has proven to be an important factor for the quality judgments of synthesized speech (Arden, 1997), all speech files have been normalized to an active speech level of -26 dB below the overload point of the digital system, when measured at the position of the codec. This level is commonly encountered in European telephone networks.

5.4.1.2 Transmission Channel Characteristics

The speech material generated in this way has been processed by the transmission channel simulation described in Section 4.2. Because the experiment was carried out in a listening-only mode, only the parameters which affect the listening situation could be investigated. The following parameters of the simulation system were adjusted for the speech file processing:

- The effect of continuous narrow-band circuit noise, Nc.
- The effect of continuous wideband noise at the receiving terminal side, $Nfor$.
- The effect of quantizing distortion of different signal-to-noise ratios Q.
- The effect of ambient room noise.
- The effects of low bit-rate coding, both single codecs as well as asynchronous combinations of two codecs (tandems).
- The combination of narrow-band circuit noise Nc with the IS-54 cellular codec.

The exact parameter settings are given in Table 5.1. All other parameters of the transmission channel simulation were set to their default values given in Table 2.4. It should be noted that the settings for $Nc = -100$ dBm0p and $Nfor = -100$ dBmp are theoretical values; in reality, the noise floor of the simulation system (expected to be in the order of -80 dBmp) may become dominant here. However, noise levels below -70 dBmp should hardly be noticeable in the telephone handset listening situation.

The pre-recorded source speech files originating from the four voices (naturally recorded or synthesized speech) were processed throughout the simulation, and then recorded on the output side of the simulation, at the position of the right bin in Figure 4.1. The processed files were played back to the test subjects via a standard wireline handset. The electro-acoustic transmission characteristics

Table 5.1. Parameter settings for the test connections of experiments 5.1 and 5.2.

No.	*Nc* (*dBm0p*)	*Nfor* (*dBmp*)	*Codec / MNRU*	*Pr* (*dB(A)*)
1	-100	-100	–	35
2	-100	-100	G.711	35
3	-70	-100	G.711	35
4	-60	-100	G.711	35
5	-50	-100	G.711	35
6	-40	-100	G.711	35
7	-30	-100	G.711	35
8	-70	-70	G.711	35
9	-70	-60	G.711	35
10	-70	-50	G.711	35
11	-70	-40	G.711	35
12	-70	-30	G.711	35
13	-70	-64	G.726	35
14	-70	-64	G.728	35
15	-70	-64	G.729	35
16	-70	-64	IS-54	35
17	-70	-64	G.726*G.726	35
18	-70	-64	IS-54*IS-54	35
19	-70	-64	G.729*IS-54	35
20	-70	-64	MNRU, $Q = 30$ dB	35
21	-70	-64	MNRU, $Q = 20$ dB	35
22	-70	-64	MNRU, $Q = 15$ dB	35
23	-70	-64	MNRU, $Q = 10$ dB	35
24	-70	-64	MNRU, $Q = 5$ dB	35
25	-70	-64	MNRU, $Q = 0$ dB	35
26	-100	-100	IS-54	35
27	-70	-100	IS-54	35
28	-60	-100	IS-54	35
29	-50	-100	IS-54	35
30	-40	-100	IS-54	35
31	-30	-100	IS-54	35
32	-70	-64	G.711	35
33	-70	-64	G.711	50
34	-70	-64	G.711	60
35	-70	-64	G.711	70

of this listening handset were adjusted with the help of the RLR' filter to the one of an intermediate reference system (see Section 4.2).

In order to minimize the effect of differences in language material, 35 sentences generated from the templates in Appendix B were equally distributed to the 35 connection settings of Table 5.1. In this way, each voice/connection combination only occurs once in the whole test, and each connection is rated on different sentences and templates for each voice. Due to the maximum number of speech samples which could be presented to the listeners without fatigue, only one test sentence per connection and voice could be tested. This limitation leads to a slightly augmented variance in the test results and has to be taken into account when outliers are interpreted.

5.4.1.3 Test Design

It is expected that the overall quality ratings of listeners who do not have to concentrate on the contents of the speech samples (what is said) will reflect to a high degree the form of the speech (how it is said). This situation is not realistic for the application scenario described above. Thus, a test situation was created in which the subjects were instructed to concentrate on the topic of the sentence, in order to extract information for two of the four slots carrying information (weekday or time, location, price, and type of food). After listening once to each speech sample, subjects were either asked to provide the information they understood, or to rate different aspects of overall quality. This situation mainly corresponds to the principles described in ITU-T Rec. P.85 (1994). Due to the limitation given by the maximum number of samples which subjects can listen to without fatigue, however, it was necessary to separate the collection of information (on what was understood by the subjects) from their rating on quality aspects. The subjects' concentration on the content may partly be relaxed by this separation, but it is still supported by the instructions given to them before the test, and by the judgment form presented to them on the screen. If both experimental tasks had to be performed in the same test session, the individual sessions would have been too long.

The test was divided into two parts (5.1 and 5.2), each part consisting of four sessions (each approx. 25 min.) with intermediate breaks. In each session, test subjects listened either to naturally produced or to synthesized speech samples. The separation of synthesized and natural speech samples was chosen to make better use of the rating scales in each case, because it was foreseen that the synthesized speech would globally be rated worse, irrespective of the transmission conditions. As a consequence, the answers have to be compared in a relative way (which is in line with the aim of the reported investigations), and not in terms of absolute numbers. Five speech samples similar to the respective set in the test session were presented as anchor conditions (without rating) before each session.

In part 5.1, test subjects were asked to give judgments for the following questions:

- Overall quality: How was your overall impression of what you just heard? Excellent, good, fair, poor, bad (German: "Wie beurteilen Sie den Gesamteindruck des gerade Gehörten? Ausgezeichnet; gut; ordentlich; dürftig; schlecht").

- Listening-effort: How much effort was necessary to understand the meaning? No effort required, complete relaxation possible; attention necessary, no appreciable effort required; moderate effort required; effort required; no meaning understood with any feasible effort (German: "Welche Anstrengung war notwendig, um die Bedeutung zu verstehen? Keine Anstrengung erforderlich, entspanntes Zuhören möglich; keine wesentliche Anstrengung, aber Aufmerksamkeit erforderlich; mäßige Anstrengung erforderlich; erhebliche Anstrengung erforderlich; selbst größte Anstrengung genügt nicht zum Verstehen").

- Intelligibility: Did you find certain words hard to understand? Never; rarely; occasionally; often; all of the time (German: "Wie häufig waren Worte schlecht zu verstehen? Nie; selten; manchmal; oft; immer").

- Perceived acceptability: Do you think that this voice could be used for the described information service? Yes; no (German: "Könnte diese Stimme im beschriebenen Informationssystem verwendet werden? Ja; nein").

Subjects rated the first three questions on 5-point absolute category rating (ACR) scales with numbers from 5 to 1, and the respective labels as indicated above. For the last question, possible answers were "yes" and "no". In part 5.2, the subjects were only asked to fill in the information they understood, but not to rate the quality.

The test run was administered on a PC terminal. Test subjects controlled the playback of the samples themselves, in order to avoid any time pressure during the quality rating task (part 5.1). Lack of playback control may however be preferable for part 5.2, and would probably have lead to lower identification scores. The subjects listened once to each speech sample via a standard telephone handset, in an office room environment (background noise floor $Pr \leq 35$ dB(A); the requirements of NC25 (Beranek, 1971) were fulfilled; reverberation time was between 0.37 and 0.50 s in the frequency range of interest). For the background noise conditions 33-35, hoth-type noise simulating office environments (Hoth, 1941) was inserted in the test room. In order to produce a relatively diffuse field at the listener's position, four loudspeakers (JBL control 1C) were used in these conditions.

23 test subjects (13 male, 10 female) participated in part 5.1, and 6 subjects (4 male, 2 female) in part 5.2. 12 of them (5 in part 5.2) reported to have no experience with synthesized speech and with the type of listening experiment carried out here. Age ranged between 21 and 48 years (mean 24.2 years) for

part 5.1, and between 23 and 29 years (mean 25.2 years) for part 5.2. The subjects were paid for their service.

5.4.2 Results

For correctly interpreting the listeners' judgments and for drawing comparisons to the estimations of quality prediction models, it is important to take the characteristics of the test set-up and the judgment scales into account. The 5-point ACR quality scale was shown not to have an interval level, see Möller (2000) for a discussion. Thus, the calculation of an arithmetic mean is – strictly speaking – not meaningful. In addition, this scale shows saturation effects at the scale end points (1 and 5). The saturation is partly reflected in the S-shaped relationship between MOS and R in the E-model, and in the *arctan* function linking predicted similarity to predicted MOS in some of the signal-based comparative measures.

Nevertheless, arithmetic mean values will be calculated from all 5-point absolute category ratings, in order to be in line with the common practice in telephony, and to be able to compare the results to the estimations of quality prediction models. The mean values will be designated MOS (mean opinion score) for the mean rating on the quality scale, $\mathrm{MOS_{LE}}$ for the mean rating on the listening-effort scale, and INT for the mean rating on the intelligibility scale. The latter should not be confused with the identification rates for the variable items in the sentences which subjects were asked to write down in part 5.2 of the test. Each voice/connection combination was rated once by each subject, so that most arithmetic means have been calculated over $n = 23$ subject ratings (connections 32-35 were rated twice as often due to the test design). The variance is not depicted in the following discussion in order to give a clear representation of the test results; it is generally in the range 0.7 to 0.8 for most of the ACR ratings. A slightly higher variance is encountered for the background noise conditions, because some of the test subjects do not take the background noise into account when rating overall quality. A slightly higher variance was also found for intelligibility judgments of the synthetic voices.

Because the judgments for naturally uttered and synthesized samples have been obtained in separate sessions, a comparison between the judgments will be relative in character. This is in line with the aim of this study, namely to determine the relative degradation due to the transmission channel, and not absolute quality levels. For abstraction from absolute levels, test results are normalized before comparison.

5.4.2.1 Normalization

Several possibilities exist for normalizing the mean auditory judgments obtained from the test subjects. The ITU-T formerly recommended a normal-

Table 5.2. Values of the *topline* parameter for the individual voices.

Judgment / prediction	*Synthetic 1*	*Synthetic 2*	*Natural 1*	*Natural 2*
MOS	2.91	3.78	4.46	4.27
MOS_{LE}	3.74	4.17	4.83	4.68
INT	4.00	4.61	4.92	4.95
PESQ MOS	4.28	4.32	4.28	4.36
TOSQA MOS	4.20	4.21	4.20	4.21

ization procedure which makes use of signal-correlated noise, the so-called 'equivalent-Q method' described in Annex G of ETSI Technical Report ETR 250 (1996). It determines the equivalent level of signal-correlated noise generated by an MNRU which is necessary for provoking the same quality degradation as the speech stimulus under consideration. Unfortunately, this methodology has proven to be unsuccessful for normalizing subjective test results obtained for low bit-rate speech codecs, because the degradations generated by the MNRU are not comparable in their perceptual character to the ones originating from codecs. A similar approach for synthesized speech has been proposed by Johnston (1997), using signals which are degraded by a time-and-frequency warping (TFW) algorithm in a controlled way. This method, however, only takes the characteristics of the (synthesized) source signals, but not of the transmission channel into account. The TFW-degraded signals are neither perceptually comparable to low bit-rate type degradations nor to noise-type degradations.

For the comparison described here, two types of normalizations have been carried out. The first one is a simple linear transformation to the scale range used for the specific voice (natural speaker or synthesized speech) under consideration. The quality range predicted by the E-model (MOS $\in$ [1;4.5]) is taken as the target range, because this model performs predictions independent of the speech signal which is transmitted through the channel:

$$\mathrm{MOS_n} = \frac{\mathrm{MOS} - 1}{topline - 1} \cdot (4.5 - 1) + 1 \qquad (5.4.1)$$

The *topline* parameter is the MOS value for the "clean" channel (namely connection No. 1 in Table 5.1) with that voice. These values are given in Table 5.2. The same formula applies to $\mathrm{MOS_{LE}}$ and INT (leading to $\mathrm{MOS_{LE,n}}$ and $\mathrm{INT_n}$). It has also been used to normalize the estimations of the signal-based comparative measures PESQ and TOSQA. For these measures, the *topline* predictions are also indicated in Table 5.2.

From Table 5.2, it can be seen that the overall quality of synthesis 1 is considerably inferior to that of synthesis 2 and of both natural voices. This does not seem to be related to the intelligibility rating, which received high values for all voices, but probably due to a lack in prosodic quality. The signal-based measures predict relatively constant *topline* values, independent of the voice. In particular, nearly no difference can be observed between the naturally produced and synthetic voices. For the E-model, this is obvious as the model does not have any signals available at its input, but only parameters characterizing the transmission channel.

The second normalization technique makes use of the non-linear (S-shaped) relationship between MOS and R defined with the E-model. Each MOS value is transformed to an R value using a numeric inversion of Formula 2.4.2, and then further transformed to a normalized R_n value in the range $R_n \in [0;100]$, using

$$R_n = \frac{R}{topline_R} \cdot 100 \qquad (5.4.2)$$

with $topline_R$ being the R value corresponding to the *topline* parameter via Formula 2.4.2. Using this normalization, comparisons can be made directly on the transmission rating scale, between auditory results for synthesized and naturally produced speech, as well as to E-model predictions. However, it can only be applied to the (quality) MOS values, and not to $\mathrm{MOS_{LE}}$ or INT.

5.4.2.2 Impact of Circuit Noise

In Figure 5.1, the overall quality degradation due to wideband circuit noise at the receive side with level $Nfor$ is depicted. The figure shows the normalized $\mathrm{MOS_n}$ values which have been calculated using Formula 5.4.1. It can be seen that the relative amount of degradation due to circuit noise is similar for the naturally produced and for the synthetic voices. An ANOVA with Tuckey-HSD post-hoc test carried out for each of the circuit conditions shows no statistically significant effect of the voice, except for $Nfor = -60$ dBmp (however with no grouping of voices in the post-hoc test). Instead, there is a statistically significant ($p < 0.001$) influence of $Nfor$ for all voices; the overall quality starts to degrade statistically significantly around -50 dBmp for the synthetic voices, and around -60 dBmp for the natural voices. The amount of quality degradation corresponds quantitatively also to the effect predicted by the E-model. In contrast to the model prediction, however, the highest quality ratings are not obtained for the lowest noise levels, but for levels around $Nfor \approx -70$ dBmp. This result will probably be due to the characteristics of the individual source stimuli and not to a preference of the test subjects for higher noise levels. The effect is very small, and it is statistically significant only for natural voice 1. The E-model has been optimized for noise levels above -60 dBmp, and does not seem to accurately predict the quality for very low noise levels. As a result,

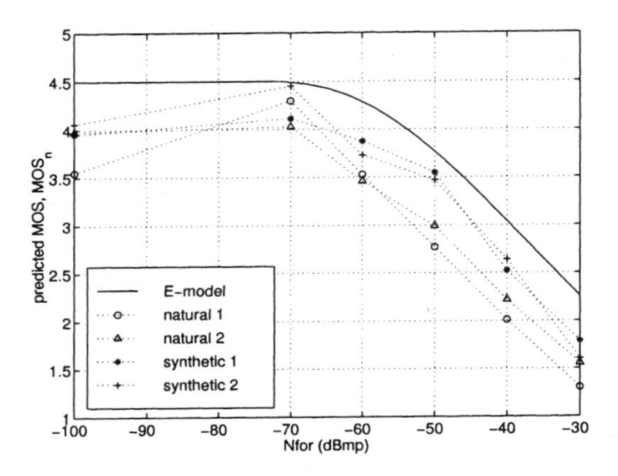

Figure 5.1. Effect of wideband noise at receive side $Nfor$. Normalized MOS_n and E-model prediction for individual voices. $Nc = -70$ dBm0p.

its estimations are too optimistic compared to the MOS_n results which have been normalized to the E-model prediction for connection No. 1 of Table 5.1 (theoretical values $Nfor = -100$ dBmp and $Nc = -100$ dBm0p, limited by the inherent noise of the simulation system).

The normalized intelligibility and listening-effort ratings show a similar behavior with respect to wide-band noise, see Figures 5.2 and 5.3. However, especially the intelligibility ratings seem to be less affected by noise levels $Nfor \leq -50$ dBmp. An ANOVA with Tuckey-HSD post-hoc test shows a statistically significant decrease of INT_n below $Nfor = -50$ dBmp for the natural voices, and below $Nfor = -40$ dBmp for the synthetic voices ($p = 0.002$ for synthetic voice 1, $p < 0.001$ for the other voices).

For higher noise levels, the synthetic voices seem to be slightly more robust than the natural voices, in the sense that the *relative* degradation due to the noise is smaller for the synthetic voices. This tendency is visible for all 3 ratings (MOS, MOS_LE and INT) but is not statistically significant. It may be due to a higher "distinctness" of the synthetic voice, which makes it more remarkable in the presence of noise. For synthetic voice 2, this relative effect turns out to be also an absolute one: MOS values for $Nfor = -50$ dBmp (2.96) and MOS (2.30), MOS_LE (3.48) and INT (4.04) values for $Nfor = -40$ dBmp are higher than the ratings for both natural voices. With the use of the second normalization technique (via R) this difference becomes even more obvious, see Figure 5.4. The overall behavior for increasing noise levels is – once again – similar to the effect predicted by the E-model.

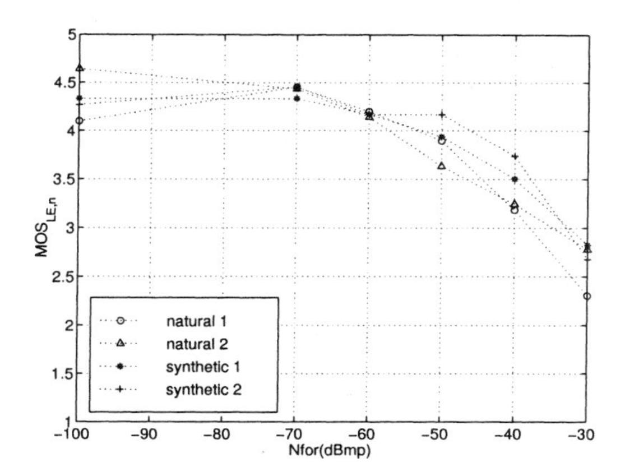

Figure 5.2. Effect of wideband noise at receive side $Nfor$. Normalized listening-effort $\mathrm{MOS_{LE,n}}$ for individual voices. $Nc = -70$ dBm0p.

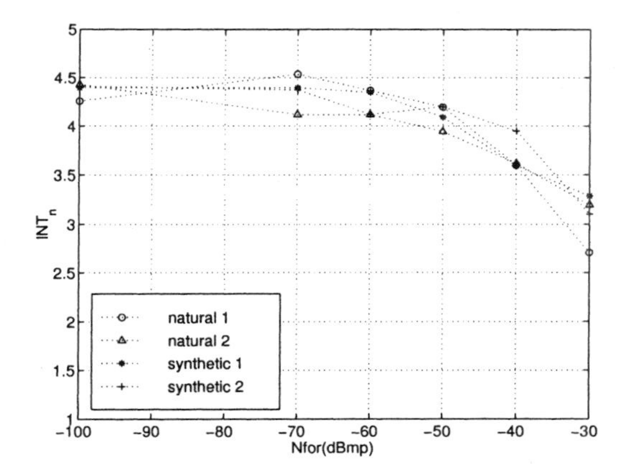

Figure 5.3. Effect of wideband noise at receive side $Nfor$. Normalized intelligibility score $\mathrm{INT_n}$ for individual voices. $Nc = -70$ dBm0p.

The auditory $\mathrm{MOS_n}$ ratings can also be compared to the normalized predictions of signal-based instrumental measures. This comparison is depicted in Figure 5.5, for the PESQ and the TOSQA model. It can be seen that both models predict the quality degradation to be very similar for natural and synthesized voices, a result which does not correspond to the observed auditory results for the higher noise levels in the test. The overall shape of the PESQ curves for increasing noise levels is similar to the one found in the auditory test, whereas the TOSQA model predicts a steeper decrease of quality than was observed in the test. This might be due to the *arctan* function linking the predicted raw

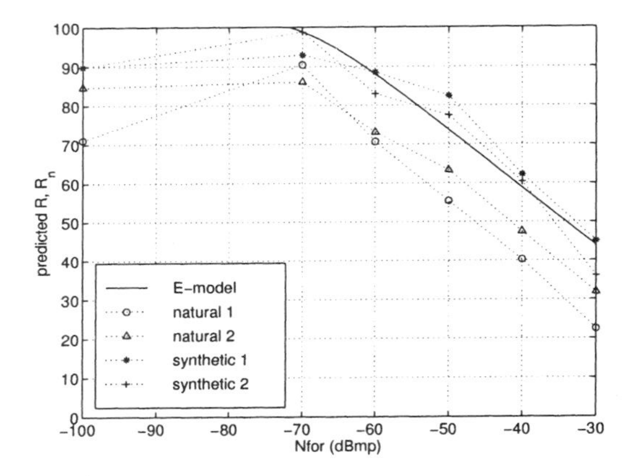

Figure 5.4. Effect of wideband noise at receive side $Nfor$. Normalized R_n and E-model transmission rating prediction for individual voices. $Nc = -70$ dBm0p.

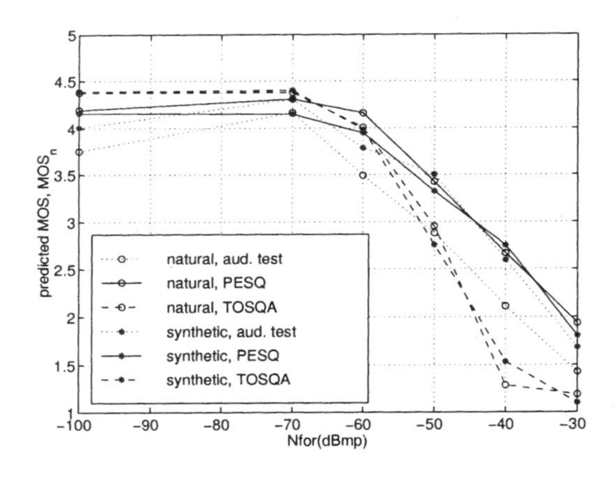

Figure 5.5. Effect of wideband noise at receive side $Nfor$. Normalized MOS_n, PESQ and TOSQA model predictions for synthetic vs. natural voices. $Nc = -70$ dBm0p.

similarities to the MOS scale. In fact, such a relationship may be very sensible at the extremities of the MOS scale. The TOSQA curves seem to be more S-shaped than the PESQ curves or those corresponding to the auditory test, with constant quality predictions for low and high noise levels and a steep step in between.

The degradation due to continuous narrow-band circuit noise of level Nc is shown in Figure 5.6. Once again, the degradations are similar for the naturally produced and for the synthetic voices. An ANOVA with Tuckey-HSD post-hoc test reveals only for $Nc = -50$ dBm0p a statistically significant effect of

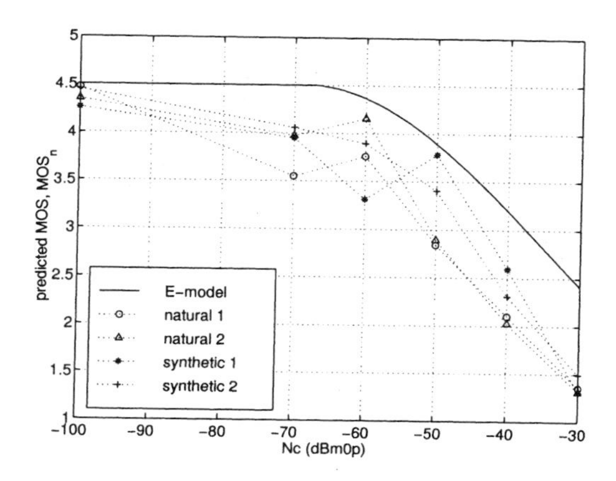

Figure 5.6. Effect of narrow-band circuit noise Nc. Normalized $\mathrm{MOS_n}$ and E-model prediction for individual voices. $Nfor = -100$ dBmp.

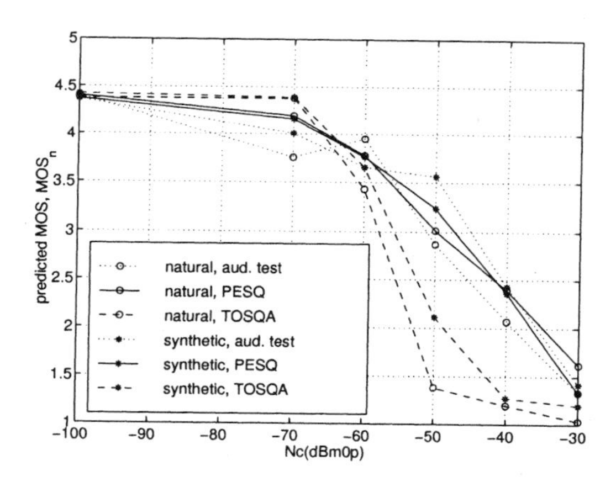

Figure 5.7. Effect of narrow-band circuit noise Nc. Normalized $\mathrm{MOS_n}$, PESQ and TOSQA model predictions for synthetic vs. natural voices. $Nfor = -100$ dBmp.

the voice ($p = 0.006$), and a grouping in synthetic and natural voices. The overall quality judgments are mainly comparable to the estimations given by the E-model. However, in contrast to the model, a remarkable MOS degradation can already be observed for very low noise levels (Nc between -100 and -60 dBm0p). This degradation is statistically significant only for natural voice 1; for all other voices, the overall quality starts to degrade significantly at narrow-band noise levels higher than -60 dBm0p. The listening-effort ($\mathrm{MOS_{LE}}$) and the intelligibility (INT) ratings are similar to those obtained for wide-band circuit noise conditions.

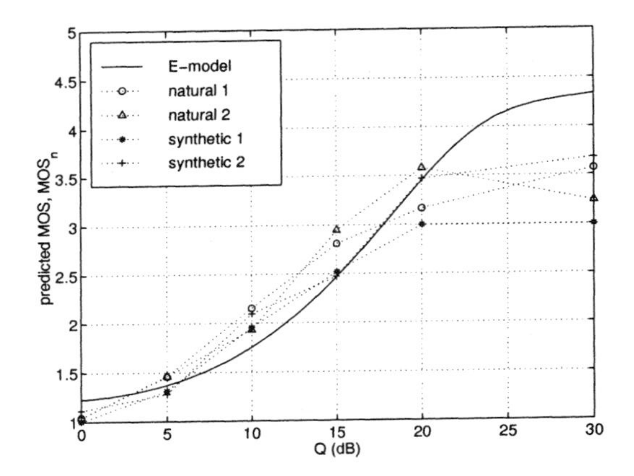

Figure 5.8. Effect of signal-correlated noise with signal-to-quantizing-noise ratio Q. Normalized MOS_n and E-model prediction for individual voices.

When comparing the results for narrow-band circuit noise, Nc, with the predictions from signal-based comparative measures, the graph is similar to the one found for wideband noise $Nfor$, see Figure 5.7. The predictions for naturally produced and synthesized speech from PESQ are close to each other, whereas the TOSQA model predicts a higher quality decrease for the naturally produced speech, an estimation which is supported by the auditory tests. As for $Nfor$, the TOSQA model predicts a very steep decrease for the MOS values with increasing noise levels, whereas the shape of the curve predicted by PESQ is closer to the one found in the auditory test. As can be expected, the scatter of the auditory test results for medium noise levels ($Nc \sim -70... - 60$ dBm0p) is not reflected in the signal-based model predictions. It will have its origin in the subjective ratings, and not in the speech stimuli presented to the test subjects.

5.4.2.3 Impact of Signal-Correlated Noise

Signal-correlated noise is perceptively different from continuous circuit noise in the sense that it only affects the speech signal, and not the pauses. Its effects on the overall quality ratings are shown in Figure 5.8. Whereas slight individual differences for the voices are discovered (not statistically significant in the ANOVA), the overall behavior for synthetic and natural voices is very similar. This can be seen when the mean values for synthetic and natural voices are compared, see the dotted lines in Figure 5.9. The degradations are – in principle – well predicted by the E-model. However, for low levels of signal-correlated noise (high Q), there is still a significant degradation which is not predicted by the model. This effect is similar to the one observed for narrow-band circuit noise, Nc; no explanation can be given for this effect so far.

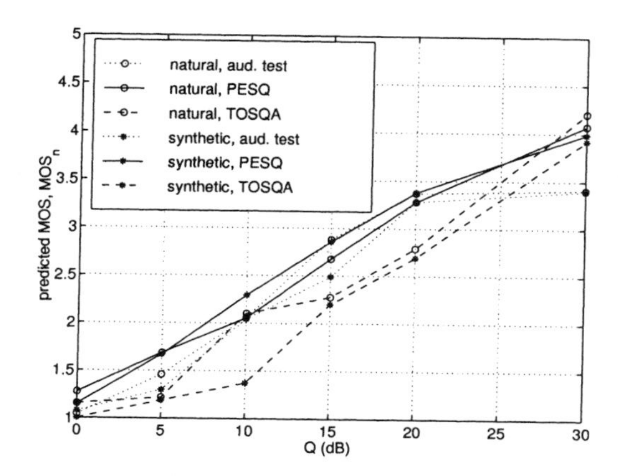

Figure 5.9. Effect of signal-correlated noise with signal-to-quantizing-noise ratio Q. Normalized $\mathrm{MOS_n}$, PESQ and TOSQA model predictions for synthetic vs. natural voices.

The predictions of the signal-based comparative measures PESQ and TOSQA do not agree very well with the auditory test results. Whereas the PESQ model estimations are close to the auditory judgments up to SNR values of $Q \approx 20$ dB, the TOSQA model estimates the signal-correlated noise impact slightly more pessimistically. This model, however, predicts a slightly lower degradation of the naturally produced speech samples, which is congruent with the auditory test. Both PESQ and TOSQA models do not predict the relatively low quality level for the highest SNR value in the test ($Q = 30$ dB), but give more optimistic estimations for these speech samples. Expressed differently, the models reach saturation (which is inevitable on the limited MOS scale) at higher SNR values than those included in the test conditions. As a general finding, both models are in line with the auditory test in that they do not predict strong differences between the naturally produced and the synthesized speech samples.

The $\mathrm{MOS_{LE}}$ and the INT values are similar in the natural and synthetic case, with slightly higher values for the natural voices. These results have not been plotted for space reasons.

5.4.2.4 Impact of Ambient Noise

Degradations due to ambient room noise are shown in Figure 5.10. The behavior slightly differs for the individual voices. In particular, the synthetic voices seem to be a little less prone to ambient noise impairments than the natural voices. Once again, this might be due to a higher 'distinctness' of the synthetic voices, which makes them more remarkable in the presence of noise. The same behavior is found for the intelligibility judgments, see Figure 5.11. For all judgments, the data point for synthetic voice 1 and $Pr = 35$ dB(A)

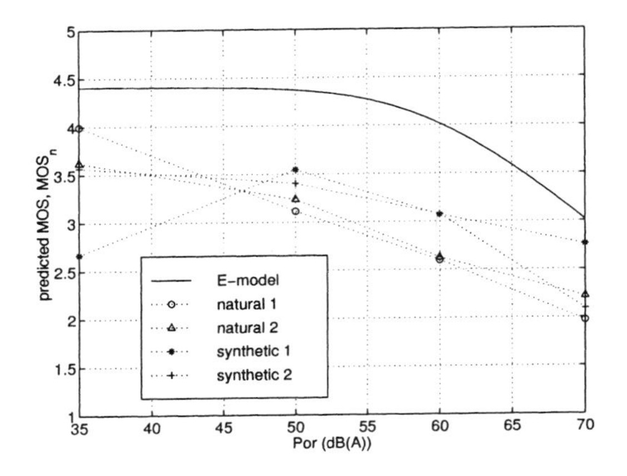

Figure 5.10. Effect of hoth-type ambient noise Pr. Normalized $\mathrm{MOS_n}$ and E-model prediction for individual voices.

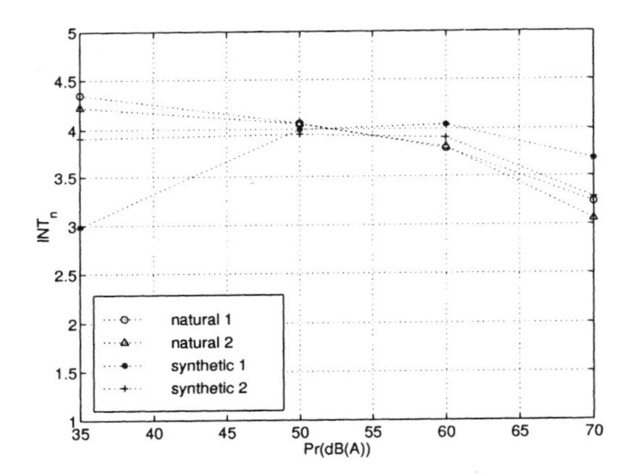

Figure 5.11. Effect of hoth-type ambient noise Pr. Normalized intelligibility score $\mathrm{INT_n}$ for individual voices.

seems to be an outlier, as it is rated particularly negative. Informal listening shows very inappropriate phone durations in two positions of the speech file, which makes this specific sample sound particularly bad. Here, the lack of optimization of the speech material discussed in Section 5.4.1.1 is noted.

5.4.2.5 Impact of Low Bit-Rate Coding

The low bit-rate codecs investigated here cover a wide range of perceptively different types of degradations. In particular, the G.726 (ADPCM) and the

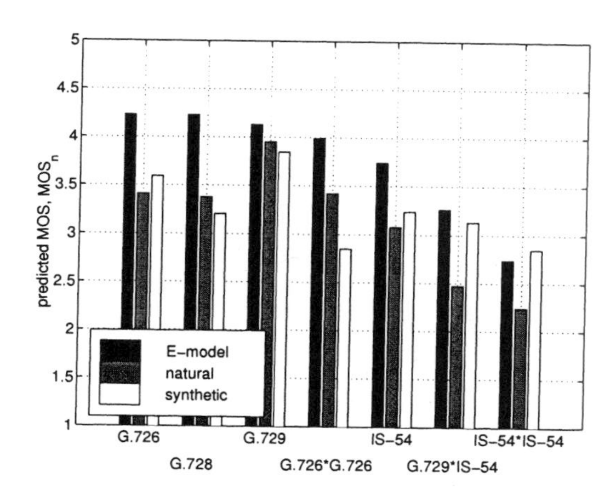

Figure 5.12. Effect of low bit-rate codecs. Normalized $\mathrm{MOS_n}$ and E-model prediction for synthetic vs. natural voices.

G.728 (LD-CELP) codecs produce an impression of noisiness, whereas G.729 and IS-54 are characterized by an artificial, unnatural sound quality (informal expert judgments).

Figures 5.12 to 5.14 show a fundamental difference in the quality judgments for natural and synthesized speech, when transmitted over channels including these codecs (mean values over the natural and synthetic voices are reproduced here for clarity reasons). Except for two cases (the G.726 and G.728 codecs, which are rated too negatively in comparison to the prediction model), the decrease in overall quality predicted by the E-model is well reflected in the auditory judgments for natural speech. On the other hand, the judgments for the synthesized speech do not follow this line. Instead, the overall quality of synthesized speech is much more strongly affected by 'noisy' codecs (G.726, G.728 and G.726*G.726) and less by the 'artificially sounding' codecs. Listening-effort and intelligibility ratings for synthesized speech are far less affected by all of these codecs (they scatter around a relatively constant value), whereas they show the same rank order for the naturally produced speech (once again, with exception of the G.726 and G.728 codec). The differences in behavior of the synthetic and the natural voices are also observed for the codec cascades (G.726*G.726 and IS-54*IS-54) compared to the single codecs: Whereas for the G.726 tandem mainly the synthetic voices suffer from the cascading, the effect is more dominant for the natural voices with the IS-54 cascade.

The observed differences may be due to differences in quality dimensions perceived as degradations by the test subjects. Whereas the 'artificiality' di-

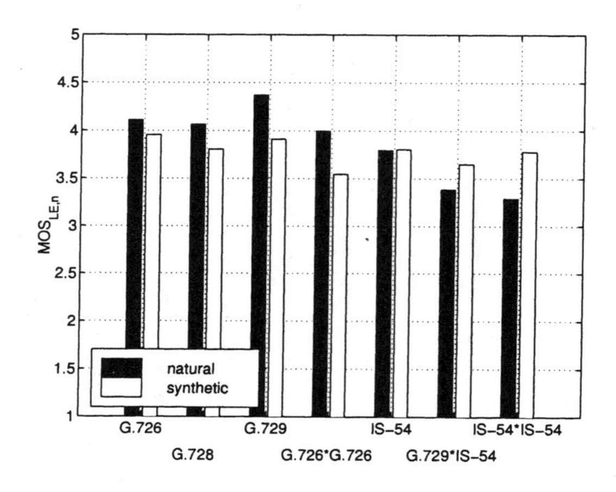

Figure 5.13. Effect of low bit-rate codecs. Normalized listening-effort $MOS_{LE,n}$ for synthetic vs. natural voices.

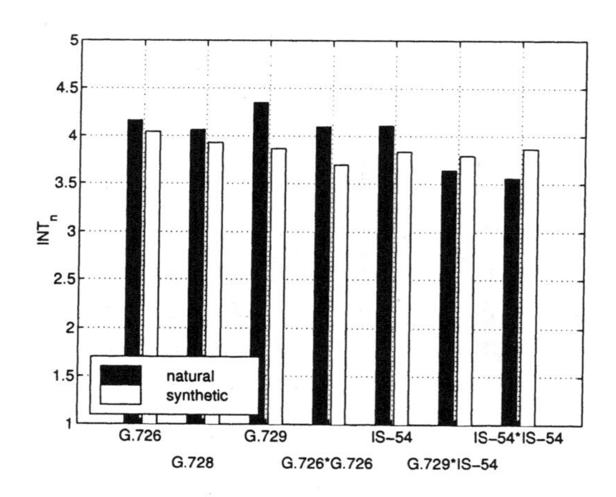

Figure 5.14. Effect of low bit-rate codecs. Normalized intelligibility score INT_n for synthetic vs. natural voices.

mension introduced by the G.729 and IS-54 codecs is an additional degradation for the naturally produced speech, this is not the case for synthesized speech, which already carries a certain degree on artificiality. It is not yet clear why the G.726 and G.728 transmission circuits result in particularly low quality, an effect which does not correspond to the prediction of the E-model. Other investigations carried out by the author in a working group of the ITU-T (Möller, 2000) also suggest that the model predictions are too optimistic for these codecs when considered in isolation, i.e. without tandeming.

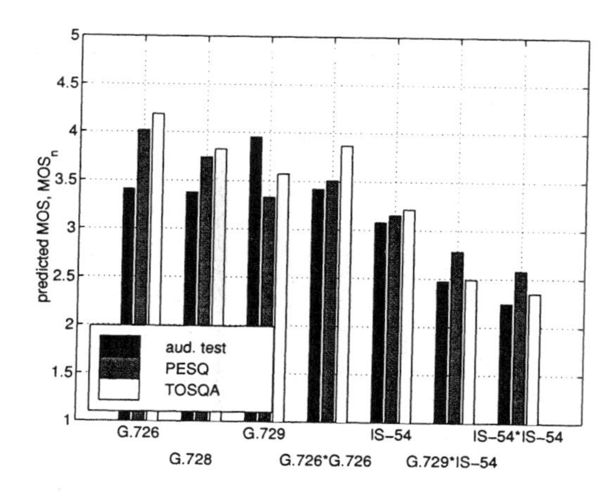

Figure 5.15. Effect of low bit-rate codecs. Normalized MOS_n, PESQ and TOSQA model predictions for natural voices.

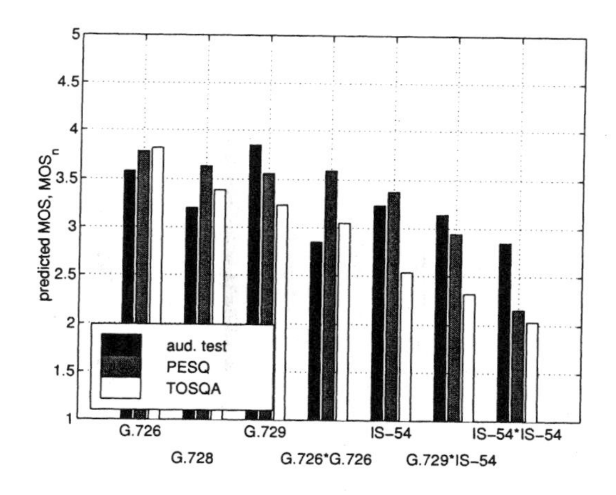

Figure 5.16. Effect of low bit-rate codecs. Normalized MOS_n, PESQ and TOSQA model predictions for synthetic voices.

Signal-based comparative measures like PESQ and TOSQA have been developed in particular for predicting the effects of low bit-rate codecs. A comparison to the normalized auditory MOS_n values is shown in Figure 5.15 for the natural voices. Whereas for the IS-54 codec and its combinations the predicted quality is in good agreement with both models' predictions, the differences are bigger for the G.726, G.728 and G.729 codecs. As was found for the E-model, the G.726 and G.728 codecs are rated significantly worse in the auditory test compared to the model predictions. On the other hand, the G.729 codec is rated better than the predictions of both PESQ and TOSQA suggest. In all cases, either both models predict the codec degradations too optimistically or too pessimistically. Thus, no advantage can be obtained when calculating the mean of the PESQ and TOSQA model predictions.

The picture is different for the synthesized voices, see Figure 5.16. The quality rank order predicted by the E-model (i.e. the bars ordered with respect to decreasing MOS values) is also found for the PESQ and TOSQA predictions, but it is not well reflected in the auditory judgments. In all, the differences between the auditory test results and the signal-based model predictions is larger for the synthesized than for the naturally produced voices. For the three 'noisy' codec conditions G.726, G.728 and G.726*G.726, both PESQ and TOSQA predict quality more optimistically than was judged in the test. For the other codecs the predictions are mainly more pessimistic. This supports the assumption that the overall quality of synthesized speech is much more strongly affected by 'noisy' and less by the 'artificially sounding' codecs.

5.4.2.6 Impact of Combined Impairments

For combinations of circuit noise and low bit-rate distortions, synthetic and natural voices behave similarly. This can be seen in Figure 5.17, showing the combination of the IS-54 cellular codec with narrow-band circuit noise (mean values for synthetic vs. natural voices are depicted). Again, the quality for low noise does not reach the optimum value (the value predicted by the E-model). This observation has already been made for the other circuit noise conditions. In high-noise-level conditions, the synthetic voices are slightly less affected by the noise than the natural voices. This finding is similar to the one described in Section 5.4.2.2.

With the help of the normalization to the R scale, the additivity of different types of impairments postulated by the E-model can be tested. Figure 5.18 shows the results after applying this transformation. It can be seen that the slope of the curve for higher noise levels is well in line with the results for the natural voices. The synthesized voices seem to be more robust under these conditions, although the individual results scatter significantly.

For low noise levels, the predictions of the E-model are once again too optimistic. This will be due to the unrealistically low theoretical noise floor level ($Nfor = -100$ dBmp) of this connection, for which the E-model predictions even exceed 100 as the limit of the R scale under normal (default) circuit conditions. The optimistic model prediction can also be observed for the judgment of the codec alone, depicted in Figure 5.12. In principle, however, the flat model curve for the lower noise levels is well in agreement with the results both for synthetic and natural voices. Thus, no specific doubts arise as to the validity of adding different impairment factors to obtain an overall transmission rating. Of course, the limited findings do not validate the additivity property as

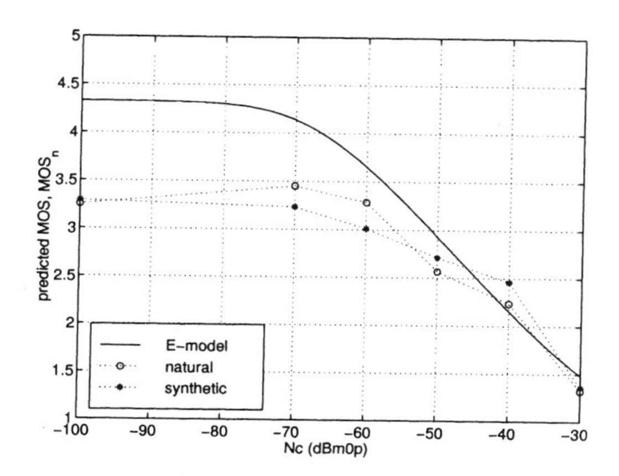

Figure 5.17. Effect of narrow-band circuit noise Nc and the IS-54 codec. Normalized MOS_n and E-model prediction for synthetic vs. natural voices.

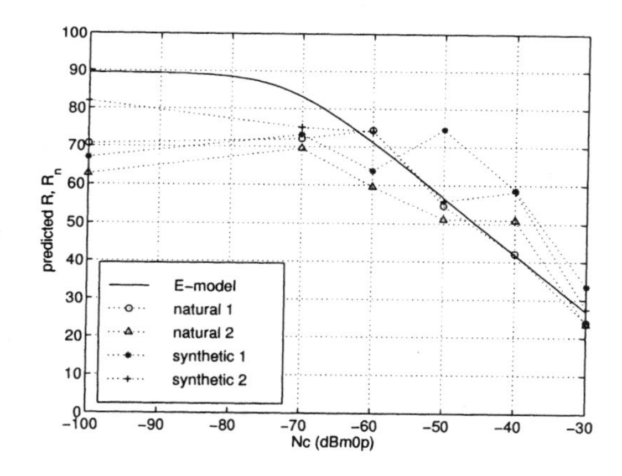

Figure 5.18. Effect of narrow-band circuit noise Nc and the IS-54 codec. Normalized R_n and E-model transmission rating prediction for individual voices.

a whole. Other combinations of impairments will have to be tested, and more experiments have to be carried out in order to reduce the obvious scatter in the results.

5.4.2.7 Acceptability Ratings

The ratings on the 'perceived acceptability' question in part 5.1 of the test have to be interpreted with care, because acceptability can only finally be assessed with a fully working system (for a definition of this term see Möller, 2000). Nevertheless, acceptability judgments are interesting for the develop-

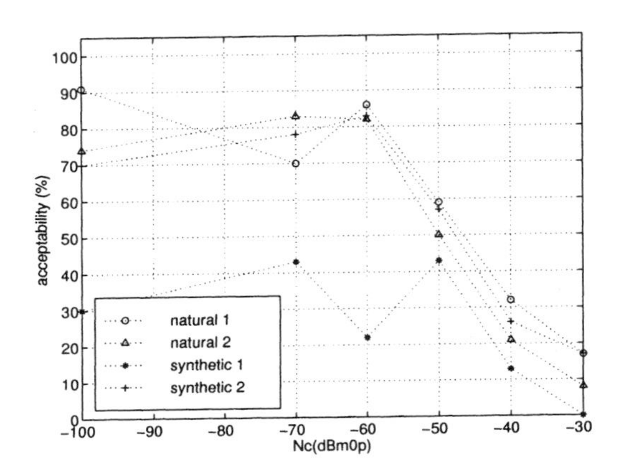

Figure 5.19. Effect of narrow-band circuit noise Nc. Perceived acceptability ratings for individual voices.

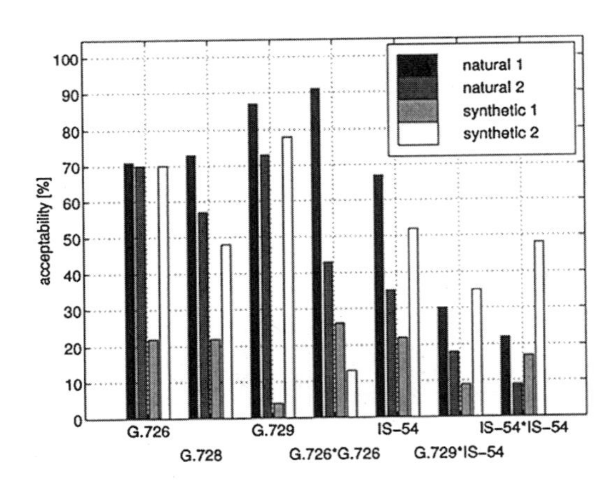

Figure 5.20. Effect of low bit-rate codecs. Perceived acceptability ratings for individual voices.

ers, because they show whether a synthetic voice is acceptable in a specific application context.

As an example, Figure 5.19 shows the overall (not normalized) level of perceived acceptability for noisy transmission channels. It can be seen that synthetic voice 2 mostly ranges in between the natural voices, whereas synthetic voice 1 is rated considerably worse. Interestingly, the highest perceived acceptability level for the three better voices seems to be reached at a moderate noise floor of $Nc \approx -60$ dBm0p, and not for the lowest noise levels (except natural voice 1 and $Nc = -100$ dBm0p). Thus, under realistic transmission characteristics, these voices seem to be more acceptable for the target application scenario then for (unrealistic) low-noise scenarios. The influence of the transmission channel on the perceived acceptability ratings for the natural voices as well as for synthetic voice 2 is very similar. The according voices seem to be acceptable up to a noise level of $Nc = -60$ dBm0p. The results for synthetic voice 1 seem to be too low to be acceptable at all in this application scenario.

A second example for the perceived acceptability ratings is depicted in Figure 5.20. Once again, the synthetic voice 2 reaches a perceived acceptability level which is in the order of magnitude of the two natural voices. Whereas the level is lower than both natural voices for the 'noisy' G.728 and the G.726*G.726 codecs, it is higher than natural voice 2 for the 'artificially sounding' codecs G.729 and IS-54, and higher than both natural voices for the G.729*IS-54 and IS-54*IS-54 tandems. Apparently, synthetic voice 2 is relatively robust against artificially sounding codecs, and more affected by noisy codecs. This supports the finding that the perceptual type of degradation which is introduced by the transmission channel has to be seen in relation to the perceptual dimensions of the carrier voice. When both are different, the degradations seem to be accumulated, whereas similar perceptive dimensions do not further impact the acceptability judgments.

5.4.2.8 Identification Scores

In part 5.2 of the test, subjects had to identify the two variable pieces of information contained in each stimulus and write this information down on the screen form. The responses have been compared to the information given in the stimuli. This evaluation had to be carried out manually, because upper/lowercase changes, abbreviations (e.g. German "Hbf" for "Hauptbahnhof") and misspellings had to be counted as correct responses. The scores only partly reflect intelligibility; they cannot easily be related to segmental intelligibility scores which have to be collected using appropriate test methods.

In nearly all cases, the identification scores reached 100% of correct answers. Most of the errors have been found for synthetic voice 1, which also showed the lowest intelligibility rating, cf. Table 5.2. Only for three stimuli more than one error was observed. In two of these stimuli, the location information was not identified by 5 or all of the 6 test subjects. Thus, it can be assumed that the particular character of the speech stimuli is responsible for the low identification scores. In principle, however, all voices allow the variable parts of the information contained in the template sentences to be identified.

The results show that the identification task cannot discriminate between the different transmission circuit conditions and voices. This finding may be partly due to the playback control which was given to the test subjects. Time pressure during the identification task may have revealed different results. A comparison to the perceived "intelligibility" ratings shows that although the test subjects occasionally judged words hard to understand, their capacity to extract the important information is not really affected.

5.4.2.9 Discussion

In the experiment reported here, the overall quality levels of natural and synthetic voices differed significantly, and in particular the levels reached by the two synthetic voices. Nevertheless, the relative amount of degradation introduced by the transmission channel was observed to be very similar, so general trends can be derived from the normalized judgments.

For most of the tested degradations, the impact on synthesized speech was similar to the one observed on naturally produced speech. This result summarizes the impact of narrow-band and wideband circuit noise, of signal-correlated noise, as well as of ambient room noise. More precisely, the synthetic voices seem to be slightly less affected by high levels of uncorrelated noise compared to the natural voices. This difference – though not statistically significant in most cases – was observed for overall quality, intelligibility and listening-effort judgments. It was explained with a higher 'distinctness' of the synthetic voice which might render it more remarkable in the presence of noise. However, it is not clear how this finding can be brought in line with a potentially higher cognitive load which has been postulated for synthetic voices, e.g. by Balestri et al. (1992).

The situation is different for the degradations caused by low bit-rate codecs. Whereas – with two exceptions – the quality ranking of different codecs as is estimated by the E-model, and partly also by the signal-based comparative measures PESQ and TOSQA, is well in line with the judgments for naturally produced speech, the synthetic voices seem to be affected differently. The quality impact seems to depend on the perceptual type of impairment which is introduced by a specific codec. When the codec introduces noisiness, it seems to degrade the synthetic voice additionally, whereas 'artificially sounding' codecs do not add a significant degradation.

Nearly no differences in intelligibility and listening-effort ratings could be observed for the codecs included in the tests. At least the intelligibility ratings seem to be in contrast to the findings of Delogu et al. (1995). In their experiments, the differences in segmental intelligibility were higher for synthesized speech when switching from good transmission conditions (high quality) to telephonic ones. The reason might be that – in the experiment reported here – no comparison to the wideband channel was made, and that the intelligibility judgments obtained from the subjects do not reflect segmental intelligibility. Thus, the 'perceived intelligibility' seems to be less affected by the transmission channel than the intelligibility measured in a segmental intelligibility test.

5.4.3 Conclusions from the Experiment

Two specific questions were addressed in the described experiment. The first one has to be answered in a differentiated way. Noise-type degradations seem to impact the quality of naturally produced and synthesized speech by roughly the same amount. However, there was a tendency observed that synthesized speech might be slightly more robust against high levels of uncorrelated noise. For codec-type degradations, the impact seems to depend on the perceptual type of degradation which is linked to the specific codec. A 'noisiness' dimension seems to be an additional degradation for the synthesized speech, whereas an 'artificiality' dimension is not – probably because it is already included in the auditory percept related to the source speech signal.

The second question can partly be answered in a positive way. All in all, the predictions of the transmission rating model which was investigated here (the E-model) seem to be in line with the auditory test results, both for naturally produced as well as for synthesized speech. Unfortunately, the model's estimations are misleading for very low noise levels, a fact which results in too optimistic predictions when such a channel is taken as a reference for normalization. When the overall quality which can be reached with a specific network configuration is over-estimated, problems may arise later on in the service operation. It has to be admitted, however, that such low noise levels are generally not assumed in the network planning process. The signal-based model PESQ provides a good approximation of the quality degradation to be expected from circuit noise, whereas the S-shaped characteristic of TOSQA underestimates the quality at high noise levels. These levels, however, are fortunately not realistic for normal network configurations. The degradations due to signal-correlated noise are poorly predicted by every model, especially for high SNRs. The situation for codec degradations has to be differentiated between the naturally produced and the synthesized speech: Whereas the degradations on the former are – with the exception of the G.726 and G.728 codec – adequately predicted by all models, the degradations on synthesized speech are not well predicted by any investigated model. This finding might be explained with the degradation dimensionality introduced by the low bit-rate codecs under consideration.

The results which could be obtained in this initial experiment are obviously limited. In particular, a choice had to be made with respect to the synthetic voices under consideration. Two typical concatenative (diphone) synthesizers, which show perceptual characteristics typical for such approaches, were chosen here. The situation will be different for formant synthesizers – especially with respect to coding degradations, but perhaps also for noise degradations, taking into account that such systems normally reach a lower level of intelligibility. The quality of speech synthesized with unit-selection approaches will depend on the size and coverage of the target sentences in the inventory. Thus, the quality will be time-variant on a long-term level. As the intelligibility and overall quality level which can be achieved with unit-selection is often higher than the one of diphone synthesizers, the differences observed in the reported experiment may become smaller in this case. It is not yet clear how different coding schemes of the synthesizer's inventory will be affected by the transmission channel degradations. The synthesizers in the reported experiment used a linear 16 bit PCM coding scheme or a vector-quantized LPC with a parametrized glottal waveform. Other coding schemes may be differently affected by the transmission channel characteristics.

A second limitation results from the purely listening-only test situation. In fact, it cannot be guaranteed that the same judgments would be obtained in a conversational situation. Experiments carried out by the author (Möller, 2000), however, do not raise any specific doubts that the relative quality degradation will be similar. Some of the degradations affecting the conversational situation do not apply to interactions with spoken dialogue systems. For example, talker echo with synthetic voice is only important for potential barge-in detectors of SDSs, and not on a perceptual level. Typical transmission delays will often be surpassed by the reaction time of the SDS. Here, the estimations for acceptable delay from prediction models like the E-model might be used as a target for what is desirable in terms of an overall reaction time, including system reaction and transmission delay.

Obviously, not all types of degradations could be tested in the reported experiment. In particular, the investigation did not address room acoustic influences (e.g. when listening to synthetic voice with a hands-free terminal), or time-variant degradations from lost packets or fading radio channels. These degradations are still poorly investigated, also with respect to their influence on naturally produced speech. They are important in mobile networks and will also limit the quality of IP-based voice transmission. Only few modelling approaches take these impairments into account so far. The E-model provides a rough estimation of packet loss impact in its latest version (ITU-T Delayed Contribution D.44, 2001; ITU-T Rec. G.107, 2003), and the signal-based comparative measures have also been tested to provide valid prediction results for this type of time-variant impairment.

5.5 Summary

In this chapter, the quality of synthesized speech has been addressed in a specific application scenario, namely an information server operated over the telephone network. In such a scenario, quality assessment and evaluation has to take into account the environmental and the contextual factors exercising an influence on the speech output quality, and subsequently on usability, user satisfaction, and acceptability.

The contextual factors have to be reflected by the design of evaluation experiments. In this way, such experiments can provide highly valid results for the future application to be set up. The requirements for such functional testing have been defined, and an exemplary evaluation for the restaurant information system used in the last chapter has been proposed. As will happen in many evaluations carried out during the set-up of spoken dialogue systems, the resources for this evaluation were limited. In particular, only a laboratory test with a limited group of subjects could be carried out, and no field test or survey with a realistic group of potential future users of the system. In spite of these limitations, interesting results with respect to the influence of the environmental factors were obtained.

The type of degradation which is introduced by the transmission channel was shown to determine whether synthesized speech is degraded by the same amount than naturally produced speech. For noise-type degradations (narrow-band and wideband circuit noise, signal-correlated noise), the amount of degradation is similar in both cases. However, synthesized speech seemed to be slightly more remarkable in high uncorrelated noise conditions. For codec-type degradations, the dimensionality of the speech and the transmission channel influences have to be taken into account. When the codec introduces an additional perceptive dimension (such as noisiness), the overall quality is impacted. When the dimensionality is already covered in the source speech signal (such as artificiality), then the quality is not further degraded, at least not by the same amount as would be expected for naturally produced speech.

The estimations provided by quality prediction models which have originally been designed for naturally produced speech can serve as an indication of the amount of degradation introduced by the transmission channel on synthesized speech. Two types of models have been investigated here. The E-model relies on the parametric description of the transmission channel, and thus does not have any information on the speech signals to be transmitted as an input. It nevertheless provides adequate estimations for the relative degradation caused by the transmission channel, especially for uncorrelated noise. The signal-based comparative measures PESQ and TOSQA are also capable of estimating quality of transmitted synthesized speech to a certain degree. All models, however, do not adequately take into account the different perceptive dimensions caused by the source speech material and by the transmission channel. In addition, they are only partly able to accurately predict the impact of signal-correlated noise.

The test results have some implications for the designers of telecommunication networks, and for speech synthesis providers. Whereas in most cases networks designed for naturally produced speech will transmit synthesized speech with the same amount of perceptual degradation, the exact level of quality will depend on the perceptual quality dimensions. These dimensions depend on the speech signal and the transmission channel characteristics. Nevertheless, rough estimations of the amount of degradation may be obtained with the help of quality prediction models like the E-model. The overall quality level is however estimated too optimistically, due to misleading model predictions for very low noise levels. In conclusion, no specific doubts arise as to whether telephone networks which are carefully designed for transmitting naturally produced speech will also enable an adequate transmission of synthesized speech.

Chapter 6

QUALITY OF SPOKEN DIALOGUE SYSTEMS

Investigations on the performance of speech recognition and on the quality of synthesized speech in telephone environments like the ones reported in the previous two chapters provide useful information on the influence of environmental factors on the system's speech input and output capability. They are, however, limited to these two specific modules, and do not address the speech understanding, the dialogue management, the application system (e.g. the database), and the response generation. Because the other modules may have a severe impact on global quality aspects of the system and the service it provides, user-orientated quality judgments can only be obtained when all system components operate together. The quality judgments will then reflect the performance of the individual components in a realistic functional situation.

The experiments described in this chapter take such a wholistic view of the system. They are not particularly limited to the dialogue management component for two obvious reasons. Firstly, users can only interact with the dialogue manager via the speech input and output components. The form of both speech input from the user and speech output from the system cannot, however, be separated from its communicative function. Thus, speech input and output components will always exercise an influence on the quality perceived by the user, even when they show perfect performance. Secondly, the quality which is attributed to certain dialogue manager functionalities can only be assessed in the realistic environment of *non-perfect* other system components. For example, an explicit confirmation strategy may be perceived as lengthy and boring in case of perfect speech recognition capabilities, but may prove extremely useful when the recognition performance decreases. Thus, quality judgments which are obtained in a set-up with non-realistic neighboring system components will not be valid for the later application scenario.

In order to estimate the impact of the mentioned module dependencies on the overall quality of the system, it will be helpful to describe the relationships between quality factors (environmental, agent, task, and contextual factors) and quality aspects in terms of a relational network. Such a network should ideally be able to identify and quantify the relationships, e.g. by algorithmically describing how and by what amount the capabilities and the performance of individual modules affect certain quality aspects. The following relationship can be taken as an example: Transmission impairments obviously affect the recognition performance, and their impact has been described in a quantitative way with the help of the E-model, see Section 4.5. Now, further relationships can be established between ASR performance (expressed e.g. by a WER or WA) on the one side, and perceived system understanding (which is the result of a user judgment) on the other. Perceived system understanding is one aspect of speech input quality, and it will contribute to communication and task efficiency, and to the comfort perceived by the user, as has been illustrated in the QoS taxonomy. These aspects in turn will influence the usability of the service, and finally the user's satisfaction. If it is possible to follow such a concatenation of relations, predictors for individual quality aspects can be established, starting either from system characteristics (e.g. a parametric description of the transmission channel) or from interaction parameters.

The illustrated goal is very ambitious, in particular if the relationships to be established shall be generic, i.e. valid for a number of different systems, tasks and domains. Nevertheless, even individual relationships will give light on how users perceive and judge the quality of a complex service like the one offered via an SDS. They will form a first basis for modelling approaches which allow quality to be addressed in an analytic way, i.e. via individual quality aspects. Thus, a first step will be to establish predictors for individual quality aspects. Such predictors may then be combined to predict quality on a global level, e.g. in terms of system usability or user satisfaction. From this perspective, the goal is far less ambitious than that of predicting overall quality directly from individual interaction parameters, as is proposed by the PARADISE framework discussed in Section 6.3. Prediction of individual quality aspects may carry the additional advantage that such predictors might be more generic in their prediction, i.e. that they may be applied to a wider range of systems.

It is the aim of the experiments described underneath to identify quality aspects which are relevant from a user's point of view and to relate them to interaction parameters which can be collected during laboratory tests. A prototypical example SDS will be used for this purpose, namely the BoRIS system for information about the restaurants in the area of Bochum, Germany. The system has been set up by the author as an experimental prototype for quality assessment and evaluation. Speech recognition and speech synthesis components which can be used in conjunction with this system have already been investigated in Chapters 4 and 5. Now, user interactions with the fully working system will be addressed, making use of the mentioned speech output components, and replacing the ASR module by a wizard simulation in order to be able to control its performance. The experimental set-up of the whole system will be described in Section 6.1.

A number of subjective interaction experiments have been carried out with this system. They generally involve the following steps to be performed:

- Set-up and running of laboratory interactions with a number of test subjects, under controlled environmental and contextual conditions.
- Collection of instrumentally measurable parameters during the interactions.
- Collection of user quality ratings after each interaction, and after a complete test session.
- Transcription of the dialogues.
- Annotation of dialogue transcriptions by a human expert.
- Automatic calculation of interaction parameters.
- Data analysis and quality modelling approaches.

The first steps serve the purpose of collecting interaction parameters and related quality judgments for specific system configurations. These data will be analyzed with respect to the interrelations among interaction parameters and quality judgments, and between interaction parameters and quality judgments, see Section 6.2.

The starting point of the analysis carried out here is the QoS taxonomy which has already been used for classifying quality aspects and interaction parameters, see Sections 3.8.5 and 3.8.6. In this case, it will be used for selecting interaction parameters and judgment scales which refer to the same quality aspect. The analysis of correlation data will highlight the relationships between interaction parameters and perceived quality, but also the limitations of using data from external (instrumental or expert) sources for describing perceptive effects. Besides this, it serves a second purpose, namely to analyze the QoS taxonomy itself. These analyses will be described in detail in Section 6.2.4.

Both interaction parameters and subjective judgments reflect the characteristics of the specific system. In the experiments, a limited number of system characteristics were varied in a controlled way, in order to quantify the effects of the responsible system components. Such a parametric setting is possible for the speech recognizer (using a wizard-controlled ASR simulation), for the speech output (using either naturally recorded or synthesized speech, or combinations of both), and for the dialogue manager (selecting different confirmation strategies). Effects of the respective system configurations on both interaction parameters and subjective ratings are analyzed, and compared to data reported in the literature, see Section 6.2.5. Other effects are a result of the test set-up (e.g. training effects) and will be discussed in Section 6.2.6.

In the final Section 6.3, analysis results will be used to define new prediction model approaches. Starting from a review of the most widely used PARADISE model and its variants, a new approach is proposed which aims at finding predictors for individual quality aspects first, before combining them to provide predictions of global quality aspects. Such a hierarchical model is expected to provide more generic predictions, i.e. better extrapolation possibilities to unknown systems and new tasks or domains. Although the final proof of this claim remains for further study, the obtained results will be important for everyone interested in estimating quality for selecting and optimizing system components. They provide evidence that an analytic view of quality aspects – as is provided by the QoS taxonomy – can fruitfully be used to enhance current state-of-the-art modelling approaches.

6.1 Experimental Set-Up

In the following sections, results from three subjective interaction experiments with the BoRIS restaurant information system will be discussed. The experiments have been carried out with slightly differing system versions during the period 2001-2002. Because the aim of each experiment was different, also the evaluation methods varied between the experiments. In particular, the following aims have been accomplished:

- Experiment 6.1: Scenario, questionnaire and test environment design and set-up; analysis of the influence of different system parameters on quality. This experiment is described in detail by Dudda (2001), and part of the results have been published in Pellegrini (2003).

- Experiment 6.2: Questionnaire design and investigation of relevant quality aspects. This experiment is described in Niculescu (2002).

- Experiment 6.3: Analysis and validation of the QoS taxonomy; analysis of the influence of different system configurations on quality aspects; analysis and definition of existing and new quality prediction models. The experiment is described in Skowronek (2002), and some initial results have been published in Möller and Skowronek (2003a,b).

Experiments 6.1 and 6.3 follow the steps mentioned in the introduction, allowing for a comparison between interaction parameters and subjective judgments. Experiment 6.2 is limited to the collection of subjective judgments, making use of guided interviews in order to optimally design the questionnaire.

6.1.1 The BoRIS Restaurant Information System

BoRIS, the "Bochumer Restaurant-Informations-System", is a mixed-initiative prototype spoken dialogue system for information on restaurants in the area of Bochum, Germany. It has been developed by the author at the Institut dalle Molle d'Intelligence Artificielle Perceptive (IDIAP) in Martigny, Switzerland, and at the Institute of Communication Acoustics (IKA), Bochum. The first ideas were derived from the Berkeley restaurant project (BeRP), see Jurafski et al. (1994). The dialogue structure was developed at Ecole Polytechnique Fédérale de Lausanne (EPFL), Switzerland (Rajman et al., 2003). Originally, the system was designed for the French language, and for the restaurants in Martigny. This so-called "Martigny Restaurant Project" (MaRP) was set up in the frame of the Swiss-funded InfoVox project. Later, the system has been adapted to the German language, and to the Bochum restaurant environment.

The system architecture follows, in principle, the pipelined structure depicted in Figure 2.4. System components are either available as fully autonomously operating modules, or as wizard simulations providing control over the module characteristics and their performance. The following components are part of BoRIS:

- Two alternatives for speech input: A commercially available speech recognizer with keyword-spotting capability (see Section 4.3), able to recognize about 395 keywords from the restaurant information domain, including proper names; or a wizard-based ASR simulation relying on typed input from the wizard, see Section 6.1.2.

- A rough keyword-matching speech understanding module. It consists of a list of canonical values which are attributed to each word in the vocabulary. On the basis of the canonical value, the interpretation of the user input in the dialogue context is determined.

- A finite-state dialogue model, see below.

- A restaurant database which can be accessed locally as a text file, or through the web via an HTML interface. The database contains around 170 restaurants in Bochum and its surroundings. Searches in this database are based on pattern matching of the canonical values in the attribute-value pairs.

- Different speech generation possibilities: Pre-recorded speech files for the fixed system messages, be they naturally produced or with TTS; and naturally-produced speech or full TTS capabilities for the variable restaurant information turns. This type of speech generation makes an additional response generation unnecessary, except for the variable restaurant information and the confirmation parts where a simple template-filling approach is chosen.

The system has been implemented in the Tcl/Tk programming language on the Rapid Application Developer platform provided by the CSLU Toolkit, see Section 2.4.3 (Sutton et al., 1996, 1998). This type of implementation implies that no strict separation between application manager and dialogue manager exists, a fact which is tolerable for the purpose of a dedicated experimental prototype. The standard platform has been amended by a number of specific functions like text windows for typed speech input and text output display, a display for internal system variables (e.g. recognized user answer, current and historical canonical slot values, state-related variables, database queries and results), windows for selecting different confirmation strategies, wizard control options, etc. The exchange of data between the dialogue manager and the speech recognition and TTS modules is performed in a blackboard way via files.

The system can be used either with a commercial speech recognizer, or with a wizard-based speech recognition simulation. For the commercial ASR module, an application-specific vocabulary has been built on the basis of initial WoZ experiments. Because the other characteristics of the recognizer are not accessible to the author, feature extraction and acoustic models have been kept in their default configuration. The recognition simulation has been developed by Skowronek (2002). It is based on a full transcription of the user utterances which has to be performed by the wizard (or an additional assistant) during the interactions. The simulation tool generates typical recognition errors on this transcription in a controlled way. Details on the simulation tool are given in Section 6.1.2. Using the simulation, it becomes possible to adjust the system's ASR performance to a pre-defined value, within a certain margin. A disadvantage is, however, that the wizard does not necessarily provide an error-free transcription. In fact, Skowronek (2002) reports that in some cases words in the user utterances are substituted by others with the same meaning. This shows that the wizard does not really act as a human "recognizer", but that higher cognitive levels seem to be involved in the transcription task.

The system is able to give information about the restaurants in Bochum and the surrounding area, more precisely the names and the addresses of restaurants which match a user query. It does not permit, however, a reservation in a selected restaurant, nor does it provide more detailed information on the menu or opening hours. The task is described in terms of five slots containing AVPs which characterize a restaurant: The type of food (*Foodtype*), the location of the restaurant (*Location*), the day (*Date*) and the time (*Time*) the user wants to eat out, and the price category (*Price*). Additional slots are necessary for the dialogue management itself, e.g. the type of slot which is addressed in a specific user answer, and logical operations ("not", "except", etc.). On these slots, the system performs a simple keyword-match in order to extract the semantic content of a user utterance. It provides a rough help capability by indicating its functionality and potential values for each slot. On the other hand, it does not understand any specific "cancel" or "help" keywords, nor does it allow user barge-in.

It is the task of the dialogue module to collect the necessary information from the user for all slots. In the case that three or fewer restaurant solutions exist, only some of the slots need to be filled with values. The system follows a mixed-initiative strategy in that it also accepts user information for slots which the system did not ask for. Meta-communication and clarification dialogues are started in the case that an incoherence in the user utterance is detected (non-understanding of a user answer, user answer is out of context, etc.). Different confirmation strategies can be selected: Explicit confirmation of each piece of information understood by the system (Skowronek, 2002), implicit confirmation with the next request for a specification, or summarizing confirmation. The latter two strategies are implemented with the help of a specialized HTML page, see Dudda (2001). In the case that restaurants exist which satisfy the requirements set by the user, BoRIS indicates names and addresses of the restaurants in packets of maximally three restaurants at a time. If no matching restaurants exist, BoRIS offers the possibility to modify the request, but provides no specific information as to the reason for the negative response. The dialogue structure of the final module used in experiment 6.3 is depicted in Appendix C, Figures C.1 to C.3.

On the speech generation side, BoRIS makes use of pre-recorded messages for the fixed system utterances, and messages which are concatenated according to a template for the variable restaurant information utterances and for the confirmation utterances. Both types of prompts can be chosen either from pre-recorded natural speech, or from TTS. Natural prompts have been recorded from one male and one female non-expert speaker in an anechoic environment, using a high-quality AKG C 414 B-ULS microphone. Synthesized speech prompts were generated with a TTS system developed at IKA. It consists of the symbolic text pre-processing unit SyRUB (Böhm, 1993) and the synthesizer IKAphon (Köster, 2003). F_0 and phone length modelling is performed as described by Böhm. The inventory consists of variable-length units which are concatenated as described by Kraft (1997). These units have been recorded from a professional male speaker, and are stored in a linear 16 bit PCM coding scheme. Because the restaurant information and the confirmation prompts are concatenated from several individual pieces without any prosodic manipulation, they show a slightly unnatural melody. This fact has to be taken into account in the interpretation of the according results.

Test subjects can interact with the BoRIS system via a telephone link which is simulated in order to guarantee identical transmission conditions. This telephone line simulation system has already been described in Section 4.2. For the experiments reported in this chapter, the simulation system has been set to its default transmission parameter values given in Table 2.4. A handset telephone with an electro-acoustic transfer characteristic corresponding to a modified IRS (ITU-T Rec. P.830, 1996) is used by the test subjects. On the wizard's side, the speech signal originating from the test subjects can be monitored via headphone, and the speech originating from the dialogue system is directly forwarded to the transmission system, without prior IRS filtering. All interactions can be recorded on DAT tape for a later expert evaluation.

The BoRIS system is integrated in an auditory test environment at IKA. It consists of three rooms: An office room for the test subject, a control room for the experimenter (wizard), and a room for the set-up of the telephone line simulation system. During the tests, subjects only had access to the office room, so that they would not suspect a wizard being behind the BoRIS system. This procedure is important in order to maintain the illusion of an automatically working system for the test subject. The office room is treated in order to limit background noise, which was ensured to satisfy the requirements of NC25 (Beranek, 1971, p. 564-566), corresponding to a noise floor of below 35 dB(A). Reverberation time is between 0.37 and 0.50 s in the frequency range of speech. The room fulfills the requirements for subjective test rooms given in ITU-T Rec. P.800 (1996).

6.1.2 Speech Recognition Simulation

In order to test the influence of speech recognition performance on different quality aspects of the service, the recognition rate of the BoRIS system should be adjustable within certain limits. This can be achieved with the help of a recognition simulation which is based on an on-line transcription of each user utterance by a wizard, or better – as has been done in experiment 6.3 – by an additional assistant to the wizard. A simple way to generate a controlled number of recognition errors on this transcription would be to substitute every tenth, fifth, fourth etc. word by a different word (substitution with existing words or with non-words, or deletion), leading to an error rate of 10%, 20%, 25% etc. This way, which has been chosen in experiment 6.1, does however not lead to a realistic distribution of substituted, deleted and inserted words. In particular, sequence effects may occur due to the regularity of the errors, as has clearly been reported by Dudda (2001).

To overcome the limitations, Skowronek (2002) designed a tool which is able to simulate recognition errors of an isolated word recognizer in a more realistic and scalable way. This tool considerably facilitates the wizard's work and generates error patterns which are far more realistic, leading to more realistic estimates of the individual interaction parameters related to speech input. The basis of this simulation is a confusion matrix C_{ref} which has been measured with the recognizer under consideration, containing the correctly identified word counts in the main diagonal cells, and the confused word counts in the other cells. This matrix has been generated as a part of the experiments of Chapter 4. It has been amended by an additional row and column for the inserted and deleted words. The matrix corresponds to a reference recognition rate E_{ref} (percentage of correctly identified words) which can be calculated by

$$E_{ref} = \frac{\sum_i C_{ref}(i,i)}{\sum_i \sum_j C_{ref}(i,j)} \tag{6.1.1}$$

In this and in the following equations, the index i refers to the rows which contain the reference words, and j to the columns of the matrix which contain the "recognized" output words.

The matrix now has to be scaled in order to reach a (simulated) target recognition performance E_x, by up-scaling the elements of the main diagonal and lowering the other ones when $E_x > E_{ref}$, or by doing the opposite when $E_x < E_{ref}$. The corresponding scaling will be different for each matrix element $C_x(i,j)$ of the scaled matrix C_x. It has to satisfy the following boundary conditions which result from the limiting recognition rates:

(a) *Target recognition rate 100%*: $E_x = E_{100} = 1$
In this case, all elements outside the main diagonal are added to the values in the main diagonal, and the out-of-diagonal values are then set to zero.

$$C_{100}(i,i) \stackrel{!}{=} \sum_j C_{ref}(i,j) \tag{6.1.2}$$

$$C_{100}(i,j)|_{j \neq i} \stackrel{!}{=} 0 \tag{6.1.3}$$

(b) *Target recognition rate* $E_x = E_{ref}$:
In this case, no change in the confusion matrix takes place.

$$C_x(i,i) \stackrel{!}{=} C_{ref}(i,i) \tag{6.1.4}$$

$$C_x(i,j)|_{j \neq i} \stackrel{!}{=} C_{ref}(i,j)|_{j \neq i} \tag{6.1.5}$$

(c) *Target recognition rate 0%*: $E_x = E_0 = 0$
In this case, all elements of the main diagonal have to be set to zero, and their counts have to be distributed to the out-of-diagonal elements. The following method is used to achieve this goal:
All elements of a row i, with exception of element (i,i), are multiplied with a factor $\alpha(i)$. $\alpha(i)$ is determined in a way that the sum of all scaled elements in the row is identical to the sum of all elements in the reference matrix row. It follows:

$$\sum_j C_{ref}(i,j)|_{j \neq i} \cdot \alpha(i) \stackrel{!}{=} \sum_j C_{ref}(i,j) \tag{6.1.6}$$

Thus

$$\alpha(i) = \frac{\sum_j C_{ref}(i,j)}{\sum_j C_{ref}(i,j) - C_{ref}(i,i)} \tag{6.1.7}$$

This results in the following boundary conditions for the target recognition rate 0%:

$$C_0(i,i) \stackrel{!}{=} 0 \tag{6.1.8}$$

$$C_0(i,j)|_{j \neq i} \stackrel{!}{=} \alpha(i) \cdot C_{ref}(i,j)|_{j \neq i} \tag{6.1.9}$$

The scaling function for each matrix element has to satisfy these three boundary conditions. In addition, it should be monotonously increasing with respect to E_x for the elements of the main diagonal, and monotonously decreasing for the elements out of the main diagonal. Here, the following approach for $C_x(i,j)$ is chosen:

$$C_x(i,j) = a_0(i,j) + a_1(i,j) \cdot E_x^{a_2(i,j)} \tag{6.1.10}$$

Consideration of the boundary conditions leads to the following equation system for determining the parameters $a_0(i,j)$, $a_1(i,j)$ and $a_2(i,j)$ of the scaling function:

$$E_x = E_{100} = 1: \quad a_0(i,j) + a_1(i,j) \cdot 1^{a_2(i,j)} \stackrel{!}{=} C_{100}(i,j) \tag{6.1.11}$$

$$E_x = E_{ref}: \quad a_0(i,j) + a_1(i,j) \cdot E_{ref}^{a_2(i,j)} \stackrel{!}{=} C_{ref}(i,j) \tag{6.1.12}$$

$$E_x = E_0 = 0: \quad a_0(i,j) + a_1(i,j) \cdot 0^{a_2(i,j)} \stackrel{!}{=} C_0(i,j) \tag{6.1.13}$$

The solution of this equation system is as follows:

$$a_0(i,j) = C_0(i,j) \tag{6.1.14}$$

$$a_1(i,j) = C_{100}(i,j) - C_0(i,j) \tag{6.1.15}$$

$$a_2(i,j) = log_{E_{ref}}\left(\frac{C_{ref}(i,j) - C_0(i,j)}{C_{100}(i,j) - C_0(i,j)}\right) \tag{6.1.16}$$

$$= \frac{ln\left(\frac{C_{ref}(i,j) - C_0(i,j)}{C_{100}(i,j) - C_0(i,j)}\right)}{ln(E_{ref})} \tag{6.1.17}$$

Finally, the elements of the scaled confusion matrix have to be rounded to integer values in order to represent word confusion counts.

It should be noted that this approach will only be possible if the final element of the confusion matrix is not zero, otherwise two boundary conditions have the same value of zero:

$$C_0(i,j) = 0 \tag{6.1.18}$$

$$C_{ref}(i,j) = 0 \tag{6.1.19}$$

As this last element describes how often an empty word is recognized as an empty word, an arbitrary number can be chosen.

On the basis of the scaled confusion matrix C_x, the individual rows are used to build a randomized word sequence in which the frequency of the individual words corresponds to their counts in the matrix. Such a sequence is built for each word, and for each simulated recognition rate. For each word which occurs in the wizard's transcription, the sequence determines whether it will be correctly recognized, or whether it will be substituted by a different word, or deleted. Word insertions are simulated using the last (insertion) row of the scaled confusion matrix, and by generating an inserted word probability sequence.

The simulation is able to generate recognition errors for an isolated word recognizer. In the experiments reported in Chapter 4, a relatively stable difference in recognition performance between a continuous and an isolated word recognizer was observed. Table 4.2 shows this difference to be in the area of 10...12% (comparison between the Swiss-French and the German recognizer). Thus, when the target recognition rate should be reached with a continuous speech recognizer, then an accordingly higher target rate E_x has to be chosen.

Several words have been added to the BoRIS vocabulary after the confusion matrix had been generated. Thus, a solution has to be found for the additional words which are not contained in the reference confusion matrix C_{ref}, in a way that the target error rate is reached for all words of the wizard's transcription, and not only for the words in the confusion matrix. A decision on the correct or incorrect recognition of the additional words is taken depending on the current simulated recognition rate: If it is higher than the target value, then the word is incorrectly "recognized"; if it is lower than the target recognition rate, then the word is correctly "recognized". The overall numbers of substitutions, deletions and insertions determine how the incorrectly "recognized" words are treated, i.e. whether they are substituted, inserted or deleted.

The described recognition simulation behaves in the way which is defined by the confusion matrix, i.e. the probability of confusion between words. It does not, however, take user and speaking style effects (dialects of the user, type of articulation, etc.) into account, for which additional information and an appropriate description would be needed in order to implement them in a simulation. However, the simulation developed here is far more realistic than a simple regular deletion of individual words with the percentage of the recognition rate. In particular, it becomes possible for the recognizer to insert information which may lead to a confusion in the dialogue flow. This effect will be very important in real-life dialogues. Other ways to simulate recognition performance in Wizard-of-Oz scenarios have recently been proposed by Trutnev et al. (2004).

6.1.3 Measurement of Interaction Parameters

During each interaction, a log file is automatically created by BoRIS. It is labelled with an identifier for the system configuration, for the test subject, and for the scenario which was given to the subject. Each log file contains the following information:

- Information on the current system configuration: Target recognition rate, speech output for variable and fixed system messages, type of confirmation.
- Start and end time stamp for each dialogue.
- Start and end time stamp for each system and user utterance. For the system utterances, this time stamp is generated automatically; for the user utterances, it is generated manually by a button-click by the wizard.
- The number of system and user turns.
- The text of each system message.
- In case of an ASR simulation (see the previous section), the wizard's transcription of each user utterance.
- The recognized user utterance, or in the case of an ASR simulation the generated recognition pattern.
- The AVPs which the system is able to extract from each recognized user utterance.

On the basis of this log file information, a number of instrumentally measurable interaction parameters can be calculated directly, namely all duration-related parameters, and all parameters related to the number of turns and words in each turn. Also the number of system questions, system error messages, and the number of ASR rejections can be calculated automatically by defining the respective system messages beforehand and counting them in the log file. Other interaction parameters require an annotation of the dialogue by a human expert. The Tcl/Tk-based annotation tool which has been described in Section 3.8.4 is used for this purpose. Appendix D.3 discusses the individual decisions which have to be taken by the expert at each step of the annotation. The evaluation is first carried out on each individual dialogue, and then in a second phase the metrics referring to the whole set of dialogues are calculated.

From the log file and the expert annotation, a relatively complete set of interaction parameters can be obtained for each dialogue. The following parameters are generated by the annotation tool:

- Dialogue duration (DD)
- System turn duration (STD), user turn duration (UTD)

- System response delay (SRD), user response delay (URD)
- # TURNS, # SYSTEM TURNS, # USER TURNS
- # SYSTEM WORDS, # USER WORDS
- Number of in-vocabulary words (# USER WORDS IV)
- Number of words per system turn ($WPST$) and per user turn ($WPUT$)
- # ASR REJECTIONS
- # SYSTEM ERROR MESSAGES
- # SYSTEM QUESTIONS, # USER QUESTIONS
- System answers: $AN{:}CO$, $AN{:}PA$, $AN{:}IC$, $AN{:}FA$
- $DARPA_s$, $DARPA_{me}$
- # HELP REQUESTS
- # BARGE-INS
- # CANCEL ATTEMPTS
- System correction turns and rate (SCT, SCR), user correction turns and rate (UCT, UCR)
- Contextual appropriateness: $CA{:}AP$, $CA{:}IA$, $CA{:}TF$, $CA{:}IC$
- Parsed user utterances: $PA{:}CO$, $PA{:}PA$, $PA{:}FA$
- Implicit recovery (IR)
- Task success labels in order to calculate TS_{bin} and TS_{ord}
- Attribute-Value-Pairs: c_{AVP}, s_{AVP}, d_{AVP}, i_{AVP}, N_{AVP}, and n_{AVP} (the latter being the number of correctly not set AVPs)
- Information content (IC)
- Understanding accuracy (UA)
- Kappa coefficient (κ), calculated for each dialogue and for a set of dialogues belonging to the same system configuration
- Recognized words: c_w, s_w, d_w, i_w (isolated and continuous recognition)
- Word error rate (WER) and word accuracy (WA), each for isolated and for continuous recognition

- Average number of errors per sentence ($\overline{NES}$), average word error per sentence ($\overline{WES}$), both calculated for isolated and for continuous recognition

Two specific points have to be noted. Speech recognition – both with the commercial recognizer and with the simulation – is based on keyword spotting. In this case, it is not obvious whether the ASR performance metrics related to a continuous speech recognizer (including insertions) or the corresponding metrics for an isolated word recognizer (ignoring insertions) are more meaningful in describing the respective behavior. In order to collect as much information as possible, both classes of metrics have been calculated in the present case.

A second decision is necessary on how to count the task-success-related labels S, SCs, SCu, $SCuCs$, SN, FS and FU. Two coding approaches have been chosen here: For a binary decision of task success (TS_{bin}), FS and FU are assigned a value of zero, and to the rest of the labels (S, SCs, SCu, $SCuCs$, SN) a value of one. As a second way, an ordinally scaled variable TS_{ord} is calculated, assigning the following rank orders to the labels: 0 for FU, 1 for FS, 2 for SN, 3 for $SCuCs$, 4 for SCu, 5 for SCs, and 6 for S. This rank order has obviously been chosen in an intuitive way. Nevertheless, it seems to provide a more fine-graded way of interpreting the task-related success labels assigned by the wizard.

6.1.4 Questionnaire Design

In order to obtain information about the perceptive quality features, a questionnaire has to be completed by the subjects at the beginning of a test session, after each interaction (except experiment 6.2), and at the end of a test session. The questionnaires have been developed during the course of the experiments, so the individual questions which have been asked in experiments 6.1, 6.2 and 6.3 are not identical. They are partly based on the literature survey (see Section 3.8.6), and partly on own considerations of the individual quality aspects which should be covered. Details on the design of the individual questionnaires can be found in Dudda (2001), Niculescu (2002) and Skowronek (2002).

It is the aim of a questionnaire to obtain information in a realistic and unbiased way. Thus, only very limited information about the system should be given to the test subjects prior to the experiment. The information should be identical for all subjects, a requirement which can best be satisfied via a written instruction sheet. A translation of the instructions which were used in the experiments is reproduced in Appendix D.1. Test subjects are asked to imagine a realistic usage situation. Nevertheless, most of the subjects will feel to be in a test situation, which differs in several respects from the later usage situation (e.g. by the motivation for using the system, time and money constraints, the fear of being judged and not being the judging subject, etc.).

The choice of questions for each questionnaire depends on the aim of the specific test. In general, completeness and appropriateness of the individual questions can be enhanced by carrying out qualitative investigations in a first step. Dudda (2001) offered open answer possibilities to the subjects in a pre-test to experiment 6.1. She used the experiences gained in this pre-test to optimize the evaluation procedure, but did not detect any specific missing items in the questionnaire.

Niculescu (2002) started experiment 6.2 with a qualitative interview in order to gain insight into the relevant quality aspects. The interview was carried out in a standardized way with eight test subjects whose answers were recorded and transcribed. The results of this qualitative interview were then condensed to formulate closed questions or statements which can be rated quantitatively in the final test session. The resulting questionnaire consists of two parts (A and B), one given before and one after the interactions to be carried out by the test subjects. Part B is reproduced in Appendix E.1, and part A is similar to the one used in experiment 6.3, see Appendix E.2. The questionnaire allows the system behavior and system capabilities to be investigated diagnostically, but is not focussed on specific system versions (which is in accordance with the aim of the experiment). Questions are often formulated as statements, in order to provoke a pronounced opinion of test subjects who would not judge in a very decided way. Additional guidelines on how to formulate such questions (briefness, use of simple words, no foreign words, reference to one fact only, etc.) can be found in the respective psychometric literature, see e.g. Bortz (1995). The questions cover all eleven quality aspects which are part of Niculescu's taxonomy, see Section 2.2.3. Relationships between questions and aspects are described in detail in Niculescu (2002). Her experimental results, which are not reproduced here because they mainly refer to the specific system under investigation and not to general quality assessment aspects, show that the questionnaire can fruitfully be used to assess individual aspects of cooperativity.

The questionnaire of experiment 6.3 will serve the analysis of the QoS taxonomy. For this aim, it would be preferable to have at least one question directed to each taxonomy item (feature, aspect and category) in the questionnaire[1]. Unfortunately, this would result in more than 120 questions, which are far too much for a single test session. In addition, it does not seem possible to formulate a simple question for each aspect or category of the taxonomy. Thus, a compromise was selected, based on the experiences gained in experiments 6.1 and 6.2. The final questionnaire consists of three parts (A, B and C), see Appendix E.2. Part A collects background information on the subjects (user

[1]With respect to Grice's maxims of cooperative behavior, this idea has recently been followed in an experiment documented by Meng et al. (2003), using four questions related to the maxims of quality, quantity, relation, and manner.

Table 6.1. BoRIS system configurations addressed in experiment 6.1, see Dudda (2001).

No.	*Recogn. rate (%)*	*Speech output*		*Confirmation*
		Fixed	*Variable*	
1	100	female	female	–
2	100	TTS	female	–
3	100	female	female	explicit
4	100	female	female	implicit
5	100	female	female	summary 1
6	83	female	female	–
7	75	female	female	–
8	67	female	female	–
9	100	female	TTS	–
10	100	TTS	TTS	–
11	100	female	female	summary 2

factors), and on their demands and ideas of the system under test, in a more-or-less unprimed way. Part B reflects the spontaneous impressions of the subjects directly after each call. Part C refers to the final impression at the end of the test session, reflecting all experience gained with the system so far. Section 6.2.6 will discuss how far the impressions after each interaction influence the final integrative judgment.

6.1.5 Test Design and Data Collection

The three experiments with the BoRIS system have mainly been carried out in a Wizard-of-Oz style. The wizard had the task of instructing the test subjects, controlling the smooth run of the experiment, selecting the respective system configuration, and starting and stopping each interaction with the system. In addition, the wizard (experiments 6.1 and 6.2) or an additional assistant (experiment 6.3) had to transcribe the user utterances for the ASR simulation. Whereas in experiment 6.1 only a simple "leave-every-nth-word-out" ASR error generation strategy was available, the ASR simulation described in Section 6.1.2 was implemented for experiment 6.3. Making use of an ASR simulation provided a higher flexibility of system versions to be tested, and guaranteed sufficient speech recognition performance which otherwise would have been too low, due to the lack of training data. The ASR simulation also provided a meaningful way to extract metrics which are related to continuous speech recognizers.

Eleven different system configurations were tested in experiment 6.1, see Table 6.1. They differed with respect to the simulated recognition rate (between 67 and 100%), the speech output provided by the system, and the confirmation

Table 6.2. BoRIS system configurations addressed in experiment 6.3, see Skowronek (2002).

No.	*Recogn. rate (%)*	*Speech output*		*Confirmation*
		Fixed	*Variable*	
1	100	female	female	–
2	100	male	female	–
3	100	female	TTS	–
4	100	TTS	TTS	–
5	70	female	female	–
6	100	female	female	explicit
7	90	female	female	explicit
8	80	female	female	explicit
9	70	female	female	explicit
10	60	female	female	explicit

strategy applied in the dialogue. Speech output either consisted of a natural pre-recorded female voice, or of TTS samples with a male voice. These options could be chosen separately for the fixed system messages and confirmations on the one hand, and for the variable restaurant information parts on the other. In addition to an explicit and an implicit confirmation of each piece of information, also two types of summarizing confirmations were tested. Strategy "summary 1" consists in a summary of all slot values before the restaurant information is given to the user. Strategy "summary 2" only indicates the summary in cases where the user wants to modify a previous request, so that the user is aware of the current search criteria of the system.

In experiment 6.2, the system was held constant, and only one configuration (no. 9 of Table 6.1) was tested. The experiment consisted of two parts: A qualitative oral interview with a limited number of subjects, in order to gain insight into the quality features perceived by the test subjects; and a quantitative evaluation via a written questionnaire, which was amended by an additional interview. The interview and the questionnaire reflect the subject's opinion before and after the test. No judgments were collected between the individual interactions in experiment 6.2.

For experiment 6.3, the collection of interaction parameters, the questionnaire design, and the simulation of recognition errors were optimized. Variable parameters in the system configuration were once again the recognition rate, the speech output provided by the system, and the confirmation strategy, see Table 6.2. In addition to the female speaker, recordings of a male speaker were also used for the fixed system prompts. This configuration (no. 2) thus uses two different natural voices for the fixed messages and for the variable restaurant information utterances. A large number of interaction parameters were collected during experiment 6.3, see Section 6.1.3.

Test subjects had to carry out 6 (experiments 6.1 and 6.2) or 5 (experiment 6.3) experimental tasks which were defined via scenarios. The scenarios were chosen in a way to reach comparable results between the different test subjects, however with the aim to avoid direct priming (e.g. via the chosen vocabulary). The tasks covered the functionality provided by the system, and different situations the system would be used in. Most of the scenarios provide only a part of the information which BoRIS requires in order to search for a restaurant. The remaining requirements have to be selected spontaneously by the test subjects. When no restaurants exist which satisfy the requirements, constraint relaxations are suggested in some scenarios. In this way, it is guaranteed that a task solution exists, provided that the user accepts constraint relaxation. One task is an 'open' scenario where the subjects could define the search criteria on their own, prior to the call. The scenarios which were used in experiment 6.3 are depicted in Appendix D.2. Further details on the scenarios can be found in Dudda (2001).

The system configurations and scenarios have to be distributed to the individual dialogues to be carried out by the test subjects, so that the influence of confounding factors can be minimized (Möller, 2000). In experiment 6.3, this goal has been reached by distributing the scenarios (5) and system configurations (10) to the available dialogues on a Graeco-Latin Square of order 5 (25 cells). This Graeco-Latin Square is repeated 8 times in order to define all 200 dialogues (40 subjects x 5 interactions), see Skowronek (2002). In experiments 6.1 and 6.2, the scenarios were always given in the same order, allowing the effect of the scenario to be studied in more detail. In this case, only the system configurations have been randomized to avoid sequence effects.

The following overview shows the characteristics of the test subject group for each experiment:

- Experiment 6.1: 25 subjects (11f, 14m), 23 to 51 years (mean 28.4 years), about half of them had prior experience with speech-based information servers.

- Experiment 6.2: Qualitative interview: 8 test subjects (4f, 4m), 22 to 39 years (mean 27.5 years), mainly students and researchers of the university, with nearly no experience in using speech-based information systems. Quantitative questionnaire: 53 test subjects (23f, 30m), large majority between 20 and 29 years, mainly students of the university, only 9 of them had ever used a speech-based information server before, 15% were non-natives.

- Experiment 6.3: 40 test subjects (11f, 29m), 18 to 53 years (mean 29 years). Most of the test subjects had a certain knowledge about the town and thus about the potential location of restaurants. Criteria which were mentioned for choosing restaurants were the price (27 nominations), the type of food (33), the quality of food (33), and the friendliness of the staff (34). Location (14) and opening hours (7) seem to be less important criteria. 13 of the 40 test subjects said that they had prior experience with speech-based information servers. Their mean rating on this experience was between fair and poor (2.8 on the continuous 5-category overall quality scale).

Dudda (2001) reports some observations about the test subjects' behavior during the test. She recognized three types of behavior: (1) Self-confident natural interaction, replying by full sentences; (2) Defensive behavior, adaptive to the system in order to reach maximum task success; answers consisted partly in simple keywords which were contained in the system utterances; or (3) Spontaneous behavior, which however did not always respect the constraints given by the scenarios. Most subjects show a behavior corresponding to the categories (1) or (2). The behavior of the subjects does not seem to be linked to their expertise with speech-based information servers (Skowronek, 2002). In the open scenario, most subjects used criteria which were very similar to the ones given in the other scenarios. Thus, the open scenario in the final position of the experiment does not seem to invite new search criteria or interaction styles. Some subjects experienced problems because the system did not always provide a complete list of available options in his explanations (Niculescu, 2002). This type of problem is however not limited to the scenario-driven evaluation, but may also occur in a real-life situation. Subjects seemed to react to recognition errors, e.g. by pronouncedly clearly speaking, by shortening their spontaneous input to simple phrases, or by following the system questions, thus leading to a more system-directed dialogue.

6.2 Analysis of Assessment and Evaluation Results

In this section, experimental results will be discussed in the case that they give an insight into quality assessment and evaluation principles which go beyond the specific restaurant information system. Results which are mainly related to the BoRIS system can be found in the detailed experimental descriptions of Dudda (2001), Niculescu (2002) and Skowronek (2002). After a brief pre-analysis of the collected data, Sections 6.2.2 and 6.2.3 investigate the underlying dimensionality of quality judgments and of interaction parameters. Such an analysis may be interpreted in terms of addressed quality aspects. For the quality judgments, it will thus provide information about the dimensions underlying the user's quality percept. For the interaction parameters, it uncovers the relationship between individual parameters and will help to choose an adequate evaluation metric. A deeper analysis of the correlations between interaction parameters and quality judgments is carried out in Section 6.2.4. The aim is to give empirical evidence for the taxonomy of QoS aspects, and to describe the relationships in a quantitative way. The subsequent section analyzes the impact of the system configuration, namely of the speech recognizer, of the speech output, and of dialogue management strategies. If these influences can be captured in a quantitative way, system developers are able to optimize their systems very efficiently, by addressing the most relevant system characteristics first. A selection of test-related influences (Section 6.2.6) concludes the analysis.

6.2.1 Data Pre-Analysis

Two types of data have been collected in the experiments: Interaction parameters which can either be measured instrumentally or calculated from the expert annotation, and subjective quality judgments. The quality judgments of experiments 6.1 and 6.3 were mostly obtained on continuous ratings scales with two or five standard labels, and 2 additional separate scale end points, see Figure 3.4. These responses were coded in a linear way so that the most positive (or fastest, shortest) attribute of the normal, bold part of the scale was attributed a value of 5, and the most negative (slowest, longest) a value of 1, the extreme points corresponding to 0 and 6. In this way, it is easy to compare the results obtained on these scales to MOS scores obtained on 5-point ACR ratings scales. In the questionnaire of experiment 6.2, test subjects had to judge whether they agreed or disagreed with a defined statement on a 5-point Likert scale. Because such a scale can generally not be assumed to have an interval level, these ratings have been accumulated for the positive and the negative part of the scale (percentage of ratings in the upper/lower two categories).

All data collected in the experiments have been aggregated in an SPSS database for further analysis. Depending on the experiment, the database contains subjective ratings, interaction parameters, or both. In the following discussion, the focus of the analysis will be put on the experiment 6.3 data, because this set is the most complete one and contains all interaction parameters listed in Section 6.1.3. It consists of 197 dialogues carried out with 40 test subjects. 3 dialogues had to be excluded from the analysis because of experimental problems. Data of experiments 6.1 and 6.2 are nevertheless indicated in the case that additional information can be obtained.

The data have been checked with respect to their distribution. When all dialogues – irrespective of the system configuration – are taken together, a Kolmogorov-Smirnov test shows that nearly all of the interaction parameters and also nearly all subject responses (exception B0) do not follow a gaussian distribution ($p \leq 0.05$). However, if the individual system configurations are analyzed separately, nearly all subjective judgments and most of the interaction parameters show a gaussian distribution. Thus, the system configuration seems to exercise an influence both on the interaction parameters and on the subjects' judgments. An analysis of variance (ANOVA) shows that the system configuration has a significant effect on most interaction parameters, and on 12 of the 26 questions asked in part B of experiment 6.3. The effects of the different system configuration characteristics on interaction parameters and subjective judgments are addressed in more detail in Section 6.2.5.

Given the large set of interaction parameters, it is important to identify parameters which do not possess informative power in order to exclude them from the further analysis. The following interaction parameters show nearly constant values for all interactions, and are consequently excluded from the data set:

- # HELP REQUESTS
- # ASR REJECTIONS
- # CANCEL ATTEMPTS from the user
- CA:TF and CA:IC
- AN:IC

In addition, the parameters # BARGE-INS and # SYSTEM ERROR MESSAGES only rarely differed from zero. They were nevertheless kept in the data set, but interpretation of results should take the low frequency of their occurrence into consideration. The parameter # USER QUESTIONS happened only in 18 cases to have values different from zero. Consequently, the AN:CO, AN:PA, AN:FA, $DARPA_s$ and $DARPA_{me}$ parameters could also only be calculated for 18 dialogues. The parameter IR can only be calculated if PA:CO and PA:PA differ from zero. This was possible in 89 out of the 197 cases. All other parameters have been calculated for each dialogue, i.e. for all 197 cases. The task success measure κ has been calculated in two ways: On a per-dialogue basis (κ_{dia}), and on a per-system-configuration basis (κ_{conf}).

In principle, the mentioned parameters have informative value, which is however restricted due to their low occurrence rate in the database. Other parameters have very close relationships *by definition*, e.g. the # SYSTEM TURNS, # SYSTEM WORDS, and $WPST$. In order to reduce the amount of data to administrable quantities, the following parameters will not be discussed in the following sections:

- # SYSTEM TURNS, # USER TURNS (related to # TURNS due to the strict alternation of turns)
- # SYSTEM WORDS (related to # SYSTEM TURNS and $WPST$)
- # USER WORDS (related to # USER TURNS and $WPUT$)
- SCT (related to SCR)
- UCT (related to UCR)

- c_w, d_w, i_w, s_w (related to WA and WER)
- c_{AVP}, d_{AVP}, i_{AVP}, s_{AVP}, n_{AVP}, N_{AVP} (related to IC and UA)

All turn-related parameters (# SYSTEM ERROR MESSAGES, # SYSTEM QUESTIONS, # USER QUESTIONS, # BARGE-INS, CA:AP, CA:IA, CA:TF, CA:IC, *PA*:CO, *PA*:*PA*, *PA*:*FA*) have been normalized to the total number of turns in the dialogue. The system-answer-related parameters AN:CO, AN:*PA*, AN:IC and AN:*FA* have been normalized to # USER QUESTIONS.

6.2.2 Multidimensional Analysis of Quality Features

The quality judgments obtained from the users cover a wide range of perceptive dimensions. A factor analysis investigates the dimensionality of the perceptive quality space, and provides information about which quality dimensions are relevant from a user's point of view. The resulting factors will, however, strongly reflect the characteristics of the specific system the judgments have been collected with. Thus, when generic information is to be obtained, extracted factors should be seen in relation to results obtained with other systems, under different circumstances. In the following discussion, several examples of multidimensional analyses which have been published in the literature are cited for comparison. They address systems with speech input or with full dialogue capabilities. The results are compared to the factors which can be extracted from the experiment 6.3 data.

Love et al. (1994) designed a questionnaire for automated telephone services, based on general usability literature. It included 18 general service attributes and 4 attributes which are specific to speech-based telephone services. The questionnaire was used in a WoZ experiment where 40 users had to pay for goods over the phone. Unfortunately, the authors did not describe the system's dialogue capability in detail, but it seems that it was relatively restricted (mainly digit recognition). A factor analysis of the subjective ratings with subsequent Varimax rotation revealed 5 factors accounting for 74% of the variance. These factors have been interpreted as follows:

- Factor 1: "Quality of interface performance". Includes efficiency and reliability of the service, required improvement for the service, willingness to use the service again, user's degree of enjoyment and frustration, whether the service was confusing to use, the degree of complication and the ease of use of the service, and the degree of control the user had over the service.
- Factor 2: "Cognitive effort and stress experienced by the user". Includes the speed of the service, the stressfulness of the experience, the concentration required to use the service, whether the user felt flustered while using the service, the degree of control the user had over the service, whether the

service was felt to be confusing to use, and the user's degree of frustration and enjoyment.

- Factor 3: "User's conversational model". Includes the user's attitude to the voice and to the tones used by the service, the perceived friendliness of the service, whether the user preferred the service or a human operator, the perceived helpfulness of the service tones, the user's degree of enjoyment, and the degree of complication of the service.

- Factor 4: "Perceived fluency of the experience (ease of use I)". Includes the perceived clarity of the voice used by the service, the perceived politeness of the service, whether users knew what they were expected to do, and the degree of complication of the service.

- Factor 5: "Transparency of the interface (ease of use II)". Includes the ease of use of the service, the perceived helpfulness of the service tones, whether the users felt flustered while using the service, whether the users knew what they were expected to do, the degree of complication of the service, the users' attitude to the tones used in the service, and whether the service was confusing to use.

Factor 1 seems to be mainly related to service efficiency, usability, and user satisfaction in general. Factor 2 contains features which are subsumed under the term "comfort" of the QoS taxonomy, and partly also user satisfaction features. Factors 3 and 4 are relatively difficult to interpret, and subsume a number of features belonging to different aspects. Factor 5 is once again related to the system's usability. A recent analysis for user judgments obtained via a Danish translation of the questionnaire used in the OVID project extracted five factors which are very similar to the described ones (Larsen, 2003).

A similar multidimensional analysis was carried out by Jack et al. (1992). They collected 256 subjective judgments on a service where the user had to provide a credit card number via spoken input. The service was simulated by a wizard and provided an adjustable level of recognition performance. Subjective ratings were obtained on 22 Likert scales, and a factor analysis revealed 4 key factors:

- A "cognitive" factor, covering the degree of concentration required, and the confidence in the use of the service.

- An "efficiency" factor, covering perceived system reliability, speed, and accuracy.

- A "personality" factor, covering friendliness and politeness of the voice prompts, and the degree of control which the user perceived in the transaction.

- A "technology" factor, covering the attitude to tone prompts (beeps), spoken prompts, and the preference to a human operator.

Two of the factors ("cognitive" and "personality") cover aspects which are addressed under the comfort category of the QoS taxonomy. The "efficiency" factor seems to be mainly related to communication efficiency, although the reliability aspect also indicates service efficiency. The "technology" factor seems to address mainly the characteristics of the user group, and thus the user factors in the QoS taxonomy.

A very structured multidimensional analysis approach has been performed by Hone and Graham (2000, 2001). They used the 50-item SASSI questionnaire (see Section 3.8.6) in order to identify perceptive dimensions underlying users' quality judgments on a number of different speech-based systems. Because not all systems provided speech output capability, this aspect was explicitly disregarded in the questionnaire. Data was collected from 226 subjects and for 8 different speech applications, ranging from simple small-vocabulary telephone dialling interfaces with a press-to-talk button to in-car command systems and telephone banking applications. A factor analysis on the whole data set revealed six factors accounting for 64.7% of the variance. These factors were tentatively labelled as follows:

- System response accuracy
- Likeability
- Cognitive demand
- Annoyance
- Habitability
- Speed

The factors address aspects on different levels of the QoS taxonomy: "System response accuracy" may be related to the system's cooperativity and to service efficiency, "speed" to communication efficiency, and "cognitive demand" to the comfort category. "Likeability", "annoyance" and "habitability" address user satisfaction and usability aspects. The "likeability" factor showed the highest correlation (0.607) to overall system quality (judgment to the statement "Overall, I think this is a good system"). This finding reflects the close relationship between user satisfaction and overall system quality judged by the user.

A similar cross-system analysis was carried out during the 1997 EUROSPEECH conference, however in a less systematic way (den Os and Bloothooft, 1998). During this so-called "ELSNET Olympics", telephone-based systems for different languages, application domains and tasks were evaluated. The systems were judged by volunteering visitors of the conference in a common questionnaire, consisting of 11 aspects to be rated on Likert scales, an overall impression judgment, and some personal information about the users. Factor analysis with Varimax rotation on the whole data set resulted in 5 factors explaining 75% of the variance. The first three factors relate to system properties, and the final two to the user. These factors were interpreted as follows:

- General appreciation of the system (task completion, error recovery, appropriate reactions, error messages, and overall satisfaction).
- Functional capability of the system (functional possibilities and dialogue structure).
- Intelligibility of output speech (speech intelligibility and wording of utterances).
- Language proficiency of the user.
- User's familiarity with dialogue systems.

The first three factors were shown to be somehow independent of the system to be judged upon. The forth factor seems to be a particular user factor which results from the uncommon test population (many non-native users). The final factor is a general user factor which will play a similar role for many walk-up-and-use dialogue systems. The study had some inherent limitations, namely the specific user population, the incomplete experimental design (due to the language differences), and the differences in interpretation of the single questions for each system.

The subjective judgments which have been collected in experiment 6.3 can be factor-analyzed in order to identify the quality dimensions underlying the interactions with BoRIS. A principal component analysis with Varimax rotation and Kaiser normalization has been used for this purpose. Five factors can be extracted from the judgments collected after each interaction (part B questions). The corresponding factor loadings for the individual questions are given in Table 6.3. The factors can tentatively be interpreted in the following way:

- Factor 1: Overall impression of the system, informativeness, overall behavior of the system, interaction capability/flexibility, user satisfaction. Related questions: B0, B1, B2, B3, B4, B9, B10, B13, B14, B23.
- Factor 2: Overall impression of the system, intelligibility, friendliness, naturalness, pleasantness. Related questions: B0, B7, B16, B18, B22, B24.
- Factor 3: Cognitive effort and dialogue smoothness. Related questions: B6, B8, B19, B21.
- Factor 4: Speed and conciseness. Related questions: B15, B17, B20.

Table 6.3. Factor analysis (principal component analysis with Varimax rotation and Kaiser normalization, pairwise exclusion of missing values) of the part B questions of experiment 6.3. Rotated component matrix. Loadings higher than 0.6 are given in italics.

Question	*Component*				
	1	2	3	4	5
B0	*0.677*	*0.677*	0.332	0.078	0.168
B1	*0.871*	0.079	0.101	0.048	-0.108
B2	*0.847*	0.094	0.064	-0.002	-0.213
B3	*0.534*	0.556	0.190	-0.158	-0.197
B4	*0.674*	0.066	0.309	0.064	-0.278
B5	0.515	0.163	0.512	0.197	0.260
B6	0.033	0.491	*0.630*	-0.084	-0.141
B7	-0.003	*0.857*	0.163	-0.095	-0.075
B8	0.204	0.039	*0.726*	-0.050	-0.113
B9	*0.749*	0.014	0.318	0.049	0.131
B10	*0.625*	0.012	0.394	0.199	0.239
B11	0.579	-0.036	0.540	0.196	0.110
B12	0.241	0.394	0.153	0.184	*0.657*
B13	*0.718*	0.021	0.075	0.127	0.369
B14	*0.730*	-0.014	0.150	-0.077	0.242
B15	0.074	0.050	-0.037	*0.644*	0.144
B16	0.159	*0.739*	0.029	-0.124	0.061
B17	0.062	-0.056	0.054	*0.729*	-0.377
B18	0.011	*0.697*	0.119	0.264	0.407
B19	0.443	0.129	*0.668*	0.080	0.198
B20	0.021	-0.102	0.194	*0.766*	0.241
B21	0.382	0.223	*0.668*	0.170	0.206
B22	-0.053	*0.883*	0.027	0.042	0.108
B23	*0.753*	0.193	0.232	0.031	0.177
B24	0.369	*0.634*	0.385	0.069	0.249
B25	0.328	0.354	0.498	0.243	0.226

- Factor 5: Reaction like a human. Related question: B12.

These 5 factors explain 68.1% of the overall variance in the data. Questions B5, B11 and B25 show only moderate to low correlations (< 0.6) to all factors, and will thus contribute to the residual variance. The overall impression (B0) is mainly addressed by Factors 1 and 2. Factor 1 reflects the overall system quality from an information exchange or task point of view, whereas Factor 2 covers the surface form and the agent personality. Factor 3 seems to be related to the cognitive effort and smoothness, while Factor 4 describes interaction length and speed. Factor 5 addresses the comparability of the system behavior

Table 6.4. Factor analysis (principal component analysis with Varimax rotation and Kaiser normalization, pairwise exclusion of missing values) of the part C questions of experiment 6.3. Rotated component matrix. Loadings higher than 0.6 are given in italics.

Question	*Component*				
	1	2	3	4	5
C1	0.582	0.133	0.168	0.372	0.426
C2	0.105	0.065	0.006	-0.003	*0.815*
C3	0.215	*0.651*	-0.322	-0.130	0.260
C4	0.350	0.169	0.017	*0.777*	-0.147
C5	*0.855*	0.137	0.181	-0.097	-0.002
C6	*0.832*	0.085	0.173	0.342	0.113
C7	0.028	-0.061	0.133	*0.891*	0.226
C8	0.444	0.595	0.359	0.095	-0.090
C9	0.253	0.484	0.181	0.302	0.504
C10	0.127	*0.625*	0.321	-0.265	0.294
C11	0.182	0.574	0.415	0.397	-0.336
C12	0.559	0.305	0.317	0.123	0.461
C13	*0.688*	0.496	0.026	0.183	0.174
C14	0.589	0.529	-0.168	0.124	0.266
C15	0.396	0.153	*0.738*	0.131	0.256
C16	0.112	*0.623*	0.165	0.133	0.017
C17	0.107	0.088	*0.877*	0.058	-0.012
C18	*0.622*	0.330	0.407	0.179	0.093

to that of a human interaction partner. The fact that this aspect accounts for an individual perceptive dimension justifies taking HHI as a reference for HMI when both are carried out over the phone.

A comparison of the extracted factors with the ones found by Love et al. (1994) does not give a clear picture. Factors 1 and 2 seem to be somehow related to the "quality of interface performance" factor extracted by Love. Factor 3 seems to be related to the "cognitive effort and stress experienced by the user". Factor 4 is once again reflected in the first of Love's factors ("quality of interface performance"). The human-like reaction factor is somehow related to the "user's conversational model" from Love et al., but does not have a real equivalent there.

A similar factor analysis has been performed on the part C questions (C1 to C18) which were judged at the end of the whole test session. Five factors can be extracted which explain 73.3% of the overall variance in the data, see Table 6.4. These factors can be interpreted as follows: Factor 1 describes the system's task and communication capabilities, which result in system acceptability (medium loading on C18). Factor 2 subsumes the system personality, the degree of enjoyment when using the system, and its ease of use. Factor 3 comprises the system's added value in respect to other interfaces. Factor 4 covers the communication initiative and the system's help capability. Factor 5 only loads high on the clarity of expression. This factor does not have any corresponding term in the other analyses described so far.

The grouping in factors shows that the individual questions of the questionnaire are not independent from each other. However, only in five cases were correlations above 0.7 observed. These cases were:

- B0 and B23 ($\rho = 0.759$): The overall impression (overall quality) is obviously highly correlated with user satisfaction, and can consequently be allocated to that category.

- B1 and B2 ($\rho = 0.759$): The provision of the desired information seems to be linked to the information completeness. Apparently, test subjects are not easily able to differentiate between accuracy and completeness of the provided information. This effect may however be due to the test situation.

- B24 and B25 ($\rho = 0.757$): Both questions are related to the emotional state of the subject (personal feeling, stress).

- C5 and C6 ($\rho = 0.735$): The non-ability to answer the user's questions seems to be mostly linked to misunderstandings.

- C15 and C18 ($\rho = 0.713$): The preference for a different interface is obviously linked to the system's acceptability, i.e. the expected future use.

No correlations higher than 0.8 could be found. Thus, no apparently redundant questions seem to be contained in the questionnaire.

Summarizing the different factor analysis results, seven common factor classes observed for all systems can be identified. The first class of factors relates to *service and task efficiency* and *acceptability*. It includes the "quality of interface performance" factor (Love et al., 1994), the "system response accuracy" factor (Hone and Graham, 2000), the "general appreciation" and "functional capabilities" factors (den Os and Bloothooft, 1998), as well as Factor 1 from the analysis of part B questions, and Factors 1 and 3 from the analysis of part C questions. Closely related are the factors or the second class describing *user satisfaction*, e.g. the "annoyance" and "likability" factors from Hone and Graham (2000), and Factor 2 of the analysis of part C questions. A third class of factors describes *usability*, e.g. the "transparency of the interface" factor (Love et al., 1994), and the "habitability" factor extracted by Hone and Graham (2000). Two other classes of factors belong to the "comfort" category of the QoS taxonomy: The factors relating to *cognitive demand* ("cognitive effort and stress experienced by the user" from Love et al. (1994), the "cognitive" factor from Jack et al. (1992), the "cognitive demand" factor from Hone and Graham (2000), and Factor 3 of the part B question analysis), and those relating to the *agent personality* ("users conversational model" from Love et al. (1994), "personality" factor from Jack et al. (1992), partially Factor 2 of the part B question analysis, and Factor 3 of the part C question analysis). The sixth class covers *user factors* like attitude or experience. It is reflected in the "user's conversational model" factor from Love et al. (1994), in the "technology" factor from Jack et al. (1992), in the "user familiarity" factor extracted by den Os and Bloothooft (1998), and in Factor 4 extracted from the part C questions. The final class relates to *communication efficiency*, and includes the "efficiency" factor from Jack et al. (1992), the "speed" factor from Hone and Graham (2000), and Factor 4 of the part B question analysis.

The summary shows that *all* categories of the lower half of the QoS taxonomy relate to quality dimensions which can be extracted empirically from subjective experiments. As the described experiments address a large number of diverse systems, the finding seems to be quite generic. This is a strong argument for the completeness and the relevance of the quality aspects covered by the taxonomy.

A slightly different way for analyzing relationships between quality judgments is a cluster analysis. In principle, both cluster analysis and factor analysis are able to describe groupings (clusters) among variables (here subjective judgments) on the basis of distances or similarities of variable values in a specified data set. Commonly, a cluster analysis aims at grouping cases, but is may also be applied to the grouping of variables. A cluster analysis may be carried out in a hierarchical way, also indicating the (distance) level on which the individual variables combine to a cluster. The clustering can then be displayed via a so-called dendrogram. The dendrogram indicates the scaled distance where two or more variables combine to form a cluster, up to the point where all variables combine to a single cluster for the whole data set. Such a hierarchical cluster analysis is carried out in the following for the part B and part C questions of experiment 6.3. Interpretation of the resulting clusters is facilitated by associating the respective questions with the quality aspects of the QoS taxonomy. This association is summarized in the following figures and tables. Figure 6.1 shows the association with the categories, and Tables 6.5 and 6.6 the association with the individual quality features. The association will be used for the subsequent analysis of the QoS taxonomy in Section 6.2.4.

Figure 6.2 shows a clear grouping of the part B questions into quality aspects or sub-aspects of the QoS taxonomy. A first cluster is formed by the questions related to the personal feeling/impression (B24 and B25). The second cluster addresses dialogue smoothness (B19 and B21), and partly also the the transparency of the interaction (B19, part of the cooperativity class). The

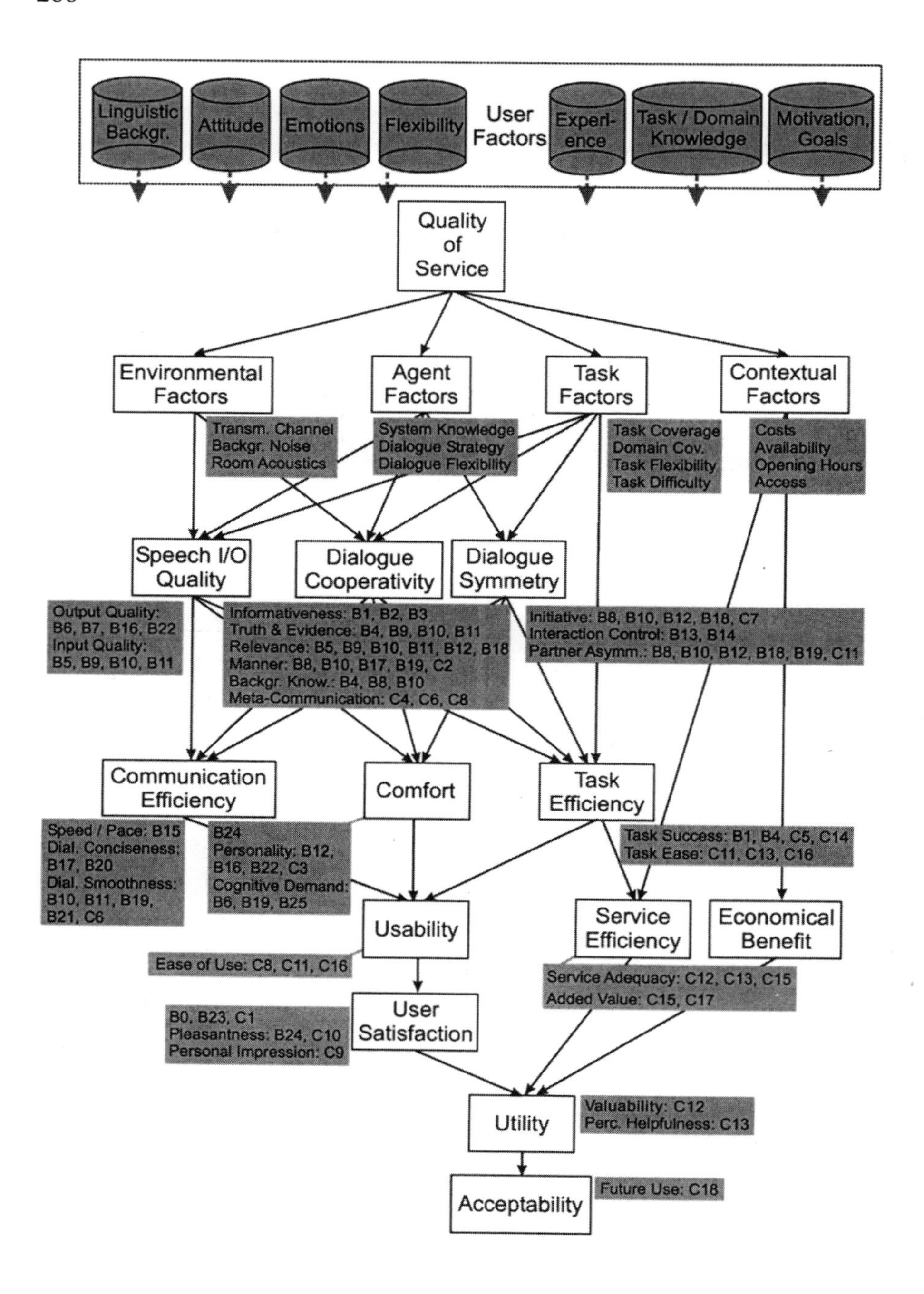

Figure 6.1. Assignment of part B and C questions of experiment 6.3 to the categories of the QoS taxonomy.

Table 6.5. Relationship between part B and C questions of experiment 6.3 and dialogue-related quality features.

Category	*Aspect*	*Quality features*
Dialogue cooperativity	Informativeness ← B1	– Accuracy / specificity of information ← B1 – Completeness of information ← B2 – Clarity of information ← B3 – Conciseness of information – System feedback adequacy
	Truth and evidence ← B4	– Credibility of information ← B4 – Consistency of information ← B4 – Reliability of information ← B4 – Perceived system reasoning ← B9, B10, B11
	Relevance	– System feedback adequacy – Perceived system understanding ← B5 – Perceived system reasoning ← B9, B10, B11 – Naturalness of interaction ← B12, B18
	Manner	– Clarity / non-ambiguity of expression ← B8, C2 – Consistency of expression ← B10 – Conciseness of expression ← B17 – Transparency of interaction ← B19 – Order of interaction ← B19 – Respect of natural information packages
	Background knowledge	– Congruence with user's task/domain knowl. ← B4 – Congruence with user experience ← B10 – Suitability of user adaptation – Inference adequacy – Interaction guidance ← B8
	Meta-comm. handling ← C8	– Repair handling adequacy ← C6 – Clarification handling adequacy ← C6 – Help capability ← C4 – Repetition capability
Dialogue symmetry	Initiative ← C7	– Interaction guidance ← B8 – Naturalness of interaction ← B10, B12, B18
	Interaction control ← B14	– Flexibility of interaction ← B13 – Perceived control capability ← B14 – Barge-in capability – Cancel capability
	Partner asymmetry	– Transparency of interaction ← B10, B19 – Transparency of task / domain coverage ← C11 – Interaction guidance ← B8 – Naturalness of interaction ← B10, B12, B18 – Respect of natural information packages

Table 6.6. Relationship between part B and C questions of experiment 6.3 and communication-, task- and service-related quality features.

Category	*Aspect*	*Quality features*
Speech I/O quality	Speech output quality	– Intelligibility ← B7 – Naturalness of speech ← B16, B22 – Listening-effort required from the user ← B6
	Speech input quality	– Perceived system understanding ← B5 – Perceived system reasoning ← B9, B10, B11
Communic. efficiency	Speed ← B15	– Perceived interaction pace – Perceived response time
	Conciseness ← B17, B20	– Perceived interaction length – Perceived interaction duration
	Smoothness ← B21	– Perceived system errors ← B11 – Repair handling adequacy ← C6 – Clarification handling adequacy ← C6 – Transparency of interaction ← B10, B19 – Congruence with user experience ← B10
Comfort ← B24	Agent personality	– Politeness ← C3 – Friendliness ← B16 – Naturalness of behavior ← B12, B22
	Cognitive demand	– Ease of communication ← B19 – Concentration required from the user ← B6 – Stress / fluster ← B25
Task efficiency	Task success ← B1	– Adequacy of task / domain coverage ← C5 – Validity of task results ← B4 – Precision of task results – Reliability of task results ← C14
	Task ease ← C16	– Perceived helpfulness ← C13 – Task guidance ← C13 – Transparency of task / domain coverage ← C11
Service efficiency	Service adequacy	– Access adequacy, availability – Modality adequacy ← C15 – Task adequacy ← C12 – Perceived service functionality ← C12, C13 – Perceived usefulness ← C13
	Added value	– Service improvement – Comparable interface ← C15, C17
Usability	Ease of use ← C8, C16	– Service operability ← C8 – Service understandability ← C11 – Service learnability ← C11
User satisfaction ← B23, B0, C1		– Pleasantness ← B24, C10 – Personal impression ← C9
Utility		– Valuability ← C12 – Perceived helpfulness ← C13
Acceptability		– Future use ← C18

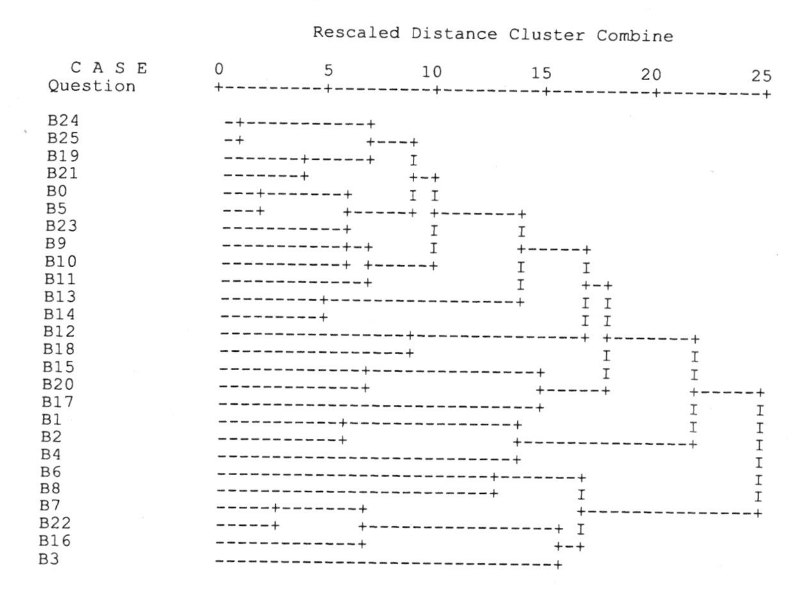

Figure 6.2. Hierarchical cluster analysis of part B question ratings in experiment 6.3. Dendrogram using average linkage between groups.

third cluster contains items of global (overall) quality importance: It consists of the overall impression when using the system (B0), the satisfaction with the interaction (B23), and the perceived system understanding (B5). The latter fact is somehow surprising, but it may be explained with the extremely high importance of this dimension for the overall flow of the interaction. The fourth cluster is formed by the system behavior questions B9, B10 and B11. It relates to the correctness and transparency of the system's interaction behavior, and also to the perceived reasoning capabilities of the system. Cluster 5 relates to the system's interaction flexibility (B13 and B14). The subsequent cluster can be subsumed under the term naturalness or comparison with human behavior (B12 and B18). Cluster 7, formed by B15 and B20, relates to the communication efficiency, and contains dialogue duration and speed. Question B17 (length of system utterances) is also related to this aspect. Cluster 8 is related to the system's informativeness (B1, B2 and B4; accuracy, completeness and truth of the provided information). Cluster 9 (B6 and B8) seems to be somehow related to the cognitive demand required from the user, and also includes the transparency of the system behavior, which seems to have a direct influence on this dimension. The next cluster contains questions B7, B16 and B22 which all relate to the system's output voice (dimensions intelligibility, friendliness and voice naturalness). The friendliness of the system thus seems to be highly related to its voice. The final dimension 'clarity of information' does not form a cluster with any of the other questions.

These clusters can now be interpreted in the QoS taxonomy. The 'personal impression' cluster is mainly related to comfort, the 'pleasantness' question (B24) to user satisfaction as well. Cluster 2 (dialogue smoothness, B19 and B21) forms one aspect of communication efficiency. The global quality aspects covered by questions B0 and B23 (Cluster 3) mainly relate to user satisfaction. The strong influence of the 'perceived system understanding' question (B5) on this dimension has already been noted. This question is however located in the speech input/output quality category of the QoS taxonomy. Cluster 4 is related to system behavior (B9, B10 and B11), and can be attributed to dialogue cooperativity, question B10 also to dialogue symmetry. The questions addressing interaction flexibility (B13 and B14) belong to the dialogue symmetry category. 'Naturalness' (B12 and B18) is once again related to both dialogue cooperativity and dialogue symmetry. These two categories cannot be clearly separated with respect to the user questions. Questions B15, B17 and B20 all reflect communication efficiency. Cluster 8, related to informativeness (B1, B2 and B4), is attributed to the dialogue cooperativity category. This is not true for Cluster 9 (B6 and B8): Whereas B8 is part of dialogue cooperativity, B6 fits best to the comfort category. Cluster 10 (B7, B16 and B22) is mainly related to the speech output quality category. However, question B16 also reflects the agent personality aspect, and thus the comfort category. The stand-alone question B3 is part of the dialogue cooperativity category.

A similar analysis can be used for the judgments on the part C questions of experiment 6.3, namely questions C1 to C18 (the rest of the questions have either free answer possibilities or are related to the user's expectations about what is important for the system). A hierarchical cluster analysis leads to the dendrogram which is shown in Figure 6.3.

Most clusters are related to the higher levels of the QoS taxonomy. The first cluster comprises C1, C9, C12, C13, C14 and C18: These questions are related to user satisfaction (overall impression, C1 and C9), the system's utility (C12, C13), task efficiency (reliability of task results, C14) and acceptability (C18). The second cluster (C8, C11) relates to the usability and the ease of using the system. Question C8 will also address the meta-communication handling capability. Cluster 3 (C2, C3) reflects the system personality (politeness, clarity of expression). Cluster 4 (C10, C16) is once again related to usability and user satisfaction (ease of use, degree of enjoyment). The fifth cluster captures the system's interaction capabilities (initiative and guidance; C4 and C7). Cluster 6 describes the system's task (task success, C5) and meta-communication (C6) capabilities. The final two questions (C15, C17) reflect the added value provided

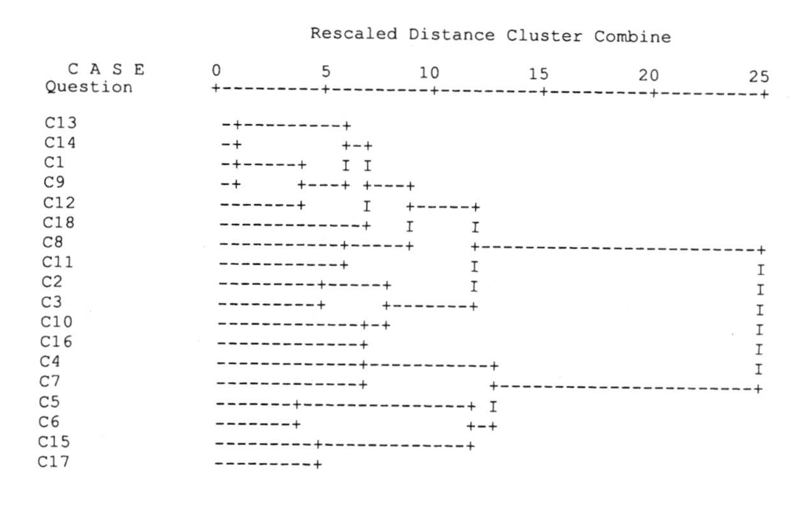

Figure 6.3. Hierarchical cluster analysis of part C question ratings in experiment 6.3. Dendrogram using average linkage between groups.

by the service, and are thus also related to the service efficiency category. Also the part C questions have been associated with the categories of the QoS taxonomy, see Figure 6.1 and Tables 6.5 and 6.6.

Similar to the factor analysis, the cluster analysis shows that many questions of part B and part C of the experiment 6.3 questionnaire group into categories which have been previously postulated by the QoS taxonomy. Part B questions can mainly be associated with the lower levels of the taxonomy, up to communication efficiency, comfort and, to some extent, task efficiency. On the other hand, part C questions mostly reflect the higher levels of the taxonomy, namely service efficiency, usability, utility and acceptability. User satisfaction is covered by both part B and part C questions. The relationship shown in Figure 6.1 will be used in Section 6.2.4 to identify subjective ratings which can be associated to specific quality aspects.

The results of multidimensional analyses give some indications on the relevance of individual quality aspects for the user, in that they show which dimensions of the perceptual space can be distinguished. The relevance may additionally be investigated by directly asking the users which characteristics of a system they rate as important or not important. This was done in Question 4 (4.1-4.15) of experiment 6.2, and Questions A8 and C22 of experiment 6.3. The data from experiment 6.2, which will be discussed here, have been ranked with respect to the number of ratings in the most positive category (n_5), and in case of equality to the accumulated positive answers to the statements (two categories close to the "agree" label, n_5 and n_4) minus the accumulated number

Table 6.7. Rank order of quality features judged to be important by the test subjects, experiment 6.2. Complete agreement corresponds to the number of ratings for n_5, accumulated agreement to the percentage of ratings for $(n_5 + n_4) - (n_1 + n_2)$.

Complete agreement (%)	*Accum. agreement (%)*	*Statement "It is important that ..."*	*Quality aspect*
83		... BoRIS always understands my utterances. (4.2)	Manner (speech understanding capability)
79		... BoRIS asks clear questions. (4.3)	Manner (comprehensibility, intelligibility)
74	94	... BoRIS is easy to handle. (4.5)	Transparency (ease of use)
74	93	... BoRIS provides adequate information to my request. (4.6)	Relevance and informativeness
53		... BoRIS repeats his questions in case I am lost. (4.1)	Meta-communication handling (repetition)
51		... it is easy to modify my input. (4.13)	Transparency (functional limits)
50		... I can continue the dialogue as long as I want, in order to be able to ask additional questions. (4.8)	Interaction control
36	72	... I can ask questions to BoRIS. (4.7)	Initiative
36	51	... I know from the beginning which type of information can be provided by BoRIS, and which not. (4.10)	Transparency (functional limits), congruence with user's background knowledge and user expectations
36	43	... I know from the beginning how I have to formulate by requests. (4.9)	Transparency (functional limits), congruence with user's background knowledge and user expectations
30	60	... the dialogue proceeds without interruptions. (4.14)	Speed and smoothness (dialogue interruptions)
30	51	... BoRIS provides a help capability. (4.15)	Transparency (help capability)
23	75	... BoRIS reacts quickly to my utterances. (4.4)	Speed and smoothness (processing speed)
23	43	... BoRIS confirms my input. (4.11)	Meta-communication handling (confirmation of user input)
13		... pauses following my input are filled by "Wait a moment, please...". (4.12)	Meta-communication handling (confirmation of pauses)

of negative answers (two categories close to the "disagree" label, n_1 and n_2). The resulting rank order is depicted in Table 6.7.

The rank order shows that manner, transparency and relevance, and partly also meta-communication handling and interaction control seem to be of major importance to the users. The result may be partly linked to the particularities of the BoRIS system (repetition capability, modification capability), but the three major aspects – manner, transparency and relevance – will be of general importance for other applications as well. They are all related to the basic communicative and functional capabilities of the system (service aspects have not been addressed by questions 4.1 to 4.15). The highest ranking is observed for the speech input and output capabilities, which is the basic requirement for the interaction with an SDS. The overall system quality seems to be largely affected by a relatively low intelligibility of the TTS speech output. Transparency subsumes the transparency of how to use the system, as well as its functional capabilities. This quality aspect seems to reflect whether the user knows what to say to the system at each step in the dialogue, in which format, as well as the system's navigation (modification, repetition and dialogue continuation) capabilities. It may result in discomfort and stress if the system is not transparent enough. Relevance can be defined on an utterance level (relevance of each utterance in the immediate dialogue context) or on a global information (task) level. In the qualitative interview, it turned out that the global information level seems to pose problems with the current BoRIS version, due, in part, to database problems, but also due to the low detail of information provided by the current system version.

The user's background knowledge and the level of experience play a role in the judgement of overall quality. The qualitative interview of experiment 6.2 shows that test subjects who had no specific idea about such a system rated it generally better than persons with a specific idea. In the questionnaire, high expectations resulted mainly in more positive quality judgments after using the system. This could clearly be observed for the judgments of the female test subjects.

6.2.3 Multidimensional Analysis of Interaction Parameters

Apart from the users' quality judgments, also the interaction parameters will be related to each other. Such relations – if they are known – can be used to define meaningful evaluation metrics, and to interpret the influences of individual system components. This section will give a brief overview about relationships which are reported in the literature and present the results of a factor and cluster analysis of the data collected in experiment 6.3. A deeper analysis with respect to the QoS taxonomy follows in the subsequent section.

A number of analyses report the obvious relationship between dialogue duration *DD* and turn-related parameters. For example, Polifroni et al. (1992) found out that the overall number of user queries correlates highly with *DD* ($r = 0.84$). The correlation between *DD* and the number of unanswered user queries was considerably lower ($r = 0.66$). The different problem-solving strategies applied in the case of misunderstandings probably have a significant impact on the duration of the interactions. Sikorski and Allen (1997) investigated the correlation between dialogue duration and recognition accuracy. The correlation turned out to be unexpectedly low ($r = 0.16$). The authors indicate three potential reasons for this finding:

- A robust parsing strategy, which makes it more important *which* words are correctly recognized than *how many*.
- Misunderstandings, i.e. the system taking an action based on erroneous understanding, seems to be more detrimental to task success than non-understanding, where both the system and the user are aware of the situation. A system which is robust in this respect (i.e. one that tries to form an interpretation even when there is low confidence in the input) can create a high variance in the effectiveness of an interaction, and thus in the length of the interaction.
- A certain amount of nondeterminism (random behavior) in the system implementation, which could not be compensated for by the small number of test subjects.

Thus, the dialogue strategy may be a determining factor of dialogue duration, although the number of turns remains an important predictor.

Several parameters indicate speech input performance on different levels. Gerbino et al. (1993) compared absolute figures for correctly understood sentences in a field test (30.4% correct, 21.3% failed, 39.7% incorrect) to the ones in a laboratory situation (72.2% correct, 11.3% failed, 16.5% incorrect). Obviously, the field test situation was considerably more difficult for the recognizer than a laboratory situation. For the field test situation, the figures can be compared to the recognition accuracy ($SA = 14.0\%$, $WA = 52.4\%$). It turns out that the understanding error rate is approximately in the middle of the word and sentence error rates.

The relation between ASR performance (WA) and speech understanding performance (CA) was also investigated by Boros et al. (1996). Both measures can differ considerably, because WA does not make a difference between functional words and filler words. Thus, perfect CA can be reached without perfect WA. On the other hand, CA may become lower than WA when words which are relevant for understanding are missing in the system's interpretation. Results from a test corpus recorded over the public telephone network however showed that WA and CA have a strong correlation, resulting in a nearly linear relationship between WA and CA. For the tested system, WA seems to be a good predictor for CA, as speech recognizer and parser collaborate smoothly. In general, it cannot however be guaranteed that an increase in ASR performance will always lead to better speech understanding capabilities. If new words are added to the ASR vocabulary, this could provoke a degradation of speech understanding performance. Investigations carried out at MIT (Polifroni et al., 1998) however showed that a decrease in word error (from 21.7% to 16.4%) also resulted in a decrease of sentence error (42.5% to 34.3%) and in speech understanding error (31.7% to 23.8%). All in all, relatively strong correlations between the ASR and speech understanding performance measures can be observed.

Speech recognition and speech understanding performance will also be related to task success. Rosset et al. (1999) illustrate the relationship between word error rate and task success for two system configurations which differ in terms of a rejection threshold for poorly recognized words. Implementation of such a threshold resulted in an increasing task success rate, especially for high word error rates. Transaction success is however not necessarily closely linked to speech understanding performance. Gerbino et al. (1993) report that their system had a task success rate of 79% with only 30.4% correctly understood sentences. Better predictors of task success may be found in the system-answer-related parameters. Goodine et al. (1992) compared the percentage of correctly resolved scenarios (as a measure of task success), the $AN{:}CO$ parameter, and $DARPA_s$. It turned out that $AN{:}CO$ was a good indicator of task success, but that the $DARPA_s$ parameter over-penalizes incorrect system answers.

During experiment 6.3, a more-or-less complete set of interaction parameters was collected. On this set, a factor analysis has been carried out, in the same way as was done for the quality judgments (principal component analysis with Varimax rotation and Kaiser normalization, missing values were replaced by means). The complete results will not be reproduced here due to space limitations; only a short summary will be given. 10 factors were extracted which accounted for 81.9% of the variance in the parameter data.

Factor 1 loads high on all speech-input related parameters (IC, UA, $\overline{NES}$, $\overline{WES}$, WER, WA, $\overline{NES_{iso}}$, $\overline{WES_{iso}}$, WER_{iso}, WA_{iso}) as well as on the parsing parameters ($PA{:}CO$ and $PA{:}FA$) and on κ_{conf}. Apparently, this factor is related to the speech input capabilities. Factor 2 loads high on the duration-related parameters DD, STD, SRD, # TURNS, $WPST$ and $WPUT$, and seems to be related to communication efficiency (additional loading on $PA{:}PA$). Factor 3 seems to be related to the system's meta-communication capabilities. It loads high on SCR, UCR, $CA{:}AP$, $CA{:}IA$, IR and $PA{:}FA$. Factor 4 is related to the system's answer capability. It has very high loadings on $AN{:}CO$, $AN{:}FA$, $DARPA_s$ and $DARPA_{me}$. Factor 5 reflects task success: Loadings are high for κ_{dia}, TS_{bin} and TS_{ord}. Interestingly, the per-configuration version of κ does not show a high loading. Apparently, the system configuration plays a significant role for determining task success. Factor 6 might be explained by the cognitive demand put on the user. It only shows high loadings on UTD and URD. The last four factors are difficult to interpret. They only show high loadings on one or two interaction parameters which are not obviously related.

```
                      Rescaled Distance Cluster Combine

  C A S E       0         5        10        15        20        25
Label           +---------+---------+---------+---------+---------+

# Sys. Err. M.   -+-----+
AN:PA            -+     +-------+
DARPA_me         -------+       +-------+
DD               ---+-----------+       I
# Turns          ---+                   I
PA:FA            -+                     I
SCR              -+---+                 I
UCR              -+   +---------+       +---------------+
CA:IA            -----+         I       I               I
NES_iso          -+-+           +---+   I               I
WES_iso          -+ +-+         I   I   I               I
WES              -+-+ +---+     I   I   I               I
WER_iso          -+   I   +-----+   +---+               I
WER              -----+   I         I                   +---------+
NES              ---------+         I                   I         I
# Barge-Ins      -------------------+                   I         I
STD              -+-+                                   I         I
SRD              -+ +-------+                           I         I
WPUT             ---+       +-----------+               I         I
# User Quest.    ---+       I           I               I         I
WPST             ---+-------+           +---------------+         I
PA:PA            ---+                   I                         I
AN:FA            -----------------------+                         I
TS_ord           -+-----------+                                   I
TS_bin           -+           +-----+                             I
AN:CO            -+-----+     I     I                             I
DARPA_s          -+     +-----+     +-----------------+           I
URD              -------+           I                 I           I
UTD              -------------------+                 I           I
WA_iso           ---+-+                               I           I
Kappa_conf       ---+ +-------+                       I           I
WA               -----+       I                       +-----------+
PA:CO            -+-+         +---+                   I
UA               -+ +-------+ I   I                   I
CA:AP            -+-+       +-+   +-------------+     I
IC               -+         I     I             I     I
IR               -----------+     I             +-----+
Kappa_dia        -----------------+             I
# Sys. Quest.    -------------------------------+
```

Figure 6.4. Hierarchical cluster analysis of interaction parameters in experiment 6.3. Dendrogram using average linkage between groups.

Links between interaction parameters can additionally be addressed by a hierarchical cluster analysis, as was performed for the subjective judgments. The resulting dendrogram is shown in Figure 6.4. The first cluster contains three parameters which are all related to meta-communication (system error messages, partially correct answers, and the DARPA error). The next cluster contains two parameters related to communication efficiency (*DD* and # TURNS). The third cluster relates once again to meta-communication, in particular to the correction capabilities (correction rates, inappropriate system utterances, and failed speech understanding). Cluster 4 contains 6 parameters related to speech recognition, and thus to the speech input quality of the system. The # BARGE-INS parameter seems to be independent of all other parameters.

The following cluster consists of 7 parameters which all seem to be related to communication efficiency: STD, SRD, $WPUT$, $WPST$ and # USER QUESTIONS all carry a direct impact on the dialogue length, and $PA{:}PA$ and $AN{:}FA$ will also contribute to lengthening of the dialogue due to subsequent clarification dialogues. The next cluster is somehow related to task efficiency. It contains the two task success measures TS_{bin} and TS_{ord}, and two parameters which reflect the number of correct system answers ($AN{:}CO$ and $DARPA_s$). The following two parameters (URD and UTD) do not form a cluster in a proper sense. They reflect the characteristics of the user, but cannot be interpreted with respect to their quality impact. The next 8 parameters all relate to speech input quality: The first group of three parameters addresses ASR performance, and the second group of five parameters addresses speech understanding performance. It is interesting to note that the κ_{conf} parameter forms a cluster with the word accuracy measures. This is an indication that the recognition rate seems to play an important role for task success, and that task success (as expressed by the κ coefficient) will depend on the target recognition rate of the system configuration under test. In the group of speech-understanding-related parameters, the $CA{:}AP$ parameter has to be noted. Apparently, appropriate system answers are related to the system's speech understanding capability. The final two parameters do not form any specific cluster. In particular, no clustering of κ_{dia} with the other task-success-related parameters can be observed.

Both cluster and factor analysis show that interaction parameters mostly address the lower level categories of the QoS taxonomy, namely speech input quality, dialogue cooperativity, communication efficiency, task efficiency, and comfort. This finding has to be placed in contrast to the higher level categories reflected in the dimensions of the user judgments, e.g. usability, service efficiency, user satisfaction and acceptability. Although individual questions (mainly part B questions) can be attributed to the lower level categories, the more wholistic user view of the service, discussed in Chapter 3, is confirmed here.

The finding may have some implications for the construction of prediction models for SDS-based services: If interaction parameters mainly address low-level categories and the user judges in high-level categories, then it might be

Table 6.8. Spearman correlation coefficients for questions and interaction parameters related to informativeness. # UQ: # USER QUESTIONS; D_s: $DARPA_s$; D_{me}: $DARPA_{me}$.

	B1	B2	B3	# UQ	D_s	D_{me}	$AN{:}CO$
B1	1.000	*0.759*	*0.503*	-0.047	*0.542*	*-0.539*	*0.542*
B2	*0.759*	1.000	*0.555*	-0.067	0.397	-0.411	0.397
B3	*0.503*	*0.555*	1.000	0.051	-0.061	0.072	-0.061
# UQ	-0.047	-0.067	0.051	1.000	-0.109	0.047	-0.109
D_s	*0.542*	0.397	-0.061	-0.109	1.000	*-0.990*	*1.000*
D_{me}	*-0.539*	-0.411	0.072	0.047	*-0.990*	1.000	*-0.990*
$AN{:}CO$	*0.542*	0.397	-0.061	-0.109	*1.000*	*-0.990*	1.000

difficult to predict global quality aspects perceived by the user from interaction parameters. Kamm et al. (1997a) already noted relatively weak correlations between users' perceptions of system quality and system performance metrics. It may be an indication that global quality aspects are not the right target to be predicted from interaction parameters, but that individual quality aspects are more adequate for this purpose. The idea will be further discussed in Section 6.3.

6.2.4 Analysis of the QoS Schematic

The factor and cluster analyses described in the previous two sections highlight the relationships amongst subjective quality judgments or interaction parameters. The extracted factors have been interpreted in the light of the QoS taxonomy introduced in Section 2.3.1, however without giving further justification for the classification it defines. In this section, the individual categories of the taxonomy will be initially addressed in isolation, showing the correlations between subjective judgments and interaction parameters. The findings will then be interpreted with respect to the prediction potential for global quality aspects like the ones addressed by questions B0 or C18.

A correlation analysis for the individual categories of the QoS taxonomy is described in the following discussion. As most of the parameters and subjective judgments do not show a gaussian distribution when accumulated over all system configurations, Spearman rank order correlation coefficients ρ have been chosen. The correlation tables contain all parameters and questions which have been attributed to a specific category (see Tables 6.5, 6.6, and Figure 6.1 for the subjective ratings, and Tables 3.1 and 3.2 for interaction parameters), as well as all additional parameters which show a correlation $\rho > 0.5$ with one of the associated questions. Correlations which are significant ($p < 0.05$) are given in italics.

Table 6.9. Spearman correlation coefficients for questions and interaction parameters related to truth and evidence. D_s: $DARPA_s$; D_{me}: $DARPA_{me}$.

	B4	B9	B10	B11	$AN{:}CO$	$AN{:}PA$	$AN{:}FA$	D_s	D_{me}
B4	1.000	*0.555*	*0.452*	*0.513*	0.079	0.205	-0.206	0.079	-0.052
B9	*0.555*	1.000	*0.630*	*0.635*	0.338	-0.164	-0.275	0.338	-0.347
B10	*0.452*	*0.630*	1.000	*0.668*	0.223	-0.118	-0.171	0.223	-0.226
B11	*0.513*	*0.635*	*0.668*	1.000	0.445	-0.305	-0.311	0.445	-0.446
$AN{:}CO$	0.079	0.338	0.223	0.445	1.000	-0.327	*-0.876*	*1.000*	*-0.990*
$AN{:}PA$	0.205	-0.164	-0.118	-0.305	-0.327	1.000	-0.170	-0.327	0.458
$AN{:}FA$	-0.206	-0.275	-0.171	-0.311	*-0.876*	-0.170	1.000	*-0.876*	*0.798*
D_s	0.079	0.338	0.223	0.445	*1.000*	-0.327	*-0.876*	1.000	*-0.990*
D_{me}	-0.052	-0.347	-0.226	-0.466	*-0.990*	0.458	*0.798*	*-0.990*	1.000

Informativeness:

The relevant questions and parameters are listed in Table 6.8. High correlations are observed between questions B1 and B2, as well as between $DARPA_s$, $DARPA_{me}$ and $AN{:}CO$. Apparently, the accuracy and the completeness of the provided information are not easy to distinguish for the test subjects. Correlation between questions and parameters is very low, with the exception of B1 which moderately correlates with $DARPA_s$, $DARPA_{me}$ and $AN{:}CO$. These parameters are however only calculated for 18 dialogues, and the correlations should be interpreted with care. # USER QUESTIONS is not correlated with any other question or parameter of the list. This corresponds to the wizard's observation that most users were asking questions in order to assess the system functionality, and not with respect to the restaurant information provided by the system.

Truth and evidence:

Four questions and five parameters are related to this aspect, see Table 6.9. All questions correlate moderately ($\rho = 0.45...0.67$). However, only question B11 also shows some (moderate) correlation to the relevant parameters. The generally low correlations may be an indication that the perception of truth by the test subjects does not necessarily require system answers to be correct from an external point of view. In fact, the test subjects have no possibility to verify the correctness of information provided by the system, except when the system gives explicit feedback on misunderstood items. The high correlations between $DARPA_s$, $DARPA_{me}$ and $AN{:}CO$ have already been noted. Also $AN{:}FA$ shows high correlations to these parameters.

Relevance:

Relevance is an aspect which is only indirectly covered in the user judgments, namely via questions related to perceived system understanding (B5),

Table 6.10. Spearman correlation coefficients for questions and interaction parameters related to relevance. # BI: # BARGE-INS.

	B5	B9	B10	B11	B12	B18	# BI
B5	1.000	*0.585*	*0.593*	*0.619*	*0.458*	*0.221*	-0.034
B9	*0.585*	1.000	*0.630*	*0.635*	*0.297*	*0.182*	-0.122
B10	*0.593*	*0.630*	1.000	*0.668*	*0.371*	*0.234*	-0.078
B11	*0.619*	*0.635*	*0.668*	1.000	*0.302*	0.130	0.053
B12	*0.458*	*0.297*	*0.371*	*0.302*	1.000	*0.583*	0.045
B18	*0.221*	*0.182*	*0.234*	0.130	*0.583*	1.000	0.083
# BI	-0.034	-0.122	-0.078	0.053	0.045	0.083	1.000

Table 6.11. Spearman correlation coefficients for questions and interaction parameters related to manner.

	B8	B10	B17	B19	C2	# TURNS	*WPST*
B8	1.000	*0.276*	0.043	*0.552*	0.070	*-0.305*	*0.318*
B10	*0.276*	1.000	0.119	*0.524*	0.138	*-0.332*	*0.312*
B17	0.043	0.119	1.000	0.102	-0.124	*-0.327*	*0.324*
B19	*0.552*	*0.524*	0.102	1.000	0.088	*-0.316*	*0.295*
C2	0.070	0.138	-0.124	0.088	1.000	-0.017	0.006
# TURNS	*-0.305*	*-0.332*	*-0.327*	*-0.316*	-0.017	1.000	*-0.874*
WPST	*0.318*	*0.312*	*0.324*	*0.295*	0.006	*-0.874*	1.000

perceived system reasoning (B9, B10 and B11), and to the naturalness of the interaction (B12, B18). Only the # BARGE-INS parameter may address this aspect. Correlations between B5, B9, B10 and B11 on the one hand, and B12 and B18 on the other, are moderately high ($\rho = 0.58...0.67$). The number of barge-ins does not correlate with any of the questions, which may however be due to the fact that this parameter is only in rare cases different from zero.

Manner:

Table 6.11 shows correlations between five questions (B8, B10, B17, B19 and C2) and two parameters (# TURNS, $WPST$) related to the manner of expression. Both interaction parameters highly correlate, but they only show weak to moderate correlations to the questions. Question C2 does not show any correlation with the part B questions. A factor analysis of all questions and

Table 6.12. Factor analysis (principal component analysis with Varimax rotation and Kaiser normalization) of the questions and parameters related to manner. Rotated component matrix.

	Component	
	1	2
# TURNS	-0.321	-0.706
WPST	0.371	0.705
B8	0.711	0.031
B10	0.728	0.213
B17	-0.099	0.768
B19	0.812	0.168
C2	0.424	-0.395

Table 6.13. Spearman correlation coefficients for questions and interaction parameters related to background knowledge. # BI: # BARGE-INS.

	B4	B8	B10	# BI
B4	1.000	*0.382*	*0.452*	0.033
B8	*0.382*	1.000	*0.276*	-0.101
B10	*0.452*	*0.276*	1.000	-0.078
# BI	0.033	-0.101	-0.078	1.000

parameters related to manner has been carried out, see Table 6.12. It reveals two factors explaining 56.2% of the variance: Factor 1 loading high on B8, B10 and B19, and tentatively labelled "transparency of the interaction", and Factor 2 loading high on B17, # TURNS and *WPST*, labelled "system utterance length". The manner aspect seems to cover at least these two dimensions.

Background knowledge:

Although Table 3.1 indicates four interaction parameters related to the background knowledge aspect, only the # BARGE-INS parameter can be used for the analysis, see the discussion in Section 6.2.1. In addition, questions B4, B8 and B10 address this aspect. No remarkable correlation can be observed, see Table 6.13. The questions indicate that background knowledge covers both the knowledge related to the task and to the interaction behavior.

Meta-communication handling:

Meta-communication is addressed by questions C4, C6, C8, and the interaction parameters # SYSTEM ERROR MESSAGES, *SCR*, and *IR* (the param-

Table 6.14. Spearman correlation coefficients for questions and interaction parameters related to meta-communication handling. # SEM: # SYSTEM ERROR MESSAGES.

	C4	C6	C8	# SEM	SCR	IR
C4	1.000	*0.516*	*0.316*	0.054	-0.064	-0.006
C6	*0.516*	1.000	*0.451*	-0.092	0.021	0.017
C8	*0.316*	*0.451*	1.000	-0.030	0.015	-0.051
# SEM	0.054	-0.092	-0.030	1.000	*0.144*	-0.175
SCR	-0.064	0.021	0.015	*0.144*	1.000	*-0.635*
IR	-0.006	0.017	-0.051	-0.175	-0.635	1.000

eters # HELP REQUESTS and # CANCEL ATTEMPTS being excluded from the analysis). Whereas the correlations between the questions are moderate, the interaction parameters do not correlate well with any of the questions. This finding might be explained by the fact that the questions are rated after the whole test session, whereas the interaction parameters are determined for each dialogue.

Dialogue cooperativity:

The dialogue cooperativity category covers all aspects analyzed so far. It may now be interesting to see which dimensions are relevant for this category, and in how far the mentioned aspects are reflected in the dimensions. Fortunately, the number of appropriate system utterances $CA{:}AP$ is, by definition, a direct measure of dialogue cooperativity. Thus, an analysis of covariance with this parameter as the dependent variable may indicate the main contributing factors to cooperativity. The result of this analysis is depicted in Figure 6.5.

Apparently, only questions B2 and B5 carry a significant influence on $CA{:}AP$, and B11 is close to the significance level. These three questions refer to different aspects of cooperativity: Whereas B2 is directly linked to the system's informativeness, B5 describes the perceived system understanding. The latter aspect is mainly attributed to the speech input/ooutput quality category, but also reflects the relevance of system messages (category cooperativity). Question B11 refers to the errors made by the system. It is related to the relevance of system messages, but in addition it depends on the background knowledge of the user, and results in meta-communication necessary for a clarification. Thus, at least the four aspects informativeness, relevance, background knowledge and meta-communication handling carry a significant contribution to dialogue cooperativity defined by the $CA{:}AP$ measure. The truth and evidence aspect may be under-estimated in the test situation. Users do not feel in a realistic situation and cannot verify the given information. It is however astonishing

Test of between-subject effects

Dependent variable: CA:AP

Source	Square sum of type III	df	Mean of squares	F	Significance
Corrected model	1.484[a]	17	0.087	3.556	0.000
Constant	1.451	1	1.451	59.108	0.000
B1	0.042	1	0.042	1.705	0.193
B2	0.116	1	0.116	4.730	0.031
B3	0.027	1	0.027	1.092	0.298
B4	0.030	1	0.030	1.241	0.267
B5	0.140	1	0.140	5.718	0.018
B8	0.013	1	0.013	0.542	0.463
B9	0.008	1	0.008	0.334	0.564
B10	0.002	1	0.002	0.079	0.778
B11	0.085	1	0.085	3.457	0.065
B12	0.007	1	0.007	0.271	0.603
B17	0.001	1	0.001	0.025	0.874
B18	0.003	1	0.003	0.115	0.735
B19	0.015	1	0.015	0.619	0.432
C2	0.069	1	0.069	2.830	0.094
C4	0.026	1	0.026	1.040	0.309
C6	0.001	1	0.001	0.030	0.862
C8	0.016	1	0.016	0.658	0.418
Error	4.199	171	0.025		
Total	126.291	189			
Corrected total variation	5.683	188			

a. R-square = 0.261 (corrected R-square = 0.188)

Figure 6.5. Univariate analysis of covariance for dialogue cooperativity. Covariate factors are part B and C questions.

that none of the manner-related questions shows a significant contribution to cooperativity. It may be the case that it is difficult for the test subjects to distinguish between the content-related manner aspect and the form-related speech input/output quality category.

A correlation analysis (which is not reproduced here) shows how $CA{:}AP$ is related to the questions and interaction parameters belonging to the individual quality aspects. High correlation levels are only obtained for SCR ($\rho = -0.754$), IR ($\rho = 0.676$), and obviously for $CA{:}IA$ ($\rho = -0.979$) which is the inverse measure. Apparently, the cooperativity of system answers is largely dependent on the system's correction and recovery strategies. This finding will have a general validity for SDSs with limited speech recognition, understanding and reasoning capabilities.

Initiative:

Question C7 directly addresses the initiative experienced by the user, and questions B8, B10, B12 and B18 describe the system behavior with respect to the user's expectations and the human background. Correlations between part B questions are relatively low, except for B12 and B18 (both related to the naturalness of the interaction). C7 does not correlate with any other question or interaction parameter. The highest correlations between questions and in-

Table 6.15. Spearman correlation coefficients for questions and interaction parameters related to initiative (1). # SQ: # SYSTEM QUESTIONS; # UQ: # USER QUESTIONS.

	B8	B10	B12	B18	C7	# TURNS
B8	1.000	*0.276*	0.136	*0.151*	-0.015	*-0.305*
B10	*0.276*	1.000	*0.371*	*0.234*	0.032	*-0.332*
B12	0.136	*0.371*	1.000	*0.583*	-0.093	-0.112
B18	*0.151*	*0.234*	*0.583*	1.000	*-0.291*	-0.059
C7	-0.015	0.032	-0.093	*-0.291*	1.000	*0.192*
# TURNS	*-0.305*	*-0.332*	-0.112	-0.059	*0.192*	1.000
$WPST$	*0.318*	*0.312*	0.065	-0.059	-0.093	*-0.874*
$WPUT$	*0.344*	*0.266*	0.102	0.138	*-0.271*	*-0.662*
# SQ	-0.039	0.064	*0.156*	*0.187*	-0.061	0.021
# UQ	-0.070	0.025	-0.048	-0.002	-0.059	-0.057
SCR	-0.080	-0.155	-0.088	0.044	-0.003	*0.462*
UCR	-0.078	*-0.142*	-0.109	0.026	0.011	*0.400*

Table 6.16. Spearman correlation coefficients for questions and interaction parameters related to initiative (2). # SQ: # SYSTEM QUESTIONS; # UQ: # USER QUESTIONS.

	$WPST$	$WPUT$	# SC	# UQ	SCR	UCR
B8	*0.318*	*0.344*	-0.039	-0.070	-0.080	-0.078
B10	*0.312*	*0.266*	0.064	0.025	*-0.155*	*-0.142*
B12	0.065	0.102	*0.156*	-0.048	-0.088	-0.109
B18	-0.059	0.138	*0.187*	-0.002	0.044	0.026
C7	-0.093	*-0.271*	0.061	-0.059	-0.003	0.011
# TURNS	*-0.874*	*-0.662*	0.021	-0.057	*0.462*	*0.400*
$WPST$	1.000	*0.602*	*-0.319*	-0.020	*-0.413*	*-0.370*
$WPUT$	*0.602*	1.000	-0.099	*0.153*	-0.018	0.033
# SQ	*-0.319*	-0.099	1.000	0.131	0.135	-0.112
# UQ	-0.020	*0.153*	-0.131	1.000	-0.070	-0.038
SCR	*-0.413*	-0.018	-0.135	-0.070	1.000	*0.926*
UCR	*-0.370*	0.033	-0.112	-0.038	*0.926*	1.000

teraction parameters are observed between B8, B10, # TURNS, $WPST$ and $WPUT$, but they are still very limited. The mentioned parameters are moderately correlated with each other, but with the exceptions of SCR and UCR no other correlations larger than 0.5 are obtained. The correlation between

Table 6.17. Spearman correlation coefficients for questions and interaction parameters related to interaction control. # BI: # BARGE-INS; D_s: $DARPA_s$; D_{me}: $DARPA_{me}$.

	B13	B14	# BI	UCR	$AN{:}CO$	D_s	D_{me}
B13	1.000	*0.677*	-0.054	*-0.184*	0.378	0.378	-0.410
B14	*0.677*	1.000	*-0.185*	-0.127	*0.573*	*0.573*	*-0.582*
# BI	-0.054	*-0.185*	1.000	-0.019	-0.120	-0.120	0.098
UCR	*-0.184*	-0.127	-0.019	1.000	0.013	0.013	0.056
$AN{:}CO$	0.378	*0.573*	-0.120	0.013	1.000	*1.000*	*-0.990*
D_s	0.378	*0.573*	-0.120	0.013	*1.000*	1.000	*-0.990*
D_{me}	-0.410	*-0.582*	0.098	0.056	*-0.990*	*-0.990*	1.000

TURNS and $WPST$ indicates that a talkative system seems to provoke more system and user turns, and also more talkative users (correlation with $WPUT$). The correlation between SCR and UCR can be explained by the way these variables are coded, see Appendix D.3.

Interaction control:

Questions B13 and B14 relate to this aspect, as well as the # BARGE-INS and UCR parameters (the other parameters of Table 3.1 have not been included in the analysis, see Section 6.2.1). The three parameters $AN{:}CO$, $DARPA_s$ and $DARPA_{me}$ have been added because of their moderate correlation with question B14. No obvious reason for this correlation can be found, but these parameters could only be calculated for 18 dialogues, and the results should consequently be interpreted with care. # BARGE-INS and UCR do not correlate with any of the interaction-control-related questions. Only between the questions a moderate correlation of $\rho = 0.68$ can be observed.

Partner asymmetry:

A number of questions relate to this aspect, namely B8, B10, B12, B18, B19 and C11, but only one interaction parameter (# BARGE-INS). Moderate correlations are observed between B8/B10 and B19, which are all related to the transparency of the dialogue, and between B12 and B18 which are related to the naturalness. These two dimensions seem to contribute to the partner asymmetry aspect. Question C11 relates to the functional capabilities of the system. Only low correlations are found for this question.

Speech output quality:

It has already been noted that no interaction parameters are known which relate to speech output quality, see Section 3.8.5. Thus, this aspect has to be investigated via subjective ratings only, namely the ones in questions B6, B7, B16 and B22. As Table 6.19 shows, the correlations are only moderate or low. This is an indication that the questions address different dimensions of

Table 6.18. Spearman correlation coefficients for questions and interaction parameters related to partner asymmetry. # BI: # BARGE-INS.

	B8	B10	B12	B18	B19	C11	# BI
B8	1.000	*0.276*	0.136	*0.151*	*0.552*	*0.315*	-0.101
B10	*0.276*	1.000	*0.371*	*0.234*	*0.524*	0.099	-0.078
B12	0.136	*0.371*	1.000	*0.583*	*0.317*	0.108	0.045
B18	*0.151*	*0.234*	*0.583*	1.000	*0.301*	0.090	0.083
B19	*0.552*	*0.524*	*0.317*	*0.301*	1.000	*0.266*	-0.046
C11	*0.315*	0.099	0.108	0.090	*0.266*	1.000	*-0.211*
# BI	-0.101	-0.078	0.045	0.083	-0.046	*-0.211*	1.000

Table 6.19. Spearman correlation coefficients for questions related to speech output quality.

	B6	B7	B16	B22
B6	1.000	*0.459*	*0.276*	*0.323*
B7	*0.459*	1.000	*0.468*	*0.550*
B16	*0.276*	*0.468*	1.000	*0.567*
B22	*0.323*	*0.550*	*0.567*	1.000

speech output quality which are independently perceivable by the test subjects. Moderate correlations are observed between B6 and B7 (listening-effort and intelligibility), and between B7, B16 and B22 (intelligibility, friendliness and naturalness). Nevertheless, it is justifiable to collect judgments on all those questions in order to better capture different speech output quality dimensions.

Speech input quality:

This aspect is addressed by a large number of interaction parameters, and by questions which relate to the perceived system understanding (B5), and those related to the perceived system reasoning (B9, B10 and B11). The correlations between the two perceptive dimensions are all moderate ($\rho = 0.59...0.66$), indicating that they are somehow related. Interestingly, the correlations between questions and interaction parameters are all very low; the highest values are observed for the *PA*:*FA* parameter ($\rho \leq 0.39$). Apparently, the perceived system understanding and reasoning is not well reflected in speech recognition or understanding performance measures. This finding is in agreement with the one made by Kamm et al. (1997a), with the correlation coefficients in the same order of magnitude.

Table 6.20. Spearman correlation coefficients for questions and interaction parameters related to speech input quality (1). # SEM: # SYSTEM ERROR MESSAGES.

	B5	B9	B10	B11	$PA{:}CO$	$PA{:}PA$	$PA{:}FA$	IC	UA
B5	1.000	*0.585*	*0.593*	*0.619*	-0.023	0.117	*-0.360*	*0.227*	0.100
B9	*0.585*	1.000	*0.630*	*0.635*	-0.074	0.078	*-0.246*	*0.185*	0.061
B10	*0.593*	*0.630*	1.000	*0.668*	-0.078	0.129	*-0.237*	*0.145*	0.039
B11	*0.619*	*0.635*	*0.668*	1.000	-0.022	0.079	*-0.387*	*0.281*	*0.156*
$PA{:}CO$	-0.023	-0.074	-0.078	-0.022	1.000	*-0.722*	*-0.376*	*0.741*	*0.865*
$PA{:}PA$	0.117	0.078	0.129	0.079	*-0.722*	1.000	0.067	*-0.586*	*-0.780*
$PA{:}FA$	*-0.360*	*-0.246*	*-0.237*	*-0.387*	*-0.376*	0.067	1.000	*-0.747*	*-0.556*
IC	*0.227*	*0.185*	*0.145*	*0.281*	*0.741*	*-0.586*	*-0.747*	1.000	*0.909*
UA	0.100	0.061	0.039	*0.156*	*0.865*	*-0.780*	*-0.556*	*0.909*	1.000
# SEM	-0.125	*-0.147*	*-0.163*	-0.108	0.035	-0.068	*0.170*	-0.124	-0.046
WA	*0.188*	0.112	0.064	*0.214*	*0.543*	*-0.386*	*-0.479*	*0.689*	*0.611*
WER	*-0.161*	-0.085	-0.038	*-0.189*	*-0.518*	*0.400*	*0.494*	*-0.689*	*-0.586*
$\overline{NES}$	-0.003	0.061	0.080	0.024	*-0.596*	*0.549*	*0.205*	*-0.522*	*-0.545*
$\overline{WES}$	*-0.181*	-0.109	-0.059	*-0.209*	*-0.528*	*0.338*	*0.549*	*-0.690*	*-0.576*
WA_{iso}	*0.230*	0.109	0.096	*0.235*	*0.563*	*-0.410*	*-0.601*	*0.792*	*0.706*
WER_{iso}	*-0.207*	-0.085	-0.072	*-0.213*	*-0.542*	*0.425*	*0.616*	*-0.792*	*-0.685*
$\overline{NES}_{iso}$	*-0.156*	-0.021	-0.030	-0.137	*-0.575*	*0.508*	*0.523*	*-0.762*	*-0.701*
$\overline{WES}_{iso}$	*-0.207*	-0.095	-0.070	*-0.216*	*-0.521*	*0.397*	*0.623*	*-0.776*	*-0.669*

There are however strong correlations between the interaction parameters. Very close relationships are found between WA, WER, $\overline{NES}$ and $\overline{WES}$, both for the continuous as well as for the isolated ASR measures. The relationships between the corresponding continuous and isolated measures are in the area of $\rho = 0.66...0.81$. On the speech understanding level, strong correlations are observed between IC and UA, and moderate correlations also to the parsing-related parameters. # SYSTEM ERROR MESSAGES is not correlated with any of the other selected parameters. For future investigations, the number of interaction parameters addressing the speech input aspect could be reduced, e.g. to the four parameters WER or WA (either continuous or isolated speech recognition), # SYSTEM ERROR MESSAGES, a parsing-related parameter, and either IC or UA. With this reduced set, the main characteristics of speech recognition and speech understanding can be captured.

Speed:

This aspect is addressed by question B15, as well as by STD, UTD, SRD, URD, and # BARGE-INS. Correlations between B15 and interaction parameters are all very low, see Table 6.22. Moderate correlations are found between UTD, SRD and URD, and also between SRD and STD. The relationship between UTD and SRD can be explained by the “processing time” needed by

Table 6.21. Spearman correlation coefficients for questions and interaction parameters related to speech input quality (2). # SEM: # SYSTEM ERROR MESSAGES.

	# SEM	WA	WER	$\overline{NES}$	$\overline{WES}$	WA_{iso}	WER_{iso}	$\overline{NES}_{iso}$	$\overline{WES}_{iso}$
B5	-0.125	0.188	*-0.161*	-0.003	*-0.182*	*0.230*	*-0.207*	*-0.156*	*-0.207*
B9	*-0.147*	0.112	-0.085	0.061	-0.109	0.109	-0.085	-0.021	-0.095
B10	*-0.163*	0.064	-0.038	0.080	-0.059	0.096	-0.072	-0.030	-0.070
B11	-0.108	*0.214*	*-0.189*	0.024	*-0.209*	*0.235*	*-0.213*	-0.137	*-0.216*
$PA{:}CO$	0.035	*0.543*	*-0.518*	*-0.596*	*-0.528*	*0.563*	*-0.542*	*-0.575*	*-0.521*
$PA{:}PA$	-0.068	*-0.386*	*0.400*	*0.549*	*0.338*	*-0.410*	*0.425*	*0.508*	*0.397*
$PA{:}FA$	*0.170*	*-0.479*	*0.494*	*0.205*	*0.549*	*-0.601*	*0.616*	*0.523*	*0.632*
IC	-0.124	*0.689*	*-0.689*	*-0.522*	*-0.690*	*0.792*	*-0.792*	*-0.762*	*-0.776*
UA	-0.046	*0.611*	*-0.586*	*-0.545*	*-0.576*	*0.706*	*-0.685*	*-0.701*	*-0.669*
# SEM	1.000	*-0.159*	*0.165*	0.065	*0.167*	*-0.181*	*0.188*	*0.172*	*0.202*
WA	*-0.159*	1.000	*-0.976*	*-0.834*	*-0.941*	*0.797*	*-0.777*	*-0.746*	*-0.770*
WER	*0.165*	*-0.976*	1.000	*0.858*	*0.965*	*-0.771*	*0.795*	*0.764*	*0.788*
$\overline{NES}$	0.065	*-0.834*	*0.858*	1.000	*0.831*	*-0.592*	*0.615*	*0.656*	*0.602*
$\overline{WES}$	*0.167*	*-0.941*	*0.965*	*0.831*	1.000	*-0.770*	*0.794*	*0.754*	*0.805*
WA_{iso}	*-0.181*	*0.797*	*-0.771*	*-0.592*	*-0.770*	1.000	*-0.980*	*-0.956*	*-0.966*
WER_{iso}	*0.188*	*-0.777*	0.795	*0.615*	*0.794*	*-0.980*	1.000	*0.976*	*0.986*
$\overline{NES}_{iso}$	*0.172*	*-0.746*	*0.764*	*0.656*	*0.754*	*-0.956*	*0.976*	1.000	*0.963*
$\overline{WES}_{iso}$	0.202	*-0.770*	*0.788*	*0.602*	*0.805*	*-0.966*	*0.986*	*0.963*	1.000

Table 6.22. Spearman correlation coefficients for questions and interaction parameters related to speed. # BI: # BARGE-INS.

	B15	STD	UTD	SRD	URD	# BI
B15	1.000	*0.169*	-0.017	-0.036	*-0.184*	*-0.141*
STD	*0.169*	1.000	*0.247*	*0.632*	*-0.281*	0.055
UTD	-0.017	*0.247*	1.000	*0.537*	*0.552*	-0.006
SRD	-0.036	*0.632*	*0.537*	1.000	0.097	-0.029
URD	*-0.184*	*-0.281*	*0.552*	0.097	1.000	-0.004
# BI	*-0.141*	0.055	-0.006	-0.029	-0.004	1.000

the wizard to transcribe the user utterances. SRD and URD may be related because a quickly responding system may also invite the user to respond quickly. For the other relations, no obvious explanation has been found. As has been observed in the other analyses, the # BARGE-INS parameter does not correlate with any of the other entities.

Table 6.23. Spearman correlation coefficients for questions and interaction parameters related to conciseness.

	B17	B20	*DD*	# TURNS	*WPST*	*WPUT*
B17	1.000	*0.331*	*-0.247*	*-0.327*	*0.324*	*0.366*
B20	*0.331*	1.000	*-0.498*	*-0.493*	*0.381*	*0.261*
DD	*-0.247*	*-0.498*	1.000	*0.950*	*-0.775*	*-0.483*
# TURNS	*-0.327*	*-0.493*	*0.950*	1.000	*-0.874*	*-0.662*
WPST	*0.324*	*0.381*	*-0.775*	*-0.874*	1.000	*0.602*
WPUT	*0.366*	*0.261*	*-0.483*	*-0.662*	*0.602*	1.000

Table 6.24. Spearman correlation coefficients for questions and interaction parameters related to dialogue smoothness. # SEM: # SYSTEM ERROR MESSAGES; # BI: # BARGE-INS.

	B10	B11	B19	B21	C6	# SEM	# BI	*UCR*	*SCR*
B10	1.000	*0.668*	*0.524*	*0.536*	*0.200*	*-0.163*	-0.078	*-0.142*	*-0.155*
B11	*0.668*	1.000	*0.630*	*0.650*	*0.319*	-0.108	0.053	*-0.224*	*-0.250*
B19	*0.524*	*0.630*	1.000	*0.739*	*0.260*	-0.129	-0.046	-0.104	-0.108
B21	*0.536*	*0.650*	*0.739*	1.000	*0.251*	-0.126	-0.030	*-0.147*	*-0.146*
C6	*0.200*	*0.319*	*0.260*	*0.251*	1.000	-0.092	*-0.301*	0.058	0.021
# SEM	*-0.163*	-0.108	-0.129	-0.126	-0.092	1.000	-0.023	*0.143*	*0.144*
# BI	-0.078	0.053	-0.046	-0.029	*-0.301*	-0.023	1.000	-0.019	-0.052
UCR	*-0.142*	*-0.224*	*-0.104*	*-0.147*	0.058	*0.143*	-0.019	1.000	*0.926*
SCR	*-0.155*	*-0.250*	-0.108	*-0.146*	0.021	*0.144*	-0.052	*0.926*	1.000

Conciseness:

The dialogue conciseness is addressed by questions B17 and B20, as well as by four interaction parameters. Only B20 is moderately correlated to *DD* and # TURNS, but B17 does not show any high correlation to the interaction parameters. This result is astonishing, because one would expect at least a correlation with *WPST*. Apparently, the length of system utterances is not directly reflected in the user's perception. A reason might be that system utterances which are interesting and new to the subjects are not perceived as lengthy. Among the interaction parameters, a high correlation is observed between the *DD* and # TURNS, and a slightly lower value between *DD*, # TURNS and *WPST*. It seems to be sufficient to extract either *DD* or the # TURNS parameter in future experiments; however, the first one has the advantage of being extracted fully instrumentally, and the latter is needed for normalization of other interaction parameters.

Table 6.25. Spearman correlation coefficients for questions related to agent personality.

	B12	B16	B22	C3
B12	1.000	*0.256*	*0.390*	*0.218*
B16	*0.256*	1.000	*0.567*	*0.377*
B22	*0.390*	*0.567*	1.000	*0.307*
C3	*0.218*	*0.377*	*0.307*	1.000

Table 6.26. Spearman correlation coefficients for questions and interaction parameters related to cognitive demand.

	B6	B19	B25	URD
B6	1.000	*0.374*	*0.403*	-0.018
B19	*0.374*	1.000	*0.535*	-0.052
B25	*0.403*	*0.535*	1.000	-0.131
URD	-0.018	-0.052	-0.131	1.000

Dialogue smoothness:
The correlations are given in Table 6.24. Whereas the part B questions all show moderate correlations to each other ($\rho = 0.52...0.74$), question C6 does not show meaningful correlations to any other question or parameter of the set. Once again, correlations between questions and interaction parameters are very low, and only between UCR and SCR can a close relationship be observed (because these parameters are related by definition, see Appendix D.3).
Agent personality:
This aspect is only addressed by subjective ratings. No specifically high correlation between the questions is noted. The only correlation value $\rho > 0.5$ is between B16 and B22, indicating that the perceived friendliness of the system is linked to its voice.
Cognitive demand:
Questions B6, B19 and B25 are related to the cognitive demand required from the user, and the parameter URD, see Table 6.26. Only the questions show moderate correlations to each other. URD is nearly independent of the questions. Apparently, it is not a good predictor for the cognitive demand or stress perceived by the user.

Test of between-subject effects

Dependent Variable: B24

Source	Square sum of type III	df	Mean of squares	F	Significance
Corrected model	215.328[a]	7	30.761	71.671	0.000
Constant	7.119	1	7.119	16.588	0.000
B6	0.871	1	0.870	2.029	0.156
B12	2.056	1	2.056	4.790	0.030
B16	4.382	1	4.382	10.209	0.002
B19	2.431	1	2.431	5.665	0.018
B22	4.659	1	4.659	10.856	0.001
B25	52.599	1	52.599	122.553	0.000
C3	1.601	1	1.601	3.731	0.055
Error	79.402	185	0.429		
Total	2394.520	193			
Corrected total variation	294.730	192			

a. R-square = 0.731 (corrected R-square = 0.720)

Figure 6.6. Univariate analysis of covariance for comfort. Covariate factors are part B and C questions.

Comfort:

Question B24 has been directly attributed to the comfort category, see Table 6.6. A univariate analysis of covariance with B24 as the dependent variable and the other questions related to comfort as the independent variables indicates the relevant features for this category. The result of this analysis is depicted in Figure 6.6. Nearly all part B questions (B12, B16, B19, B22 and B25) show a significant contribution to B24, covering about 72% of the variance. Whereas B12 and B22 relate to the naturalness of the system's voice and behavior, B16 addresses the friendliness of the system's reaction, B19 the transparency of the interaction, and B25 the stress experienced by the user. Although a high correlation between B24 and B25 has been observed (both refer to the emotional state of the subject), also naturalness, transparency and friendliness seem to contribute significantly to the comfort perceived during the interaction. Thus, if B24 is accepted as a descriptor of comfort, then the two aspects of the comfort category (agent personality and cognitive demand) have an important relationship to each other.

Task success:

Questions B1, B4, C5 and C14 relate to this aspect, as well as all task success measures (TS_{bin}, TS_{ord}, κ_{dia} and κ_{conf}). The parameters $AN{:}CO$, $DARPA_s$ and $DARPA_{me}$ have been included because their correlation to B1 exceeds 0.5. Moderate correlations exist between B1 and B4. On the other hand, the relations between questions and task success measures are all relatively low. This may be an indication that many test subjects thought they would have obtained the right information from the system, but in fact they didn't. As an example, subjects who asked for a moderately priced Italian restaurant got information about Italian restaurants in another price category. For a user, such

Table 6.27. Spearman correlation coefficients for questions and interaction parameters related to task success. D_s: $DARPA_s$; D_{me}: $DARPA_{me}$.

	B1	B4	C5	C14	TS_{bin}	TS_{ord}	κ_{dia}	κ_{conf}	$AN{:}CO$	D_s	D_{me}
B1	1.000	*0.601*	*0.314*	*0.265*	*0.240*	*0.315*	-0.043	0.070	*0.542*	*0.542*	*-0.539*
B4	*0.601*	1.000	*0.304*	*0.476*	*0.287*	*0.331*	-0.006	*0.156*	0.079	0.079	-0.052
C5	*0.314*	*0.304*	1.000	*0.396*	0.035	0.037	-0.046	-0.012	0.337	0.337	-0.311
C14	*0.265*	*0.476*	*0.396*	1.000	*0.156*	*0.150*	0.008	0.064	-0.206	-0.206	0.219
TS_{bin}	*0.240*	*0.287*	0.035	*0.156*	1.000	*0.921*	*0.270*	*0.177*	*0.742*	*0.742*	*-0.747*
TS_{ord}	*0.315*	*0.331*	0.037	*0.150*	*0.921*	1.000	*0.422*	*0.191*	*0.574*	*0.574*	*-0.584*
κ_{dia}	-0.043	-0.006	-0.046	0.008	*0.270*	*0.422*	1.000	*0.324*	-0.024	-0.024	-0.033
κ_{conf}	0.070	*0.156*	-0.012	0.064	*0.177*	*0.191*	*0.324*	1.000	0.222	0.222	-0.273
$AN{:}CO$	*0.542*	*0.079*	*0.337*	*-0.206*	*0.742*	*0.574*	-0.024	0.222	1.000	*1.000*	*-0.990*
D_s	*0.542*	*0.079*	*0.337*	*-0.206*	*0.742*	*0.574*	-0.024	0.222	*1.000*	1.000	*-0.990*
D_{me}	*-0.539*	-0.052	-0.311	0.219	*-0.747*	*-0.584*	-0.033	-0.273	*-0.990*	*-0.990*	1.000

Table 6.28. Spearman correlation coefficients for questions related to task ease.

	C11	C13	C16
C11	1.000	*0.513*	*0.362*
C13	*0.513*	1.000	*0.077*
C16	*0.362*	*0.077*	1.000

an error cannot easily be identified, also if he/she has the possibility to visit the restaurant after using BoRIS.

Among the parameters, TS_{bin} and TS_{ord} are highly correlated, as well as $AN{:}CO$, $DARPA_s$ and $DARPA_{me}$. Interestingly, the correlation between κ_{dia} and κ_{conf} is very low, as well as the correlation between the κ measures and the TS measures. Thus, both types of task success metrics seem to provide different types of information: Whereas TS always requires the full agreement of all slots determining a restaurant, κ also takes partial task success and the chance agreement into account. A moderate correlation can be observed between the DARPA measures and the TS measures.

Task ease:

This aspect is only addressed by questions C11, C13 and C16, and no interaction parameter showed a correlation higher than 0.5 to one of these questions. A moderate correlation between C11 and C13 can be observed. A service provided by the system seems to be more helpful when the user is informed about its functionality.

Table 6.29. Spearman correlation coefficients for questions related to service efficiency.

	C12	C13	C15	C17
C12	1.000	*0.687*	*0.603*	*0.377*
C13	*0.687*	1.000	*0.363*	*0.144*
C15	*0.603*	*0.363*	1.000	*0.626*
C17	*0.377*	*0.144*	*0.626*	1.000

Table 6.30. Spearman correlation coefficients for questions related to usability.

	C8	C11	C16
C8	1.000	*0.629*	*0.437*
C11	*0.629*	1.000	*0.362*
C16	*0.437*	*0.362*	1.000

Service efficiency:

This category comprises the aspects of service adequacy and added value. It is addressed by the questions C12, C13, C15 and C17, from which C12 shows moderate correlations with C13 and C15, and C15 with C17. C12, C13 and C15 all seem to be related to the perceived usefulness of the service. C15 and C17 explicitly address the preference for a comparable interface, be it another system or a human operator. No interaction parameters seem to be related to these quality aspects.

Usability:

Usability is addressed by questions C8, C11 and C16. C8 and C11 are moderately correlated; thus, if the users are adequately informed about the system's functionality, handling will be easier for them. It is surprising that C8 and C16 do not show a higher correlation. Both address the ease of handling the system. However, users may have the impression that they were responsible for interaction problems, and answer question C8 with "no" although they gave a positive answer to question C16. It is important to find question wordings which cannot be misinterpreted in this way.

User satisfaction:

User satisfaction in general is addressed by questions B0, B23 and C1. The underlying aspects pleasantness (B24, C10) and personal impression (C9) have additional related questions. Correlations between these questions are shown in Table 6.31. Because $AN{:}CO$, $DARPA_s$ and $DARPA_{me}$ have moderate

Table 6.31. Spearman correlation coefficients for questions and interaction parameters related to user satisfaction. D_s: $DARPA_s$; D_{me}: $DARPA_{me}$.

	B0	B23	C1	B24	C9	C10	$AN{:}CO$	D_s	D_{me}
B0	1.000	*0.759*	*0.291*	*0.639*	*0.321*	*0.255*	0.309	0.309	-0.355
B23	*0.759*	1.000	*0.378*	*0.541*	*0.302*	*0.165*	*0.493*	*0.493*	*-0.511*
C1	*0.291*	*0.378*	1.000	*0.157*	*0.676*	*0.207*	0.196	0.196	-0.237
B24	*0.639*	*0.541*	*0.157*	1.000	*0.207*	*0.309*	0.227	0.227	-0.219
C9	*0.321*	*0.302*	*0.676*	*0.207*	1.000	*0.408*	0.440	0.440	-0.466
C10	*0.255*	*0.165*	*0.207*	*0.309*	*0.408*	1.000	*-0.584*	*-0.584*	*0.589*
$AN{:}CO$	0.309	*0.493*	0.196	0.227	0.440	*-0.584*	1.000	*1.000*	*-0.990*
D_s	0.309	*0.493*	0.196	0.227	0.440	*-0.584*	*1.000*	1.000	*-0.990*
D_{me}	-0.355	*-0.511*	-0.237	-0.219	-0.466	*0.589*	*-0.990*	*-0.990*	1.000

correlations to B23, C9 and C10, these parameters have also been included in the table. Amongst the questions, B0 and B23 are highly correlated (both indicate the overall satisfaction), and moderate correlations can be seen for B24 with B0 and B23 (the system is rated pleasant when the user is satisfied), and C1 with C9 (the user is impressed when the overall rating is positive). Once again, correlations between part B questions (reflecting the individual interaction) and part C questions (reflecting the whole test session) are relatively low. Correlations between questions and interaction parameters are only moderate, especially to B23, C9 and C10. The degree of correlation is similar for all mentioned parameters, as their inter-correlation is very high.

In order to investigate the contribution of the individual questions to the user satisfaction indicators, an analysis of covariance is performed. B0 (overall impression) is taken as the dependent (target) variable, and all other part B questions are taken as covariate factors, except B23 which is on the same level and highly correlated with B0. The result is shown in Figure 6.7. Significant contributors to B0 are B1 (system provided information), B3 (information was clear), B5 (perceived system understanding), B6 (listening-effort) and B24 (pleasantness). B4 (truth/evidence) and B13 (system flexibility) are close-to-significant contributors. The significant contributors reflect the low-level categories speech input/output quality, cooperativity, comfort, task efficiency, and partly also dialogue symmetry. For the first category, both speech input (perceived understanding) and output (listening-effort) are relevant. In the cooperativity category, informativeness and relevance seem to be the most important aspects, followed by truth and evidence. Interestingly, communication

Test of between-subject effects

Dependent variable: B0

Source	Square sum of type III	df	Mean of squares	F	Significance
Corrected model	186.780[a]	24	7.782	16.974	0.000
Constant	0.306	1	0.306	0.667	0.412
B1	3.337	1	3.337	7.279	0.008
B2	0.811	1	0.811	1.770	0.185
B3	2.047	1	2.047	4.464	0.036
B4	1.750	1	1.750	3.817	0.053
B5	8.630	1	8.630	18.823	0.000
B6	2.357	1	2.357	5.140	0.025
B7	0.307	1	0.307	0.669	0.415
B8	0.023	1	0.023	0.050	0.823
B9	1.322	1	1.322	2.882	0.092
B10	0.024	1	0.024	0.053	0.819
B11	0.211	1	0.211	0.460	0.498
B12	0.403	1	0.403	0.878	0.350
B13	1.525	1	1.523	3.326	0.070
B14	0.397	1	0.397	0.866	0.354
B15	0.798	1	0.798	1.741	0.189
B16	0.138	1	0.138	0.301	0.584
B17	0.093	1	0.093	0.204	0.652
B18	0.077	1	0.077	0.168	0.682
B19	0.033	1	0.033	0.072	0.788
B20	0.071	1	0.071	0.155	0.695
B21	0.003	1	0.003	0.007	0.935
B22	0.193	1	0.193	0.420	0.518
B24	1.910	1	1.910	4.166	0.043
B25	0.347	1	0.347	0.758	0.385
Error	70.149	153	0.458		
Total	1582.78	178			
Corrected total variation	256.928	177			

a. R-square = 0.727 (corrected R-square = 0.684)

Figure 6.7. Univariate analysis of covariance for overall impression addressed by question B0. Covariate factors are part B questions, except B23.

efficiency is not reflected in the significant predictors to B0. In particular the speed- and conciseness-related questions B15, B17 and B20 do not provide a significant contribution to the overall user satisfaction.

Utility and acceptability:

These two categories are addressed by questions C12 and C13 (utility) and C18 (acceptability). Correlation between all these questions is moderate ($\rho = 0.57...0.69$). Apparently, the questions are related, but they are not addressing identical perceptive dimensions. In particular, the future use seems to depend on the perceived helpfulness and on the value attributed to the service. These two dimensions are however not the only influencing factors. Other dimensions for utility and acceptability should be identified in future experiments. They may be related to the economical benefit for the user, a dimension which can hardly be investigated in laboratory tests.

Significant predictors to acceptability can be identified by an analysis of covariance of the part C questions, taking question C18 as the target variable.

Table 6.32. Spearman correlation coefficients for questions related to utility and acceptability.

	C12	C13	C18
C12	1.000	*0.687*	*0.571*
C13	*0.687*	1.000	*0.630*
C18	*0.571*	*0.630*	1.000

Test of between-subject effects

Dependent variable: C18

Source	Square sum of type III	df	Mean of squares	F	Significance
Corrected model	229.813[a]	17	13.518	26.295	0.000
Constant	0.098	1	0.098	0.191	0.662
C1	0.781	1	0.781	1.519	0.219
C2	1.046	1	1.046	2.034	0.156
C3	0.361	1	0.361	0.701	0.403
C4	1.124	1	1.124	2.187	0.141
C5	2.575	1	2.575	5.009	0.026
C6	0.025	1	0.025	0.049	0.824
C7	0.707	1	0.707	1.375	0.243
C8	1.877	1	1.877	3.651	0.058
C9	0.955	1	0.955	1.858	0.175
C10	0.226	1	0.226	0.439	0.508
C11	6.147	1	6.147	11.957	0.001
C12	0.003	1	0.003	0.007	0.935
C13	1.716	1	1.716	3.338	0.069
C14	1.833	1	1.833	3.565	0.061
C15	19.581	1	19.581	38.088	0.000
C16	0.199	1	0.199	0.387	0.535
C17	0.605	1	0.605	1.178	0.279
Error	92.027	179	0.514		
Total	2777.27	197			
Corrected total variation	321.837	196			

a. R-square = 0.714 (corrected R-square = 0.687)

Figure 6.8. Univariate analysis of covariance for acceptability addressed by question C18. Covariate factors are part C questions.

Figure 6.8 shows that significant contributions to the acceptability (future use) question stem from C5 (answer capabilities), C11 (information about system capabilities), C15 (alternative interface), and close-to-significant contributions from C8 (ease of use), C13 (perceived helpfulness) and C14 (perceived reliability). The interpretation of these findings is difficult, but the informative and functional value of the system seems to be important for the test subjects.

Summary of the analyses:

In this section, the relationships between subjective judgments and interaction parameters have been analyzed on the basis of the QoS taxonomy. Because not each category or quality aspect defined in the taxonomy can be described by a single interaction parameter or subjective judgment, it is not possible to verify the taxonomy in a strict sense. Nevertheless, correlation analyses within the subjective judgments, within the interaction parameters, or between both groups reveal dependencies which can be interpreted with respect to the user's quality perceptions.

In general, the correlations between different subjective judgments associated with a single quality aspect or category are moderate. This is an indication that the questions address the correct aspect, but that it is justifiable to use several questions in order to better capture the dimensionality of the perceptive quality space. In the case that all aspects of the QoS taxonomy are interesting to the evaluator, the questionnaire could not be shortened without losing information. However, there might be information contained in the questionnaire which is not really useful for the system developer, either because it cannot be related to a specific system property, or because the addressed property cannot be changed in the system set-up. In order to get a complete picture of the aspects contributing to the quality of the service, it is nevertheless desirable to keep the questionnaire in its current form.

Particularly low correlations were found between judgments obtained for individual interactions (part B questions) and those related to the whole test session (part C questions), even if they address the same quality aspect (like B0 and C1). This is an indication of a strong learning effect, and shows that it is necessary to distinguish well between questions which are related to an individual experience and those which reflect a general feeling. Obviously, questions which refer to system configuration properties (which differ between the interactions) should be asked in part B of the questionnaire. In contrast, usability may be better addressed after a certain interaction experience. The informal comments which some test subjects made after completing the experiment showed that the overall impression of the system is often closely linked to a specific detail of the system or interaction. For example, a bad system voice or a specifically bad interaction behavior at one dialogue point may impact an otherwise positive global opinion quite strongly. Learning effects will be further addressed in Section 6.2.6.

The correlations observed between subjective judgments and interaction parameters are disappointingly low in most cases. Only in very few cases do both metrics correlate to an extent which would allow the replacement of a subjective judgment by an interaction parameter. Although the same tendency has been noted earlier, e.g. by Kamm et al. (1997a), the low correlation will be a limiting factor for predicting perceived quality from instrumentally or expert-derived parameters. Clearly speaking, it is not possible to replace questionnaires by logging interaction parameters without losing information on quality. This finding is in agreement with traditional and on-going discussions for describing the quality of transmitted or synthesized speech. For some aspects, questions and interaction parameters seem to address completely different perceptive dimensions. An example is the truth and evidence aspect: Although the system often did not understand the user's request and provided wrong information, users still rated this aspect positively. One reason for this behavior might be the unrealistic test situation, but also in real life users will not always be able to verify information which is provided by the system.

Amongst the interaction parameters, there are several ones which are highly correlated. It will thus be possible to reduce the number of parameters which have to be determined during the interaction, and consequently to reduce the effort for test run and analysis. A good starting point are the parameters related to speech input quality, where a proposal for a reduced set has been made. However, not all interaction parameters which belong to a single quality aspect can be reduced to one single parameter. For instance, two different types of parameters related to task success have been observed, namely the TS and the κ metrics. These metrics will play an important role for the quality prediction models discussed in Section 6.3.

6.2.5 Impact of System Characteristics

Experiments 6.1 and 6.3 cover a wide range of system configurations which differ with respect to the speech input component (performance of the recognizer), the dialogue management component (confirmation strategy), and the speech output (natural speech and/or synthesized speech). So far, the experimental data have been analyzed as an ensemble, disregarding the effects of the configuration the results have been obtained with. However, subjective quality judgments as well as interaction parameter values will definitely reflect the characteristics of the system. In this section, the impact of the system configuration will be analyzed separately and in more detail for the three variable parts of the system. Findings are compared to literature data, and potential implications both for quality evaluation as well as for system development are pointed out.

6.2.5.1 Impact of Speech Input

The performance of the speech input is influenced by the user who provides the speech input, by the acoustic and transmission channel characteristics (environmental factors), and obviously by the recognition and understanding components (agent factors). These influences are well reflected in the QoS taxonomy. Speech input quality, in turn, is linked to communication efficiency, comfort, and task efficiency. It can therefore be assumed that speech input performance will not only be reflected in interaction parameters and judgments directly related to the speech input quality category, but also in those of the dependant categories.

Several authors address the impact of speech input on communication efficiency and task performance. Delogu et al. (1998) compared DTMF, simple digit recognition, and more complex "audible quoting" speech input using keyword recognition. They found that keyword recognition – although showing a lower recognition score and a higher transaction time – was preferred and judged to be more enjoyable by most users compared to digit recognition or DTMF. Similar observations were made by Lai and Lee (2002) for a unified messaging system: The interaction with a speech-based system version was rated as being more satisfying, more entertaining and more natural than the one with a DTMF version, although the speech-based system was less efficient. The results from Basson et al. (1996) suggest that speech may be generally preferable to DTMF in cases where services are new and not already well established in the DTMF domain. On the other hand, it may be difficult for a spoken-dialogue-based system to compete with DTMF when the latter is already well accepted in the market.

Sanders et al. (2002) analyzed data collected in the DARPA Communicator project for potential correlations between word error rate and task success, the latter either being rated by the test subjects (perceived task success) or calculated from expert transcriptions. They found a linear relationship between word error rate and task completion for a large range of error conditions, up to a WER of 75%. The linear relationship held better for the calculated than for the perceived task success. Casali et al. (1990) investigated the influence of simulated recognition accuracy and vocabulary size on task completion time and user satisfaction for a computer data entry task. The authors varied the recognition accuracy between 99 and 91%, and found that task completion time increased significantly from 99% to 95% (17% longer duration) and to 91% (50% longer duration). User ratings on "acceptability" (combination of ratings on 12 bipolar scales) were also significantly affected by speech recognition accuracy between 91 and 99%. Unfortunately, the combination of subjective ratings did not provide any insight into the underlying quality dimensions affected from the user's point of view.

The impact of speech recognition on global user quality judgments was also addressed by Jack et al. (1992) and Larsen (1999). Jack et al. investigated a prototype telephone service where test subjects had to provide a credit card number via spoken input. They compared the mean over 22 ratings obtained on Likert scales (meant to indicate "usability") to simulated ASR performance. Results show that the mean rating rises very weakly between a recognition accuracy of 85 and 90%, and then increases significantly when approaching 100% accuracy. Perceptive quality dimensions which were most affected were the ease of use and the perceived system reliability (Foster et al., 1993). On the

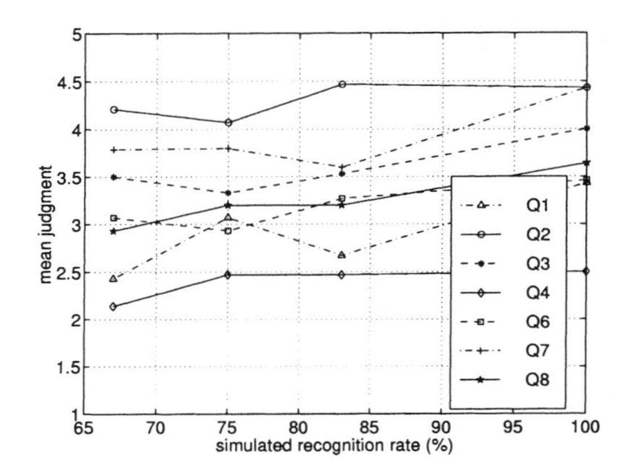

Figure 6.9. Effect of simulated recognition rate on subjective ratings of experiment 6.1. Q1: "How well did you feel understood by the system?"; Q2: "How well was the system acoustically intelligible?"; Q3: "How well did you understand what the system asked from you?"; Q4: "The system reacted like a human."; Q6: "You perceived the dialogue as pleasant...unpleasant."; Q7: "The system reacted in a friendly...unfriendly way."; Q8: "The course of the dialogue was clear...confusing.".

other hand, subjective ratings related to the system's voice characteristic and to the helpfulness of system prompts received more positive ratings when the ASR accuracy was low. The finding was explained by an increased helpfulness of the system messages in case of recognition errors. Larsen (1999) investigated the relationship between "average user attitude" (mean calculated over 26 Likert scale statements) and the speech recognition accuracy. In contrast to Jack et al., he found no significant decrease in user attitude from perfect recognition down to about 70% recognition accuracy. Apparently, the speech recognition impact is strongly dependent on the characteristics of the system as a whole, and on the actual application. Thus, an "acceptable" level of ASR performance has to be determined for each system and application anew, taking the dialogue management strategy into account.

Within the BoRIS set-up, two approaches have been made to control recognition accuracy. Dudda (2001) applied a simple replacement scheme for simulating recognition errors in experiment 6.1, see Section 6.1.2. A statistically significant effect on user judgments was only observed for question Q7 ("The system reacted in a friendly...unfriendly way.") which was answered significantly more positively in case of perfect recognition. No definite explanation for this finding can be given; it may be linked to the specific way of generating recognition errors (problems have been reported here because the regularity of misrecognized words strongly affects short user utterances), or to the regular order of scenarios used for the interactions. Other judgments on the perceived

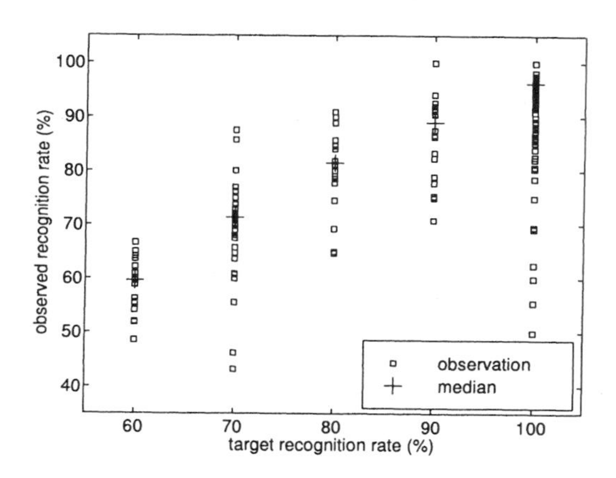

Figure 6.10. Comparison of observed and target recognition rates for the recognizer simulation in experiment 6.3.

system understanding (Q1), the intelligibility of the system voice (Q2), the impression of having understood what the system required (Q3), the human-like reaction (Q4), the pleasantness (Q6), and the transparency of the interaction (Q8) show a slight decrease with the recognition performance, see Figure 6.9. The decrease can already be observed with 83% recognition performance for Q1, Q3, Q7 and Q8, but for other ratings a recognition accuracy of 75% still seems to be acceptable. All in all, the observed influences on the quality judgments are relatively small.

In experiment 6.3, a more complex recognition simulation approach has been chosen. This approach is partly based on a previously collected confusion matrix (for the in-vocabulary words), and partly on the difference between the confusion-matrix-generated word error rate and the target error rate. Before analyzing the impact of recognition accuracy for this experiment, a comparison between the observed and the target recognition rate has to be performed. The comparison is depicted in Figure 6.10. Although the correlation between the individual observations and the target rate is not excessively high (0.78), the ensemble of data reaches the target recognition rate ($c_w/\#$ WORDS) relatively well. Median values are 0.596 for 60% target rate, 0.713 for 70%, 0.815 for 80%, 0.900 for 90% target, and 0.963 for 100% target recognition rate. The slightly larger difference for the latter value has to be explained with transcription errors from the wizard, which mainly affected non-words and filler words ("uhm", "aah", etc.) or abbreviations ("I would like" instead of "I'd like"). These transcription errors will not confuse the speech understanding component. Isolated word recognition rates ($c_{w,iso}/\#$ USER WORDS IV) have been

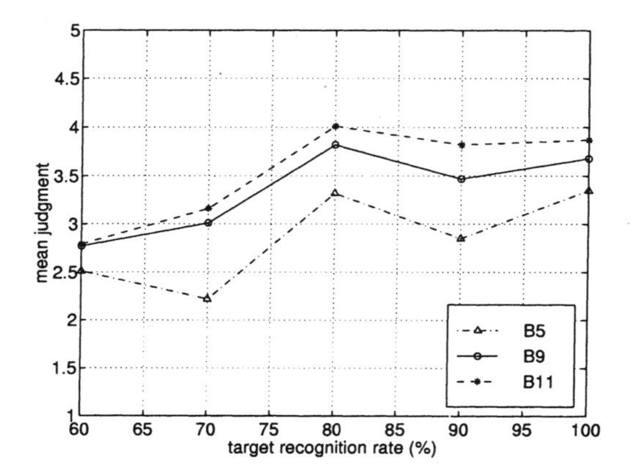

Figure 6.11. Effect of simulated recognition rate on subjective ratings on part B questions of experiment 6.3. B5: "How well did you feel understood by the system?"; B9: "In your opinion, the system processed your specifications correctly?"; B11: How often did the system make mistakes?".

calculated in order to investigate the difference between the two simulation principles (confusion matrix for the in-vocabulary words, and difference to the target rate for the others), and their interaction. A somehow larger difference for the first principle indicates that the target rate can only be reached with difficulty when recognition errors are generated purely on the basis of confusion matrices. The combination of generation principles better approaches the target rate.

Both interaction parameters and subjective quality judgments have been collected in experiment 6.3. Statistically significant differences between system configurations were determined with the help of a non-parametric Kruskal-Wallis test. The simulated recognition rate had a significant impact on the parameters # BARGE-INS ($p = 0.015$), the parsing-related parameters $PA{:}PA$ ($p = 0.027$), $PA{:}CO$ and $PA{:}FA$ ($p \leq 0.001$), the correction rates SCR and UCR ($p \leq 0.001$), the cooperativity-related parameters $CA{:}AP$ and $CA{:}IA$ ($p \leq 0.001$), the task success parameter κ_{dia} ($p = 0.004$), the understanding parameters IC and UA ($p \leq 0.001$), and obviously also the recognition-related parameters WER, WA, $\overline{WES}$ and $\overline{NES}$ (all $p \leq 0.001$, both for continuous and isolated ASR). Apparently, parameters describing speech understanding (IC, UA and PA measures) are strongly influenced by the ASR performance. SCR and UCR show that ASR performance degradation increases the need for (correction) meta-communication, and thus also affects dialogue smoothness. Interestingly, the cooperativity of system answers seems to be strongly affected by the ASR performance (by definition, the CA measures are a direct indicator of cooperativity). Task success also depends on ASR performance: Although the TS_{ord} and TS_{bin} parameters do not indicate a significant difference, they follow the same tendency as the κ_{dia} parameter.

Apart from the interaction parameters, the subjective judgments are also affected by the recognition performance. Statistically significant impacts on part B judgments were found for B5 ($p = 0.013$), and B9 ($p = 0.088$) and B11 ($p = 0.051$) were close to significance level. Figure 6.11 shows that all three aspects are rated relatively constant down to around 80% target recognition rate. Below this threshold, subjective judgments degrade quite remarkably. The observed threshold may however be system- or application-specific, or it may be linked to the recognizer simulation used in the experiment. Users seem to be able to localize the changes in the system and to attribute them to the perceived system understanding, to system errors, and to whether the system correctly processed the users' specifications. No significant impact was found for B0 (overall impression) or B23 (user satisfaction). This is astonishing, because B5 and B9 were identified as significant contributors to B0, and task success was shown to be affected by the imperfect understanding.

The results indicate that recognition performance mainly affects interaction parameters related to the categories speech input quality, cooperativity, communication efficiency, and task success. These links are well illustrated in the QoS taxonomy. Users seem to be able to localize the source of interaction problems, and attribute it to the system's understanding capability. The finding seems to be in contrast to the relatively low correlation between subjective judgments and interaction parameters related to speech input quality, see Section 6.2.4 and Kamm et al. (1997a). A reason may be that the correlations in Tables 6.20 and 6.21 have been calculated over all system configurations, differing in more than just the recognition accuracy. Narayanan et al. (1998) used a number of ASR-related performance measures to predict perceived system understanding, and reached a relatively good prediction accuracy ($R^2 = 0.89...0.90$). The models however disregarded about 10% of outliers in the data, and should consequently be interpreted with care.

6.2.5.2 Impact of Speech Output

Most spoken dialogue systems which are available on the market provide output to the user via pre-recorded messages of naturally produced speech. In this way, a relatively high level of intelligibility and naturalness can be reached, as long as the messages uttered by the system are similar to the ones which would be produced by a human operator. However, this method is not always viable, namely in case of undetermined or principally infinite vocabulary (e.g. for an email reading service), or for quickly changing messages (e.g. in a weather information service). In this case, synthesized speech from orthographic text or from concepts is necessary for generating speech output.

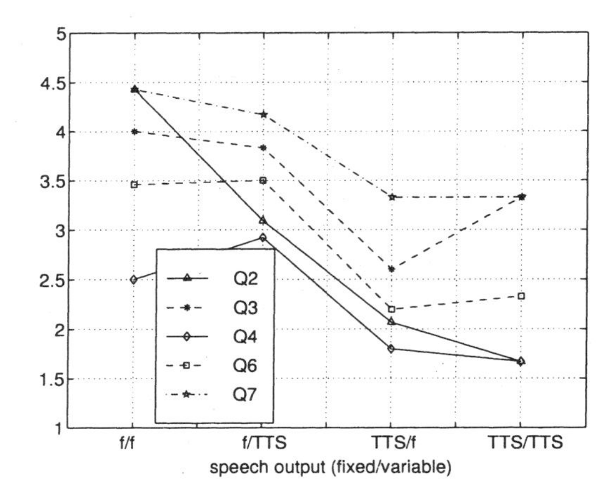

Figure 6.12. Effect of speech output configurations on subjective ratings of experiment 6.1. Q2: "How well was the system acoustically intelligible?"; Q3: "How well did you understand what the system asked from you?"; Q4: "The system reacted like a human."; Q6: "You perceived the dialogue as pleasant...unpleasant."; Q7: "The system reacted in a friendly...unfriendly way.".

Until it is absolutely necessary, however, most system designers try to avoid synthesized speech in their service, and use pre-recorded messages instead. The common assumption is that low intelligibility and naturalness would considerably impact the overall quality of the service. This assumption was confirmed in experiment 6.2, where a partly TTS-based version of the BoRIS system has been used. Although the cooperativity of the system is rated positive for the majority of the aspects addressed by the questionnaire, the overall impression of the system is still negative. Niculescu (2002) attributes this finding to the low intelligibility. Intelligibility was ranked highly important for a good system in the quantitative interview. Thus, low intelligibility may severely impact overall system quality, even if the fundamentals of cooperativity are well respected by the system.

A comparative evaluation of different speech output strategies was performed in experiments 6.1 and 6.3. In experiment 6.1, naturally produced speech from a female speaker and synthesized speech with a male unit inventory were tested. Both naturally produced and synthesized speech could be used for the fixed system messages and/or the (variable) restaurant information parts, resulting in four system configurations. As was expected, the completely natural version was rated best in most aspects, and the completely TTS-based version worst. Statistically significant differences in the subjective ratings (non-parametric Kruskal-Wallis test) were observed for Q2 ($p \leq 0.001$), Q3 ($p = 0.010$), Q4 ($p = 0.006$), Q6 ($p = 0.002$) and Q7 ($p = 0.002$), see Figure 6.12. Appar-

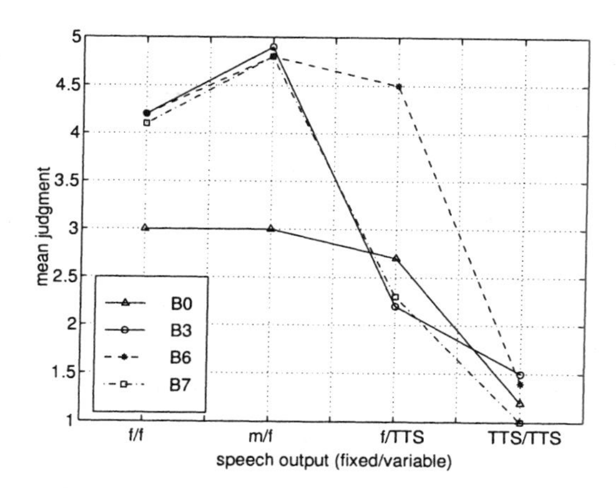

Figure 6.13. Effect of speech output configurations on subjective ratings on part B questions of experiment 6.3. B0: "Overall impression"; B3: "The information was clear/unclear"; B6: "You had to concentrate in order to understand what the system expected from you?"; B7: "How well was the system acoustically intelligible?".

ently, the system's intelligibility and the cognitive demand put on the user (i.e. whether the user understood what the system requested) as well as voice-specific characteristics (naturalness, friendliness and pleasantness) are most affected by the TTS system used in this experiment. The mixed (f/TTS and TTS/f) system configurations mainly range in between these two extreme points. Natural speech for the fixed system prompts (which is the majority of the prompts) was rated better than the opposite case on all mentioned questions. Whereas the use of TTS for the (short, but important) variable parts only slightly affects the voice and cognitive-effort related questions, it still strongly affects intelligibility. Interestingly, some aspects seem to be more prominent for the users when natural and synthesized speech are mixed, which may be a reason for the higher rating on Q4 (reaction like a human) for the f/TTS version compared to the completely natural version, and for the lower rating on Q3 (user understood what the system requested) for the TTS/f compared to the completely TTS-based version.

Similar system configurations were tested in experiment 6.3. Here, two different natural voices (male and female) were used, and the female voice was also combined with TTS (configurations 1 to 4 of Table 6.2). The choice of both voices significantly affects the parameters STD ($p = 0.002$) and URD ($p = 0.049$), as well as $WPST$ ($p = 0.041$). The influence on STD will be due to the lower speaking rate of the TTS system in comparison to the natural speakers. A significant effect on URD may indicate a higher cognitive demand put on the user when listening to synthesized speech. The number of words per

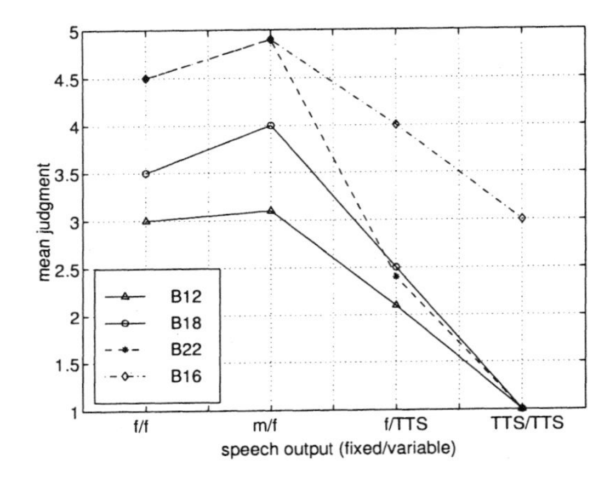

Figure 6.14. Effect of speech output configurations on subjective ratings on part B questions of experiment 6.3. B12: "The system reacted in the same way as humans do?"; B18: "You perceived the dialogue as natural/unnatural"; B22: "The system's voice was natural/unnatural"; B16: "The system reacted in a friendly/unfrienly way".

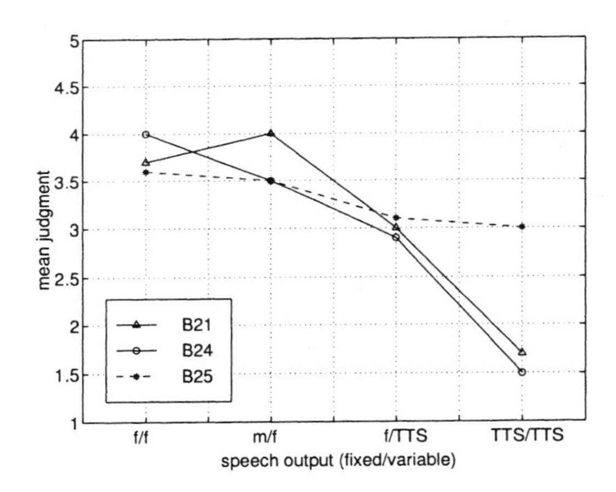

Figure 6.15. Effect of speech output configurations on subjective ratings on part B questions of experiment 6.3. B21: "The course of the dialogue was smooth/bumpy"; B24: "You perceived the dialogue as pleasant/unpleasant"; B25: "During the dialogue, you felt relaxed/stressed".

system turn is linked to the specific messages generated in the dialogue, but no explanation can be given why this number differs between the configurations.

In contrast to the restricted effects on interaction parameters, subjective ratings are affected quite drastically, see Figures 6.13 to 6.15. A statistically significant impact was found for the ratings on B0 ($p \leq 0.001$), B3 ($p \leq 0.001$),

B6 ($p \leq 0.001$), B7 ($p \leq 0.001$), B12 ($p = 0.003$), B16 ($p \leq 0.001$), B18 ($p \leq 0.001$), B21 ($p \leq 0.001$), B22 ($p \leq 0.001$), B24 ($p \leq 0.001$) and B25 ($p = 0.043$). Besides the obvious effects on listening-effort (B6) and intelligibility (B7), the clarity of the information (B3), the naturalness (B12, B18 and B22), the friendliness (B16), the smoothness (B21), the pleasantness (B24), stress (B25), and the overall impression of the system (B0) are also strongly affected. The perceptive degradation is significantly higher than it was observed for the speech recognition configurations. In nearly all cases, the use of TTS for the fixed messages only is rated significantly better than using TTS for both fixed and variable messages. A natural male voice for the fixed parts is rated slightly better in nearly all judgments; only the personal feeling judgments are not enhanced (B24 and B25). However, the overall rating B0 is identical in these two cases.

Comparing these findings with the ones made in the previous section, significant influences on interaction parameters and subjective quality ratings have been observed both on the speech input and the speech output side. Whereas the recognition rate mainly influenced interaction parameters related to the categories speech input quality, cooperativity, communication efficiency and task success, synthesized speech strongly affects perceptive dimensions which can be associated with different categories of the QoS taxonomy, down to the global overall impression of the user. Apparently, the system voice describes the system's "personality", and seems to act as a kind of "business card" for the whole system. This effect was postulated earlier and could now be quantified with experiment 6.3. Users seem to be less able to localize their perceptions with respect to speech output than to speech input.

6.2.5.3 Impact of Dialogue Strategy

The dialogue manager has a number of characteristics which strongly determine the flow of the interaction, and consequently also the quality perceived by the user. They include the distribution of initiative between the user and the system, the applied confirmation strategy, the manner of presenting the information to the user, and other characteristics which may be difficult to describe in precise terms. Together with the speech input and output characteristics, they form the "personality" of the machine agent.

In fact, the metaphor, i.e. the transfer of meaning due to similarities in the external form or function, plays an important role for the reference a user has in mind when judging quality. Dutton et al. (1999) compared three types of metaphors for a home shopping service: A menu-style metaphor, a department store metaphor, and a magazine metaphor. Influences of the used metaphor were observed both on interaction parameters and on subjective dimensions of usability. The authors found that task completion, navigation through the service, and subjective attitudes towards the service significantly ameliorated with the right metaphor, in particular when using the service several times.

The distribution of initiative between the system and the user seems to be an important characteristic as well. In experiment 6.2, test subjects were asked about their preferred initiative before the first interaction with the BoRIS system. The results show a difference between male and female test subjects: Only 9% of the female, but 40% of the male subjects preferred to ask the system themselves, and 39% of the female users (20% male) preferred to answer questions posed by the system.

Dialogue initiative was shown to carry an influence on task success in the DARPA communicator program. In general, system-directed dialogues were more successful than mixed-initiative ones. Rudnicky (2002) attributes the higher success of system-initiative to an increased recognition performance, and to the additional task structure which such systems provide to the user. In fact, system initiative reminds the user which information is still needed, and relieves memory load. The advantage was mainly visible in the constraint-gathering part of the dialogue, but system initiative was less useful in the solution navigation phase. Billi et al. (1996) compared two Italian rail information inquiry systems (RailTel and Dialogos) which differ with respect to initiative. It was observed that most users felt more comfortable with the system-initiative, isolated ASR RailTel system compared to the mixed-initiative Dialogos system providing continuous speech recognition capabilities. The authors argue that the test subjects used different quality references for both systems: RailTel was perceived as a traditional automatic telephone information service (associated parameters: system usability and service efficiency), whereas Dialogos is compared to human operator services (associated parameters: naturalness of the dialogue, naturalness of the synthetic voice). Users seem to appreciate both systems, but the parameters they take into account are different. The authors conclude that for simple services, system-driven dialogue strategies can lead to a similar level of user satisfaction as it can be reached with more sophisticated mixed-initiative systems.

Dudda (2001) investigated the effect of different confirmation strategies (explicit, implicit, summarizing confirmation, or no confirmation at all) on subjective judgments of user quality, see experiment 6.1. No clear preference for any of the strategies can be observed. In most attributes, either the implicit confirmation strategy or no confirmation were rated best. The worst ratings were mostly obtained either for the explicit or the summarizing confirmation strategy. The results seem to reflect the specific strategies applied in the test, and may also be linked to the unnatural test situation where the subjects were not really interested in obtaining the right information. Explicit confirmation resulted in the subjects being unsure about what the system expected from them (low ratings on Q3 and Q8), and their overall confidence in the system (low rating on Q10).

System configurations addressed in experiment 6.3 also differed with respect to the confirmation strategy, namely configurations 1 vs. 6 (perfect recognition) and 5 vs. 9 (70% recognition rate). Using the explicit confirmation strategy, the number of system questions raised significantly in both cases ($p = 0.017$ for 100%, $p = 0.033$ for 70% recognition rate). With perfect recognition, the implicit recovery rate also significantly increased ($p = 0.042$), whereas with lower recognition performance an additional decrease of the system turn duration, the system response delay, and the words per system turn could be observed. These effects will be linked to the specific confirmation strategy implementation. Task success could not be increased by using confirmation in both cases. The reason might be that the confirmation was only applied in the information-gathering parts of the dialogue, and not in the final navigation through the system responses which seems to cause many errors. Interestingly, the effect on the subjective judgments is relatively low. Only two questions showed a *negative* impact of applying the explicit confirmation strategy, and then only for the 70% recognition rate: B6 (concentration required) and B8 ("You knew at each point of the dialogue what the system expected from you."). Apparently, transparency suffers from an inappropriately chosen confirmation strategy when the error rate is high.

6.2.6 Test-Related Issues

Before attributing a specific effect observed in a laboratory experiment to certain system characteristics, it is important to ensure that it has not been caused by the test set-up itself. Test-related effects are difficult to control in laboratory experiments which are necessarily limited with respect to the number of dialogues which can be carried out, and to the number of test subjects. The most important ones are sequence effects when test subjects carry out several dialogues, and the effects of pre-defined scenarios. Both will be briefly addressed on the basis of data from experiments 6.1 and 6.3.

In experiment 6.3, Skowronek (2002) asked for the user's requirements of the system before and after the whole test session. Because the corresponding questions (A8.1-A8.7 and C22.1-C22.7) were identical in both cases, ratings can be directly compared to investigate the differences in requirements which will be linked to the current experience with the BoRIS system. Interestingly, the correlation between corresponding questions is not very high (between 0.175 and 0.632). A comparison of the mean ratings before (A) and after the test (C) is given in Table 6.33. In all cases where the mean differs remarkably (A8.1/C22.1, A8.2/C22.2, A8.3/C22.3, A8.4/C22.4 and A8.7/C22.7) the rating after the test is higher (i.e. the aspect is rated more important) than before the test. This increase is observed for a friendly voice (perhaps due to the low TTS quality

Table 6.33. Mean judgments on part A (A8.1-A8.7) and part C (C22.1-C22.7) questions of experiment 6.3.

Test part	*Question*						
	1	2	3	4	5	6	7
A	3.50	4.44	4.27	2.93	4.69	4.89	4.16
C	3.89	4.75	4.56	3.46	4.67	4.86	4.50

in some system configurations), for system initiative (perhaps the subjects felt uncomfortable or lost in parts of their dialogues), and for the system's help capabilities (due to the non-information about the system's functional limits).

When comparing the users' quality judgments during the five interactions carried out with BoRIS, a clearly positive tendency is found. Median values from 9 questions show a monotonously raising behavior, i.e. the judgments get more positive during the test session: B0, B1, B2, B5, B12, B14, B15, B23, B25. This list includes both questions addressing user satisfaction (B0 and B23), the accuracy and completeness of the provided information (B1 and B2), perceived system understanding (B5), the system's reaction (B12, B15), the perceived control capability (B14), and the stress experienced during the dialogue (B25). No question with a monotonously falling behavior were observed; the behavior of the other questions is heterogenous. The finding is in line with an early investigation reported by Dintruff et al. (1985). They distributed a 20-item questionnaire before and after test subjects interacted with a simulated office system capable of speech input and output. Half of the items were rated significantly differently after the experiment. The participants were more favorable towards speech technology after their test experience: Users felt significantly more active, more relaxed, happier, individualized, interested, stronger, more satisfied, and slightly more comfortable. They thought that the system was easier to use, more flexible, more appropriate, and somewhat faster than expected.

Whereas the scenarios of experiment 6.3 were equally distributed through the system configurations and positions in the experiment (Graeco-Latin square design), they were always kept in the same order in experiment 6.1. Mixing of scenarios and order allows the scenario impact to be analyzed independently of the order within the experiment. For the experiment 6.3 data, statistically significant influences of the scenario on a number of interaction parameters were observed, namely on the duration-related parameters *DD* ($p \leq 0.001$), STD ($p = 0.001$) and UTD ($p = 0.019$), the turn-related parameters # TURNS

($p \leq 0.001$) and $WPST$ ($p \leq 0.001$), the parsing-related parameters $PA{:}PA$ ($p = 0.017$) and $PA{:}FA$ ($p = 0.006$), the correction rates SCR ($p \leq 0.001$) and UCR ($p = 0.001$), the cooperativity parameters $CA{:}AP$ ($p = 0.026$) and $CA{:}IA$ ($p = 0.035$), the understanding-related parameters IC ($p = 0.005$) and UA ($p = 0.002$), and task success expressed by κ_{dia}, TS_{ord} or TS_{bin} (all $p \leq 0.001$). In contrast to that, no statistically significant effect on the user judgments to part B questions was observed.

When maintaining the scenario order in experiment 6.1, it seems that the impact of the scenario and of the system configuration is higher than the sequence learning effect. No clear tendency was found for shorter dialogues towards the end of the experiment, or for a higher transparency of the dialogue for the users. Dialogues in which the system failed to provide a restaurant resulted in particularly low ratings on perceived system understanding, in low comfort, and in low ratings about the comparability to human behavior. Confidence in the system's responses suffered in cases of failed system success, as well as in cases where the system succeeded in spotting that no answer exists. The hypotheses made in Section 3.8.3, namely that the choice of scenarios carries a significant impact on the quality judgments, is thus supported by the experimental data.

6.3 Quality Prediction Modelling

The analysis of experimental data in Section 6.2.4 showed only moderate or low correlations between interaction parameters and subjective quality judgments. The conclusion which has been drawn from this finding is that the collection of subjective quality judgments cannot be replaced by the collection of interaction parameters in order to obtain information about quality. This conclusion is in agreement with findings made in other areas of auditory quality research. In fact, the idea of predicting quality from the results of instrumental measurements has been followed for a number of years, mainly for speech transmitted via telephone channels (see Section 2.4.1.1), but also for wide-band audio quality, synthesized speech quality, and sound quality. The models developed in these areas provided useful information for specified application domains, but they do not present a generic estimation of quality which could be extrapolated to other systems or situations. Still, quality modelling proves to be important for system developers, in order to optimally design systems without depending too much on resource-consuming auditory tests. It should, however, be noted that subjective tests persist as the final reference for describing quality.

Information about the quality perceived during the interaction with an SDS is contained in interaction parameters which can be collected either instrumentally, or (in contrast to models for describing the quality of transmitted speech) with the help of expert transcriptions. From these parameters, estimations of quality, as would be perceived by the user in the addressed (test) situation, may be calculated. The corresponding algorithms will be subsumed under the term 'quality model'. In many cases, however, the parameters of such a model have to be determined anew for a specific service and the underlying system (configuration), and only the general algorithmic approach remains unchanged.

The most widely used prediction model for SDS-based services is the PARADISE framework (PARAdigm for DIalogue System Evaluation) which has been developed by Walker and co-workers at AT&T (Walker et al., 1997, 1998b, 2000a). PARADISE uses methods from decision theory (Keeney and Raiffa, 1993) in order to combine a set of task success and "dialogue cost" measures into a prediction of user satisfaction. A linear multivariate regression is performed in order to determine the contributing variables and the model coefficients. This approach will be analyzed and discussed in the following section. It forms the basis of new approaches which have been developed with the help of own experimental data, see Sections 6.3.2 and 6.3.3.

6.3.1 PARADISE Framework

Interactions with current state-of-the-art dialogue systems are generally task-orientated, i.e. the interaction has a defined task goal which the user and the system try to reach cooperatively. In most cases, the task determines system performance to a high degree, and consequently also the user's perception of quality. The PARADISE framework has been designed to compare the performance of different systems or system versions, by normalizing for task complexity. Three research goals have been addressed by the model (Walker et al., 2000a):

- Support comparison of multiple systems (or system versions) doing the same domain task.
- Develop predictive models of usability of a system as a function of a range of system properties.
- Provide a technique for making generalizations across systems, about which system properties really impact usability.

Generalization across tasks is supported by decoupling task requirements from dialogue behavior.

The aim of the PARADISE model is to predict a so-called "performance" measure which is defined as an estimation of user satisfaction. This variable will be called US_w in the following discussion. The model postulates that

> "performance can be correlated with a meaningful external criterion such as usability, and thus that the overall goal of a spoken dialogue agent is to maximize an objective related to usability. User satisfaction ratings [...] have been frequently used in the literature as an external indicator of the usability of an agent." (Walker et al., 1998b, p. 320)

Taking user satisfaction as an external validation criterion is based on the assumption that user satisfaction is predictive of other objectives (e.g. willingness to use or pay for a service) that might be more difficult to measure (Kamm and Walker, 1997). Although there will be no strict one-to-one relationship between system performance and usability, the author agrees that user satisfaction can be taken as an indicator of usability, and this is well reflected in the QoS taxonomy. It is also one contributing aspect of acceptability, as is suggested by Kamm and Walker (1997).

6.3.1.1 Model Description

In the PARADISE model, a combined "performance" metric for an SDS is derived as a weighted linear combination of task-based success measures and dialogue costs. The model assumes that the objective is to maximize user satisfaction, and that user satisfaction is composed of maximal task success and minimal dialogue costs. Task success is measured by how well the system and the user achieve the information requirements of the task by the end of the dialogue, expressed in terms of correctly filled AVMs. Dialogue costs are of two types: "Dialogue efficiency costs" which are measures of the system's efficiency in helping the user to complete the task, e.g. the number of utterances needed to complete the task; and "dialogue quality costs" which are intended to capture other aspects of the system that may have strong effects on the user's perception of the system, e.g. the number of times a user has to repeat an utterance. In the QoS taxonomy, "dialogue quality costs" would be mainly related to dialogue smoothness. The situation is depicted in Figure 6.16.

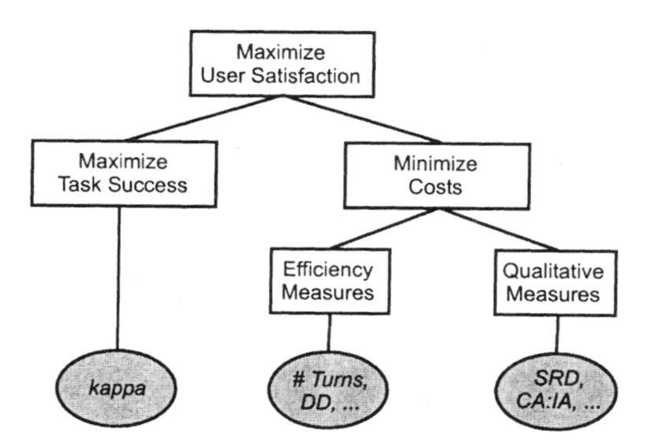

Figure 6.16. PARADISE model structure of objectives for the performance of SDSs (Walker et al., 2000a).

Both task success measures and dialogue costs can be extracted from dialogue corpora which have been collected under controlled experimental conditions.

The dialogue cost measures are either logged automatically, or they have to be annotated manually by a human expert. In order to be able to determine task success, interactions are normally guided by experimental tasks given to the test subjects, see Section 3.8.3.

The task is represented as a set of ordered pairs of attributes and their possible values (an AVM). For infinite sets of values, actual values found in the experimental data constitute the required finite set. Based on the AVM observed in the dialogue and on the scenario key AVM, a confusion matrix M can be established for a set of dialogues. In this confusion matrix, off-diagonal elements represent misunderstandings which have not been corrected in the course of the dialogue (corrected misunderstandings are included in the cost measures). From the confusion matrix M, the κ coefficient as an indicator of task success can be calculated, see Table A.9. In contrast to other measures like TS (Danieli and Gerbino, 1995), κ takes into account task complexity and corrects for chance agreement. It also reflects partial task success in a more realistic way. Using a single confusion matrix across all attributes inflates κ by reducing $P(E)$ when there is little cross-attribute confusion (Walker et al., 1997). To avoid this effect, Walker indicates that it is possible to calculate κ first for the identification of attributes, and then for the values within the attribute, or to average κ for each attribute. Task success evaluation can also be referred to a sub-dialogue level, by segmenting the whole dialogue into parts that consider certain attributes, and tagging these parts with the corresponding attribute. In this case, additional tags can be attributed to repair utterances in order to calculate cost measures.

Different cost measures have been used in conjunction with PARADISE. Usually, "dialogue efficiency" is addressed by the parameters DD, # SYSTEM TURNS and # USER TURNS, and "dialogue quality" by a mean recognition score MRS (calculated in a similar way to IC in Walker et al., 1998b, or as a mean over IC and # ASR REJECTIONS in Walker et al., 2000b), # TIME-OUT PROMPTS, # ASR REJECTIONS, # HELP REQUESTS, # CANCEL ATTEMPTS, and # BARGE-INS. These parameters can either be used as raw values, or normalized to the total number of turns.

Users' perceptions of quality are addressed by a questionnaire which is given to the test subjects after each interaction (usually via a web-based interface). One important judgment in this questionnaire is the perceived task completion $COMP$ which is collected as a binary rating to the question of whether the users thought they had completed the task or not. The other questions address very diverse aspects of user satisfaction (TTS performance, ASR performance, task ease, interaction pace, user expertise, system response, expected behavior, future use, sometimes also comparable interface) and have already been listed in Section 3.8.6. The questions are rated on 5-point ACR scales. A cumulative rating of "user satisfaction" is then computed by summing up the individual ratings, resulting in a US_w measure for each dialogue, ranging between 8 and 40 (9 to 45 for 9 questions). This measure is the target variable of the PARADISE model.

Given the definition of task success measures, dialogue cost measures, and user satisfaction, an estimation of US_w can be calculated as follows:

$$\widehat{US_w} = \alpha \cdot \mathcal{N}(\kappa) - \sum_{i=1}^{n} w_i \cdot \mathcal{N}(c_i) \tag{6.3.1}$$

In this equation, α is the weight of κ, w_i the weight for the cost measures c_i, and $\mathcal{N}$ is the Z-score normalization function

$$\mathcal{N}(x) = \frac{x - \bar{x}}{\sigma_x} \tag{6.3.2}$$

normalizing a variable x with mean value $\bar{x}$ and standard deviation σ_x to a mean of zero and a unity standard deviation. Thus, US_w as an estimation of system performance is assumed to be additively composed of maximal task success and minimal dialogue costs. The Z-score normalization is necessary because κ and c_i are calculated on different scales.

The parameters and the relevant cost measures c_i can be derived from corpus data which has been collected under controlled conditions, by applying a multivariate linear regression with US_w as the dependent variable, and normalized task success κ, normalized dialogue quality and dialogue efficiency measures as the independent variables. The regression produces a set of weighting coefficients (α and w_i) describing the relative contribution of each predictor (task success or cost measure) in accounting for the variance in the predicted variable US_w. Once the function has been determined for a particular system and set of data, it is expected to be useful for predicting user satisfaction for other user groups or system configurations, and for comparison between systems.

6.3.1.2 Application Examples

Several application examples of the PARADISE model are reported, mainly addressing systems for voice dialling, email reading, or train timetable information. Because the measures of the prediction function are normalized, the individual weighting factors can be directly interpreted as the significance of the individual quality aspects to the integral US_w rating. The amount of the variance in US_w which is accounted for by the model is defined by the R^2 variable of the regression analysis.

At AT&T, the model was applied to the email reader ELVIS, the voice dialling system Annie, and the train timetable system TOOT. Different system versions have been evaluated, which resulted in a number of different performance functions:

For ELVIS (Walker et al., 1998b, no indication of R^2):

$$\widehat{US_w} = 0.20 \cdot \mathcal{N}(COMP) + 0.45 \cdot \mathcal{N}(MRS) - 0.23 \cdot \mathcal{N}(DD) \quad (6.3.3)$$

For Annie (Kamm et al., 1998, $R^2 = 0.41$):

$$\widehat{US_w} = 0.33 \cdot \mathcal{N}(COMP) + 0.25 \cdot \mathcal{N}(MRS) - 0.33 \cdot \mathcal{N}(\#\ \text{HELP REQUESTS}) \quad (6.3.4)$$

For TOOT (Walker et al., 1998b, $R^2 = 0.47$):

$$\widehat{US_w} = 0.45 \cdot \mathcal{N}(COMP) + 0.35 \cdot \mathcal{N}(MRS) - 0.42 \cdot \mathcal{N}(\#\ \text{BARGE-INS}) \quad (6.3.5)$$

In all models, $COMP$ and MRS seem to play a dominant role. Efficiency-related measures like DD are only of secondary importance. This finding is interpreted in the sense that the qualitative system behavior associated with poor recognition performance and the predictability of system behavior are more important than the commonly assumed efficiency factors. The # HELP REQUESTS parameter is a significant predictor for the Annie system, which may be due to the fact that users with different levels of experience took part in the experiment. Interestingly, not the κ coefficient is selected by the regression algorithm for describing task success, but the binary $COMP$ rating which is collected from the test subjects. It seems that the user's perception of task success is not always a reliable indicator for the correctness of the final task solution AVM. This observation is congruent with the findings of the correlation analyses in Section 6.2.4. Other data sets for the mentioned systems gave slightly different model functions (Walker et al., 1998a).

PARADISE can also be applied to a mixed set of data obtained with two different systems. Walker et al. (1998b) report on the determination of a model function for a cumulative data set of ELVIS and TOOT dialogues. The model function results in

$$\widehat{US_w} = 0.23 \cdot \mathcal{N}(COMP) + 0.43 \cdot \mathcal{N}(MRS) - 0.21 \cdot \mathcal{N}(DD) \quad (6.3.6)$$

Apparently, the factors which appear in the individual system functions (see Formulae 6.3.3 and 6.3.5) generalize across the combined set. The function also shows that there seems to be a trade-off between recognizer performance (MRS) and efficiency (DD) which has already been reported earlier (Danieli and Gerbino, 1995). Users seem to be more concerned about smoothness aspects of the dialogue, which may be related to ASR performance and the confusion that is caused by misunderstandings, and they are not very accurate perceivers of efficiency. This result agrees with the low correlation observed between efficiency measures and efficiency-related questions in experiment 6.3.

Cross-system validity of model predictions has also been addressed in a later publication of the PARADISE authors (Walker et al., 2000a). Depending on the data sets used for training and testing, the models differed slightly in their

Table 6.34. Predictive power of the PARADISE model for different training and test data sets (Walker et al., 2000a). Percentages indicate the percentage of the dialogues which have been used for training and testing. $MRS \leq 0.95$ and $MRS > 0.95$ indicate user groups with a low or high mean recognition score, respectively.

Training set	*Training R^2*	*Test set*	*Test R^2*
Annie 90%	0.50	Annie 10%	0.40
ELVIS 90%	0.39	ELVIS 10%	0.43
TOOT 90%	0.56	TOOT 10%	0.54
All 90%	0.47	All 10%	0.50
ELVIS 90%	0.42	TOOT	0.55
ELVIS 90%	0.42	Annie	0.36
All Novices	0.47	Annie Experts	0.04
All $MRS \leq 0.95$	0.46	All $MRS > 0.95$	0.23

weighting coefficients and contributing variables, as well as in their predictive power R^2 which is usually in the range between 0.39 and 0.56. The experimental results reproduced in Table 6.34 show that some generalization seems to be possible, i.e. a model trained on one system can be used as a predictor for another system without losing prediction accuracy ($R^2 = 0.36...0.55$). Apparently, there are general predictors of user satisfaction which seem to be independent from the set of training data. On the other hand, generalization from one group of users (e.g. novices) to another (e.g. experts) proved to be impossible. The most significant contributions came in all but one case from ASR performance (*MRS*) and perceived task success (*COMP*), followed by the # ASR REJECTIONS parameter. The model accuracy starts to saturate from about 200 training dialogues onwards, suggesting that the use of larger corpora may not increase the predictive power of the model. It has to be noted that all systems have been set up and evaluated in the same laboratory (AT&T). It is thus probable that individual components and strategies of the systems were identical.

Bel et al. (2002) applied PARADISE to data collected with a multilingual telephone-based email reader. A large number of cost measures have been taken as potential input parameters, including parameters related to ASR performance (*WER*, parser error rate, *IC*, # BARGE-INS, # ASR REJECTIONS), language identification performance (accuracy rate), text verification performance, and to the overall dialogue (*DD*, # SYSTEM TURNS, # USER TURNS, # TIME-OUT PROMPTS, # HELP REQUESTS, the percentage of turns in which the user apparently was lost in the dialogue, etc.). Subjective ratings were collected on 12 Likert scales, and a summed mean rating was taken as the target variable of the model. Depending on the system configuration, the percentage of turns during which the user apparently was lost in the dialogue, # BARGE-INS and *DD* were among the significant predictors. Only for one system configuration was the *COMP* judgment a significant predictor. The authors noted that it was not possible to apply a model which has been trained on Castilian Spanish system data to data which was collected with a Catalan system.

At LIMSI, two French information systems (the ARISE train timetable information system and the PARIS-SITI tourist information system) were evaluated with PARADISE (Bonneau-Maynard et al., 2000). Three different ways to describe user satisfaction were taken as the model's target variable: A mean rating on a 10-point scale of "overall system performance" (SAT), a mean over 7 ratings on scales similar to the ones used by Walker (US_w), or a combination of SAT, US_w and task success ($QComb$). Cost measures included word accuracy, literal understanding accuracy, contextual understanding accuracy, a speech generation error rate, a history management error rate, a user repetition rate, and the task completion rate. Relatively stable predictors of SAT across sets of subjects and tasks were the contextual understanding accuracy, the word accuracy, and the user repetition rate. Contextual understanding was a more significant predictor than literal understanding in all cases. For the PARIS-SITI system, $QComb$ gave the best explanation of the variance ($R^2 = 0.44$, 0.42 for US_w and 0.37 for SAT), whereas the opposite was true for the ARISE system ($R^2 = 0.43$ for SAT, 0.42 for $QComb$, and 0.34 for US_w). Smeele and Waals (2003) also observed an increase of the predicted variance when augmenting the judgments on the scales used by Walker (US_w) with an additional rating on "overall quality" (rise of R^2 from 0.47 to 0.61). Apparently, it is important how user satisfaction is calculated from the users' judgments. The target variable of the PARADISE model will be further addressed in Section 6.3.2.

6.3.1.3 Variations

Apart from the standard application of the multivariate regression analysis, the idea underlying PARADISE has been applied in a number of variations. Assuming that the predictors derived for the whole dialogue generalize for the prediction of local performance, a model function derived from the global dialogue can be used on the sub-dialogue level. Walker et al. (1997) state that this assumption has a sound theoretical basis in dialogue structure theory. They propose to apply PARADISE on the sub-dialogue level in order to compare specific strategies which relate to the dialogue sub-tasks. Bel et al. (2002) propose to apply PARADISE for predicting component-specific user judgments, for example judgments related to the dialogue manager or the ASR component. This would allow for a better insight into the correlation between the users' perceptions of the behavior of the different components and the component-specific measures. However, the approach assumes that the user is able to judge the individual components separately, an assumption which is certainly not really satisfied.

Instead of the standard multivariate linear regression, other mathematical methods can be used to estimate the relevant predictors of user satisfaction. The PARADISE authors compare linear regression models to tree regression models (Walker et al., 2000b). Both methods reach a similar predictive power, in terms that the proportion reduction in error, PRE, comparable to R^2 for continuous dependent variables, is similar to that observed for the linear regression models (0.37...0.56). For the prediction of dialogue duration on the basis of speech-understanding-related parameters, Higashinaka et al. (2003) found a slightly better performance with non-linear regression models.

Hastie et al. (2002a) used classification and regression trees (CARTs) for determining relevant predictors to user satisfaction. Input variables in their study were dialogue act tags which were automatically extracted from the dialogue with the DATE tool (see Section 3.8.4), as well as other cost measures. Target variables were either a binary or ternary measure of perceived task success (user judgment), or an accumulated user satisfaction rating (similar to US_w). Perceived task completion can be predicted very accurately (85...92% correct predictions) from the DATE tags. Depending on whether automatically or manually determined task success is taken as an input, US_w can be predicted with an accuracy of R^2 =0.23...0.38. An advantage of CARTs is that they can easily be interpreted with respect to the relevant predictors. For the example given in Hastie et al. (2002a), task success, the type of phone access (influencing ASR performance) and the dialogue length were important predictors of US_w. An analysis via Feature Using Frequencies showed that task success accounted for 21% of the discriminatory power of the CART, efficiency measures for 47%, and qualitative measures for 32%. The authors also used CARTs in conjunction with DATE tags to automatically identify problematic dialogues (Hastie et al., 2002b). These dialogues are thought to be particularly useful for identifying dialogue system problems.

Walker et al. (2001) describe a similar approach for predicting user satisfaction for the year 2000 DARPA Communicator corpus. Input parameters were either standard measures similar to the ones indicated in Section 6.3.1.1, or an augmented set including DATE tags. The best model based on standard measures accounts for 37% of the variance in US_w (predictors were task success, DD, STD, WA), whereas the augmented input parameter set increased the predicted variance to 42%. It was observed that – in addition to exact scenario completion – tags related to single subtasks are significant predictors. This shows that the solution of single subtasks may positively influence user satisfaction, although the overall task is not solved.

Beringer et al. (2002b,a) applied an adapted version of the PARADISE model (called PROMISE) to the evaluation of the multimodal SmartKom system. Due to the rather unprecise task definition given to the users (e.g. to search for a TV movie) and to the lack of a fixed reference (e.g. TV guides), the determination of task success turned out to be a major problem. Instead of using standard AVM confusion measures, task success is scored on the basis of "information bits" which can be assigned to different aspects of the proposed task solution. These information bits represent superordinate concepts of a task element, e.g. the title, genre and channel of a TV movie. They have to be carefully selected and weighted by hand before the evaluation starts. The number of information bits can vary within one completed task, but they have to define a task unambiguously. For example, the title of a movie is sufficient in the case that the user mentions this in the first instance, whereas more bits are necessary when the users start by selecting the movie by channel, time, etc. Hjalmarsson (2002) used a similar task concept for the evaluation of the multimodal AdApt system, counting the number of correctly exchanged concepts with a task success "point". In this way, tasks requiring several concepts to be exchanged are ranked higher than simple tasks involving only one concept. A second difficulty in applying PARADISE to multimodal systems consists in the relation of the user's input modalities, which are similar in functionality and thus may interfere with each other.

6.3.1.4 Discussion

Despite its simplicity, the PARADISE model has proven to be a valuable tool for describing the impact of individual system characteristics (which are reflected in the interaction parameters) on the users' perception of quality. The prediction function which is defined by the model can be used to determine which aspects of a system need improvement, and tentatively also in which direction user satisfaction will develop in the case of system enhancement. It may also serve as a feedback to the dialogue system, adapting the dialogue strategy as a function of the estimated user satisfaction (Walker et al., 1998b; Walker, 2004). If metrics representing dialogue strategies are included in the qualitative measures, then the significant predictors of user satisfaction may indicate which dialogue strategies are optimal, and facilitate the choice of appropriate strategies. However, because the individual predictors of US_w somehow depend on the characteristics of specific system components, a change in one component will not leave the model's prediction function unaffected. For example, if the mean recognition score is a significant predictor in the case of poor ASR performance, then it might be far less important when ASR performance increases to a level which makes other dialogue aspects more prominent. Integral judgments like user satisfaction will always depend on the interaction of all system components during the dialogue.

It is interesting to note that nevertheless the model reaches a certain degree of cross-system and cross-domain validity. The cited investigations carried out by the PARADISE authors show that nearly the same amount of variance can be captured when the model is trained on data from a different system which has nevertheless been obtained under the same circumstances. However, when the user group changes (e.g. from novices to experts), prediction accuracy will suffer significantly. Cross-laboratory application of PARADISE models have not yet been reported. It has to be noted that the model achieves best performance when $COMP$ instead of κ is used for describing task success. Because $COMP$ has to be determined with the help of test subjects, the final aim of a quality prediction model – namely to become independent of user judgments – is lost in this way.

A main limitation of PARADISE is that it tries to predict an amalgam of user satisfaction which is calculated as a simple mean over user judgments on quality aspects as different as TTS quality, perceived speech recognition and understanding, task-related issues, or usability and acceptability. Calculation of a mean implies that each question has the same weight for user satisfaction. Both empirical and theoretical evidence for this assumption is lacking. As a consequence, the target variable of the model is highly questionable. This topic will be addressed in the set of experiments reported below.

Additional limitations of the model result from the simple structure of Formula 6.3.1 and from the model's input parameters (Walker et al., 1998b). Due to the simple linear combination of task success and dialogue costs, no interaction terms can be accounted for by the model. Other tentative indicators of user satisfaction have been proposed, e.g. a product of task success and the average number of concepts per turn (Hacioglu and Ward, 2002), but they have still not been validated against experimental data. The κ coefficient of task success does not adequately handle the quality of the task solution. Namely, some solutions might be better than others in the perception of the user, even when they have higher dialogue costs. The current measures of task success do not distinguish between providing a wrong solution, or not providing any solution at all. The limitations in describing task success may be an important reason why the $COMP$ judgment regularly achieves better prediction performance than the κ measure.

6.3.2 Quality Prediction Models for Experiment 6.3 Data

The PARADISE model establishes a simple relationship between a set of input parameters (task success and "dialogue costs") and a target variable (predicted "user satisfaction") via a linear regression function. In order to analyze the impact of input and output parameters in more detail, this paradigm will now be applied to the experiment 6.3 data. Several possible combinations of input and output parameters will be compared with respect to the amount of variance in the data which is covered by the model (R^2), as well as with respect to the significant predictors of the target variable. The following constellations will be tested:

- *Target parameters*: The target parameter shall be an indicator of user satisfaction. In the PARADISE model, a mean over the subjective ratings for 8 or 9 questions is used for this purpose (US_w). These questions roughly correspond to questions B5, B7, B8, B10, B15, B20, C15, C16 and C18 of the experiment 6.3 questionnaire. Thus, they comprise of judgments addressing the individual system configurations (part B questions), and others reflecting the overall experience with the system (part C questions). In the QoS taxonomy, questions B0, B23 and C1 have been classified in the category "user satisfaction". In order to investigate the differences between these targets, the following parameters have been chosen:
  - Arithmetic mean over the judgments on the 9 questions cited above, US_w
  - Judgment on question B0 (overall impression)
  - Judgment on question B23 (user satisfaction)
  - Arithmetic mean over the judgments on questions B0 and B23, MEAN (B0,B23)
  - Arithmetic mean over all judgments on part B questions, MEAN(B)
- *Input parameters reflecting "dialogue costs"*: Input parameters are interaction parameters which may be measured either instrumentally, or with the help of experts. Parameters which are frequently used by the PARADISE authors include # BARGE-INS, *DD*, # ASR REJECTIONS, # HELP REQUESTS, # SYSTEM TURNS, # USER TURNS, and a mean recognition score which is either equivalent to IC, or calculated as a mean of IC and # ASR REJECTIONS. The parameters # ASR REJECTIONS and # HELP REQUESTS were not included in the data analysis, see Section 6.2.1. A time-out capability was not implemented in the current system version, and thus the # TIME-OUT PROMPTS parameter cannot be used here. Because of the strict alternation of system and user turns, # SYSTEM TURNS and # USER TURNS are perfectly correlated and can be replaced by a general # TURNS parameter. Two sets of input parameters were finally used for modelling:
  - Set 1: Contains the parameters used by Walker, provided that they could be meaningfully collected in the experiment. This set includes *DD*, IC, # TURNS, and # BARGE-INS.
  - Set 2: Contains all available parameters which have been analyzed in the previous sections, excluding the variables which did not prove to

be informative in the data pre-analysis. This set includes DD, STD, UTD, SRD, URD, # TURNS, $WPST$, $WPUT$, # BARGE-INS, # SYSTEM ERROR MESSAGES, # SYSTEM QUESTIONS, # USER QUESTIONS, AN:CO, AN:PA, AN:FA, $DARPA_s$, $DARPA_{me}$, PA: CO, PA: PA, PA:FA, SCR, UCR, CA:AP, CA:IA, IR, IC, UA, WA, WER, $\overline{NES}$, $\overline{WES}$, WA_{iso}, WER_{iso}, $\overline{NES_{iso}}$, $\overline{WES_{iso}}$.

In both sets, the turn-related parameters have been normalized to the overall number of turns (or for the AN parameters to the number of user questions), as is described in Section 6.2.1.

- *Input parameters reflecting task success*: Although the basic formula of the PARADISE model contains κ as a mandatory input parameter, it has been often replaced by the user judgment on task completion, $COMP$, in the practical application of the model. This $COMP$ parameter roughly corresponds to a binary version of the judgment on question B1. For the analysis of the experiment 6.3 data, the following options for describing task success have been chosen:
  - Task success calculated on the basis of the AVM, either on a per-dialogue level (κ_{dia}), or on a per-configuration level (κ_{conf}).
  - Task success measures based on the overall solution, TS_{bin} and TS_{ord}.
  - User judgment on question B1.
  - A binary version of B1 (B1_{bin}), calculated by assigning a value of 0 for a rating B1 ≤ 3.0 and a value of 1 for a rating B1 > 3.0.

It should be noted that B1 and B1_{bin} are calculated on the basis of user judgments. Thus, using one of these parameters as an input to a prediction model is not in line with the general idea of quality prediction, namely to become independent of direct user ratings.

Apart from the input and output parameters, the choice of the regression approach carries an influence on the resulting model. A linear multivariate analysis like the one used in PARADISE has been chosen here. The choice of parameters which are included in the regression function depends on the amount of available parameters. For set 1, a forced inclusion of all four parameters has been chosen. For set 2, a stepwise inclusion method is more appropriate, because of the large number of input parameters. The stepwise method sequentially includes the variables with the highest partial correlation values with the target variable (forward step), and then excludes variables with the lowest partial correlation (backward step). In case of missing values, the corresponding cases have been excluded from the analysis for the set 1 data (listwise exclusion). For set 2, such an exclusion would lead to a relatively low

Table 6.35. Multivariate regression models for the experiment 6.3 data, using input parameters of set 1. Indicated are the R^2 values for different combinations of target variables and task-success-related input parameters. Listwise exclusion of missing cases, forced inclusion of all input parameters.

Target parameter	*Task success input parameter*					
	κ_{dia}	κ_{conf}	TS_{bin}	TS_{ord}	B1	$B1_{bin}$
US_w	0.139	0.138	0.150	0.154	0.245	0.236
B0	0.154	0.153	0.164	0.165	0.445	0.386
B23	0.122	0.120	0.194	0.191	0.481	0.410
MEAN(B0, B23)	0.150	0.149	0.190	0.189	0.517	0.445
MEAN(B)	0.162	0.156	0.194	0.197	0.517	0.440

number of valid cases; therefore, the missing values have been replaced by the corresponding mean value instead.

In Table 6.35, the amount of variance covered by models with set 1 input parameters is listed for different target variables and task-success-related input parameters. When the κ coefficient is used for describing task success, the amount of covered variance is very low ($R^2 = 0.12...0.16.$). Unfortunately, the authors of PARADISE do not provide example models which include the κ coefficient. Instead, their models rely on the $COMP$ measure. Making use of the $B1_{bin}$ parameter (which is similar to $COMP$) increases R^2 to 0.24...0.45, which is in the range of values given in the literature. The model's performance can be further increased by using the non-simplified judgment on question B1 for describing task success. In this case, R^2 reaches 0.52, a value which is amongst the best of Table 6.34. Task success measures which are based on the overall solution (TS_{bin} and the modified version TS_{ord}) provide slightly better estimations than κ_{dia} and κ_{conf}, but they are not competitive with the subject-derived measures B1 and $B1_{bin}$. Apparently, the PARADISE model performs best when it is not completely relying on interaction parameters, but when subjective estimations of task success are included in the estimation function. This finding is in line with comparative experiments described by Bonneau-Maynard et al. (2000). When using subjective judgments of task success instead of the κ coefficient, the amount of predicted variance raised from 0.41 to 0.48.

Comparing the performance for the different target variables, US_w seems to be least predictable. The amount of covered variance is significantly lower than in the experiments described by Walker. The relatively low number of input parameters in set 1 may be a major reason for this finding. Prediction accuracy significantly raises when B0 or B23 are taken as the target parameter, and with B1 or $B1_{bin}$ describing task success. A further improvement is observed when

Table 6.36. Influence of different task-success-related input parameters on multivariate regression models for the experiment 6.3 data, using input parameters of set 1. Listwise exclusion of missing cases, forced inclusion of all input parameters.

Model parameters			*Significant predictors*	R^2
Target	*Dialogue cost input*	*Task success input*		
US_w	set 1	κ_{dia}	-0.022 *DD* - 0.359 # TURNS -0.088 # BARGE-INS -0.020 IC -0.019 κ_{dia}	0.139
US_w	set 1	κ_{conf}	-0.013 *DD* -0.364 # TURNS -0.089 # BARGE-INS -0.022 IC -0.005 κ_{conf}	0.138
US_w	set 1	TS_{bin}	-0.055 *DD* -0.339 # TURNS -0.093 # BARGE-INS -0.067 IC + 0.118 TS_{bin}	0.150
US_w	set 1	TS_{ord}	- 0.010 *DD* -0.360 # TURNS -0.100 # BARGE-INS -0.068 IC -0.132 TS_{ord}	0.154
US_w	set 1	B1	- 0.049 *DD* -0.236 # TURNS -0.074 # BARGE-INS -0.045 IC + 0.339 B1	0.245
US_w	set 1	$B1_{bin}$	-0.014 *DD* -0.290 # TURNS -0.059 # BARGE-INS -0.035 IC + 0.320 $B1_{bin}$	0.236
MEAN(B)	set 1	κ_{dia}	-0.062 *DD* - 0.353 # TURNS -0.065 # BARGE-INS -0.035 IC -0.080 κ_{dia}	0.162
MEAN(B)	set 1	κ_{conf}	-0.019 *DD* -0.373 # TURNS -0.069 # BARGE-INS + 0.006 IC +0.014 κ_{conf}	0.156
MEAN(B)	set 1	TS_{bin}	-0.096 *DD* -0.329 # TURNS -0.076 # BARGE-INS -0.060 IC + 0.208 TS_{bin}	0.194
MEAN(B)	set 1	TS_{ord}	-0.018 *DD* -0.367 # TURNS -0.087 # BARGE-INS -0.055 IC + 0.215 TS_{ord}	0.197
MEAN(B)	set 1	B1	-0.097 *DD* -0.124 # TURNS -0.039 # BARGE-INS -0.027 IC + 0.630 B1	0.517
MEAN(B)	set 1	$B1_{bin}$	-0.031 *DD* -0.235 # TURNS -0.015 # BARGE-INS -0.007 IC + 0.550 $B1_{bin}$	0.440

the target parameter is calculated as a mean over several ratings, namely as MEAN(B0, B23) or MEAN(B). The model's performance is equally high in these cases. Apparently, the smoothing of individual judgments which is inherent to the calculation of the mean has a positive effect on the model's prediction accuracy.

Table 6.36 shows the significant predictors for different models determined using the set 1 "dialogue cost" parameters and different task-success-related parameters as the input. Target variables are either the US_w or the MEAN(B) parameter. For US_w, most significant dialogue cost contributions come from # TURNS (with a negative sign), and partly also from the # BARGE-INS parameter (negative sign). *DD* and IC only play a subordinate role in predicting US_w. For the task-success-related parameters, a clear order can be observed: B1 and $B1_{bin}$ have a dominant effect on US_w (both with a positive sign), TS_{bin} and TS_{ord} only a moderate one (the first with a positive and the second with a

Table 6.37. Influence of different target parameters on multivariate regression models for the experiment 6.3 data, using input parameters of set 1. Listwise exclusion of missing cases, forced inclusion of all input parameters.

Model parameters			*Significant predictors*	R^2
Target	*Dialogue cost input*	*Task success input*		
US_w	set 1	TS_{bin}	-0.055 *DD* -0.339 # TURNS -0.093 # BARGE-INS -0.067 IC + 0.118 TS_{bin}	0.150
B0	set 1	TS_{bin}	-0.216 *DD* -0.168 # TURNS -0.155 # BARGE-INS + 0.000 IC + 0.117 TS_{bin}	0.164
B23	set 1	TS_{bin}	-0.280 *DD* -0.077 # TURNS -0.140 # BARGE-INS - 0.026 IC + 0.293 TS_{bin}	0.194
MEAN(B0,B23)	set 1	TS_{bin}	-0.262 *DD* -0.125 # TURNS -0.153 # BARGE-INS -0.009 IC + 0.219 TS_{bin}	0.190
MEAN(B)	set 1	TS_{bin}	-0.096 *DD* -0.329 # TURNS -0.076 # BARGE-INS -0.060 IC + 0.208 TS_{bin}	0.194
US_w	set 1	$B1_{bin}$	-0.014 *DD* -0.290 # TURNS -0.059 # BARGE-INS -0.035 IC + 0.320 $B1_{bin}$	0.236
B0	set 1	$B1_{bin}$	-0.235 *DD* -0.023 # TURNS -0.102 # BARGE-INS + 0.021 IC + 0.497 $B1_{bin}$	0.386
B23	set 1	$B1_{bin}$	-0.184 *DD* + 0.001 # TURNS -0.076 # BARGE-INS + 0.064 IC + 0.533 $B1_{bin}$	0.410
MEAN(B0,B23)	set 1	$B1_{bin}$	-0.189 *DD* -0.038 # TURNS -0.093 # BARGE-INS + 0.050 IC + 0.559 $B1_{bin}$	0.445
MEAN(B)	set 1	$B1_{bin}$	-0.031 *DD* -0.235 # TURNS -0.015 # BARGE-INS -0.007 IC + 0.550 $B1_{bin}$	0.440

negative sign), and κ_{dia} and κ_{conf} are nearly irrelevant in predicting US_w. For MEAN(B) as the target, the situation is very similar. Once again, # TURNS is a persistent predictor (always with a negative sign), and *DD*, IC and # BARGE-INS only have minor importance. The task-success-related input parameters show the same significance order in predicting MEAN(B): B1 and $B1_{bin}$ have a strong effect (positive sign), TS_{bin} and TS_{ord} a moderate one (also positive sign), and κ_{dia} and κ_{conf} are not important predictors. Apparently, the PARADISE model is strongly dependent on the type of the input parameter describing task success.

The prediction results for different target variables are depicted in Table 6.37, both for the expert-derived TS_{bin} parameter and for the user-derived $B1_{bin}$ parameter describing task success. The most important contributors for the prediction of US_w are # TURNS (negative sign) and the task-success-related parameter. For predicting B0, also *DD* and # BARGE-INS (both negative sign) play a certain role. B23 and MEAN(B0, B23) seem to be better predicted from

Table 6.38. Multivariate regression models for the experiment 6.3 data, using input parameters of set 2. Indicated are the R^2 values for different combinations of target variables and task-success-related input parameters, and in brackets the number of parameters included in the respective model. *: Task success parameter was not significant, and consequently not included in the model. Missing cases are replaced by the mean value, stepwise inclusion of parameters.

Target parameter	*Task success input parameter*					
	κ_{dia}	κ_{conf}	TS_{bin}	TS_{ord}	B1	$B1_{bin}$
US_w	0.409 (9*)	0.409 (9*)	0.409 (9*)	0.409 (9*)	0.349 (4)	0.429 (8)
B0	0.259 (6*)	0.259 (6*)	0.281 (7)	0.280 (7)	0.505 (7)	0.507 (9)
B23	0.144 (3*)	0.144 (3*)	0.289 (7)	0.332 (9)	0.464 (4)	0.382 (2)
MEAN(B0, B23)	0.279 (8*)	0.279 (8*)	0.294 (7)	0.303 (7)	0.546 (6)	0.430 (2)
MEAN(B)	0.297 (4*)	0.297 (4*)	0.350 (6)	0.467 (8)	0.585 (5)	0.623 (10)

DD and the task-success-related parameter; here, the # TURNS parameter is relatively irrelevant. For predicting MEAN(B), the most significant contributions come from $B1_{bin}$, # TURNS and TS_{bin}. As may be expected, the different target parameters related to user satisfaction require different input parameters for an adequate prediction. Thus, the models established by the multivariate regression analysis are only capable of predicting different indicators of user satisfaction to a limited extent.

The number of input parameters in set 1 is very restricted (four "dialogue cost" parameters and one task-success-related parameter). Taking the set 2 parameters as an input, it can be expected that more general aspects of quality are covered by the resulting models. An overview of the achievable variance coverage is given in Table 6.38. In general, the coverage is much better than was observed for set 1. Using the interaction parameters TS_{ord} or TS_{bin} for describing task success, R^2 raises to 0.28...0.47, depending on the target parameter. With B1 or $B1_{bin}$, an even better coverage can be reached ($R^2 \sim 0.35...0.62$). As was observed for the set 1 data, it seems to be important to include subject-derived estimations of task success in the prediction function. Expert-derived parameters like κ_{dia}, κ_{conf}, TS_{bin} or TS_{ord} are far less efficient in predicting indicators of user satisfaction. Interestingly, the κ_{dia} and κ_{conf} parameters are never selected by the stepwise inclusion algorithm. Thus, the low importance of these parameters in the prediction function (see Table 6.36) is confirmed for the augmented set of input parameters. Overall, the prediction functions include a relatively large number of input parameters. However, the amount of variance covered by the function does not seem to be strictly related to the number of input parameters, as the results in the final row or column of Table 6.38 show.

Taking subject-derived estimations of task success as an input, the best prediction results can be obtained for B0 and MEAN(B). Prediction functions with a good data coverage can be obtained especially in the latter case. The R^2 values in these cases exceed the best results reported by Walker et al. (2000a), see Table 6.34. However, it has to be noted that a larger number of input parameters are used in these models. For the set 2 data, no clear tendency towards better results for the prediction of smoothed arithmetic mean values (US_w, MEAN(B), MEAN(B0,B23)) can be observed. In summary, the augmented data set leads to far better prediction results, with a wider coverage of the resulting prediction functions.

Table 6.39 shows the resulting prediction functions for different task-success-related input parameters. The following parameters seem to be stable contributors to the respective targets:

- Measures of communication efficiency: Most models include either the $WPST$ and SRD parameters (positive sign), STD (negative sign), or # TURNS (negative sign). The latter two parameters seem to indicate a preference for shorter interactions, whereas the positive sign for the $WPST$ parameter indicates the opposite, namely that a talkative system would be preferred. A higher value for SRD is in principle linked to longer user utterances which require an increased processing time from the system/wizard. No conclusive explanation can be drawn with respect to the communication efficiency measures.

- Measures of appropriateness of system utterances: All prediction functions contain the $CA{:}AP$ parameter with a positive sign. Two models of Table 6.39 also contain $CA{:}IA$ (positive sign), which seems to rule out a part of the high effect of $CA{:}AP$ in these functions. In any case, dialogue cooperativity proves to be a significant contributor to user satisfaction.

- Measures of task success: The task-success-related parameters do not always provide an important contribution to the target parameter, except for B1 which is in both cases a significant contributor. In the model estimated from the first four input parameter sets (identical model), task success is completely omitted.

- Measures of initiative: Most models contain the # SYSTEM QUESTIONS parameter, with a positive sign. Apparently, the user likes systems which take a considerable part of the initiative. Only one model contains the # USER QUESTIONS parameter.

- Measures of meta-communication: Two parameters are frequently selected in the models. The $PA{:}PA$ parameter (positive sign) indicates that partial

Table 6.39. Influence of different task-success-related input parameters on multivariate regression models for the experiment 6.3 data, using input parameters of set 2. Missing cases are replaced by the mean value, stepwise inclusion of parameters.

Model parameters			*Significant predictors*	R^2
Target	*Dialogue cost input*	*Task success input*		
US_w	set 2	κ_{dia}	1.103 $WPST$ + 0.308 $PA{:}PA$ + 0.218 # SYSTEM QUESTIONS + 0.317 $CA{:}AP$ + 0.287 SCR -0.144 # BARGE-INS -1.020 STD + 0.447 SRD + 0.153 WA_{iso}	0.409
US_w	set 2	κ_{conf}	1.103 $WPST$ + 0.308 $PA{:}PA$ + 0.218 # SYSTEM QUESTIONS + 0.317 $CA{:}AP$ + 0.287 SCR -0.144 # BARGE-INS -1.020 STD + 0.447 SRD + 0.153 WA_{iso}	0.409
US_w	set 2	TS_{bin}	1.103 $WPST$ + 0.308 $PA{:}PA$ + 0.218 # SYSTEM QUESTIONS + 0.317 $CA{:}AP$ + 0.287 SCR -0.144 # BARGE-INS -1.020 STD + 0.447 SRD + 0.153 WA_{iso}	0.409
US_w	set 2	TS_{ord}	1.103 $WPST$ + 0.308 $PA{:}PA$ + 0.218 # SYSTEM QUESTIONS + 0.317 $CA{:}AP$ + 0.287 SCR -0.144 # BARGE-INS -1.020 STD + 0.447 SRD + 0.153 WA_{iso}	0.409
US_w	set 2	B1	0.326 B1 + 0.333 $PA{:}PA$ + 0.264 $CA{:}AP$ + 0.131 # USER QUESTIONS	0.349
US_w	set 2	$B1_{bin}$	0.260 $B1_{bin}$ + 0.251 $PA{:}PA$ + 0.751 $CA{:}AP$ + 0.964 $WPST$ + 0.169 # SYSTEM QUESTIONS + 0.607 $CA{:}IA$ -0.987 STD + 0.447 SRD	0.429
MEAN(B)	set 2	κ_{dia}	-0.270 # TURNS + 0.564 $CA{:}AP$ + 0.414 SCR + 0.208 $PA{:}PA$	0.297
MEAN(B)	set 2	κ_{conf}	-0.270 # TURNS + 0.564 $CA{:}AP$ + 0.414 SCR + 0.208 $PA{:}PA$	0.297
MEAN(B)	set 2	TS_{bin}	-0.258 # TURNS + 0.510 $CA{:}AP$ + 0.365 SCR + 0.267 $PA{:}PA$ + 0.214 TS_{bin} + 0.132 # SYSTEM QUESTIONS	0.350
MEAN(B)	set 2	TS_{ord}	-0.277 # TURNS + 0.262 TS_{ord} +0.224 $PA{:}PA$ + 0.540 $CA{:}AP$ + 0.385 SCR -1.564 STD + 0.806 SRD + 1.037 $WPST$	0.467
MEAN(B)	set 2	B1	0.612 B1 + 0.224 $PA{:}PA$ + 0.258 $CA{:}AP$ + 0.153 # SYSTEM QUESTIONS -0.164 IR	0.585
MEAN(B)	set 2	$B1_{bin}$	0.493 $B1_{bin}$ -0.141 # TURNS + 0.171 $PA{:}PA$ + 0.873 $CA{:}AP$ + 0.630 $CA{:}IA$ -1.421 STD -0.145 IR + 1.008 $WPST$ + 0.689 SRD + 0.115 # SYSTEM QUESTIONS	0.623

system understanding seems to be a relevant factor for user satisfaction. The SCR parameter is an indicator for corrected misunderstandings. It is always used with a positive sign.

Table 6.40. Influence of different target parameters on multivariate regression models for the experiment 6.3 data, using input parameters of set 2. Missing cases are replaced by the mean value, stepwise inclusion of parameters.

Model parameters			*Significant predictors*	R^2
Target	*Dialogue cost input*	*Task success input*		
US_w	set 2	TS_{bin}	1.103 $WPST$ + 0.308 $PA{:}PA$ + 0.218 # SYSTEM QUESTIONS + 0.317 $CA{:}AP$ + 0.287 SCR -0.144 # BARGE-INS -1.020 STD + 0.447 SRD + 0.153 WA_{iso}	0.409
B0	set 2	TS_{bin}	-0.249 # TURNS + 0.389 $CA{:}AP$ + 0.225 $PA{:}PA$ -0.177 # BARGE-INS + 0.232 SCR + 0.176 # SYSTEM QUESTIONS + 0.157 TS_{bin}	0.281
B23	set 2	TS_{bin}	-0.327 DD + 0.287 TS_{bin} + 0.242 $PA{:}PA$ -0.163 $WPUT$ -0.392 $CA{:}IA$ + 0.292 SCR -0.149 # BARGE-INS	0.289
MEAN(B0,B23)	set 2	TS_{bin}	-0.279 DD + 0.244 TS_{bin} +0.206 $PA{:}PA$ -0.403 $CA{:}IA$ -0.174 # BARGE-INS + 0.284 SCR + 0.130 # SYSTEM QUESTIONS	0.294
MEAN(B)	set 2	TS_{bin}	-0.258 # TURNS + 0.510 $CA{:}AP$ + 0.365 SCR + 0.267 $PA{:}PA$ + 0.214 TS_{bin} + 0.132 # SYSTEM QUESTIONS	0.350
US_w	set 2	$B1_{bin}$	0.260 $B1_{bin}$ + 0.251 $PA{:}PA$ + 0.751 $CA{:}AP$ + 0.964 $WPST$ + 0.169 # SYSTEM QUESTIONS + 0.607 $CA{:}IA$ -0.987 STD + 0.447 SRD	0.429
B0	set 2	$B1_{bin}$	0.477 $B1_{bin}$ -0.166 # TURNS + 0.123 # SYSTEM QUESTIONS + 0.179 $PA{:}PA$ + 0.122 $CA{:}AP$ -1.294 STD +0.644 SRD + 0.837 $WPST$ + 0.137 WA	0.507
B23	set 2	$B1_{bin}$	0.543 $B1_{bin}$ - 0.199 DD	0.382
MEAN(B0,B23)	set 2	$B1_{bin}$	0.570 $B1_{bin}$ -0.223 DD	0.430
MEAN(B)	set 2	$B1_{bin}$	0.493 $B1_{bin}$ -0.141 # TURNS + 0.171 $PA{:}PA$ + 0.873 $CA{:}AP$ + 0.630 $CA{:}IA$ -1.421 STD -0.145 IR + 1.008 $WPST$ + 0.689 SRD + 0.115 # SYSTEM QUESTIONS	0.623

The prediction functions differ for the mentioned target parameters, see Table 6.40. Apart from the parameters listed above, new contributors are the dialogue duration (negative sign), the # BARGE-INS parameter (negative sign), and in two cases the word accuracy as well. Whereas the first parameter underlines the significant influence of communication efficiency, the latter introduces speech input quality as a new quality aspect in the prediction function. Two models differ significantly from the others, namely the ones for predicting B23 and MEAN(B0, B23) on the basis of $B1_{bin}$ and the set 2 input parameters. The models are very simple (only two input parameters), but reach a relatively

Table 6.41. Influence of training and test data sets on multivariate regression models for the experiment 6.3 data, using 90% of the dialogues for training and 10% for testing. Indicated are R^2, and in brackets the number of parameters included in the models. Input parameters of set 2, TS_{bin} or $B1_{bin}$ for describing task success, and MEAN(B) as the target variable. Missing cases are replaced by the mean value, stepwise inclusion of parameters.

Training Set No.	*Training data R^2*		*Test data R^2*	
	TS_{bin}	$B1_{bin}$	TS_{bin}	$B1_{bin}$
1	0.342 (5)	0.631 (10)	0.158	0.454
2	0.327 (5)	0.627 (10)	0.404	0.312
3	0.368 (7)	0.552 (6)	0.380	0.748
4	0.417 (7)	0.539 (5)	0.960	0.035
5	0.320 (6)	0.512 (5)	0.145	0.006
6	0.344 (6)	0.509 (6)	0.086	0.314
7	0.370 (6)	0.642 (9)	0.099	0.058
8	0.357 (6)	0.511 (5)	0.216	0.345
9	0.365 (6)	0.609 (10)	0.028	0.133
10	0.343 (6)	0.634 (11)	0.149	0.642
mean	0.355	0.577	0.263	0.305

high amount of covered variance. The relatively high correlation between B1 and B23 ($\rho = 0.642$) may be responsible for this result.

The R^2 values given so far reflect the amount of variance in the *training* data covered by the respective model. However, the aim of a model is to allow for predictions of new, unseen data. Experiments have been carried out to train a model on 90% of the available data, and to test it on the remaining 10% of data. The sets of training and test data can be chosen either in a purely randomized way, i.e. selecting a randomized 10% of the dialogues for testing (comparable to the results reported in Table 6.34), or in a per-subject way, i.e. selecting a randomized set of 4 of the 40 test subjects for testing. The latter way is slightly more independent, as it prevents within-subject extrapolation. Both analyses have been applied ten times, and the amount of variance covered by the training and test data sets (R^2 values) is reported in Tables 6.41 and 6.42.

It turns out that the models show a significantly lower predictive power for the test data than for the training data. The performance on the training data is comparable to the one observed in Table 6.40, namely $R^2 = 0.350$ using TS_{bin} and $R^2 = 0.623$ using $B1_{bin}$ as the input parameter related to task success. For a purely randomized set of unseen test data, the mean amount of covered variance drops to 0.263 with TS_{bin}, and to 0.305 with $B1_{bin}$. The situation is similar when within-subject extrapolation is excluded: Here, the mean R^2 drops to 0.198 with TS_{bin}, and to 0.360 with $B1_{bin}$. In contrast to what has been reported

Table 6.42. Influence of training and test data sets on multivariate regression models for the experiment 6.3 data, using 36 test subjects for training and 4 subjects for testing. Indicated are R^2, and in brackets the number of parameters included in the models. Input parameters of set 2, TS_{bin} or $\mathrm{B1}_{bin}$ for describing task success, and $\mathrm{MEAN(B)}$ as the target variable. Missing cases are replaced by the mean value, stepwise inclusion of parameters.

Left-out	*Training data R^2*		*Test data R^2*	
Speakers	TS_{bin}	$\mathrm{B1}_{bin}$	TS_{bin}	$\mathrm{B1}_{bin}$
1-4	0.387 (6)	0.523 (5)	0.000	0.022
5-8	0.323 (5)	0.510 (6)	0.398	0.505
9-12	0.336 (6)	0.636 (10)	0.279	0.354
13-16	0.352 (5)	0.544 (5)	0.083	0.018
17-20	0.387 (8)	0.600 (8)	0.304	0.486
21-24	0.370 (7)	0.603 (9)	0.024	0.848
25-28	0.354 (6)	0.504 (5)	0.091	0.145
29-32	0.359 (6)	0.668 (10)	0.382	0.203
33-36	0.332 (6)	0.522 (6)	0.192	0.766
37-40	0.341 (5)	0.512 (5)	0.225	0.254
mean	0.354	0.562	0.198	0.360

by Walker et al. (see Table 6.34), the model predictions are more limited to the training data. Several reasons may be responsible for this finding. Firstly, the differences between system versions seem to be larger in experiment 6.3 than in Walker et al. (2000a). Although different functionalities are offered by the systems at AT&T, it is to be expected that the components for speech input and output were identical for all systems. Secondly, the amount of available training data is considerably lower for each system version of experiment 6.3. Walker et al. showed saturation from about 200 dialogues onwards, but these 200 dialogues only reflected three instead of ten different system versions. Finally, several of the parameters used in the original PARADISE version only have limited predictive power for experiment 6.3, e.g. the # BARGE-INS, # ASR REJECTIONS and # HELP REQUESTS parameters, see Section 6.2.1. It can be expected that a linear regression analysis on parameters which are only different from zero in a few cases, will not lead to an optimally fitting curve.

The interaction parameters and user judgments which form the model input have been collected with different system versions. In order to capture the resulting differences in perceived quality, it is possible to build separate prediction models for each system configuration. In this way, model functions for different system versions can be compared, as well as the amount of variance which is covered in each case. Table 6.43 shows models derived for each of the ten system versions of experiment 6.3, as well as the overall model derived

Table 6.43. Influence of the system configuration on multivariate regression models for the experiment 6.3 data, using input parameters of set 1, TS_{bin} for describing task success, and MEAN(B) as the target variable. Listwise exclusion of missing cases, forced inclusion of all input parameters.

Conf. No.	*Significant predictors*	R^2
1	-0.895 *DD* -0.054 # TURNS + 0.581 # BARGE-INS -0.731 IC + 0.300 TS_{bin}	0.609
2	0.502 *DD* -0.623 # TURNS + 0.033 # BARGE-INS -0.298 IC -0.073 TS_{bin}	0.152
3	1.594 *DD* -1.666 # TURNS -0.477 # BARGE-INS -0.211 IC -0.075 TS_{bin}	0.380
4	0.535 *DD* -0.863 # TURNS -0.451 # BARGE-INS + 0.254 IC + 0.464 TS_{bin}	0.504
5	-0.317 *DD* -0.336 # TURNS + 0.157 # BARGE-INS -0.148 IC + 0.516 TS_{bin}	0.369
6	0.152 *DD* -0.926 # TURNS -0.007 IC + 0.263 TS_{bin}	0.581
7	0.664 *DD* -1.235 # TURNS -0.248 IC + 0.373 TS_{bin}	0.406
8	0.006 *DD* -0.749 # TURNS + 0.105 # BARGE-INS -0.181 IC + 0.471 TS_{bin}	0.538
9	1.018 *DD* -1.803 # TURNS -0.283 # BARGE-INS -0.052 IC + 0.691 TS_{bin}	0.735
10	1.111 *DD* -1.622 # TURNS + 0.174 # BARGE-INS + 0.384 IC -0.221 TS_{bin}	0.542
all	-0.096 *DD* -0.329 # TURNS -0.076 # BARGE-INS -0.060 IC + 0.028 TS_{bin}	0.194

for all system versions, using set 1 and TS_{bin} as input parameters. Except for configurations 6 and 7, where the # BARGE-INS parameter is constantly zero, all models include the same input parameters. It turns out that the individual models attribute different degrees of importance (coefficient values) to each input parameter. Unfortunately, the coefficient values cannot easily be interpreted with respect to the specific system configuration. The speech-input-related parameter IC does not show a stronger effect if ASR performance decreases (configurations 6 to 10), nor does the extensive use of TTS have an interpretable effect on the prediction function. The amount of variance covered by the models also differs significantly between the system configurations ($R^2 = 0.15...0.74$). Apparently, the system configuration has a strong influence on *how* and *how well* a prediction model is able to estimate parameters related to user satisfaction.

The same analysis has been carried out for the augmented set of input parameters (set 2 and TS_{bin}). The results are given in Table 6.44. Once again, the amount of covered variance differs significantly ($R^2 = 0.26...0.84$) between the system configurations. Some of the configurations for which set 1 fails to provide an adequate model basis (e.g. configuration 2) can be well covered by the augmented set 2. Input parameters which are frequently included in the prediction function are those related to dialogue cooperativity ($CA{:}AP$ with a positive sign, $CA{:}IA$ with a negative sign), communication efficiency (STD with positive sign, # TURNS with a negative sign), task success (TS_{bin} with a positive sign), and meta-communication handling (SCR with a positive sign).

Table 6.44. Influence of the system configuration on multivariate regression models for the experiment 6.3 data, using input parameters of set 2, TS_{bin} for describing task success, and MEAN(B) as the target variable. Missing cases are replaced by the mean value, stepwise inclusion of parameters.

Conf. No.	*Significant predictors*	R^2
1	-1.380 $PA{:}CO$ + 0.851 UA	0.531
2	1.391 $CA{:}AP$ + 0.937 SCR + 0.641 SRD + 0.349 # SYSTEM QUESTIONS	0.840
3	0.514 SRD	0.264
4	0.711 UA + 0.429 $\overline{NES_{iso}}$ -0.367 # USER QUESTIONS	0.563
5	0.797 STD + 0.331 TS_{bin}	0.631
6	-0.717 # TURNS	0.514
7	0.632 STD + 0.392 TS_{bin}	0.463
8	-0.724 $CA{:}IA$ -0.368 $PA{:}CO$	0.698
9	-0.686 # TURNS + 0.588 TS_{bin}	0.641
10	0.659 $CA{:}AP$ -0.345 # TURNS	0.690
all	-0.258 # TURNS + 0.510 $CA{:}AP$ + 0.365 SCR + 0.267 $PA{:}PA$ + 0.214 TS_{bin} + 0.132 # SYSTEM QUESTIONS	0.350

The contradicting tendencies for the communication-efficiency-related parameters have already been discussed above. Interestingly, speech-input-related parameters are also included in the performance functions, but partly in an opposite sense: UA with a positive sign, $\overline{NES_{iso}}$ with a positive sign, $PA{:}CO$ with a negative sign, and $PA{:}PA$ with a positive sign. No explanation for this finding can be given so far. In conclusion, the regression model functions proved to be highly dependent on the system configuration under test. Thus, generalizability of model estimations – as reported in Section 6.3.1.2 – seems to be very limited for the described experiment. The large differences in the system configurations of experiment 6.3 may be responsible for this finding. Although the systems described by Walker et al. (2000a) differ with respect to their functionality, is is possible that the underlying components and their performance are very similar. Further cross-laboratory experiments are necessary to thoroughly test how generic quality prediction models are.

In the case that system characteristics are known beforehand (which is normally true for system developers), this information can be included in the input parameter set. Because the regression analysis is not able to handle nominally scaled variables with more than two distinct values, the system information has to be coded beforehand. Five coding variables were used for this purpose:

- $conf\_type$: 0 for no confirmation, 1 for explicit confirmation.
- rec_rate: Target recognition rate in percent (already given on an ordinal scale).

Table 6.45. Multivariate regression models for the experiment 6.3 data, using input parameters of set 2 augmented by system-specific information. Indicated are the R^2 values for different combinations of target variables and task-success-related input parameters, and in brackets the number of set 2 parameters included in the respective model. *: Task success parameter was not significant, and consequently not included in the model. Missing cases are replaced by the mean value, stepwise inclusion of parameters, system-specific parameters are always included in the models.

Target parameter	*Task success input parameter*					
	κ_{dia}	κ_{conf}	TS_{bin}	TS_{ord}	B1	$B1_{bin}$
US_w	0.445 (9*)	0.445 (9*)	0.445 (9*)	0.445 (9*)	0.400 (4)	0.465 (8)
B0	0.376 (6*)	0.376 (6*)	0.400 (7)	0.397 (7)	0.575 (7)	0.530 (9)
B23	0.188 (3*)	0.188 (3*)	0.327 (7)	0.386 (9)	0.478 (4)	0.410 (2)
MEAN(B0, B23)	0.356 (8*)	0.356 (8*)	0.376 (7)	0.390 (7)	0.580 (6)	0.497 (2)
MEAN(B)	0.482 (4*)	0.482 (4*)	0.527 (6)	0.554 (8)	0.689 (5)	0.663 (10)

- *voc_m*: 1 for natural male voice uttering the fixed system turns, 0 otherwise.
- *voc_s*: 1 for synthetic male voice uttering the fixed and variable system turns, 0 otherwise.
- *voc_f*: 1 for natural female voice uttering the variable system turns, 0 otherwise.

These variables completely describe the system configuration with respect to the speech output characteristic, the recognition rate, and the confirmation strategy. Table 6.45 shows that the amount of covered variance can be increased for all models in this case (comparison to Table 6.38). The number of input parameters is only increased by the system information; other parameters of set 2 remain unchanged.

The influence of individual parameters coding the system-specific information is depicted in Table 6.46, for different target parameters. In all cases, the most important system information seems to be coded in the *voc_s* and *voc_f* parameters. As has been observed in the analyses of Section 6.2.5.2, the speech output component seems to be the one with the highest impact on overall system quality and user satisfaction. However, speech-output-related information is not covered in any of the interaction parameters. Thus, the increase in overall model coverage can be explained by the new aspect which is introduced with the additional input parameters. In most cases, the *voc_s* parameter carries a negative coefficient, showing that synthetic speech leads to lower user satisfaction scores. In only a few cases the *rec_rate* parameter has a coefficient with a value higher than 0.1 (always with a positive sign). Apparently, the recog-

Table 6.46. Influence of different target parameters on multivariate regression models for the experiment 6.3 data, using input parameters of set 2 augmented by system-specific information. Missing cases are replaced by the mean value, stepwise inclusion of parameters, system-specific parameters are always included in the models.

Model parameters		*Significant predictors*	R^2
Target	*Task success input*		
US_w	TS_{bin}	0.813 $WPST$ + 0.292 $PA{:}PA$ + 0.169 # SYSTEM QUESTIONS + 0.320 $CA{:}AP$ + 0.272 SCR -0.170 # BARGE-INS -0.528 STD + 0.199 SRD + 0.117 WA_{iso} + 0.012 voc_s + 0.094 voc_m + 0.201 voc_f + 0.052 rec_rate + 0.049 $conf\_type$	0.445
B0	TS_{bin}	-0.291 # TURNS + 0.386 $CA{:}AP$ + 0.199 $PA{:}PA$ -0.160 # BARGE-INS + 0.213 SCR + 0.127 # SYSTEM QUESTIONS + 0.166 TS_{bin} -0.267 voc_s + 0.028 voc_m + 0.150 voc_f + 0.036 rec_rate -0.100 $conf\_type$	0.400
B23	TS_{bin}	-0.333 DD + 0.297 TS_{bin} + 0.199 $PA{:}PA$ -0.109 $WPUT$ -0.414 $CA{:}IA$ + 0.298 SCR -0.154 # BARGE-INS -0.107 voc_s + 0.068 voc_m + 0.073 voc_f + 0.021 rec_rate + 0.056 $conf\_type$	0.327
MEAN(B0, B23)	TS_{bin}	-0.307 DD + 0.252 TS_{bin} +0.185 $PA{:}PA$ -0.429 $CA{:}IA$ -0.165 # BARGE-INS + 0.286 SCR + 0.068 # SYSTEM QUESTIONS -0.209 voc_s + 0.042 voc_m + 0.121 voc_f + 0.035 rec_rate -0.025 $conf\_type$	0.376
MEAN(B)	TS_{bin}	-0.339 # TURNS + 0.501 $CA{:}AP$ + 0.343 SCR + 0.220 $PA{:}PA$ + 0.228 TS_{bin} + 0.022 # SYSTEM QUESTIONS -0.295 voc_s + 0.089 voc_m + 0.167 voc_f + 0.066 rec_rate + 0.035 $conf\_type$	0.527
US_w	$B1_{bin}$	0.257 $B1_{bin}$ + 0.266 $PA{:}PA$ + 0.806 $CA{:}AP$ + 1.087 $WPST$ + 0.135 # SYSTEM QUESTIONS + 0.689 $CA{:}IA$ -1.118 STD + 0.483 SRD + 0.171 voc_s + 0.058 voc_m + 0.219 voc_f + 0.112 rec_rate + 0.049 $conf\_type$	0.465
B0	$B1_{bin}$	0.466 $B1_{bin}$ -0.155 # TURNS + 0.120 # SYSTEM QUESTIONS + 0.178 $PA{:}PA$ + 0.139 $CA{:}AP$ -0.387 STD + 0.201 SRD + 0.244 $WPST$ + 0.116 WA -0.149 voc_s -0.039 voc_m + 0.183 voc_f + 0.017 rec_rate -0.145 $conf\_type$	0.530
B23	$B1_{bin}$	0.531 $B1_{bin}$ - 0.202 DD -0.059 voc_s -0.019 voc_m + 0.143 voc_f + 0.113 rec_rate + 0.009 $conf\_type$	0.410
MEAN(B0, B23)	$B1_{bin}$	0.550 $B1_{bin}$ -0.232 DD -0.147 voc_s -0.024 voc_m + 0.184 voc_f + 0.123 rec_rate -0.042 $conf\_type$	0.497
MEAN(B)	$B1_{bin}$	0.477 $B1_{bin}$ -0.130 # TURNS + 0.170 $PA{:}PA$ + 0.667 $CA{:}AP$ + 0.456 $CA{:}IA$ -0.720 STD -0.123 IR + 0.593 $WPST$ + 0.351 SRD + 0.058 # SYSTEM QUESTIONS -0.078 voc_s + 0.029 voc_m + 0.220 voc_f + 0.130 rec_rate + 0.041 $conf\_type$	0.663

nition rate does not have a direct impact on user satisfaction. This finding is congruent with the ones made in Section 6.2.5.1. The $conf\_type$ parameter shows coefficients with positive and negative signs, indicating that there is no clear preference with respect to the confirmation strategy.

6.3.3 Hierarchical Quality Prediction Models

Following the idea of the PARADISE model, the regression analyses carried out so far aim at predicting high-level quality aspects like overall user satisfaction. The target values for these aspects were either chosen according to the classification given by the QoS taxonomy (B0 and B23), or calculated as a simple arithmetic mean over different quality aspects. In this way, no distinction is made between the quality aspects and categories of the QoS taxonomy, and their interrelationships are not taken into account. Even worse, different aspects like perceived system understanding, TTS intelligibility, dialogue conciseness, or acceptability are explicitly mixed in the US_w variable.

In order to better incorporate knowledge about quality aspects, related interaction parameters as well as interrelationships between aspects, new modelling approaches are presented in the following which are based on the QoS taxonomy. In a first step, the taxonomy serves to define target variables for individual quality aspects and categories. The targets are the arithmetic mean values over all judgments belonging to the respective aspect or category (see Figure 6.1), namely the judgments on part B and C questions obtained in experiment 6.3. Tables 6.47 and 6.48 show the definitions of target variables (noted t_x for the target x) for each quality aspect and category. Input parameters to the following models consist of the set 2 interaction parameters, augmented by the four interaction parameters (not user judgments!) on task success, namely TS_{ord}, TS_{bin}, κ_{dia} and κ_{conf}. This augmented set will be called set 3 in the following discussion.

In a first approach, the different quality categories listed in Table 6.48 are modelled on the basis of the complete set 3 data. A standard multivariate regression analysis with stepwise inclusion of parameters and replacement by the mean for missing values is used for this purpose. The resulting models are shown in Table 6.49. It can be seen that for several quality categories the amount of covered variance is similar or even exceeds the one observed for the global quality predictors, see the first four rows of Table 6.38 (results for pure interaction parameters as the input). The best prediction results are obtained for communication efficiency, dialogue cooperativity, comfort, and task efficiency. Usability, service efficiency, utility and acceptability resist a meaningful prediction, probably because they are only addressed by the judgments on part C questions, which do not reflect the characteristics of the individual system configurations.

Table 6.47. Target variable definitions for all quality aspects of the QoS taxonomy.

Quality aspect	*Target name*	*Target definition*
Informativeness	t_inform	MEAN(B1,B2,B3)
Truth & evidence	t_truth	MEAN(B4,B9,B10,B11)
Relevance	$t\_relevance$	MEAN(B5,B9,B10,B11,B12,B18)
Manner	t_manner	MEAN(B8,B10,B17,B19,C2)
Background knowledge	t_backgr	MEAN(B4,B8,B10)
Meta-comm. handling	t_meta	MEAN(C4,C6,C8)
Initiative	$t\_initiative$	MEAN(B8,B10,B12,B18,C7)
Interaction control	t_inter	MEAN(B13,B14)
Partner asymmetry	$t\_partner$	MEAN(B8,B10,B12,B18,B19,C11)
Speech input quality	t_sp_input	MEAN(B5,B9,B10,B11)
Speech output quality	t_sp_output	MEAN(B6,B7,B16,B22)
Speed	t_speed	B15
Conciseness	$t\_concise$	MEAN(B17,B20)
Smoothness	t_smooth	MEAN(B10,B11,B19,21,C6)
Agent personality	t_agent	MEAN(B12,B16,B22,C3)
Cognitive demand	t_cognit	MEAN(B6,B19,B25)
Task success	t_ts	MEAN(B1,B4,C5,C14)
Task ease	$t\_taskease$	MEAN(C11,C13,C16)
Ease of use	t_ease	MEAN(C8,C11,C16)
Service adequacy	$t\_adequacy$	MEAN(C12,C13,C15)
Added value	t_added	MEAN(C15,C17)
Pleasantness	$t\_pleasant$	MEAN(B24,C10)
Personal impression	$t\_impression$	C9
Valuability	t_value	C12
Perc. helpfulness	$t\_helpful$	C13
Future use	t_future	C18

The predictors chosen by the algorithm give an indication on the interaction parameters which are relevant for each quality category. Independent of the parameter definition, dialogue cooperativity receives the strongest contribution from the $CA{:}AP$ parameter. This shows that indeed contextual appropriateness is the dominating dimension of cooperativity. Other relevant predictors are the system's meta-communication capability (SCR) and task success (TS_{ord}). Dialogue symmetry also seems to be dominated by the appropriateness of system utterances. The significant predictors are very similar to the ones observed for cooperativity. Apparently, there is a close relationship between these two categories, which can partly be explained by the considerable overlap of questions in both categories, see Table 6.47. The speech input/output quality category cannot be well predicted. This is mainly due to the absence of speech-output-

Table 6.48. Target variable definitions for all quality categories of the QoS taxonomy.

Quality category	*Target name*	*Target definition*
Dialogue cooperativity	t_coop	MEAN(t_inform,t_truth,$t\_relevance$, t_manner,t_backgr,t_meta)
Dialogue symmetry	t_symm	MEAN($t\_initiative$,t_inter,$t\_partner$)
Speech I/O quality	t_speech	MEAN(t_sp_input,t_sp_output)
Comm. efficiency	t_commun	MEAN(t_speed,$t\_concise$,t_smooth)
Comfort	$t\_comfort$	MEAN(t_agent,t_cognit,B24)
Task efficiency	$t\_taskeff$	MEAN(t_ts,$t\_taskease$)
Usability	t_usabil	t_ease
Service efficiency	$t\_service$	MEAN($t\_adequacy$,t_added)
User satisfaction	t_satis	MEAN($t\_pleasant$,$t\_impression$,B0,B23,C1)
Utility	t_util	MEAN(t_value,$t\_helpful$)
Acceptability	t_accept	t_future

related interaction parameters. Only the speech input aspect of the category is covered by the interaction parameters of set 3. However, these parameters were not identified as relevant predictors by the algorithm. This finding underlines the fact that information may be lost when different quality aspects are mixed as a target variable of the regression algorithm.

Communication efficiency is the category which can be predicted best from the experimental data. As may be expected, the most important predictors are $WPST$ (positive sign), # TURNS (negative sign), STD (negative sign), and DD (positive sign). The apparent contradiction in the signs has already been observed above. It seems that the users prefer to have few turns, but that the system turns should be as informative as possible (high number of words), even if this increases the overall dialogue duration. The comfort experienced by the user seems to be largely dominated by the STD parameter. However, part of this effect is ruled out by the $WPST$ parameter which influences predicted comfort in the opposite direction, with a positive sign. Further influencing factors on comfort are SRD which is correlated to long user utterances (the more the user is able to speak, the higher the comfort), as well as $CA{:}AP$ (appropriate system utterances increase comfort). Task efficiency can be predicted to a similar degree as comfort. The most important contributors are UCR, $CA{:}AP$, and TS_{ord}. Interestingly, the κ_{dia} parameter gives a negative contribution. As observed in the last section, κ coefficients do not seem to be reliable indicators of perceived task success. Apart from the user satisfaction category, which can be predicted to a similar degree and with similar parameters as observed in Table 6.40, all other target variables do not allow for satisfactory predictions.

Table 6.49. Multivariate regression models for different quality categories based on the experiment 6.3 data, using input parameters of set 3. Missing cases are replaced by the mean value, stepwise inclusion of parameters.

Model parameters		*Significant predictors*	R^2
Target	*Input*		
Dialogue cooperativity	set 3	0.370 TS_{ord} + 0.264 *PA:PA* + 0.571 $CA{:}AP$ + 0.404 SCR -0.245 # TURNS -0.134 # BARGE-INS -0.161 κ_{dia}	0.415
Dialogue symmetry	set 3	0.539 $CA{:}AP$ + 0.276 *PA:PA* + 0.207 TS_{bin} + 0.286 SCR -0.148 # BARGE-INS -0.158 # TURNS	0.307
Speech I/O quality	set 3	0.317 *PA:PA* + 1.246 $CA{:}AP$ + 0.960 $CA{:}IA$ + 0.151 TS_{ord} + 0.161 # SYSTEM QUESTIONS -0.145 IR	0.250
Communication efficiency	set 3	-0.669 # TURNS + 0.711 *WPST* + 0.300 $CA{:}AP$ + 0.289 *PA:PA* -0.152 $AN{:}PA$ -0.495 STD + 0.204 SCR + 0.218 # SYSTEM QUESTIONS + 0.146 *WA* + 0.482 *DD*	0.505
Comfort	set 3	0.207 *PA:PA* + 0.975 $CA{:}AP$ + 0.211 SCR -0.282 *DD* + 0.641 $CA{:}IA$ -2.026 STD + 1.054 *SRD* + 1.208 *WPST* + 0.168 TS_{ord}	0.414
Task efficiency	set 3	0.271 STD + 0.235 *PA:PA* -0.196 # BARGE-INS + 0.433 $CA{:}AP$ + 0.458 UCR + 0.126 # SYSTEM QUESTIONS -0.242 κ_{dia} + 0.383 TS_{ord} + 0.155 WA_{iso}	0.397
Usability	set 3	0.199 *WPUT* -0.147 # BARGE-INS + 0.174 TS_{bin} + 0.166 *PA:PA*	0.123
Service efficiency	set 3	0.151 *PA:PA* -0.225 # BARGE-INS +0.227 *WPST* + 0.170 # SYSTEM QUESTIONS	0.121
User satisfaction	set 3	-0.159 # TURNS + 0.251 TS_{bin} + 0.214 *PA:PA* + 0.465 $CA{:}AP$ + 0.352 UCR -0.197 # BARGE-INS + 0.185 # SYSTEM QUESTIONS	0.282
Utility	set 3	0.214 STD -0.212 # BARGE-INS	0.084
Acceptability	set 3	-0.231 # BARGE-INS + 0.180 *WPST* -0.153 # SYSTEM ERROR MESSAGES + 0.150 *PA:PA*	0.122

In the literature, only few examples of predicting individual quality aspects are documented. In the frame of the EURESCOM project MIVA, Johnston (2000) described a simple regression analysis for predicting single quality dimensions from instrumentally measurable interaction parameters. He found relatively good simple predictors for ease of use, learnability, pleasantness, effort required to use the service, correctness of the provided information, and perceived duration. However, no R^2 values have been calculated, and the number of input interaction parameters is very low. Thus, it has to be expected that the derived models are relatively specific to the system they have been developed for.

The models in Table 6.49 show that the interaction parameters assigned beforehand to a specific quality aspect or category (see Figure 6.1) are not always the most relevant predictors. Nevertheless, an approach will be presented in the following discussion to include some of the knowledge contained in the QoS taxonomy in a regression model. A 3-layer hierarchical structure, reflecting the quality aspects, quality categories, and the global target variables, is used in an initial approach. This structure is depicted in Figure 6.17. On the first layer, quality aspect targets (see Table 6.47) are predicted on the basis of the previously assigned interaction parameters (see Tables 3.1 and 3.2). On the second layer, quality category targets (see Table 6.48) are predicted on the basis of the predictions from layer 1 (indicated p_x for category x), and in one case (contextual appropriateness) amended by additional interaction parameters which have been directly assigned to this quality category. On the third layer, the 5 target variables used in the last section are predicted on the basis of the predictions for layer 2. All regression models are determined by forced inclusion of all mentioned input parameters, and by replacing missing values by the respective means. Figure 6.17 shows the input and output parameters of each layer and for each target, and the resulting amount of covered variance, R^2, for each prediction. It should be noted that only those quality aspects and categories for which interaction parameters have been assigned can be modelled in this way.

It turns out that a meaningful prediction of individual quality aspects is only possible in rare cases. Reasonable R^2 values are observed for speech input quality ($R^2 = 0.316$), conciseness and smoothness ($R^2 = 0.308$), and partly also for manner ($R^2 = 0.202$) and task success ($R^2 = 0.167$). All other aspects cannot be predicted on the basis of the assigned interaction parameters. One reason will be the limited number of parameters which are attributed in some cases. However, the amount of covered variance is not strictly related to the number of input parameters, as the predictions for conciseness and smoothness show. When the predicted values of the first layer are taken as an input to predict quality categories on layer 2, the prediction accuracy is not completely satisfactory. All R^2 values are far below a direct prediction on the basis of all set 3 parameters, cf. Table 6.49. Only communication efficiency can be predicted with an R^2 value of 0.323. The reason for the comparatively low amount of covered variance will be linked to the restricted number of input parameters for each category.

On the highest level (layer 3), prediction accuracy on the basis of layer 2 predictions turns out to be lower than for the direct modelling in most cases, compare Figure 6.17 and Table 6.38. It should be noted that the hierarchical model includes all input parameters by force, according to the hierarchical structure. If a comparable forced-inclusion approach is chosen for the models of Table 6.40, the amount of covered variance increases to $R^2 = 0.535$ for US_w,

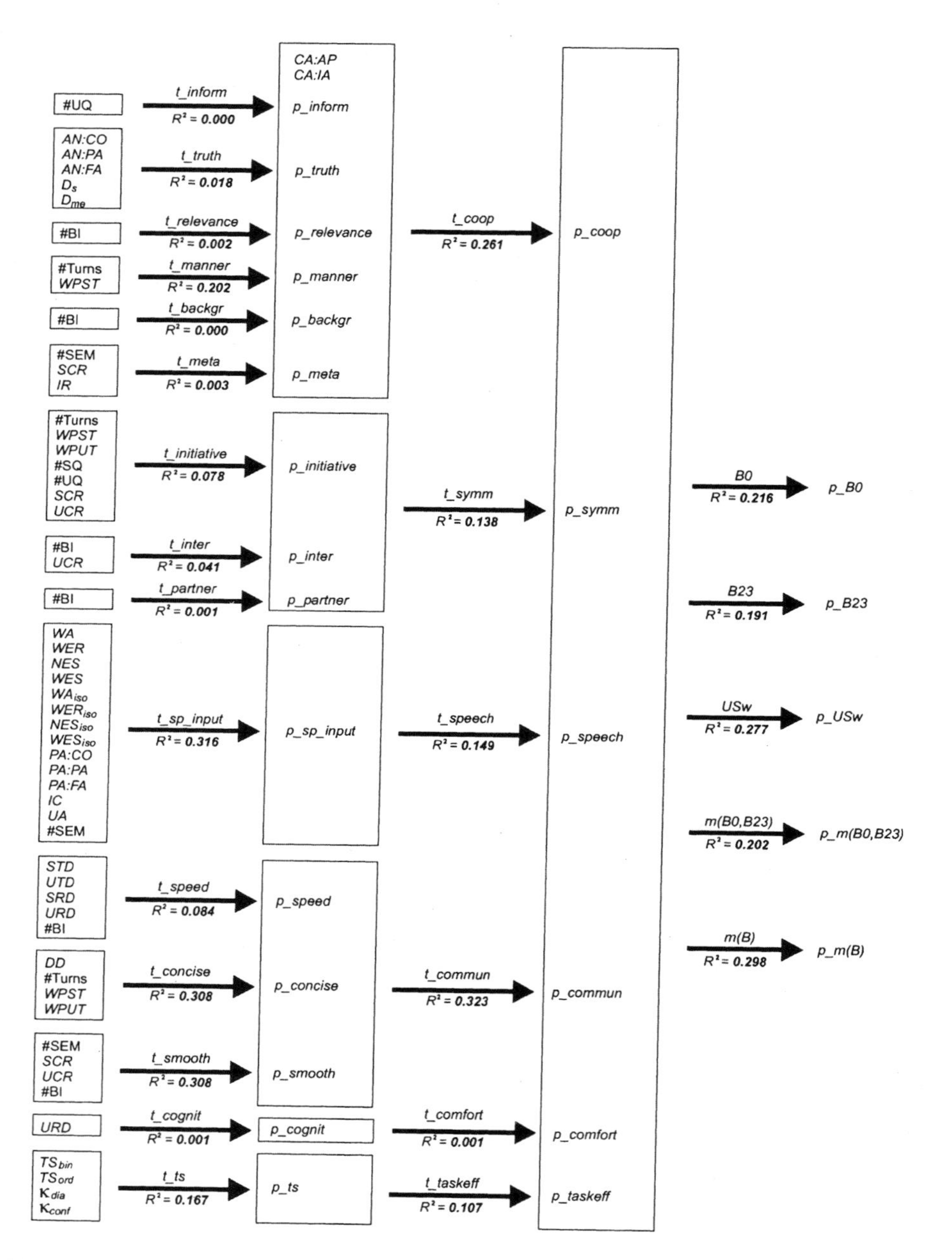

Figure 6.17. 3-layer hierarchical multivariate regression model for experiment 6.3 data. Input parameters are indicated in the black boxes. Missing cases are replaced by the mean value, forced inclusion of all input parameters. # UQ: # USER QUESTIONS; D_s: $DARPA_s$; D_{me}: $DARPA_{me}$; # BI: # BARGE-INS; # SEM: # SYSTEM ERROR MESSAGES; # SQ: # SYSTEM QUESTIONS; # UQ: # USER QUESTIONS.

$R^2 = 0.442$ for B0, $R^2 = 0.440$ for B23, $R^2 = 0.456$ for MEAN(B0,B23), and $R^2 = 0.559$ for MEAN(B). These values show that the amount of variance which can be covered by a regression model strongly depends on the choice of available input parameters. It also shows that a simple hierarchical model structure, as was used here, does not lead to better results for predicting global quality aspects.

As an alternative, a 2-layer hierarchical structure has been chosen, see Figure 6.18. In this structure, the first layer for predicting quality aspects is skipped, due to the low prediction accuracy (low amount of covered variance) which has been observed for most quality aspect targets. For predicting communication efficiency, comfort and task efficiency, the predictions for cooperativity, dialogue symmetry and speech input/output quality are taken as input variables, together with additional interaction parameters which have been assigned to these categories. In this way, the interdependence of quality categories displayed in the QoS taxonomy is reflected in the model structure. On the basis of the predictions for all six quality categories, estimations of global quality aspects are calculated, as in the previous example.

A comparison between the prediction results of Figures 6.17 and 6.18 shows that the amount of variance which is covered increases for all six predicted quality categories. The increase is most remarkable for the categories in the lower part of the QoS taxonomy, namely communication efficiency, comfort, and task efficiency. Apparently, the interrelations indicated in the taxonomy have to be taken into account when perceptive quality dimensions are to be predicted. Still, the overall amount of covered variance is lower than the one obtained for direct estimation on the basis of all set 3 parameters, see Table 6.49. It is also slightly lower when predicting global quality aspects like user satisfaction, e.g. in comparison to Table 6.40 (except for MEAN(B)).

The reasons for this finding may be threefold: (1) Either incorrect target values (here: mean over all questions related to a quality aspect or category) were chosen; or (2) incorrect input parameters for predicting the target value were chosen; or (3) the aspects or categories used in the taxonomy are not adequate for quality prediction. Indeed, the choice of input parameters has proven to carry a significant impact on quality prediction results. It is difficult to decide whether the quality categories defined in the taxonomy are adequate for a prediction, and whether the respective target variables are adequate representatives for each category. The example of speech output quality shows that quality aspects which are not at all covered by instrumentally or expert-derived interaction parameters may be nevertheless very important for the user's quality perception. Further investigations will be necessary to choose optimum target variables. Such variables will have to represent a compromise between the informative value for the system developer, the types of questions which can be answered by the user, and the interaction parameters available for model input.

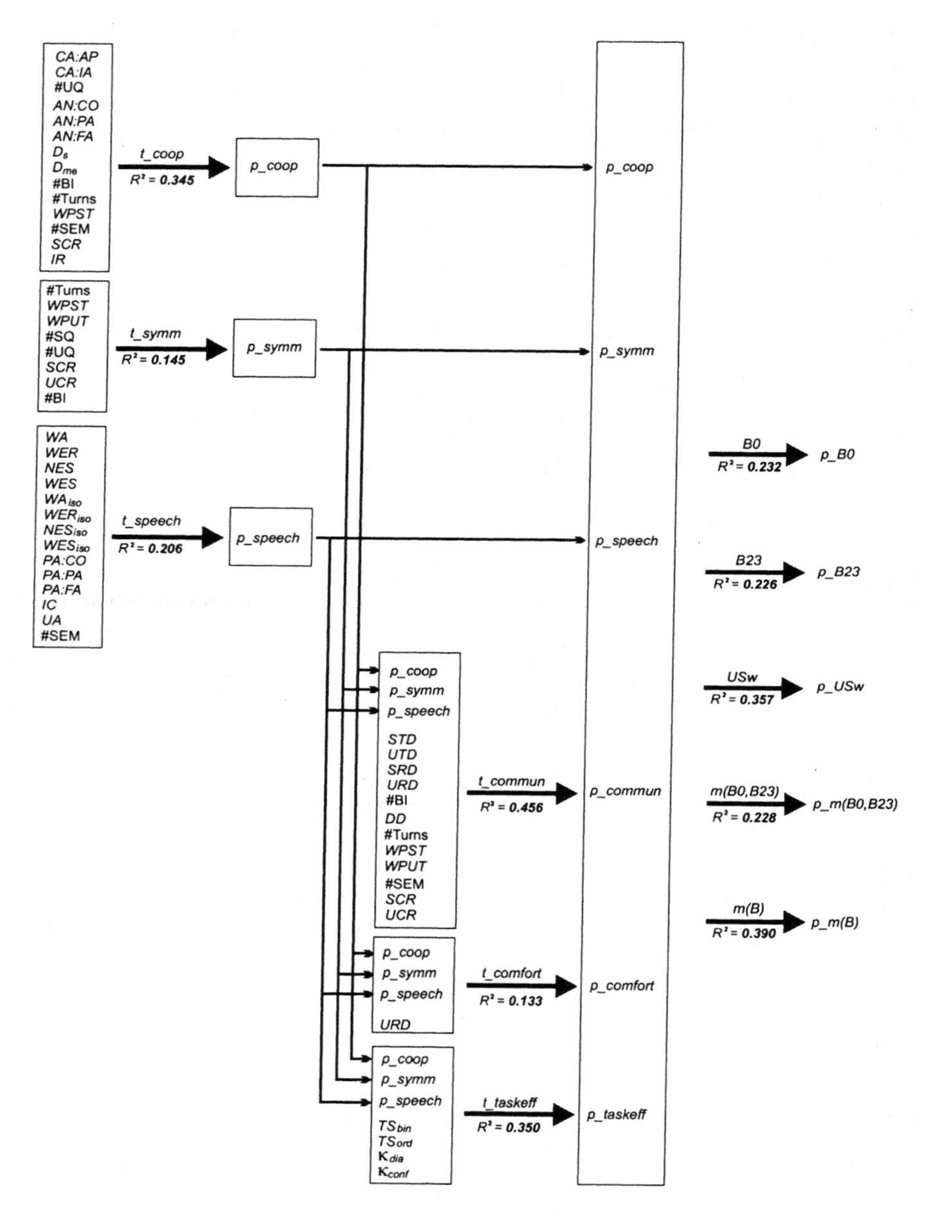

Figure 6.18. 2-layer hierarchical multivariate regression model for experiment 6.3 data. Input parameters are indicated in the black boxes. Missing cases are replaced by the mean value, forced inclusion of all input parameters. # UQ: # USER QUESTIONS; D_s: $DARPA_s$; D_{me}: $DARPA_{me}$; # BI: # BARGE-INS; # SEM: # SYSTEM ERROR MESSAGES; # SQ: # SYSTEM QUESTIONS; # UQ: # USER QUESTIONS.

For the models calculated in Section 6.3.2, the amount of covered variance was highly dependent on the system configuration. As an example, the 2-layer hierarchical model has been calculated separately for configurations 1 and 2 of experiment 6.3, see Figures 6.19 and 6.20. It can be seen that the R^2 values still differ considerably between the two configurations, depending on the prediction target. In both cases, good variance coverage is reached for communication efficiency, task efficiency, and MEAN(B). Communication efficiency in particular can be predicted in a nearly ideal way. It should however be noted that the number of input parameters for this category is very high, and the amount of target data is very restricted (20 dialogues for each connection). Thus, the optimization problem may be an easy one, even for linear regression models.

6.3.4 Conclusion of Modelling Approaches

The described modelling approaches perform a simple transformation of instrumentally or expert-derived interaction parameters in mean user judgments with respect to specific quality dimensions, or in global quality aspects like user satisfaction. The amount of variance which can be covered in most cases does not exceed 50%. Consequently, there seems to be a significant number of contributors to perceived quality which are not covered by the interaction parameters. For some quality aspects – like speech output quality – this fact is obvious. However, other aspects which seem to be well captured by the respective interaction parameters – like perceived system understanding – are still quite difficult to predict. Thus, there is strong evidence that direct judgments from the users are still the only reliable way for collecting information about perceived quality. A description via interaction parameters can only be an additional source of information, e.g. in the system optimization phase.

Because the traditional modelling approaches like PARADISE do not distinguish between different quality dimensions, it was hoped that the incorporation of knowledge about quality aspects into the model structure would lead to *better* or *more generic* results. At least the first target could not be reached by the proposed – admittedly simple – hierarchical structures. Although the 2-layer model which reflects the interrelationships between quality categories shows some improvements with respect to the 3-layer model, both approaches still do not provide any advantage in prediction accuracy with respect to a simple straight-forward approach. An increase in genericness is difficult to estimate, namely on the basis of experimental data which has been collected with a single system. All models – hierarchical as well as straight-forward, PARADISE-style ones – proved to be highly influenced by the system configuration. This will be a limiting factor of model usability: In order to estimate which level of quality can be reached with an improved system version, quality prediction models should at least be able to extrapolate to higher recognition rates, other speech

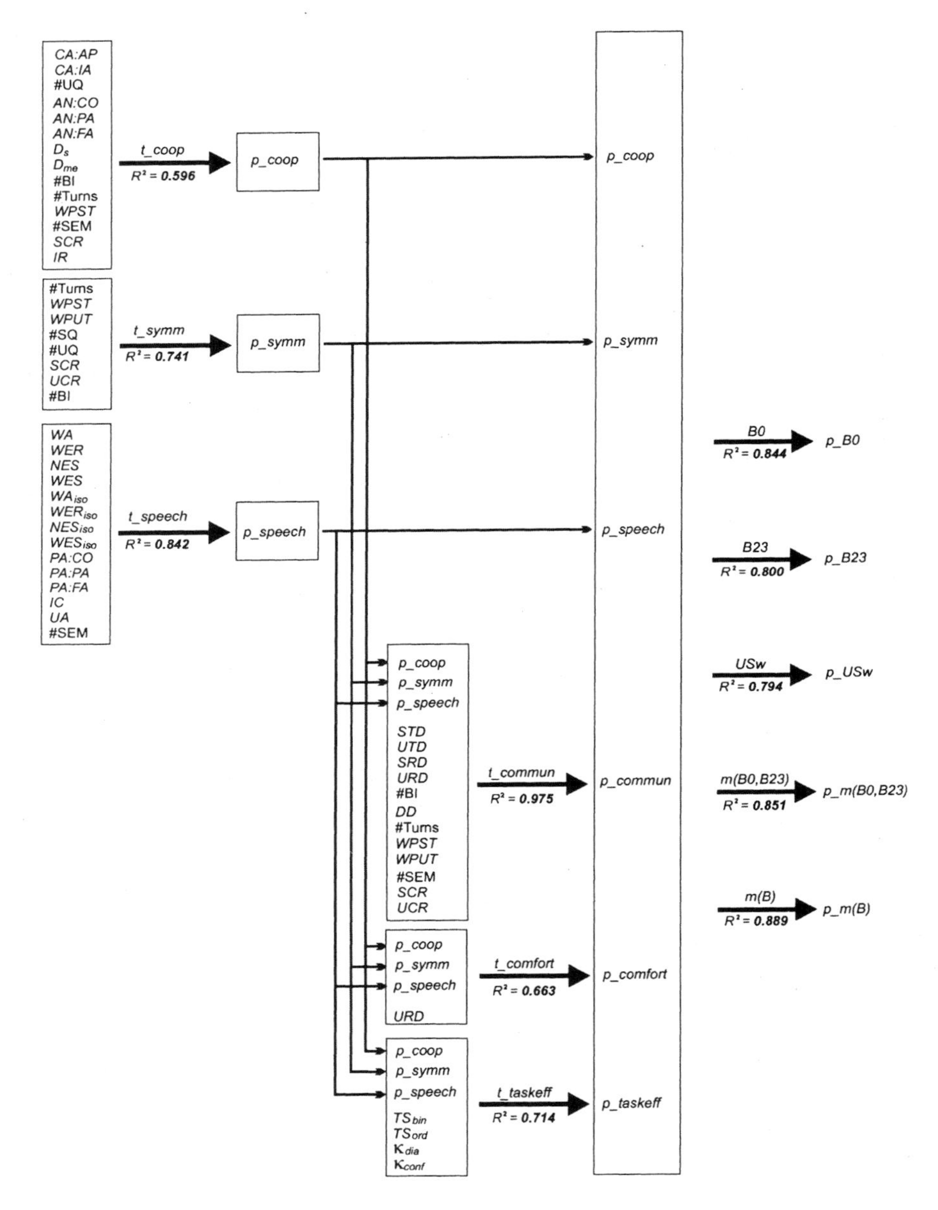

Figure 6.19. 2-layer hierarchical multivariate regression model for experiment 6.3 data, system configuration 1 of Table 6.2. Input parameters are indicated in the black boxes. Missing cases are replaced by the mean value, forced inclusion of all input parameters. # UQ: # USER QUESTIONS; D_s: $DARPA_s$; D_{me}: $DARPA_{me}$; # BI: # BARGE-INS; # SEM: # SYSTEM ERROR MESSAGES; # SQ: # SYSTEM QUESTIONS; # UQ: # USER QUESTIONS.

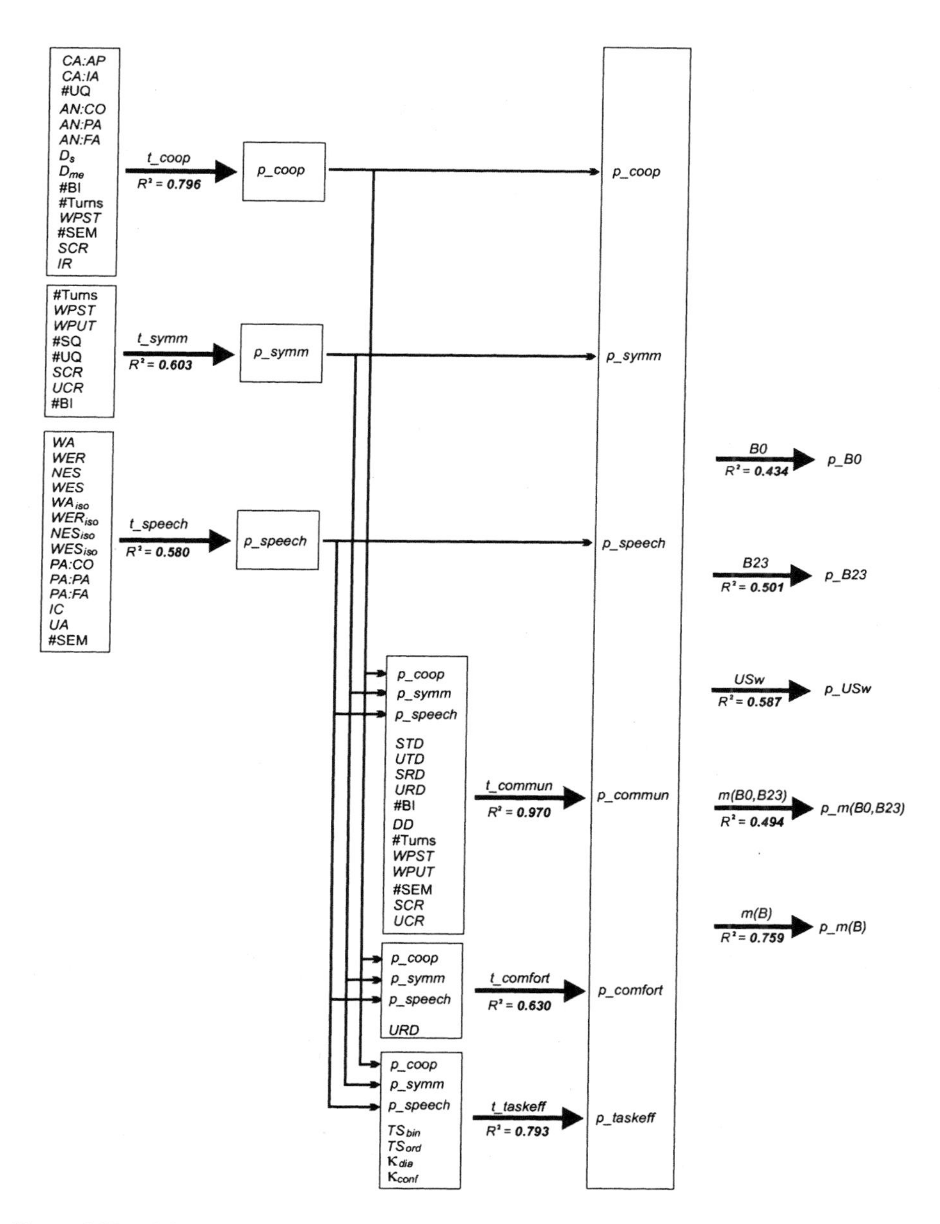

Figure 6.20. 2-layer hierarchical multivariate regression model for experiment 6.3 data, system configuration 2 of Table 6.2. Input parameters are indicated in the black boxes. Missing cases are replaced by the mean value, forced inclusion of all input parameters. # UQ: # USER QUESTIONS; D_s: $DARPA_s$; D_{me}: $DARPA_{me}$; # BI: # BARGE-INS; # SEM: # SYSTEM ERROR MESSAGES; # SQ: # SYSTEM QUESTIONS; # UQ: # USER QUESTIONS.

output principles, or different confirmation strategies. It remains questionable in how far this goal can be reached by the simple linear regression models which were discussed here.

Due to the sparse data which was available for defining the models and to the large differences between system versions, testing on an independent set of experimental data leads to a loss in prediction accuracy in most cases. This is a principled limitation of the described experiments. Unfortunately, the situation seems to be very similar in other laboratories: Only in the AT&T investigations (Walker et al., 2000a) and in the MIVA project (Johnston, 2000) have tentative cross-data validations been reported. It is up to future experiments to thoroughly test the reliability of the proposed models.

6.4 Summary

Spoken dialogue systems are relatively complex entities which rely on the interaction of a number of system components. Although it is useful to assess the performance of individual components in isolation, e.g. by investigating their input-output-relationship, results obtained in this way will provide only limited information on how the respective system component will perform in the setting of the overall system, and how it will contribute to the quality perceived by the user. In the present chapter, new experimental results were described which take a wholistic view of the system. The aim was to quantify the impact of individual system characteristics on different quality dimensions perceived by the user. A typical prototype system offering information about restaurants was taken as an example. This system has been set up by the author for the present evaluation experiments. It can be configured in a flexible way, simulating different levels of recognition performance, different confirmation strategies, and different types of speech output.

In order to determine the quality of the offered service, human test subjects interacted with the system, and rated the quality they experienced during the interaction. In parallel, the interactions have been logged, and the log files were submitted to an expert for evaluation. The evaluation resulted in a set of interaction parameters which reflect the same system settings as the subjective quality judgments do. The relationship between both has been investigated in detail via correlation and covariance analyses. In most cases, the observed correlations between quality judgments and interaction parameters were disappointingly low. Only in few cases do both types of metrics correlate to an extent which would justify renouncing on quality judgments obtained from the test subjects, and replacing them by collected interaction parameters. Such an approach would inevitably lead to a loss of information, because the interaction parameters do not reflect the quality perceived by the test subjects.

Via a multidimensional analysis, it is possible to identify and to describe the quality dimensions perceived by the user. A number of analyses reported in the literature as well as own experimental investigations show that subjective judgments mainly address the high-level categories of the QoS taxonomy (usability, service efficiency, user satisfaction, and acceptability), whereas the interaction parameters are closer related to the low-level categories, namely speech input quality, cooperativity, communication and task efficiency. This finding has some implications for the definition of quality prediction models: If both address different quality aspects, it will be difficult to predict one on the basis of the other. The classification in different categories of the QoS taxonomy however explains the low correlation between judgments and parameters.

Although the individual relationships are not very strong, it may nevertheless be possible to estimate quality judgments on the basis of a combination of input parameters. The PARADISE model follows this line, estimating an amalgam of judgments termed "user satisfaction" on the basis of a fixed set of interaction parameters. A multivariate linear regression analysis is used to estimate the prediction function coefficients. The predicted "user satisfaction" values cover only a limited amount of variance (around 40% to 50%) observed in the subjective judgment data. An analysis on the basis of own experimental data shows that the prediction accuracy is even lower when purely instrumentally and expert-derived parameters are used as an input; only subjective judgments of task success help to maintain an acceptable level of prediction accuracy. In addition, the prediction functions and the prediction accuracy strongly depend on the system configuration. Thus, it may be difficult to extrapolate quality predictions for systems or services other than those that predictors have been obtained from. This result is partly in contradiction to observations reported in the literature. It may be linked to the large range of system characteristics covered by the experimental set-up.

A major limitation of the PARADISE approach is that it tries to predict an amalgam of user satisfaction, without regard for potential quality dimensions. This disadvantage may be overcome with the help of the QoS taxonomy defined in Chapter 2. On the basis of the quality aspects and categories identified in the taxonomy, predictors of individual quality aspects have been derived. Target variables defined for each quality category can be predicted with a similar degree of precision as was observed for the global quality aspects. However, prediction accuracy decreases when the predictors are used as an input to further prediction models for the high-level categories, like user satisfaction. Two types of these so-called hierarchical quality prediction models have been proposed. Although they promised to be more generic because they better reflect underlying quality dimensions, the overall prediction accuracy of the PARADISE model could not be increased. Potential reasons for the still relatively low prediction accuracy are the inherent limitations of currently available interaction parameters, the generally low correlation between interaction parameters and perceived quality, and the sparsity of data to derive models from. So far, input and output variables have been selected from the QoS taxonomy, in a relatively straight-forward way. It is estimated that a high potential for optimization exists, namely with respect to the target variables (predicted quality dimensions), the set of input parameters, as well as the mathematical function used for the prediction.

Nevertheless, the taxonomy of QoS aspects holds a thorough analysis on the basis of experimental data. So far, it has proven to be useful in at least three ways: (1) Questionnaires for obtaining information about quality aspects perceived by the user can be constructed in a very efficient way; (2) quality aspects which are addressed by interaction parameters or subjective judgments can be identified, and relationships between corresponding metrics can be established; and (3) information which is necessary for system optimization can be obtained. In this way, the taxonomy supports a very efficient and well-structured way of system set-up and optimization.

The experimental data with respect to the different system configurations is an interesting application example of the QoS taxonomy. On both the speech input and the speech output side, significant influences on interaction parameters and subjective quality ratings were observed. Whereas the recognition rate mainly influenced interaction parameters related to the categories speech input quality, cooperativity, communication efficiency, and task success, synthesized speech strongly affects perceptual dimensions which can be associated with different categories of the QoS taxonomy, up to the overall global impression of the user. Apparently, the system voice describes the system's "personality", and seems to act as a kind of "business card" of the system. This effect has been postulated earlier and could now be quantified with the experimental data. Users seem to be less able to "localize" their perceptions with respect to speech output than to speech input. The confirmation strategy only showed weak effects on both interaction parameters and subjective quality judgments.

Chapter 7

FINAL CONCLUSIONS AND OUTLOOK

Spoken dialogue systems form the common basis for a number of information and transaction services which are increasingly offered via telephone networks. They possess speech recognition, speech interpretation, dialogue managing, and speech output capabilities, and thus enable a more-or-less "natural" bi-directional spoken interaction with an underlying application program. The increasing success in the market shows that these systems can be used with profit for both the end users and the service operators, provided that they are sufficiently well designed. It is to be expected that the range of potential application programs will grow massively in the near future, because many web-based services can also efficiently be offered via speech. The use of markup languages (VoiceXML, SALT) and specific application servers will facilitate such cross-media portability.

Speech-based human-machine interaction over the phone follows principles which are different from the ones underlying human-to-human telephone conversations. The machine partner has only limited understanding, reasoning, and expression capabilities, and a limited knowledge of the domain and the "world" the interaction takes place in. The human user adapts to these limitations, e.g. by simplifying the language, or restricting the communication topics. In general, users are able to understand a certain amount of limitations, and to accept them in order to reach the communicative goal. A successful interaction between the unequal partners is thus possible for limited tasks, provided that both behave cooperatively. In fact, cooperativity turned out to be a key requirement for high-quality human-machine interaction. Consequently, cooperativity principles can and should be used for a successful system design. The notion of cooperative behavior should not, however, be interpreted in the same way as is common for human-to-human interactions. The limitations of the machine agent have to be taken into account as well as the user's ability to adapt to these limitations. Bernsen et al. (1998) provided a very valuable theory in this respect, which can be directly applied to SDS development and optimization. This theory forms a basis for the detailed analysis of quality aspects governing the interaction with telephone-based spoken dialogue systems which was presented here.

Applying a recent definition of quality to this type of service (Jekosch, 2000), the quality perceived by a user when interacting with a spoken dialogue system will strongly reflect the expectations he or she has in the specific situation. Quality implies a human perception and judgment process, and consequently has to be measured subjectively, i.e. by directly asking human test subjects. The experiments described in this book provide examples of how such measurements can be carried out. They lead to estimations of perceived quality dimensions (quality features) which are somehow linked to the quality elements, i.e. the physical or algorithmic characteristics of the system and the interaction scenario, including the transmission channel. A new taxonomy of quality of service aspects has been proposed, and provides a generic structure for the relevant quality elements (grouped into factors) and quality features (grouped into aspects and subsequent categories). It is in agreement with other definitions of quality and quality aspects, as was illustrated in Section 2.3. This taxonomy proved to be useful in at least four respects: For classifying quality features and interaction parameters; for setting up evaluation metrics and questionnaires; for interpreting experimental quality measurement results; and for defining quality prediction models.

The link between perceived quality features and physical or algorithmic quality elements is not always obvious and straight-forward, because spoken dialogue systems are particularly complex and consist of a number of interconnected components. Two fundamental ways of addressing their quality have been discussed. Technology-centered quality assessment and evaluation tries to capture the performance or quality of individual system components, isolated from their final function in the overall system setting. The user-centered evaluation tries to capture the performance or quality of system components in the overall setting, and is the final reference for describing the quality of the system and the offered service as a whole. Both types of evaluation should go hand in hand, because they provide complementary information to the system developer. In Chapter 3, an overview of the respective assessment and evaluation methods has been presented.

Two system components are of primary importance in this respect, namely those which directly deal with the acoustic speech signals. For current state-of-the-art systems, their performance now seems to be sufficiently high to enable a relatively natural interaction via spoken language. Automatic speech recognition (ASR) is already commonly used in telephony applications. Synthesized speech still lacks a certain degree of naturalness for some applications where the "apparent quality" is of primary importance; for other applications, its quality is already sufficiently high. Assessment and evaluation examples for both components have been presented in Chapters 4 and 5, focussing mainly on the effect of the transmission channel, which is a limiting factor of system robustness. However, such a technology-centered evaluation quickly reaches its limits when components with strong interrelations to the other components are to be assessed, namely the dialogue manager and the response generation components. Here, interaction parameters will give a rough indication of the component's behavior in the overall system setting. They are however not able to replace subjective quality judgments.

The relationship between technology-centered evaluation and user-centered evaluation has been addressed in an additional set of experiments (Chapter 6). A prototype SDS which can be configured in a flexible way has been designed and implemented for these tests. During subjective interactions with the system, both interaction parameters showing certain aspects of system (component) performance and user quality judgments have been collected. A detailed statistical analysis shows that the two sets of data are only moderately correlated. This leads to the conclusion that user judgments on specific quality aspects are still the only way to capture the relevant information to describe quality, and that user-centric evaluation cannot be replaced by collecting interaction parameters. Both quality judgments and interaction parameters are strongly affected by the configuration of the SDS: Whereas speech input performance is well reflected in the interaction parameters and can be interpreted in this way by the user, speech output carries a general impact on global quality aspects, including the overall user satisfaction. Apparently, the system's voice seems to reflect the perceived "personality" of the machine interaction partner.

Quality aspects play an important role in all phases of system specification, design, set-up, and operation. In order to take full advantage of quality design principles, they should be taken into account as early as possible in the system design process. Design guidelines based on the cooperativity principle, as have been proposed by Bernsen and Dybkjær, provide qualitative information about which system behavior is desirable, and which is not. On the other hand, quality prediction models discussed in Chapters 4 to 6 provide quantitative information about the expected system performance under certain conditions (linked e.g. to the transmission channel), or even about the quality experienced by the user when interacting with the system. Such predictions should however be used with care, as they are still very limited in their predictive power.

Quality prediction models which can be used to describe the performance or quality of spoken dialogue systems and their components mainly address the effects of environmental and agent factors, and only to a limited extent the effects of task factors. Their aim is to estimate a target variable related to performance (e.g. word error rate or task success) or quality (e.g. overall user satisfaction) on the basis of instrumentally or expert-derived input parameters. Different types of models have been examined with respect to their predictive power in the

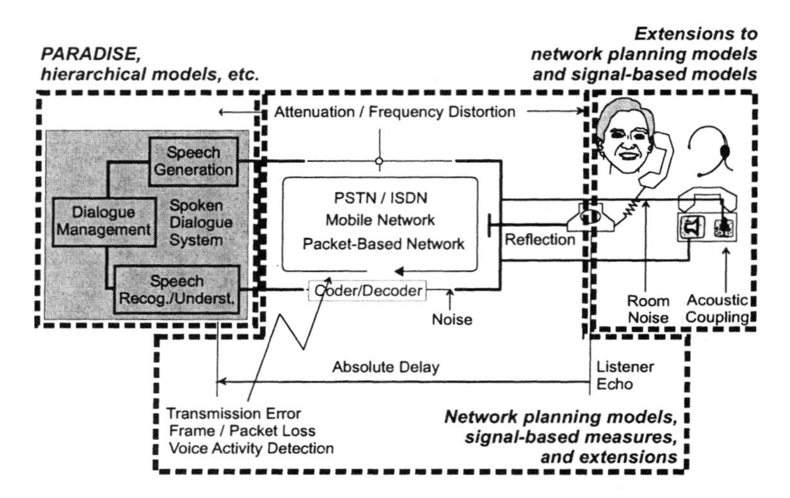

Figure 7.1. Overview of quality prediction models for the interaction of a human user with a spoken dialogue system over an impaired transmission channel.

SDS context, and an overview is depicted in Figure 7.1. Transmission channel impacts are mainly covered by the signal-based comparative measures like PESQ or TOSQA, or by parametric network planning models like the E-model. These models have been developed for predicting the transmission channel effect on human-to-human speech quality, in a listening-only or a conversational situation. As the investigations in Chapters 4 and 5 show, they may also be successfully used for predicting the transmission channel impact on speech recognizer performance and on synthesized speech originating from the SDS. Still, both types of models have their limitations, namely that they cover the acoustic transmission characteristics of user interfaces in a much simplified way, and only for traditional (handset) interfaces. Room acoustic influences, which become important when the user interacts with the system via a hands-free terminal, are not yet taken into account. The same holds true for time-variant transmission channels, e.g. in mobile or IP-based telephony, or when other time-variant signal processing equipment is integrated in the network (e.g. echo cancellers or voice activity detectors). Initial modelling approaches are underway to cover these effects, but they are mainly limited to the human-to-human communication situation (Raake, 2004; ITU-T Delayed Contribution D.44, 2001; Rix et al., 2003).

When quality prediction models are applied to synthesized speech, the characteristics of both the transmission channel and the source speech material have to be taken into account. For the user, it seems to be important whether a certain quality dimension is already contained in the source material, or whether it represents a new degradation which can be attributed to the transmission channel. As the currently available quality prediction models for transmission channel effects (both signal-based and network planning models) have only been optimized for natural speech, their predictive power for synthesized speech is still limited. It has to be noted that all transmission-related quality prediction models have been used here to predict the additional degradation caused by the channel, and not absolute quality numbers. Instrumental measures for predicting the overall quality of synthesized speech are still not available, despite some first (and very limited) approaches in this direction.

Quality design based on prediction models is relatively well established for telecommunication networks and is used in the system development phase (e.g. testing of newly developed codecs with signal-based comparative measures), the network-planning and set-up phase (e.g. using the E-model), and in the operation phase (using monitoring models, see Möller and Raake, 2002). In contrast to this, the design of SDS still requires a number of simulate-test-revise or implement-test-revise cycles. It is obvious that this procedure is highly resource-demanding, both with respect to time and money. Thus, it seems to be highly desirable to follow the lines of telecommunication network design, and to develop appropriate prediction models which provide estimations of quality in the early phases of the system design process. It is important that such models are agreed upon by system developers, network operators, and service providers, because the latter will be responsible for an adequate quality to be delivered to the user, however without being able to fully control the underlying systems.

The most widely used quality prediction model which directly addresses the agent factors is the PARADISE framework. It estimates a mean value of satisfaction-related user judgments on the basis of instrumentally or expert-derived interaction parameters. The detailed analysis presented in Chapter 6 shows that these predictions cover only a limited amount of variance in the subjective ratings (typically 30 to 50%). They heavily rely on task success estimations obtained from the user, and are strongly affected by the system configuration which the model parameters have been derived from. Still, a certain degree of cross-system and cross-task extrapolation seems to be possible. It has to be noted that all PARADISE models derived so far have been used within one laboratory, for one specific set of systems. Cross-laboratory validation of the approach is necessary, but requires a standardized set of input and output parameters to be defined.

The amount of variance in the data which can be predicted with PARADISE ($R^2 \sim 0.30 ... \ 0.50$) is small compared to the accuracy of prediction models for transmission channel effects (correlations r in the area of 0.90). This is due to the fact that PARADISE aims at predicting the quality of the whole interaction, whereas transmission-related models only address the degradation of the interaction quality due to the transmission channel. The number of influencing factors is considerably higher in the first case, a fact which becomes apparent when the taxonomies for both situations are compared (Möller, 2002a). Another reason for the limited predictive power of PARADISE is the set of interaction parameters used as an input. Currently available parameters still cover only simple aspects of the dialogue interaction. For example, no parameters related to speech output quality and no adequate parameters for describing ill-structured tasks are available. Many tasks which are nowadays mainly performed by human operators or not offered at all via spoken language are ill-structured, because they involve multiple related topics or topic jumps. Examples are integrated tourist services, TV program information, product order hotlines, or smart-home systems. These tasks cannot always be described by a simple attribute-value matrix, and require new measures of speech understanding and task success.

In order to enhance quality prediction models for agent factors, different ways can be taken. Besides extending the set of input parameters (e.g. to speech-output-related parameters and other measures of task success), other model functions could be used. Walker and colleagues experimented with classification and regression trees, but did not observe a significant increase in predictive power. Non-linear regression models or neural networks are also possible, but it remains to be studied in how far the genericness of model predictions can be increased by simply using better mathematical models mapping input to output parameters. From a scientific point of view, more sophisticated model structures are needed which better take the quality dimensions perceived by the user into account. By integrating knowledge about the human quality perception and judgment process, it can be expected to obtain models with better extrapolation possibilities, both to other system configurations and to other tasks and domains. A simple hierarchical approach has been proposed in Chapter 6, but it still has a high potential for optimization, e.g. with respect to the target variables.

In the discussion, quality prediction results have been used in a purely analytic way, describing the impact of the different factors on perceived quality. However, it is also possible to adapt the SDS behavior on-line, depending on the performance of the system components. In a first step, indicators of performance may be calculated on the basis of instrumental measurements. For example, speech recognition accuracy can be estimated on-line by measuring the parameters of the transmission channel during the interaction, and calculating expected recognition performance using a non-intrusive version of the (modified) E-model (see ITU-T Rec. P.562, 2000, and Section 4.5). The estimated parameters can be used to adapt specific system components. For example, the speech recognizer can be adapted by restricting the vocabulary in case of poor transmission conditions, or by selecting acoustic models which are particularly trained on the transmission channel characteristics. The dialogue manager may change the dialogue strategy depending on ASR accuracy, and switch to a purely system-directed dialogue when the confidence in the recognized word sequence is low. Speech output can be adapted to generate Lombard speech in the case that background noise is high. These are just examples of how performance and quality estimations can fruitfully be used in adaptive dialogue systems.

The presented experimental investigations are obviously limited. They only addressed the standard modules which are needed in every type of system with full speech input and output capabilities. In particular, speaker recognition was not covered in the experimental settings. The identification of the speaker will however be a key requirement for security-critical systems, offering e.g. account transaction and personal information capabilities. For such services, the perception of the user with respect to security and privacy may be a dominant factor of service acceptability. The investigations were further limited to native users of the considered language. In a multilingual context, systems will have to cope with non-native users and foreign accents (Schaden, 2003), and will have to offer multilingual speech input and output capabilities. These aspects are a further topic of current scientific research.

Although this book only addressed unimodal dialogue systems which are accessed over telephone networks, quality assessment results and modelling approaches described here may be transferred to other systems involving speech interaction. Examples include smart-home systems or navigation systems for which transmission channel and room acoustic influences play a similarly important role. In the future, the present work for the spoken modality should be extended to other modalities. Multimodal dialogue systems are already close to the market, and the problems they have to face will be even more complex than the ones observed with SDSs (Cohen and Oviatt, 1995). Modality theory may form a theoretical basis for addressing the additional modalities (Bernsen and Dybkjær, 1999). An overview of current state-of-the-art multimodal dialogue systems was given during a recent ISCA workshop (Dybkjær et al., 2002). Beringer et al. (2002b) also presented a first approach to transfer the PARADISE framework to multimodal systems (SmartKom), and described the difficulties which are linked to the interaction of modalities. It is hoped that the present work will serve as a basis to cope with some of the problems, and will help to set up systems which really satisfy the demands of their human users.

GLOSSARY

Acceptability of Service (AoS) Multidimensional property of a service, describing how readily a customer will use the service. Represented as the ratio of the number of potential users to the quantity of the target group. (EURESCOM Project P.807 Deliverable 1, 1998)

Accuracy Capability of a product to provide the right or agreed results or effects with the needed degree or precision. (ISO Standard ISO/IEC 9126-1, 2001, p. 8, for software products)

Agent Dialogue participant, e.g. a dialogue system or a human user. (Gibbon et al., 2000, p. 373)

Barge-In The ability for the human to speak over a system prompt or system output. (Gibbon et al., 2000, p. 382)

Communication Intentionally controlled transfer or common-making of information between a sender and a receiver via a channel, using signals or sign systems. (Lewandowski, 1994)

Communication Media Communication means (material or device) which is used in the interaction. (Fraser, 1997, p. 601)

Communication Modality The way a communicating agent conveys information to a communication partner (human or machine). E.g. intonation, gaze, hand gestures, body gestures, facial expressions. (Gibbon et al., 2000, p. 438)

Communication Mode Perception sense which allows for a communication (vocal, visual, auditive, tactile, olfactive). (Fraser, 1997, p. 601)

Desired Composition Totality of features of individual expectations and/or relevant demands and/or social requirements. (Jekosch, 2000)

Dialogue A type of discourse taking place between two or more human participants or between human participants and a computer (Gibbon et al., 2000, p. 400). As an evaluation unit: One of several possible paths through the dialogue structure.

Dialogue Act See speech act.

Dialogue History A system-internal record of what has happened in a dialogue so far. The dialogue history provides the immediate context within which interaction takes place. (Gibbon et al., 2000, p. 400)

Domain A variable defining the type of dialogue according to its subject-matter, e.g. travel, transport, appointment scheduling, etc. (Gibbon et al., 2000, p. 403)

Effectiveness The accuracy and completeness with which specified users can achieve specified goals in particular environments. (ETSI Technical Report ETR 095, 1993, p. 13)

Efficiency The effort and resources expended in relation to the accuracy and completeness with which specified users can achieve specified goals. (Definition similar to ETSI Technical Report ETR 095, 1993)

Exchange A pair of contiguous and related turns, one spoken by each party in the dialogue. (Fraser, 1997, p. 569)

Feature Recognized and nameable characteristic of a unit. (Jekosch, 2000)

Functionality Capability of a product to provide functions which meet stated and implied needs when the product is used under specified conditions. (ISO Standard ISO/IEC 9126-1, 2001, p. 7, for software products)

Grammar A set of rules that define how the words in a language can follow each other. This can include information about the probability that a sequence of words occurs. (Gibbon et al., 1997, p. 843)

Interactive Dialogue System Computer system with which humans interact on a turn-by-turn basis. (Fraser, 1997, p. 564)

Interaction Communication of information between two agents, in which (except for the special case of the initial turn) an agent's contribution at any given point can be construed as a response to the previous turn or turns. (Gibbon et al., 1997, p. 844-845)

Language Model A formal representation of the structure (usually the sentence structure) of a natural language. (Gibbon et al., 2000, p. 430)

Operability Capability of a product to enable the user to operate and control it. (ISO Standard ISO/IEC 9126-1, 2001, p. 9, for software products)

Perceived Composition Totality of features of a unit. Signal for the identity of the unit visible to the perceiver. (Jekosch, 2000)

Performance The ability of a unit to provide the function it has been designed for.

Prosody The totality of ways to structure discourse, including accent, stress, intonation, distribution of pauses, and increase/decrease of pitch. (Lewandowski, 1994).

Quality Result of appraisal of perceived composition of a unit with respect to its desired composition. (Jekosch, 2000)

Quality Aspect A namable component of the quality of a unit. A quality aspect contains several quality features, and a quality category consists of one or several quality aspects.

Quality Category See quality aspect.

Quality Element Contribution to the quality (a) of a material or immaterial product as the result of an action/activity or a process in one of the planning, execution or usage phases, (b) of an action or of a process as the result of an element in the course of this action or process. (Definition for "element of quality" from DIN 55350 according to Jekosch, 2000)

Quality Feature A recognized and designated characteristic of a unit that is relevant to the unit's quality. (Jekosch, 2000)

Quality of Service (QoS) The collective effect of service performance which determine the degree of satisfaction of a user of the service. (ITU-T Rec. E.800, 1994)

Reliability Capability of a product to maintain a specified level of performance when used under specified conditions. (ISO Standard ISO/IEC 9126-1, 2001, p. 8, for software products)

Safety Capability of a product to achieve acceptable levels of risk of harm people, business, software, property or the environment in a specified context of use. (ISO Standard ISO/IEC 9126-1, 2001, p. 13, for software products)

Security Capability of a product to protect information and data so that unauthorized persons or systems cannot read or modify them and authorized persons or systems are not denied access to them. (ISO Standard ISO/IEC 9126-1, 2001, p. 8, for software products)

Service A set of functions offered to a user. (Definition similar to ITU-T Rec. E.800, 1994)

Speech Act Basic unit of conversational theory, according to Searle (1969). Can be categorized into assertives (commit the speaker to something being the case), directives (represents attempts by the speaker to get the hearer to do something), commissives (commit the speaker to some future course of action), expressives (express the psychological state with respect to the state of affairs specified in the propositional contents), and declaratives (bring about some alteration in the status or condition of the referred object solely in virtue of the fact that the declaration has been successfully performed), see Searle (1979). All speech acts have a propositional content, i.e. the state of affairs addressed. Sometimes called "dialogue acts" or "moves".

Spoken Dialogue System (SDS) A variety of interactive dialogue system in which the primary mode of communication is spoken natural language. (Definition for spoken language dialogue systems according to Gibbon et al., 2000, p. 477)

Spoken Language Dialogue Oral dialogue. A complete spoken verbal interaction between two parties, each of whom is capable of independent actions. A dialogue is composed of a sequence of steps which are related and build on each other. (Fraser, 1997, p. 568)

Suitability Capability of a product to provide an appropriate set of functions for specified tasks and user objectives. (ISO Standard ISO/IEC 9126-1, 2001, p. 8, for software products)

System An integrated composite that consists of one or more processes, hardware, software, facilities and people, that provides a capability to satisfy a stated need or objective. (ISO Standard ISO/IEC 9126-1, 2001)

Task All the activities which a user must develop in order to attain a fixed objective in some domain (Gibbon et al., 1997, p. 850). In the evaluation context: A piece of work which is given to the test subjects, e.g. using a pre-defined scenario.

Task-Orientated Dialogue A dialogue concerning a specific subject, aiming at an explicit goal (such as resolving a problem or obtaining specific information). (Fraser, 1997, p. 568)

Transaction The part of a dialogue devoted to a single high-level task (for example, making a travel booking or checking a bank account balance). A transaction may be coextensive with a dialogue, or a dialogue may consist of more than one transaction. (Fraser, 1997, p. 568)

Turn Utterance. A stretch of speech, spoken by one party in a dialogue, from when this party starts speaking until another party definitely takes over (Bernsen et al., 1998). A stretch of speech may contain several linguistic acts or actions. A dialogue consists of a sequence of turns produced alternatively by each party. (Fraser, 1997, p. 568)

Unit Material or immaterial object under observation. (Definition from DIN 55350 according to Jekosch, 2000)

Usability Suitability of a system or service to fulfill the user's requirements. Includes effectiveness and efficiency of the system and results in user satisfaction. (Möller, 2000)

User Satisfaction The user's response to interaction with a service (ISO Standard ISO/IEC 9126-1, 2001, p. 13, for software products). Indicator of the perceived usefulness and usability of a service for the intended user group in a specific context of use. Includes communication and task efficiency and the comfort experienced with the service.

Utility Usability of a system or service in relation to the financial costs. (Möller, 2000)

Utterance See turn.

Appendix A
Definition of Interaction Parameters

Table A.1. Dialogue- and communication-related parameters (1). System transparency: bb (black box), gb (glass box). Addressed interaction partner or component: s (system), u (user), SU (speech understanding component), LU (language understanding component), DM (dialogue manager). Measurement method: instr. (instrumental), exp. (expert-based). Interaction level: word, utt. (utterance), dial. (dialogue).

Name	*Definition*	*Unit*	*Syst. transp.*	*Int. partn.*	*Meas. method*	*Int. level*
DD: Dialogue duration	Average duration of a dialogue in (s), see e.g. Fraser (1997), Cookson (1988), Goodine et al. (1992), or Polifroni et al. (1992).	s	bb	s + u	instr.	dial.
TD (*STD*, *UTD*): Turn duration (system turn duration/ user turn duration)	Average duration of one turn in a dialogue in (s): – *STD*: From the user stopping speaking to the system stopping speaking. – *UTD*: From the system stopping speaking to the user stopping speaking. (Fraser, 1997)	s	bb	s, u	instr.	utt.
RD (*SRD*, *URD*): Response delay (system response delay, user response delay)	Average delay of a system or user response in (s): – *SRD*: From the user stopping speaking to the system starting speaking. – *URD*: From the system stopping speaking to the user starting speaking (or push-to-talk button press). (Price et al., 1992)	s	bb	s, u	instr.	utt.

Table A.2. Dialogue- and communication-related parameters (2). For a legend see Table A.1.

Name	*Definition*	*Unit*	*Syst. transp.*	*Int. partn.*	*Meas. method*	*Int. level*
# TURNS (# SYSTEM TURNS, # USER TURNS)	Average number of turns uttered by the system or by the user in a dialogue. (Walker et al., 1998b)	1	bb	s + u	instr./ exp.	dial.
# WORDS (# SYSTEM WORDS, # USER WORDS)	Average number of words uttered by the system or by the user in a dialogue.	1	bb	s/u	instr./ exp.	dial.
WPST, *WPUT*: Words per system/user turn	Average number of words per system or user turn. (Cookson, 1988)	1	bb	s/u	instr.	utt.
# USER QUESTIONS	Average number of questions from the user, per dialogue. (Polifroni et al., 1992; Goodine et al., 1992)	1	bb	s + u	exp.	utt.
# SYSTEM QUESTIONS	Average number of questions from the system, per dialogue.	1	bb	s + u	instr./ exp.	utt.

Table A.3. Dialogue- and communication-related parameters (3). For a legend see Table A.1.

Name	*Definition*	*Unit*	*Syst. transp.*	*Int. partn.*	*Meas. method*	*Int. level*
QD: Query density	Average number of new concepts introduced per user query. Being N_d the number of dialogues, $N_q(i)$ the total number of user queries in the ith dialogue, and $N_u(i)$ the number of unique concepts correctly understood by the system in the ith dialogue (a concept is not counted to $N_u(i)$ if the system has already understood it from a previous utterance): $$QD = \frac{1}{N_d} \sum_{i=1}^{N_d} \frac{N_u(i)}{N_q(i)}$$ (Glass et al., 2000)	1 (%)	gb	s, u	exp.	dial.
CE: Concept efficiency	Average number of turns necessary for each concept to be understood by the system. Being N_d the number of dialogues, $N_u(i)$ the number of unique concepts correctly understood by the system in the ith dialogue, and $N_c(i)$ the total number of concepts in the ith dialogue (a concept is counted whenever it was uttered by the user and was not already understood by the system): $$CE = \frac{1}{N_d} \sum_{i=1}^{N_d} \frac{N_u(i)}{N_c(i)}$$ (Glass et al., 2000)	1 (%)	gb	s, u	exp.	dial.

Table A.4. Meta-communication-related parameters (1). For a legend see Table A.1.

Name	*Definition*	*Unit*	*Syst. transp.*	*Int. partn.*	*Meas. method*	*Int. level*
# HELP REQUESTS	Average number of help requests from the user, per dialogue. (Walker et al., 1998b)	1	bb	u	exp.	utt.
# TIME-OUT PROMPTS	Average number of time-out prompts due to no response from the user, per dialogue. (Walker et al., 1998b)	1	gb	s + u	instr.	utt.
# ASR REJECTIONS	Average number of ASR rejections per dialogue. ASR rejections lead to system re-prompts, indicating that the system is unable to extract anything from the user utterance. (Walker et al., 1998b)	1	gb	s + u	instr.	utt.
# SYSTEM ERROR MESS.	Average number of diagnostic error messages generated by the system, per dialogue. (Price et al., 1992)	1	bb	s	instr./ exp.	utt.
# BARGE-INS	Average number of barge-in attempts from the user, per dialogue. (Walker et al., 1998b)	1	bb	u	exp.	utt.
# CANCEL ATTEMPTS	Average number of cancel attempts from the user, per dialogue. Sometimes divided into # CORRECTION REQUESTS and # START-OVER REQUESTS from the user. (Kamm et al., 1998; San-Segundo et al., 2001a)	1	bb	u	exp.	utt.

Table A.5. Meta-communication-related parameters (2). For a legend see Table A.1.

Name	*Definition*	*Unit*	*Syst. transp.*	*Int. partn.*	*Meas. method*	*Int. level*
Correction rate (CR, SCR, UCR); Also inverse: Turn correction ratio (TCR, $STCR$, $UTCR$)	Percentage of all turns in a dialogue primarily concerned with rectifying a "trouble" (thus not contributing new propositional content, and interrupting the dialogue flow, excluding turns which are concerned with correcting a user's misunderstanding about the system's capabilities). – SCT: Number of system correction turns (rectifying user troubles). – UCT: Number of user correction turns (rectifying system troubles). – SCR: Percentage system correction turns (rectifying user troubles) of all system turns. – UCR: Percentage user correction turns (rectifying system troubles) of all user turns. (Fraser, 1997; Simpson and Fraser, 1993; Gerbino et al., 1993; Danieli and Gerbino, 1995)	1 (%)	bb	s, u	exp.	utt.

Table A.6. Meta-communication-related parameters (3). For a legend see Table A.1.

Name	*Definition*	*Unit*	*Syst. transp.*	*Int. partn.*	*Meas. method*	*Int. level*
IR: Implicit recovery	Capacity of the system to recover from utterances which have partially failed during recognition or understanding. Determined by labelling all utterances as to whether they have been correctly parsed or not; all incorrectly – but not completely failingly – parsed user utterances (see definition of PA:PA) are further examined as to whether the system response was appropriate (see definition of CA); these cases are regarded as implicit recovery: $IR = \frac{\text{\# utterances with appropriate system answer}}{\text{\# of utterances with concept errors}}$ (Danieli and Gerbino, 1995)	1 (%)	gb	s: DM	exp.	utt.

Table A.7. Cooperativity-related parameters. For a legend see Table A.1.

Name	*Definition*	*Unit*	*Syst. transp.*	*Int. partn.*	*Meas. method*	*Int. level*
CA: Contextual appropriateness	Percentage of system utterances which are judged to be appropriate in their immediate dialogue context. Determined by labelling utterances according to whether they violate one or more of Grice's maxims for cooperativity: – CA:TF: Total failure (no linguistic response) – CA:AP: Appropriate (not violating Grice's maxims, not unexpectedly conspicuous or marked in some way) – CA:IA: Inappropriate (violating one or more of Grice's maxims) – CA:AI: Appropriate/inappropriate (experts cannot reach agreement on contextual appropriateness) – CA:IC: Incomprehensible (content cannot be discerned) (Simpson and Fraser, 1993; Fraser, 1997; Gerbino et al., 1993); similar to the appropriateness classification given by Hirschman et al. (1993)	1 (%)	bb	s	exp.	utt.

Table A.8. Task-related parameters (1). For a legend see Table A.1.

Name	*Definition*	*Unit*	*Syst. transp.*	*Int. partn.*	*Meas. method*	*Int. level*
TS: Task success; Also: Transaction success	Percentage of dialogues in which the user has reached his/her goal, i.e. the system provided the information or action desired, provided that this information is available or the action possible. The following categories are differentiated: – $TS{:}S$: Succeeded (tasks for which answers exist) – $TS{:}SCs$: Succeeded with constraint relaxation by the system – $TS{:}SCu$: Succeeded with constraint relaxation by the user (tasks involving unknown objects) – $TS{:}SCuCs$: Combination of both types of $TS{:}SC$ – $TS{:}SN$: Succeeded in spotting that no answer exists (tasks involving unknown objects, tasks with no answer) – $TS{:}Fs$: Failed because of the system's behavior: due to system inadequacies – $TS{:}Fu$: Failed because of the user's behavior: due to non–cooperative user behavior (Fraser, 1997; Danieli and Gerbino, 1995; Simpson and Fraser, 1993)	1 (%)	bb	s + u	exp.	dial.

Table A.9. Task-related parameters (2). For a legend see Table A.1.

Name	*Definition*	*Unit*	*Syst. transp.*	*Int. partn.*	*Meas. method*	*Int. level*
Task solution	Existence of a task solution reported by the user, in percent of all tasks. (Goodine et al., 1992; Polifroni et al., 1992)	1 (%)	bb	s + u	exp.	dial.
Solution correctness	Correctness of a task solution as reported by the user, in percent of all tasks. (Goodine et al., 1992; Polifroni et al., 1992)	1 (%)	bb	s + u	exp.	dial.
Solution quality	For scalable tasks: Judgment on the quality of the reported solution. E.g., the quality of a shortest possible routing task can be measured in terms of the resulting route length. (Sikorski and Allen, 1997)		bb	s + u	instr.	dial.
κ: Kappa coefficient	Percentage of task completion according to the kappa statistics. Determined on the basis of the correctness of the result AVM of a dialogue with respect to the scenario (key) AVM. A confusion matrix $M(i,j)$ can be set up for the attributes in the result and in the key, with T the number of counts in M, and t_i the sum of counts in column i of M. Then $$\kappa = \frac{P(A) - P(E)}{1 - P(E)}$$ with $P(A)$ the proportion of times that the AVM of the actual dialogue and the key agree, $P(A) = \sum_{i=1}^{n} \frac{M(i,i)}{T}$; $P(E)$ can be estimated from the proportion of times that they are expected to agree by chance, $P(E) = \sum_{i=1}^{n} (\frac{t_i}{T})^2$ (Carletta, 1996; Walker et al., 1997)	1 ($\leq$ 1)	bb	s + u	exp.	dial. or set of dial.

Table A.10. Speech-input-related parameters (1). For a legend see Table A.1.

Name	*Definition*	*Unit*	*Syst. transp.*	*Int. partn.*	*Meas. method*	*Int. level*
WA: Word accuracy; WER: Word error rate	Percentage of words which have been correctly identified, based on the orthographic form of the hypothesized and the reference utterance. Requires the availability of expert transcriptions, and an agreed alignment method (e.g. using DTW); after the alignment, all substitutions s_w, insertions i_w (not defined for isolated word recognition) and deletions d_w are counted, and referred to the total number of words $W = c_w + s_w + d_w$ (c_w being the number of correctly recognized words): $WA = 1 - \frac{s_w + i_w + d_w}{W} = 1 - WER = \frac{c_w - i_w}{c_w + s_w + d_w}$ with the word error rate WER being defined as $WER = \frac{s_w + i_w + d_w}{W}$ (Simpson and Fraser, 1993; van Leeuwen and Steeneken, 1997)	1 (%)	gb	s: ASR	instr./ exp.	word

Table A.11. Speech-input-related parameters (2). For a legend see Table A.1.

Name	*Definition*	*Unit*	*Syst. transp.*	*Int. partn.*	*Meas. method*	*Int. level*
SA: Sentence accuracy; SER: Sentence error rate	Percentage of entire sentences which have been correctly identified. Similar to WA, but on the sentence level. Denoting S the total number of sentences, and s_s, i_s and d_s the number of substituted, inserted and deleted sentences, then $$SA = 1 - \frac{s_s + i_s + d_s}{S} = 1 - SER$$ with the sentence error rate SER being defined as $$SER = \frac{s_s + i_s + d_s}{S}$$ (Simpson and Fraser, 1993)	1 (%)	gb	s: ASR	instr./ exp.	utt.

Table A.12. Speech-input-related parameters (3). For a legend see Table A.1.

Name	*Definition*	*Unit*	*Syst. transp.*	*Int. partn.*	*Meas. method*	*Int. level*
NES: Number of errors per sentence	Related to the word error rate, but calculated on a per sentence level. Being $s_w(k)$, $i_w(k)$ and $d_w(k)$ the number of substituted, inserted and deleted words in sentence k, and S the overall number of sentences, then $$NES(k) = s_w(k) + i_w(k) + d_w(k)$$ The average $\overline{NES}$ can be calculated as follows: $$\overline{NES(k)} = \frac{\sum_{k=1}^{S} NES(k)}{S} = \frac{WER \cdot W}{S}$$ (Strik et al., 2001)	1	gb	s: ASR	instr./ exp.	utt.
WES: Word error per sentence	Related measure to NES, but normalized to the number of words in sentence k, $w(k)$ $$WES(k) = \frac{NES(k)}{w(k)}$$ The average $\overline{WES}$ can be calculated as follows: $$\overline{WES} = \frac{\sum_{k=1}^{S} WES(k)}{S}$$ (Strik et al., 2001)	1	gb	s: ASR	instr./ exp.	word

Table A.13. Speech-input-related parameters (4). For a legend see Table A.1.

Name	*Definition*	*Unit*	*Syst. transp.*	*Int. partn.*	*Meas. method*	*Int. level*
HC: Recognition success metrics	Recognition success metrics after prior classification of user utterances into (1) speech vs. non-speech, and (2) for speech into in-vocabulary, embedded, related to a keyword, or OOV. Each input can be either correctly accepted or rejected, or incorrectly accepted or rejected. The following four measures can be calculated from the raw data: $HC_{U1} = \frac{\text{\# in-voc. correctly recognized}}{\text{\# in-voc.}}$ $HC_{S1} = \frac{\text{\# in-voc. correctly recognized}}{\text{\# in–voc.}} + \frac{\text{\# in-voc. correctly rejected}}{\text{\# in-voc.}}$ $HC_{U2} = \frac{\text{\# correctly recognized}}{\text{\# foreground speech}}$ $HC_{S2} = \frac{\text{\# in-voc./emb./related correctly recognized}}{\text{\# foreground speech}} + \frac{\text{\# in-voc./emb./related correctly rejected}}{\text{\# foreground speech}}$ (Kamm et al., 1997a)	1 (%)	gb	s: ASR	instr./ exp.	utt.

Table A.14. Speech-input-related parameters (5). For a legend see Table A.1.

Name	*Definition*	*Unit*	*Syst. transp.*	*Int. partn.*	*Meas. method*	*Int. level*
AN: Number of user questions correctly/ incorrectly/ partially/ failed to be answered	Total number of questions from the user which are – correctly (AN:CO) – incorrectly (AN:IC) – partially correctly (AN:PA) – not at all (AN:FA) answered by the system, per dialogue. (Polifroni et al., 1992; Goodine et al., 1992; Hirschman and Pao, 1993)	1	bb	s	exp./ instr.	utt.
DARPA measures	Measures according to the US DARPA speech understanding initiative, augmented with partially answered utterances by Skowronek (2002). All user queries in a dialogue are labelled as to whether they have been answered correctly, partially correctly, incorrectly, or not at all. Three measures are defined: $DARPA_s = \frac{AN:CO - AN:IC}{\#\text{ USER QUESTIONS}}$ $DARPA_e = \frac{AN:FA + 2 \cdot AN:IC}{\#\text{ USER QUESTIONS}}$ $DARPA_{me} = \frac{AN:FA + 2 \cdot (AN:IC + AN:PA)}{\#\text{ USER QUESTIONS}}$ (Polifroni et al., 1992; Goodine et al., 1992; Skowronek, 2002)	1 (%)	bb	s	exp./ instr.	utt.

Table A.15. Speech-input-related parameters (6). For a legend see Table A.1.

Name	*Definition*	*Unit*	*Syst. transp.*	*Int. partn.*	*Meas. method*	*Int. level*
PA: Number of correctly (*PA:CO*) / partially (*PA:PA*) / failingly (*PA:FA*) parsed utterances	Evaluation of the number of concepts (AVPs) in an utterance which have been extracted by the system: – *PA:CO*: All concepts of a user utterance are correctly understood by the system. – *PA:PA*: Not all but at least one concept of a user utterance is correctly understood by the system. – *PA:FA*: No concepts of a user utterance are correctly understood by the system. Used by Danieli and Gerbino (1995) to calculate IR.	1 (%)	gb	s: SU/ LU	instr./ exp.	utt.

Table A.16. Speech-input-related parameters (7). For a legend see Table A.1.

Name	*Definition*	*Unit*	*Syst. transp.*	*Int. partn.*	*Meas. method*	*Int. level*
IC: Information content; Also: Concept accuracy; CER: Concept error rate; Also: keyword error rate	Percentage of correctly identified semantic units (concepts) from a user utterance. Describes the system's ability to extract propositional content from user utterances. Concepts are defined as attribute-value-pairs, AVPs, of total number N_{AVP} in the utterance. The observed AVPs are compared with human-annotated reference AVPs, and substitutions, insertions and deletions are marked: $$IC = 1 - \frac{s_{AVP} + i_{AVP} + d_{AVP}}{N_{AVP}}$$ $$CER = \frac{s_{AVP} + i_{AVP} + d_{AVP}}{N_{AVP}}$$ The correct AVPs are determined from a reference AVP for the utterance by an expert. (Gerbino et al., 1993; Simpson and Fraser, 1993; Boros et al., 1996; Billi et al., 1996)	1 (%)	gb	s: SU/ LU	instr./ exp.	utt.
UA: understanding accuracy	Percentage of utterances in which all semantic units (AVPs) are correctly extracted, cf. IC: $$UA = \frac{PA:CO}{\#\ \text{USER TURNS}}$$ (Zue et al., 2000)	1 (%)	gb	s: SU/ LU	instr./ exp.	utt.

Appendix B
Template Sentences for Synthesis Evaluation, Exp. 5.1 and 5.2

Original German Version

1 Sie möchten also am *weekday foodtype* essen gehen? (*weekday*: Montag, Dienstag, Mittwoch, Donnerstag, Freitag, Samstag, Sonntag; *foodtype*: vegetarisch, italienisch, französisch, griechisch, spanisch, orientalisch, asiatisch)

2 Das Restaurant *location* hat am *weekday* Ruhetag. (*location*: am Schauspielhaus, in der Innenstadt, am Hauptbahnhof, am Stadtpark, am Kunstmuseum, am Stadion, am Opernhaus; *weekday*: Montag, Dienstag, Mittwoch, Donnerstag, Freitag, Samstag, Sonntag)

3 Wann möchten Sie *location foodtype* essen gehen? (*location*: am Schauspielhaus, in der Innenstadt, am Hauptbahnhof, am Stadtpark, am Kunstmuseum, am Stadion, am Opernhaus; *foodtype*: vegetarisch, italienisch, französisch, griechisch, spanisch, orientalisch, asiatisch)

4 Das Lokal *price* und öffnet um *time* Uhr. (*price*: ist billig, ist preiswert, ist teuer, hat gehobene Preise, ist in der unteren Preisklasse, ist in der mittleren Preisklasse, ist in der oberen Preisklasse; *time*: dreizehn, sieben, fünfzehn, achtzehn, zwanzig, vierzehn, siebzehn)

5 Die Gerichte im *foodtype* Restaurant beginnen bei *price* Mark. (*foodtype*: vegetarischen, italienischen, französischen, griechischen, spanischen, orientalischen, asiatischen; *price*: fünfzehn, zwanzig, vierzig, achtzehn, dreißig, dreizehn, siebzehn)

English Translation

1 So you would like to eat *foodtype* food on *weekday*? (*weekday*: Monday, Tuesday, Wednesday, Thursday, Friday, Saturday, Sunday; *foodtype*: vegetarian, Italian, French, Greek, Spanish, oriental, asian)

2 The restaurant *location* is closed on *weekday*. (*location*: at the theater, in town center, at the main station, at the city park, at the art museum, at the stadium, at the opera house; *weekday*: Monday, Tuesday, Wednesday, Thursday, Friday, Saturday, Sunday)

3 When would you like to eat *foodtype* food *location*? (*foodtype*: vegetarian, Italian, French, Greek, Spanish, oriental, asian; *location*: at the theater, in town center, at the main station, at the city park, at the art museum, at the stadium, at the opera house)

4 The restaurant *price* and opens at *time*. (*price*: is cheap, is good value, is expensive, has high prices, is in the lower price category, is in the middle price category, is in the upper price category; *time*: one p.m., seven o'clock, three p.m., six p.m., eight p.m., two p.m., five p.m.)

5 The menu in the *foodtype* restaurant starts at *price* DM. (*foodtype*: vegetarian, Italian, French, Greek, Spanish, oriental, asian; *price*: fifteen, twenty, forty, eighteen, thirty, thirteen, seventeen)

Appendix C
BoRIS Dialogue Structure

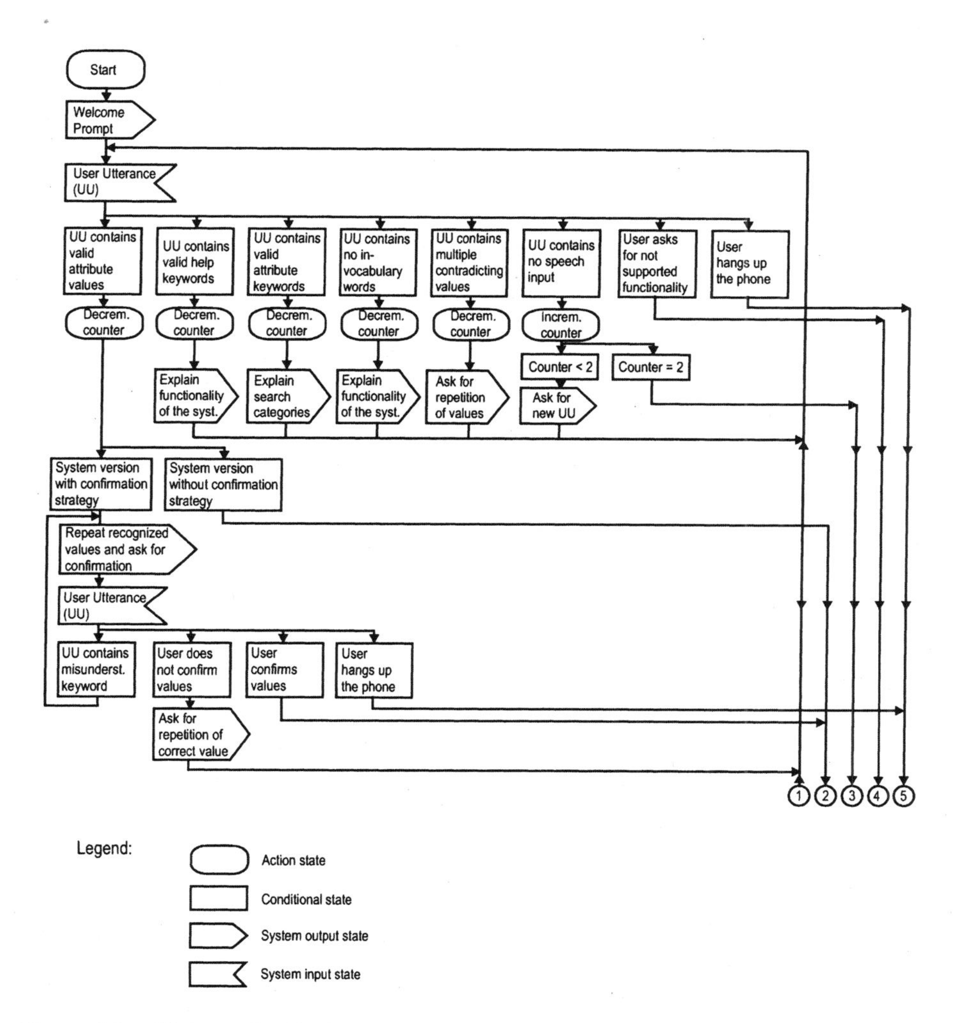

Figure C.1. Dialogue flow in the BoRIS restaurant information system of experiment 6.3, part 1.

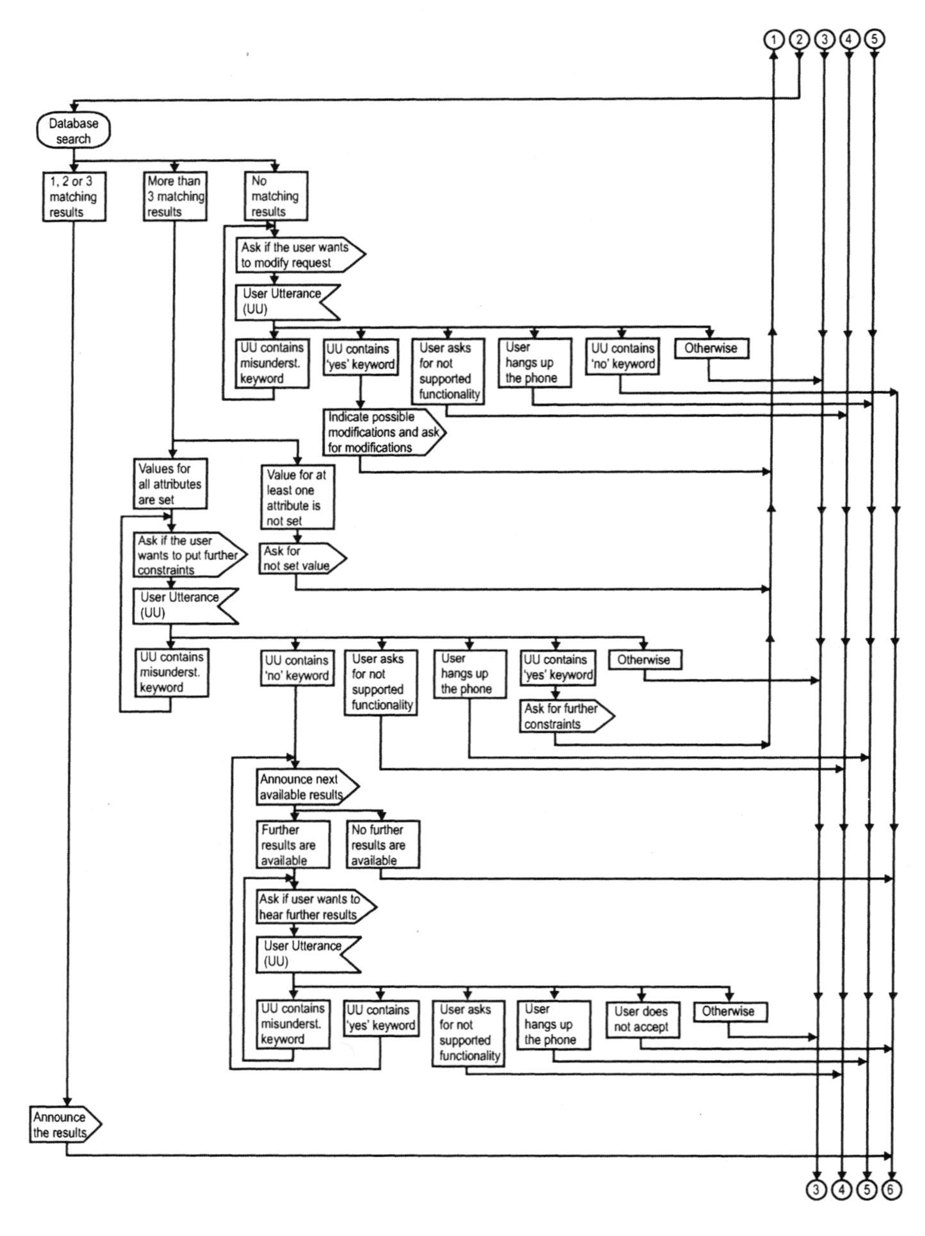

Figure C.2. Dialogue flow in the BoRIS restaurant information system of experiment 6.3, part 2. For a legend see Figure C.1.

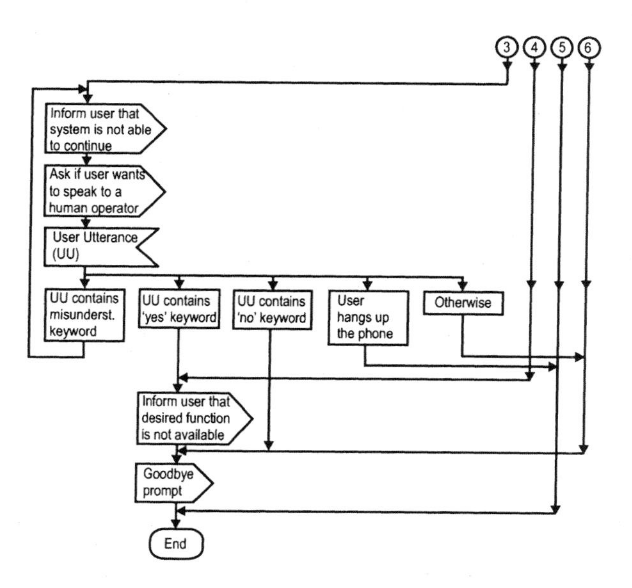

Figure C.3. Dialogue flow in the BoRIS restaurant information system of experiment 6.3, part 3. For a legend see Figure C.1.

Appendix D
Instructions and Scenarios

D.1 Instructions to Test Subjects

BoRIS

Dear participant!

Thank you for taking the time to do this experiment!

During the next hour you will get to know BoRIS via the telephone: The Bochum restaurant information system.

This test will show how you experience a telephone call with BoRIS. For this aim, we ask you to call BoRIS five times. Before each call you will get a small task. At the end of **each** telephone call, we ask you to write down what you think about the system. You can do this easily by filling out a questionnaire.

Before the test starts, we would like to ask you to answer the questions given on the following pages. For the test evaluation, we need some personal information from your side, information which will be treated anonymously of course.

At the end of the **whole** experiment, we ask you to give an overall judgment about all the calls you had with BoRIS.

For some assessments you will find the following scale:

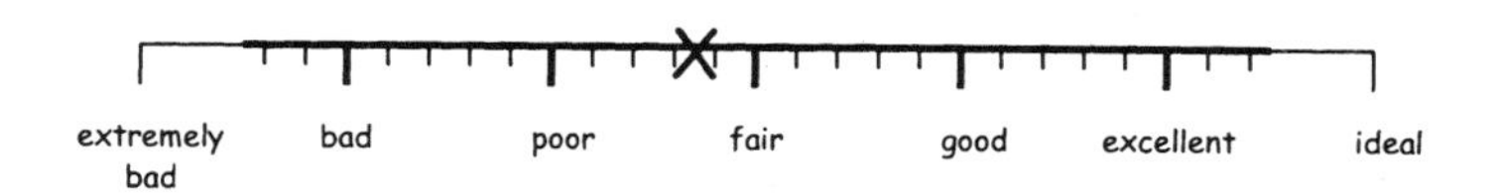

Usually, your judgment should be in the range between bad and excellent. In case of an unpredictable extreme judgment, you can use the thinly drawn edges of the scale as well. Please also use the spaces between the grid marks, as depicted above.

Assess the system in a very self-confident way and remember during the whole test session:

Not **you** are tested, but **you** test our system!

And now: Have a lot of fun!

D.2 Scenarios

Dialogue no. ___

You would like to know where you can eat duck. Please ask BoRIS.

Restaurant name(s):

Dialogue no. ___

You plan to go out for a Greek dinner on Tuesday night in Grumme.

Price: - |—X—|———|———|———| +

Restaurant name(s):

If BoRIS is unable to indicate a restaurant, please change the following specification:

You want to have the dinner in Weitmar.

Restaurant name(s):

Dialogue no. ___

You plan to have your lunch break in a Chinese restaurant downtown.

Price: - |——|——|——|—x—| +

Restaurant name(s):

If BoRIS is unable to indicate a restaurant, please change the following specification:

The price.

Restaurant name(s):

Dialogue no. ___

Please gather your information from the following hints:

Price: - |—|—*—|—| +

Type of food:

Location:

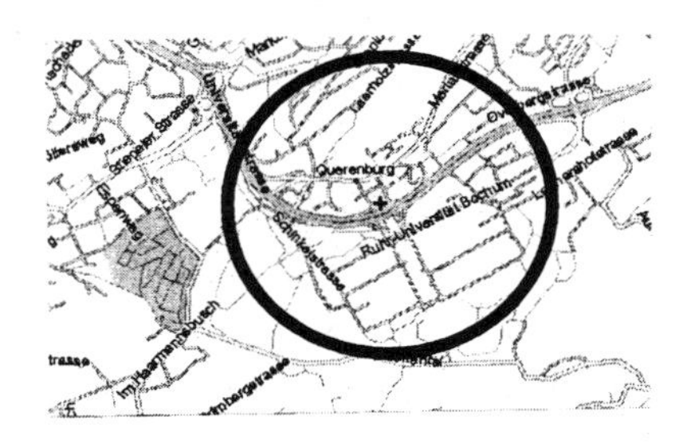

Restaurant name(s):

If BoRIS doesn't find a matching restaurant, please change the following:

Price: - |X|—|—|—| +

Restaurant name(s):

Dialogue no. ___

You plan to eat out in Bochum. Because your favorite restaurant is closed for holidays, ask BoRIS for a restaurant.

Please write down first which specifications you want to give to BoRIS.

If BoRIS is unable to find a matching restaurant, please search for an alternative until BoRIS indicates at least one restaurant.

Restaurant name(s):

D.3 Instructions for Expert Evaluation

The following guidelines describe the steps an evaluation expert has to perform in order to analyze and annotate an interaction with the BoRIS restaurant information system, see experiments 6.1 to 6.3. A number of criteria are given which have to be judged upon in each step, and it is illustrated how these criteria have to be interpreted in the context of the restaurant information task. It has to be noted that the criteria and recommendations are not strict rules. Instead, the evaluation expert often has a certain degree of freedom for interpretation. In order to take a decision in an individual case, the expert should consider the objective of the criteria, and the course of the interaction up to the specific point. In the case that a certain interpretation is chosen, the expert should try to adhere to this interpretation in order to reach consistent results for all dialogues in the analysis set.

The analysis and annotation procedure consists of the following steps:

1. Scenario definition
2. Dialogue execution
3. Transcription
4. Barge-in labelling
5. Task AVM analysis
6. Task success labelling
7. Contextual appropriateness labelling
8. System correction turn labelling
9. User correction turn labelling
10. Cancel attempt labelling
11. Help request labelling
12. User question labelling
13. System answer labelling
14. Speech understanding labelling
15. Automatic calculation of speech-recognition-related measures.
16. Automatical calculation of further interaction parameters.

The following guidelines focus on steps where the expert has to take a judgment on a specific interaction aspect (Steps 1 and 3 to 14). Practical information on the operation of the CSLU-based WoZ workbench and of the expert evaluation tool are given in Skowronek (2002).

Step 1: Scenario Definition

The scenario for the restaurant information task consists of six slots for attribute-value pairs, namely *Task*, *Foodtype*, *Date*, *Time*, *Price* and *Location*. The field *Task* can take two different values: "Get information" where the aim of the dialogue is to obtain information about restaurants, or "unknown" where the user asks for a task which is not supported by the system, e.g. a reservation. The expert has to interpret relative date specifications like today, tomorrow, etc. as follows:

- Today, tomorrow, the day after tomorrow etc. $\Rightarrow$ the corresponding weekday.
- Now, in a little while etc. $\Rightarrow$ the corresponding day and time.

This interpretation corresponds to the canonical values used by the language understanding component of the system. The following expressions should not be changed in this way because they are out of the understanding capability of the system:

- During the week, weekdays, weekend, etc. ⇒ leave unchanged.

In the case that no specifications for a slot are given in the scenario definition, the according slot should be left undetermined. The same principle applies to the free scenario.

Step 3: Transcription

The system utterances are automatically logged during the interaction. Thus, only the user utterances have to be transcribed by the expert, in the case that no transcription has been produced during the interaction (which is the case for simulated recognition). The user utterance transcription has to include literally everything that has been articulated during a user's turn, including laughing, talking to himself, etc. In this way, it will reflect the input of the system in a real-life environment.

The expert has to type the transcription into the according field of the evaluation tool. All letters (including German "Umlaute") and punctuation marks are allowed. Linebreaks are generated automatically, but they can also be enforced by pressing the return key. However, it has to be ensured that no empty lines are transcribed, except when the whole user utterance is empty. Scrolling over several lines is possible.

Step 4: Barge-In Labelling

This step refers to the user utterances only. A barge-in attempt is counted when the user intentionally addresses the system while the system is still speaking. In this definition, user utterances which are not intended to influence the course of the dialogue (laughing, speaking to himself/herself) are not counted as barge-ins. They are treated as spontaneous reactions which are not intended to influence the course of the dialogue.

All barge-in attempts are labelled by setting the according radio button in the expert evaluation tool. The barge-in utterance will not be transcribed until the user repeats it when the turn is on the user again.

Step 5: Task AVM Analysis

The "Scenario AVM" is specified by the scenario and consists of six attribute-value pairs for the slots *Task*, *Date*, *Foodtype*, *Time*, *Price* and *Location*.

During the course of the interaction, it may happen that the user changes one or several of the specifications given in the scenario, either by adding further constraints, by omitting to give constraints, or by changing the constraint values. Such a change may happen either on the user's own authority, or because the system requested to do so. In both cases the "Scenario AVM" has to be amended, resulting in a "Modified Scenario AVM" and in a "Changed AVM".

- In a first step, the attributes of the user query which differ from the specification given in the scenario have to be identified. These attributes and the corresponding values are written down in the according "Modified Scenario AVM".
- If the user voluntarily sets the value of an attribute to a neutral value (e.g. by saying "don't know", "doesn't matter", etc.), the value "neutral" has to be set in the AVM. However, in the case that the user has no possibility to specify the value (e.g. because the system did not ask him/her to do so), the AVM remains unchanged at this point. This guideline assumes that the user would have provided the missing information but the system prematurely directed the dialogue in a different way.

- In the case that the user specifies a value for an attribute that is not indicated in the scenario, this value has to be included in the "Modified Scenario AVM", independently of whether the system asked for it or not.
- In the case that the system asks the user to modify attribute values during the interaction (e.g. because it did not find a matching restaurant), such modifications should be included in the "Changed AVM".
- In the case that the user changes an attribute value which was previously specified without being asked to do so by the system, two situations have to be distinguished:
  - If the user changes a specification spontaneously, by intuition, this modification should be handled in the "Modified Scenario AVM".
  - If the user changes a specification because the system obviously did not process his/her first specification attempt, this modification should be handled by the "Changed AVM".

  This principle is in accordance with the definition of user correction turns, see below. When the expert would rate such an utterance as a user correction turn, then the modification should be handled by the "Changed AVM".
- When the modification occurs in an explicit confirmation situation (e.g. as a response to a system confirmation utterance like "Do you really want to eat out in ... ?"), then it should be handled by the "Modified Scenario AVM".

The expert is only allowed to provide values which are in the system's vocabulary (so-called "canonical values"). Other values, although they might be specified by the user, should not be introduced in the AVMs. This rule corresponds to a system-orientated point of view.

All AVMs are amended by an additional slot, namely the one with the restaurants which match the specified attribute values. This slot is automatically calculated from the system database.

Example:
Scenario AVM:

Attribute	*Value*
Task	Get information
Date	–
Foodtype	Greek
Time	–
Price	–
Location	Center

Dialogue:

S: ...
U: "*Today*, I'd like to eat out in a Greek restaurant downtown."
S/U: ...
S: "I'm sorry. There's no restaurant that matches your query.
Would you like to change you query?"
U: "Yes, please."
S: "You can change the type of food, the preferred price range, ..."
What is your modification?"
U: "Well, I want *Italian* food."
S: ...

Remark: Current weekday is Friday.

The resulting "Modified Scenario AVM" looks as follows:

Attribute	*Value*
Task	Get information
Date	*Friday*
Foodtype	Greek
Time	–
Price	–
Location	Center

The "Changed AVM" is as follows:

Attribute	*Value*
Task	Get information
Date	*Friday*
Foodtype	*Italian*
Time	–
Price	–
Location	Center

Step 6: Task Success Labelling

Task success describes whether the aim of the dialogue was achieved or not. In order to judge upon task success, the expert makes use of the restaurant slot of the "Modified Scenario AVM", the restaurant slot of the "Changed AVM", and the transcription of the dialogue.

The following categories are available:

1 S: Succeed
The aim of the dialogue is achieved and the user obtains the requested information, provided that the requested information is covered by the system functionality. The aim of the dialogue can be regarded as achieved when all restaurants given by the system are contained in the restaurant slot of the "Modified Scenario AVM". However, not all restaurants of the "Modified Scenario AVM" need to be included in the system answer.

2 SCu: Succeed with constraint relaxation by the user
The user made a query which is covered by the functionality of the system, but the system cannot find a matching restaurant in the database. The user follows the system request for modification, and the system is able to provide an answer for the modified request. This case is regarded as a constraint relaxation by the user. The restaurants given by the system all have to be contained in the "Changed AVM". However, not all restaurants of the "Changed AVM" need to be included in the system answer.

3 SCs: Succeed with constraint relaxation by the system
The aim of the dialogue is achieved although the user was not able to provide all specifications. All restaurants provided by the system have to be part of the "Modified Scenario AVM". However, not all restaurants of the "Modified Scenario AVM" need to be included in the system answer.

4 *SCuCs*: Succeed with constraint relaxation by the user and by the system
Combination of the previous two cases: The aim of the dialogue is achieved after the user had to change his/her query, but nevertheless the user was unable to give all specifications. The aim of the dialogue can be regarded as achieved if all restaurants provided by the system are contained in the "Changed AVM". However, not all restaurants of the "Changed AVM" need to be contained in the system answer.

5 *SN*: Succeed in spotting that no answer exists
The user made a query which can be covered by the current system functionality, however the system cannot find a matching restaurant in the database. The user does not follow the system request for modification, and the interaction ends by spotting that no answer exists. This case is regarded as a success in spotting that no answer exists (*SN*). The *SN* category also has to be chosen when the system informs the user that his/her request is outside the current system functionality, e.g. when the user asks for a reservation. All the "Modified Scenario AVM", the "Changed AVM" and the system response do not contain any restaurants in this case.

6 *FS*: Failed because of system behavior
The system provides an answer which is neither contained in the "Modified Scenario AVM" nor in the "Changed AVM", or it finishes the interaction prematurely although the user behaves cooperatively.

7 *FU*: Failed because of user behavior
The aim of the dialogue cannot be achieved because the user behaves uncooperatively, e.g. by giving permanently senseless answers. The interaction is also classified as *FU* in the case that the user finishes the interaction prematurely (e.g. by simply hanging up), irrespective of the user's motivation.

A first proposal for task success is made by the evaluation tool. This proposal might have to be modified by the expert, notably in the case of unexpected termination of the dialogue. In such a case, the expert tool might not be able to provide any meaningful proposal for task success.

Step 7: Contextual Appropriateness Labelling

The expert has to judge each system utterance with respect to its appropriateness in the current dialogue context. The following criteria serve as a basis for this judgment:

- Informativeness: Make your contribution as informative as required (for the current purposes of the exchange); do not make your contribution more informative than is required.
- Truth and evidence: Do not say what you believe to be false; do not say for which you lack adequate evidence.
- Relevance: Be relevant.
- Manner: Be perspicuous; avoid obscurity of expression; avoid ambiguity; be brief; be orderly.

The following categories are available:

1 *AP*: Appropriate
None of the four criteria violated.

2 *IA*: Inappropriate
One or more of the four criteria is violated, e.g. in the following cases:

- The system asks a counterquestion or ignores a user query completely.
  Example:
  U: "Which types of food are available?"
  S: "At the moment only the following functions can be provided by BoRIS: ..."

- The system provides an uncooperative answer which needlessly lengthens or aborts the dialogue, e.g. by ignoring a specification given by the user.
  Example:
  S: "Would you like to modify your query?"
  U: "Yes, then I would like to eat a pizza."
  S: "Please indicate your modification."

- The system provides a perceivably over-informative answer, i.e. it gives more information than the user requested.
  Example:
  S: "Would you like to hear more restaurants?"
  U: "Yes, one more please."
  S: "Restaurant 1, restaurant 2, restaurant 3, ..."

- The system provides a perceivably under-informative answer, i.e. it gives less information than the user requested.
  Example:
  U: "Which types of food may I choose?"
  S: "The possibilities are: Chinese, Greek, German or Italian."
  Note: The system knows more than 20 types of food.

It should be decided in the scenario context whether a system answer is over- or under-informative.

- In the case that the user asks for a specific type of food (e.g. pizza, duck, etc.) and the system only proposes different nationalities (e.g. Chinese, Italian, etc.), this answer should be classified as under-informative, because the system ignores these specific types of food.
- In the case that the user is looking for a nationality of food that the system knows but does not mention following a user request, then this answer also has to be classified as being under-informative.
- If the user is looking for a type of food which is outside the system knowledge, then the respective answer should be classified as appropriate.

In his classification, the expert may also make use of the "Modified Scenario AVM" and the "Changed AVM". The expert should consider whether the answer is *perceivably* over- or under-informative to the user.

3 *IC*: Incomprehensible
The system answer is incomprehensible in the current dialogue context.

4 *TF*: Total failure
The system fails to give an answer.

In the case of repeated system questions, the expert has to assess whether the repeated asking is appropriate or not.

- Questions are appropriate if they have the purpose of building a common ground of knowledge between the system and the user. For example, if the system asks to confirm a user

utterance, then this confirmation question has to be classified as AP. This is particularly important in an explicit confirmation strategy.
Example:
U: "I'd like to eat Greek food."
S: "Thus, you want to eat Greek food?" $\Rightarrow AP$

- On the other hand, a system question should be classified as IA when the system asks for an information which was already previously specified and confirmed in the dialogue.
  Example:
  S: "Where would you like to go?" $\Rightarrow AP$
  U: "To the town center."
  S: ...
  U: ...
  S: "Could you repeat the desired location of the restaurant?" $\Rightarrow IA$
  U: "As I already said: Downtown."

- The expert should take into account whether the repeated question would occur in a comparable dialogue between humans. In the case that the system repeats the question several times without obvious reason (of misunderstanding), then the repeated questions should be classified as IA.
  Example:
  S: "Where would you like to go?"
  U: "To the town center."
  S: "Sorry, I didn't understand you. Where would you like to go?" $\Rightarrow AP$
  U: "To the town center."
  but:
  S: "Where would you like to go?"
  U: "To the town center."
  S: "Where would you like to go?" $\Rightarrow IA$
  U: "To the town center."

Step 8: System Correction Turn Labelling

A system correction turn (SCT) is defined as a system turn which is primarily concerned with resolving a "trouble". Such a turn interrupts the normal flow of the dialogue and provides no additional information with respect to the aim of the dialogue. A "trouble" may be caused by speech recognition or understanding errors, or by illogical, contradictory, or undefined user utterances. An important exception is the case when the user asks for information or action which is not supported by the current system functionality. The utterance in which the system informs the user about this limitation is not considered as a system correction turn.

In order to judge on system correction turns, the expert needs to know the exact source of the problem. A system utterance should be labelled as a SCT when the expert assumes that the system tries to correct an error or a contradiction in the flow of the dialogue. The information of the system about empty results in the database is not considered as a correction turn, because it provides previously unknown task-related information to the user. The same principle applies to explicit confirmation questions, because they serve the set-up of a common ground of knowledge between the system and the user.

Step 9: User Correction Turn Labelling

Analogously to a SCT, a user correction turn (UCT) is a user utterance which is primarily concerned with rectifying some kind of "trouble". If the user made a mistake and corrects his/her own specifications, this turn should be counted as a UCT. If the user provides an answer to a SCT, then this turn should not be counted as a confirmation, but rather as a UCT. The underlying assumption in this case is that a problem might have occurred before, resulting in the SCT. All subsequent user turns concerned with resolving the same problem should be counted as user correction turns.

In case of an explicit confirmation from the user, this utterance should only be counted as a UCT if the user has to change a specification which was interpreted by the system in a wrong way. If, however, the user has to repeat information because the system was not able to extract it from a previous utterance, this repetition should be classified as a UCT.

Step 10: Cancel Attempt Labelling

A user turn is classified as a cancel attempt if the user tries to restart the dialogue from the beginning, or if he/she explicitly wants to step one or several levels backwards in the dialogue hierarchy.

Step 11: Help Request Labelling

A user turn is classified as a help request if the user explicitly asks for help. This request may be formulated as a question or as a statement.

Example:
User: "Could you repeat the possibilities I have?"

However, if the dialogue control is with the system, and if the system gives hints on available options on its own, then the preceding user utterance should not be classified as a help request.

Step 12: User Question Labelling

A user turn is classified as a user question if the user asks a question with the aim of putting the dialogue goal forward. Example: "What types of food may I choose from?" However, users sometimes ask for feedback on whether the system is still present (e.g. "Hello?"). This question should not be classified as a user question because it is not meant to put the dialogue goal forward. It has to be interpreted as a spontaneous user reaction in the consequence of excessive response delay from the system, or of transmission impairments perceived by the user. If the user asks a question and directly answers it on his own, this question should neither be classified as a user question. User questions may also be help requests.

Step 13: System Answer Labelling

System answers can be classified according to the following categories:

1 Correct (AN:CO)
The system answer contains all information which was requested by the user.
Example:
U: "What days may I choose?"
S: "The possibilities are: The five weekdays, Saturday, Sunday, or no preference."

2 Incorrect (AN:IC)
The system answer is wrong.

Example:

U: "What types of food are available?"

S: "The possibilities are: The five weekdays, Saturday, Sunday, or no preference."

3 Partially correct $AN{:}PA$

The system answer is only partially correct.

Example:

U: "What types of food are available?"

S: "You can choose between the following types: Italian, French Greek, Chinese, or no preference."

Note: There are far more types of food in the database.

4 Failed ($AN{:}FA$)

The system ignored or didn't process the user question.

Example:

U: "What types of food are available?"

S: "On which day are you planning to go?"

If the user asks for a restaurant which cannot directly be given because of lacking information, then an according system question should be classified as a correct "answer". This is consistent with the classification of contextual appropriateness.

Example:

U: "Where may I get Greek food tonight?"

S: "How much would you like to spend per person?"

Step 14: Speech Understanding Labelling

The speech understanding capability of the system is defined by seven slots (attribute-value pairs) which can be extracted from each user utterance: Five slots describing the specifications for a restaurant (*Date*, *Foodtype*, *Time*, *Price* and *Location*) and two additional slots which are necessary for language understanding (a *Logical* slot for describing negations and logical operations, and a *Field* slot for addressing a specific attribute). The expert has to search for any attribute-related values in a user utterance, and assess whether the canonical meaning representation used by the system has been correctly identified or not.

The following classifications shall be used by the expert:

1 Not set

When no value for an attribute is contained in the user utterance, and the system leaves the according slot empty, then the "not set" value should be assigned to the according slot.

2 Correct

The attribute value contained in the user utterance has the same meaning (canonical representation) as was set by the system.

3 Substituted

The attribute value contained in the user utterance has a different meaning (canonical representation) than was set by the system.

4 Inserted

The system assigned a meaning (canonical representation) for an attribute which was not given in the user utterance.

5 Deleted

The system assigned no meaning (canonical representation) for an attribute which was given in the user utterance.

The following specific cases have to be noted:

- The user utterance may contain several values for a single attribute. In this case, the logical operator between the attribute values is decisive for the required interpretation:

  - Logical OR: Both values have to be set by the system.
    Example:
    User: "I'd like to have Greek or Italian food."
    System: *Foodtype* Greek, Italian
    Expert: *Foodtype* Greek, Italian
    Classification: *Foodtype* ⇒ Correct

  - The user corrects himself/herself. In this case only the corrected value should be set by the system.
    Example:
    User: "I'd like to have Greek food. No, I prefer Italian."
    System: *Foodtype* Italian
    Expert: *Foodtype* Italian
    Classification: *Foodtype* ⇒ Correct

- The user utterance contains a denial.
  Example:
  User: "I don't like Greek food."
  System: *Foodtype* African, Italian...
  Expert: *Foodtype* African, Italian...
  Classification: *Foodtype* ⇒ Correct

- In each case that the system sets values for a specific attribute which are not contained in the user utterance (but other values for the attribute are contained in the utterance), then this is classified as a substitution. Substitutions are also assigned when the system sets more or less values than contained in the user utterance.

- The following examples show the difference between substitution, insertion, and deletion classifications:
  Example 1:
  User: "I'd like to have Greek or Italian food."
  System: *Foodtype* Greek *Time* –
  Expert: *Foodtype* Greek, Italian *Time* –
  Classification: *Foodtype* ⇒ Substituted *Time* ⇒ Not set
  Example 2:
  User: "I'd like to have Greek or Italian food."
  System: *Foodtype* Greek, Italian *Time* Evening
  Expert: *Foodtype* Greek, Italian *Time* –
  Classification: *Foodtype* ⇒ Correct *Time* ⇒ Inserted
  Example 3:
  User: "I'd like to have Greek or Italian food."
  System: *Foodtype* – *Time* –
  Expert: *Foodtype* Greek, Italian *Time* –
  Classification: *Foodtype* ⇒ Deleted *Time* ⇒ Not set

In contrast to the classification of task AVMs, the expert does not have to check whether the words uttered by the user are contained in the system vocabulary. For example, if the user gives a specification with an unambiguous meaning which is contained in the database, however using expressions which are outside the system vocabulary, than this meaning representation should be used by the evaluation expert.

Example:

User:	"The price is not important."	
	Note: The expression "not important" is not part of the system vocabulary.	
System:	*Price*	–
Expert:	*Price*	Neutral
Classification:	*Price*	⇒ Deleted

This interpretation is in line with the assessment of the speech understanding component. The performance of the ASR component is described in terms of WA, WER, etc., which are directly extracted from the expert transcriptions.

Appendix E
Questionnaires

E.1 Questionnaire for Experiment 6.2

Original German Version

1. Ihr Qualitätsurteil über BoRIS:

1.1 Wie würden Sie Ihren Gesamteindruck von BoRIS einschätzen?

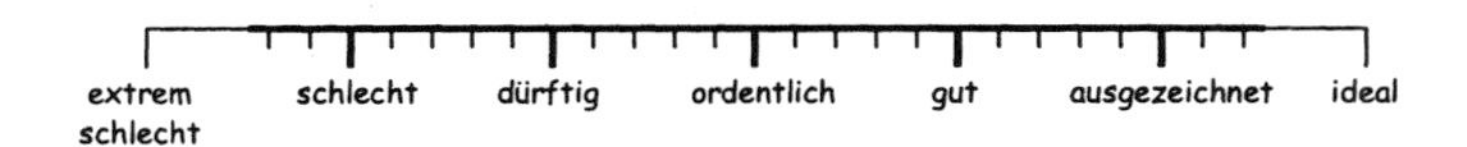

1.2 Wie würden Sie BoRIS einem Freund beschreiben?

-
-
-
-
-

1.3 Was würden Sie bei BoRIS verbessern?

-
-
-
-

2. Bitte beurteilen Sie die folgenden Aussagen über Ihre Erfahrung mit BoRIS:

2.1 BoRIS hat genau so funktioniert, wie ich es mir vorgestellt habe.

trifft zu ☐ ☐ ☐ ☐ ☐ trifft nicht zu

2.2 BoRIS bietet umfangreiche Informationen über Bochumer Restaurants.

trifft zu ☐ ☐ ☐ ☐ ☐ trifft nicht zu

2.3 Ich bekam immer die gewünschte Antwort auf meine Eingaben.

trifft zu ☐ ☐ ☐ ☐ ☐ trifft nicht zu

2.4 Ich spreche lieber mit einem Menschen als mit BoRIS .

trifft zu ☐ ☐ ☐ ☐ ☐ trifft nicht zu

2.5 Ich konnte mich äußern wie in einem normalen Gespräch mit einem Menschen.

trifft zu ☐ ☐ ☐ ☐ ☐ trifft nicht zu

2.6 Ich bin davon überzeugt, dass BoRIS mir richtige Informationen geliefert hat.

trifft zu ☐ ☐ ☐ ☐ ☐ trifft nicht zu

2.7 Das Gespräch mit BoRIS hat schnell zur gewünschten Information geführt.

trifft zu ☐ ☐ ☐ ☐ ☐ trifft nicht zu

2.8 Der Anruf bei BORIS hat sich gelohnt.

trifft zu ☐ ☐ ☐ ☐ ☐ trifft nicht zu

2.9 Jeder kann mit BoRIS ohne Probleme umgehen.

trifft zu ☐ ☐ ☐ ☐ ☐ trifft nicht zu

2.10 Durch das Gespräch mit BoRIS fühlte ich mich :

wohl	☐	☐	☐	☐	☐	unwohl
sicher	☐	☐	☐	☐	☐	unsicher
ruhig	☐	☐	☐	☐	☐	aufgeregt
zufrieden	☐	☐	☐	☐	☐	unzufrieden

2.11 Ich habe die Fragen von BoRIS beantwortet:

stichwortartig ☐ in ganzen Sätzen ☐ teils-teils ☐

3. Was Sie bei BoRIS sehr gestört hat:

3.1 BoRIS hat plötzlich das Gespräch unterbrochen und ist spontan zu einem vorherigen Menupunkt zurückgekehrt.

trifft zu ☐ ☐ ☐ ☐ ☐ trifft nicht zu

3.2 BoRIS hat das Gespräch oft mittendrin abgebrochen.

trifft zu ☐ ☐ ☐ ☐ ☐ trifft nicht zu

3.3 BoRIS konnte die Restaurantadressen nicht wiederholen.

trifft zu ☐ ☐ ☐ ☐ ☐ trifft nicht zu

3.4 BoRIS bietet keine detaillierteren Informationen über die gewünschten Restaurants.

trifft zu ☐ ☐ ☐ ☐ ☐ trifft nicht zu

3.5 Ich musste oft meine Eingaben wiederholen.

trifft zu ☐ ☐ ☐ ☐ ☐ trifft nicht zu

3.6 Ich konnte mir die Auswahlkriterien am Anfang des Gesprächs nicht merken.

trifft zu ☐ ☐ ☐ ☐ ☐ trifft nicht zu

3.7 Die Pausen zwischen meinen Eingaben und den Antworten von BoRIS waren zu lang.

trifft zu ☐ ☐ ☐ ☐ ☐ trifft nicht zu

3.8 Die Ansage der Restaurantnamen und der Adressen war nicht deutlich.

trifft zu ☐ ☐ ☐ ☐ ☐ trifft nicht zu

3.9 Der Weg bis zu der gewünschten Information war umständlich.

trifft zu ☐ ☐ ☐ ☐ ☐ trifft nicht zu

3.10 Es war mir nicht klar, wie ich meine Eingaben modifizieren soll, um schnell die gewünschte Information zu bekommen.

trifft zu ☐ ☐ ☐ ☐ ☐ trifft nicht zu

4. Für einen guten Gesprächsverlauf mit BoRIS ist es sehr wichtig,

4.1 dass BoRIS seine Fragen wiederholt, wenn ich nicht mehr weiter weiß.

trifft zu ☐ ☐ ☐ ☐ ☐ trifft nicht zu

4.2 dass BoRIS meine Eingaben immer gut versteht.

trifft zu ☐ ☐ ☐ ☐ ☐ trifft nicht zu

4.3 dass BoRIS klare Fragen stellt.

trifft zu ☐ ☐ ☐ ☐ ☐ trifft nicht zu

4.4 dass BoRIS schnell auf meine Eingaben reagiert.

trifft zu ☐ ☐ ☐ ☐ ☐ trifft nicht zu

4.5 dass BoRIS einfach zu bedienen ist.

trifft zu ☐ ☐ ☐ ☐ ☐ trifft nicht zu

4.6 dass BoRIS passende Informationen auf meine Anfragen gibt.

trifft zu ☐ ☐ ☐ ☐ ☐ trifft nicht zu

4.7 dass ich BoRIS Fragen stellen kann.

trifft zu ☐ ☐ ☐ ☐ ☐ trifft nicht zu

4.8 dass ich - nachdem ich die gewünschte Information bekommen habe - das Gespräch weiterführen kann, solange ich noch zusätzliche Fragen habe.

trifft zu ☐ ☐ ☐ ☐ ☐ trifft nicht zu

4.9 dass ich von Anfang an weiß, wie ich meine Eingaben formulieren soll.

trifft zu ☐ ☐ ☐ ☐ ☐ trifft nicht zu

4.10 dass ich von Anfang an weiß, welche Art von Informationen mir BoRIS liefern kann und welche nicht.

trifft zu ☐ ☐ ☐ ☐ ☐ trifft nicht zu

4.11 dass ich von BoRIS eine Bestätigung meiner Eingaben bekomme.

trifft zu ☐ ☐ ☐ ☐ ☐ trifft nicht zu

4.12 dass die Pausen nach meinen Eingaben von einen "Warten Sie einen Moment" ausgefüllt werden.

trifft zu ☐ ☐ ☐ ☐ ☐ trifft nicht zu

4.13 dass es einfach ist, die Eingaben zu modifizieren.

trifft zu ☐ ☐ ☐ ☐ ☐ trifft nicht zu

4.14 dass das Gespräch ohne Unterbrechungen verläuft.

trifft zu ☐ ☐ ☐ ☐ ☐ trifft nicht zu

4.15 dass BoRIS mit einer Hilfefunktion ausgestattet ist.

trifft zu ☐ ☐ ☐ ☐ ☐ trifft nicht zu

English Translation

1. Your quality judgment about BoRIS:

1.1 How would you describe your overall impression of BoRIS?

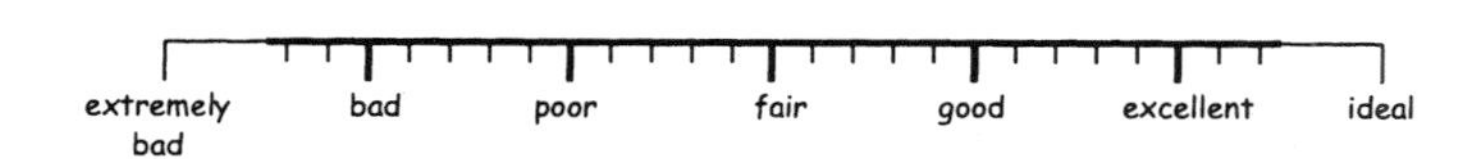

1.2 How would you describe BoRIS to a friend?

-
-
-
-
-

1.3 Which characteristics of BoRIS would you like to enhance?

-
-
-
-

2. Please judge the following statements reflecting your experience with BoRIS:

2.1 BoRIS worked exactly the way I expected.

yes ☐ ☐ ☐ ☐ ☐ no

2.2 BoRIS provides extensive information about restaurants in Bochum.

yes ☐ ☐ ☐ ☐ ☐ no

2.3 I always obtained the desired information.

yes ☐ ☐ ☐ ☐ ☐ no

2.4 I prefer to speak to a human being instead of BoRIS .

yes ☐ ☐ ☐ ☐ ☐ no

2.5 I could express myself like in a normal conversation with a human being.

yes ☐ ☐ ☐ ☐ ☐ no

2.6 I am convinced that BoRIS provided the right information to me.

yes ☐ ☐ ☐ ☐ ☐ no

2.7 The dialogue with BoRIS quickly lead to the desired information.

yes ☐ ☐ ☐ ☐ ☐ no

2.8 Calling BoRIS is worthwhile.

yes ☐ ☐ ☐ ☐ ☐ no

2.9 Everyone can handle BoRIS without any problems.

yes ☐ ☐ ☐ ☐ ☐ no

2.10 Due to the dialogue with BoRIS, I felt:

happy	☐	☐	☐	☐	☐	unhappy
sure	☐	☐	☐	☐	☐	unsure
quiet	☐	☐	☐	☐	☐	excited
satisfied	☐	☐	☐	☐	☐	dissatisfied

2.11 I answered the questions from BoRIS in the following way:

shorthand ☐ full sentences ☐ partly-partly ☐

3. The following characteristics of BoRIS were very disturbing to me:

3.1 BoRIS suddenly interrupted the dialogue and spontaneously returned to a previous menu point.

yes ☐ ☐ ☐ ☐ ☐ no

3.2 BoRIS often aborted the call in the middle of the dialogue.

yes ☐ ☐ ☐ ☐ ☐ no

3.3 BoRIS was not able to repeat the addresses of the restaurants.

yes ☐ ☐ ☐ ☐ ☐ no

3.4 BoRIS could not provide detailed information on the selected restaurants.

yes ☐ ☐ ☐ ☐ ☐ no

3.5 I had to repeat my utterances frequently.

yes ☐ ☐ ☐ ☐ ☐ no

3.6 I could not remember the selection criteria given at the beginning of the dialogue.

yes ☐ ☐ ☐ ☐ ☐ no

3.7 The pauses between my requests and the answer from BoRIS were too long.

yes ☐ ☐ ☐ ☐ ☐ no

3.8 The pronunciation of the restaurant names and addresses was unclear.

yes ☐ ☐ ☐ ☐ ☐ no

3.9 The way for obtaining the desired information was tedious.

yes ☐ ☐ ☐ ☐ ☐ no

3.10 It was unclear to me how I could modify my request in order to quickly obtain the desired information.

yes ☐ ☐ ☐ ☐ ☐ no

4. For a good dialogue with BoRIS, it is very important...

4.1 that BoRIS repeats his questions in case that I am lost.

yes ☐ ☐ ☐ ☐ ☐ no

4.2 that BoRIS always understands my utterances.

yes ☐ ☐ ☐ ☐ ☐ no

4.3 that BoRIS asks clear questions.

yes ☐ ☐ ☐ ☐ ☐ no

4.4 that BoRIS reacts quickly to my utterances.

yes ☐ ☐ ☐ ☐ ☐ no

4.5 that BoRIS is easy to handle.

yes ☐ ☐ ☐ ☐ ☐ no

4.6 that BoRIS provides adequate information to my request.

yes ☐ ☐ ☐ ☐ ☐ no

4.7 that I can ask questions to BoRIS.

yes ☐ ☐ ☐ ☐ ☐ no

4.8 that I can continue the dialogue as long as I want, in order to be able to ask additional questions.

yes ☐ ☐ ☐ ☐ ☐ no

4.9 that I know from the beginning how I have to formulate my requests.

yes ☐ ☐ ☐ ☐ ☐ no

4.10 that I know from the beginning which type of information can be provided by BoRIS, and which not.

yes ☐ ☐ ☐ ☐ ☐ no

4.11 that BoRIS confirms my input.

yes ☐ ☐ ☐ ☐ ☐ no

4.12 that the pauses following my input are filled by "Wait a moment, please..."

yes ☐ ☐ ☐ ☐ ☐ no

4.13 that it is easy to modify my input.

yes ☐ ☐ ☐ ☐ ☐ no

4.14 that the dialogue proceeds without interruptions.

yes ☐ ☐ ☐ ☐ ☐ no

4.15 that BoRIS provides a help capability.

yes ☐ ☐ ☐ ☐ ☐ no

E.2 Questionnaire for Experiment 6.3

Original German Version

Teil A

Persönliche Daten

Geschlecht:	☐ weiblich	☐ männlich
Alter:	_____ Jahre	
Beruf / Ausbildung:	____________________	
Herkunftsregion / -stadt:	____________________	
Jetziger Wohnort:	____________________	

1.0 Wie häufig gehen Sie durchschnittlich auswärts essen?

___ mal in der Woche ___ mal im Monat ___ mal im Jahr

2.0 Wie erkundigen Sie sich nach einem Restaurant, wenn Sie in einem fremden Ort essen gehen möchten (mehrfache Nennung möglich)?

1.1. Magazine	☐	1.6. Tipp von Bekannten	☐
1.2 Werbebroschüren	☐	1.7. Anruf bei einem Sprach-Informationssystem	☐
1.3 Stadtführer	☐	1.8 Sonstiges: ______________	☐
1.4 Gelbe Seiten	☐		
1.5 Internet	☐		

3.0 Was ist Ihnen wichtig, wenn Sie sich für ein Restaurant entscheiden (mehrfache Nennung möglich) ?

3.1 Preis	☐	3.6 Äußeres Ambiente	☐
3.2 Art der Küche	☐	3.7 Öffnungszeiten	☐
3.3 Qualität des Essens	☐	3.8 Schnelligkeit der Bedienung	☐
3.4 Breitgefächertes Speiseangebot	☐	3.9 Freundlichkeit des Personals	☐
3.5 Ort des Restaurants	☐	3.10 Sonstiges:___________	☐

4.0 Haben Sie bereits ein automatisches sprachbasiertes Auskunftssystem benutzt?

☐ ja ☐ nein

4.1. wenn ja, bei welcher Gelegenheit?______________________________

____

4.1.1 Wie würden Sie Ihre Erfahrung damit charakterisieren?

extrem schlecht | schlecht | dürftig | ordentlich | gut | ausgezeichnet | ideal

5.0 Haben Sie Erfahrung mit einem Sprache verstehenden System?

☐ ja ☐ nein

5.1. wenn ja, welcher Art? ______________________________

6.0 Haben Sie Erfahrung mit maschinell erzeugter Sprache?

☐ ja ☐ nein

6.1. wenn ja, bei welcher Gelegenheit? ______________________________

7.0 Welche Informationen über ein Restaurant möchten Sie von einem Auskunftssystem erfragen können?

____

8.0 Wenn Sie bei einem Restaurant-Auskunftssystem anrufen, wie wichtig ist es Ihnen...

8.1 ein normales Gespräch zu führen wie mit einem Menschen?

wichtig nicht wichtig

8.2 von einer freundlichen Stimme bedient zu werden?

wichtig nicht wichtig

8.3 dem System Fragen stellen zu können?

wichtig nicht wichtig

8.4 vom System Fragen über Ihre Vorlieben gestellt zu bekommen?

wichtig nicht wichtig

8.5 schnell die gewünschte Information zu bekommen?

wichtig nicht wichtig

8.6 das System leicht bedienen zu können?

wichtig nicht wichtig

8.7 vom System erklärt zu bekommen, wie es Ihnen helfen kann?

wichtig nicht wichtig

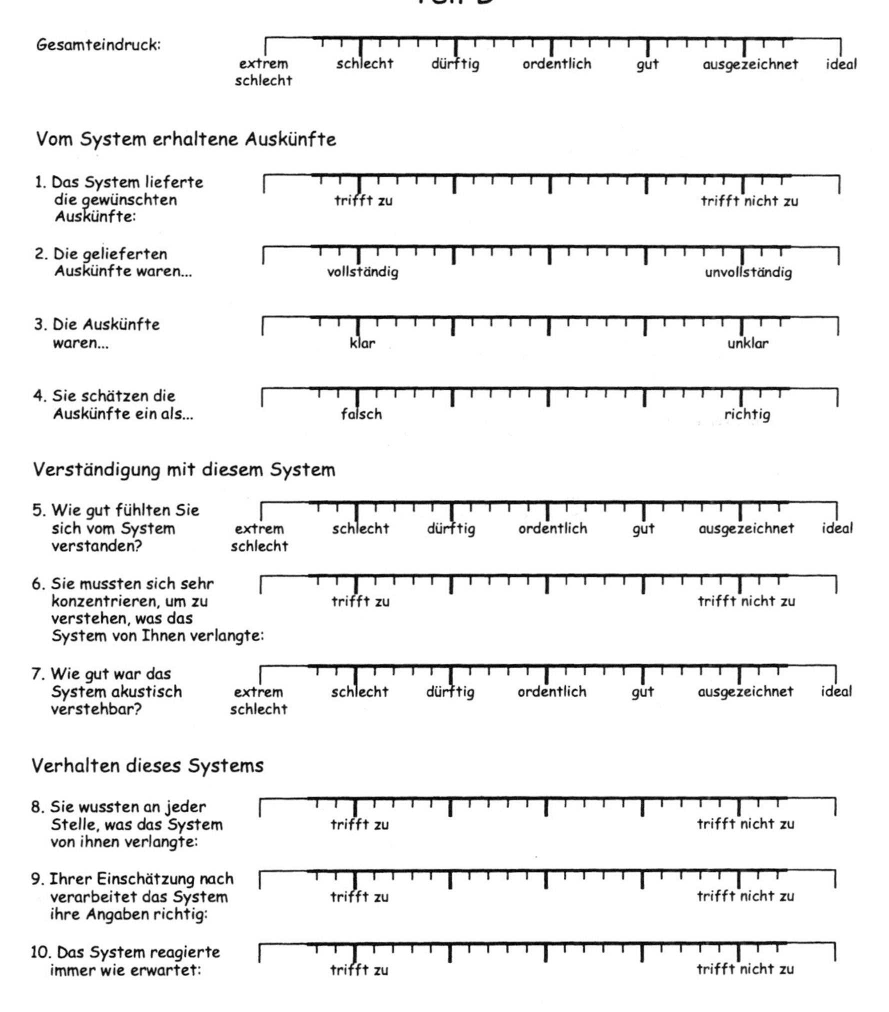

Teil B

Gesamteindruck: extrem schlecht — schlecht — dürftig — ordentlich — gut — ausgezeichnet — ideal

Vom System erhaltene Auskünfte

1. Das System lieferte die gewünschten Auskünfte: trifft zu — trifft nicht zu
2. Die gelieferten Auskünfte waren... vollständig — unvollständig
3. Die Auskünfte waren... klar — unklar
4. Sie schätzen die Auskünfte ein als... falsch — richtig

Verständigung mit diesem System

5. Wie gut fühlten Sie sich vom System verstanden? extrem schlecht — schlecht — dürftig — ordentlich — gut — ausgezeichnet — ideal
6. Sie mussten sich sehr konzentrieren, um zu verstehen, was das System von Ihnen verlangte: trifft zu — trifft nicht zu
7. Wie gut war das System akustisch verstehbar? extrem schlecht — schlecht — dürftig — ordentlich — gut — ausgezeichnet — ideal

Verhalten dieses Systems

8. Sie wussten an jeder Stelle, was das System von ihnen verlangte: trifft zu — trifft nicht zu
9. Ihrer Einschätzung nach verarbeitet das System ihre Angaben richtig: trifft zu — trifft nicht zu
10. Das System reagierte immer wie erwartet: trifft zu — trifft nicht zu

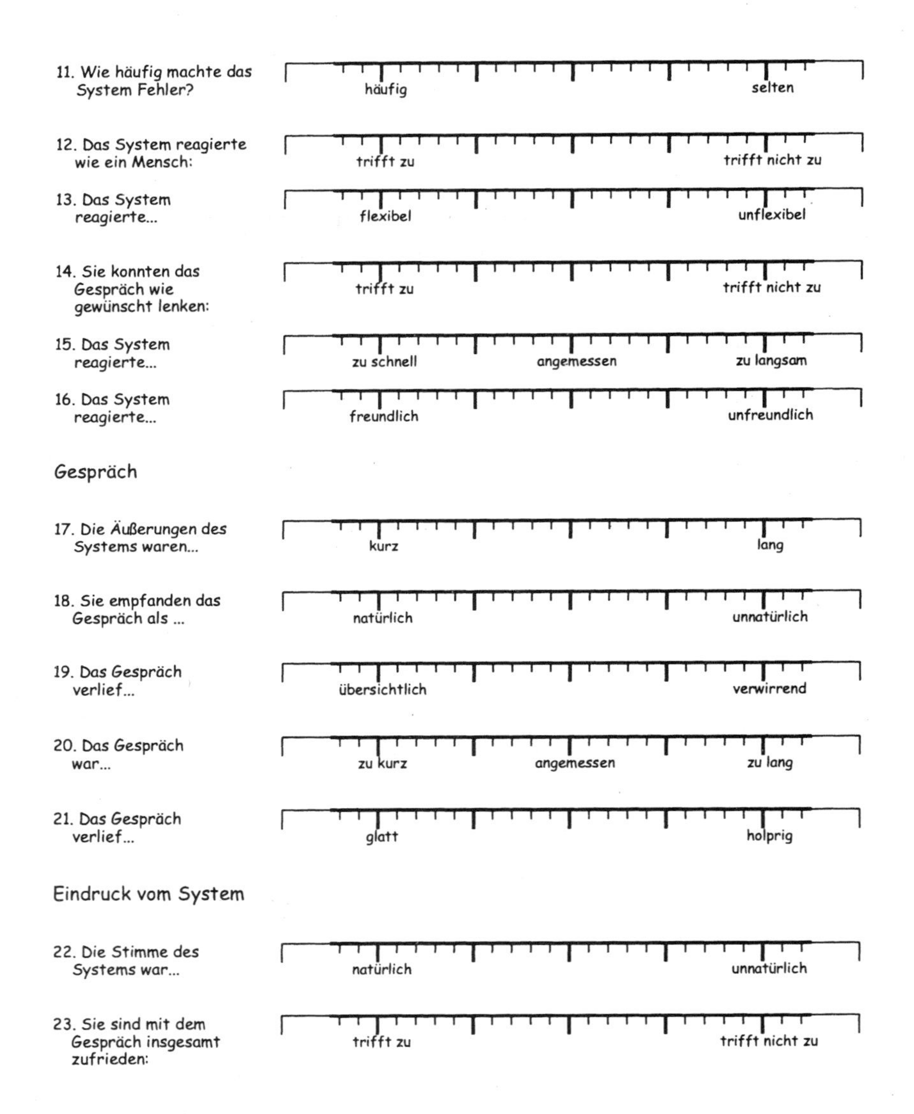
11. Wie häufig machte das System Fehler?
häufig
selten
12. Das System reagierte wie ein Mensch:
trifft zu
trifft nicht zu
13. Das System reagierte...
flexibel
unflexibel
14. Sie konnten das Gespräch wie gewünscht lenken:
trifft zu
trifft nicht zu
15. Das System reagierte...
zu schnell
angemessen
zu langsam
16. Das System reagierte...
freundlich
unfreundlich
Gespräch
17. Die Äußerungen des Systems waren...
kurz
lang
18. Sie empfanden das Gespräch als ...
natürlich
unnatürlich
19. Das Gespräch verlief...
übersichtlich
verwirrend
20. Das Gespräch war...
zu kurz
angemessen
zu lang
21. Das Gespräch verlief...
glatt
holprig
Eindruck vom System
22. Die Stimme des Systems war...
natürlich
unnatürlich
23. Sie sind mit dem Gespräch insgesamt zufrieden:
trifft zu
trifft nicht zu

Persönliche Wirkung

24. Sie empfanden das Gespräch als...

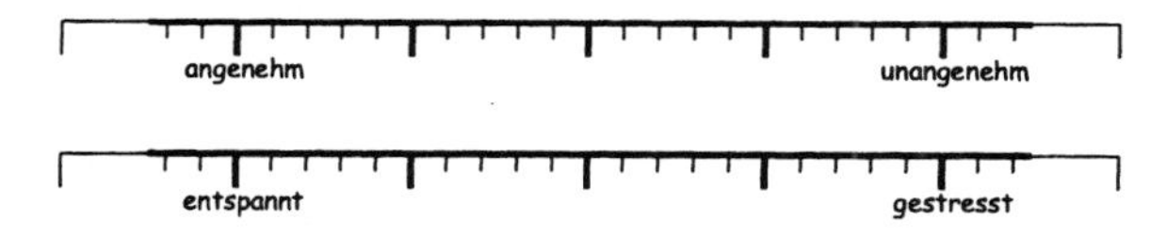

25. Sie fühlten sich während des Gespräches...

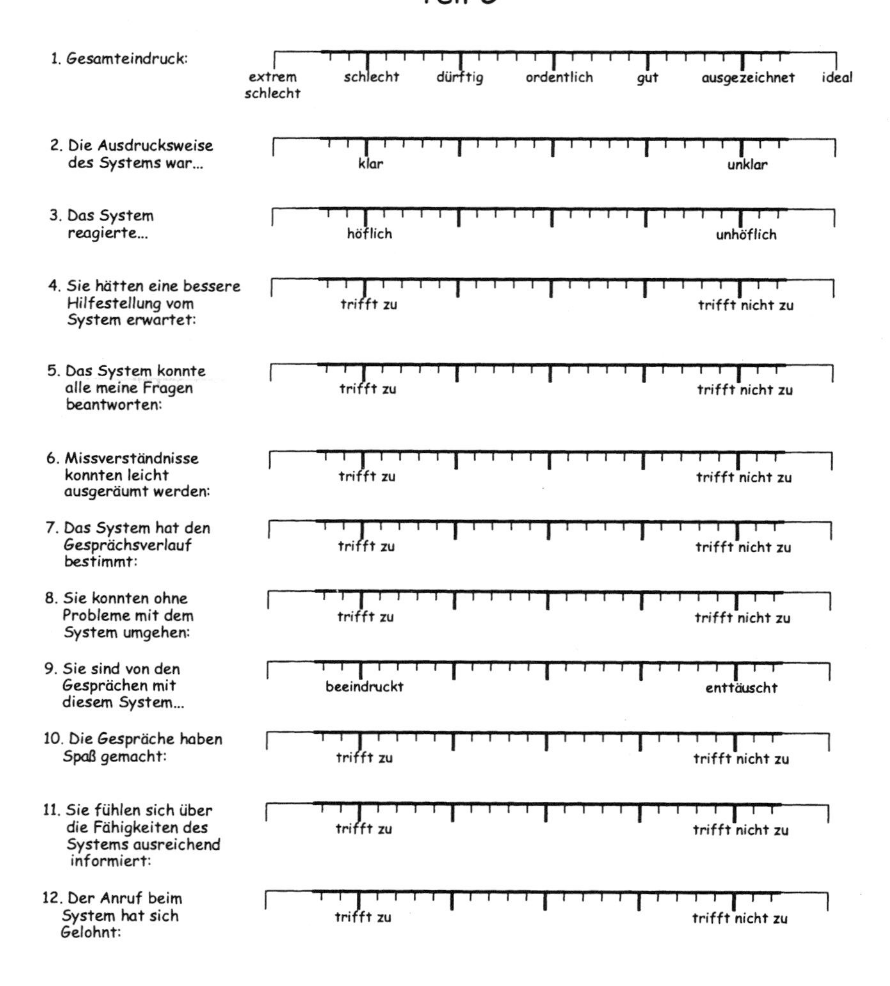
Teil C

1. Gesamteindruck: extrem schlecht – schlecht – dürftig – ordentlich – gut – ausgezeichnet – ideal

2. Die Ausdrucksweise des Systems war... klar – unklar

3. Das System reagierte... höflich – unhöflich

4. Sie hätten eine bessere Hilfestellung vom System erwartet: trifft zu – trifft nicht zu

5. Das System konnte alle meine Fragen beantworten: trifft zu – trifft nicht zu

6. Missverständnisse konnten leicht ausgeräumt werden: trifft zu – trifft nicht zu

7. Das System hat den Gesprächsverlauf bestimmt: trifft zu – trifft nicht zu

8. Sie konnten ohne Probleme mit dem System umgehen: trifft zu – trifft nicht zu

9. Sie sind von den Gesprächen mit diesem System... beeindruckt – enttäuscht

10. Die Gespräche haben Spaß gemacht: trifft zu – trifft nicht zu

11. Sie fühlen sich über die Fähigkeiten des Systems ausreichend informiert: trifft zu – trifft nicht zu

12. Der Anruf beim System hat sich Gelohnt: trifft zu – trifft nicht zu

13. Sie empfanden die Auskunftsmöglichkeit als...	hilfreich	nicht hilfreich
14. Sie schätzen das System ein als...	vertrauenswürdig	zweifelhaft
15. Sie verwenden lieber eine andere Auskunftsquelle:	trifft zu	trifft nicht zu
16. Die Bedienung des Systems war...	einfach	kompliziert
17. Sie bevorzugen eine mit einem Menschen besetzte Auskunftsstelle:	trifft zu	trifft nicht zu
18. Sie würden dieses System noch einmal Benutzen:	trifft zu	trifft nicht zu

19. Was hat Ihnen am System gefallen?

____

____

20. Was hat Sie am System gestört?

____

____

21. Haben Sie Verbesserungsvorschläge für das System?

____

____

Nachdem Sie nun Erfahrungen mit BoRIS gemacht haben, bitten wir Sie, die folgenden Fragen noch einmal zu beantworten.

Wenn Sie bei einem Restaurant-Auskunftssystem anrufen, wie wichtig ist es Ihnen...

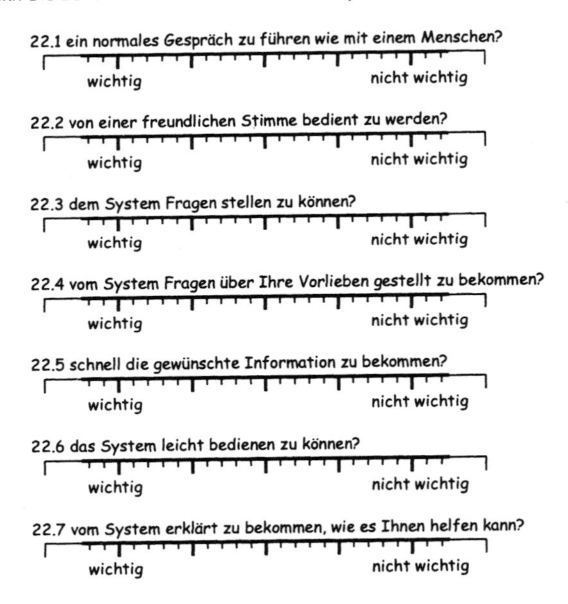

22.1 ein normales Gespräch zu führen wie mit einem Menschen?

wichtig — nicht wichtig

22.2 von einer freundlichen Stimme bedient zu werden?

wichtig — nicht wichtig

22.3 dem System Fragen stellen zu können?

wichtig — nicht wichtig

22.4 vom System Fragen über Ihre Vorlieben gestellt zu bekommen?

wichtig — nicht wichtig

22.5 schnell die gewünschte Information zu bekommen?

wichtig — nicht wichtig

22.6 das System leicht bedienen zu können?

wichtig — nicht wichtig

22.7 vom System erklärt zu bekommen, wie es Ihnen helfen kann?

wichtig — nicht wichtig

Vielen Dank für Ihre Mühe!

English Translation

Part A

Personal data

Gender:	☐ *female*	☐ *male*
Age:	____ *years*	
Profession / education:	________________	
Region / city of birth:	________________	
Current residence:	________________	

1.0 How often do you eat out on an average?

___ times a week ___ times a month ___ times a year

2.0 How would you search for a restaurant when you are in a foreign place (multiple choices possible)?

1.1. Magazines	☐	1.6 Tips from friends	☐
1.2 Commercial flyers	☐	1.7 Calling a automatic speech-based system	☐
1.3 City guide	☐	1.8 Other: ____________	☐
1.4 Telephone book	☐		
1.5 Internet	☐		

3.0 What is important for you when you decide on a restaurant (multiple choices possible)?

3.1 Price	☐	3.6 Ambience	☐
3.2 Food type	☐	3.7 Opening hours	☐
3.3 Food quality	☐	3.8 Service speed	☐
3.4 Variety of food offered	☐	3.9 Service friendliness	☐
3.5 Location	☐	3.10 Other: ____________	☐

4.0 Have you ever used an automatic speech-based information system?

☐ yes ☐ no

4.1. If yes, on which occasion? ________________________________

4.1.1 How would you characterize your experience with it?

extremely bad — bad — poor — fair — good — excellent — ideal

5.0 Do you have experience with a speech understanding system?

☐ yes ☐ no

5.1. If yes, what kind of system? ____________________

6.0 Do you have experience with synthesized speech?

☐ yes ☐ no

6.1. If yes, on which occasion? ____________________

7.0 What information about a restaurant do you want to get from an information system?

____

8.0 If you would call a restaurant information system, how important is it...

8.1 to have a normal conversation just like with a human?

important unimportant

8.2 to be served by a friendly voice?

important unimportant

8.3 that you can ask questions to the system?

important unimportant

8.4 that the system asks questions to you about your preferences?

important unimportant

8.5 to get the desired information quickly?

important unimportant

8.6 that the system can be used easily?

important unimportant

8.7 to get help from the system?

important unimportant

Part B

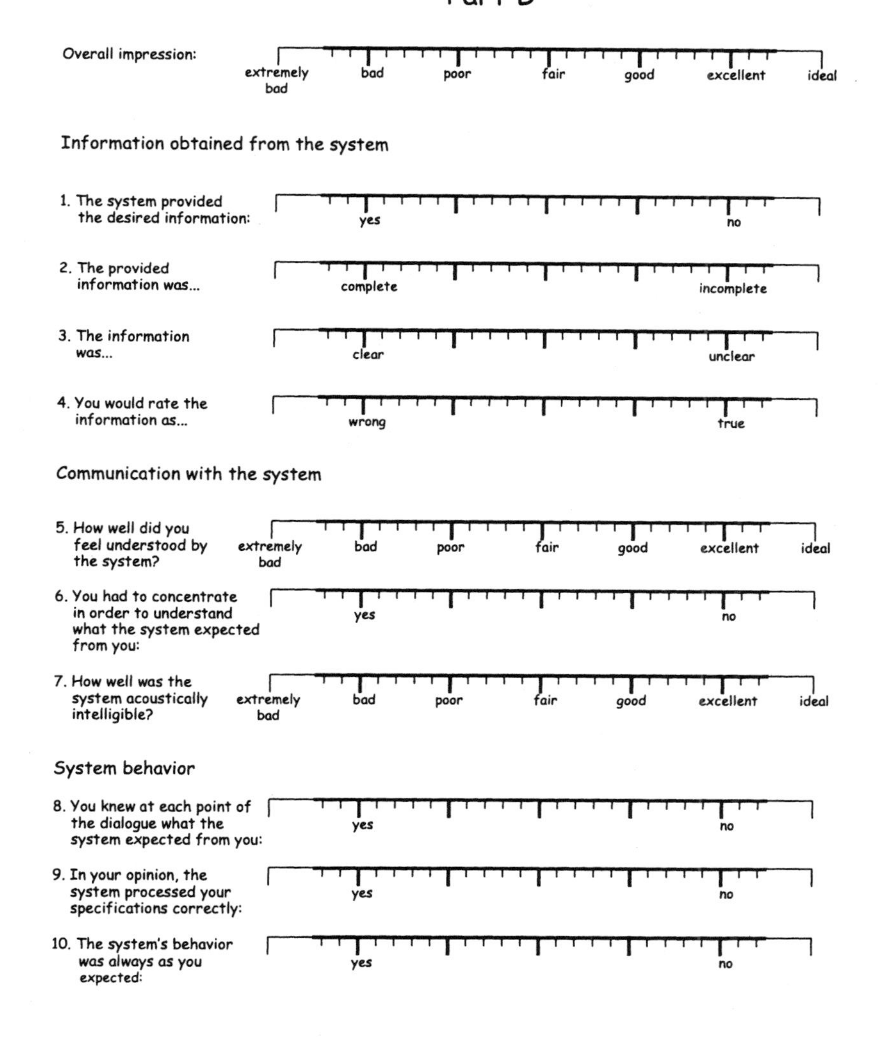

Overall impression: extremely bad — bad — poor — fair — good — excellent — ideal

Information obtained from the system

1. The system provided the desired information: yes — no

2. The provided information was... complete — incomplete

3. The information was... clear — unclear

4. You would rate the information as... wrong — true

Communication with the system

5. How well did you feel understood by the system? extremely bad — bad — poor — fair — good — excellent — ideal

6. You had to concentrate in order to understand what the system expected from you: yes — no

7. How well was the system acoustically intelligible? extremely bad — bad — poor — fair — good — excellent — ideal

System behavior

8. You knew at each point of the dialogue what the system expected from you: yes — no

9. In your opinion, the system processed your specifications correctly: yes — no

10. The system's behavior was always as you expected: yes — no

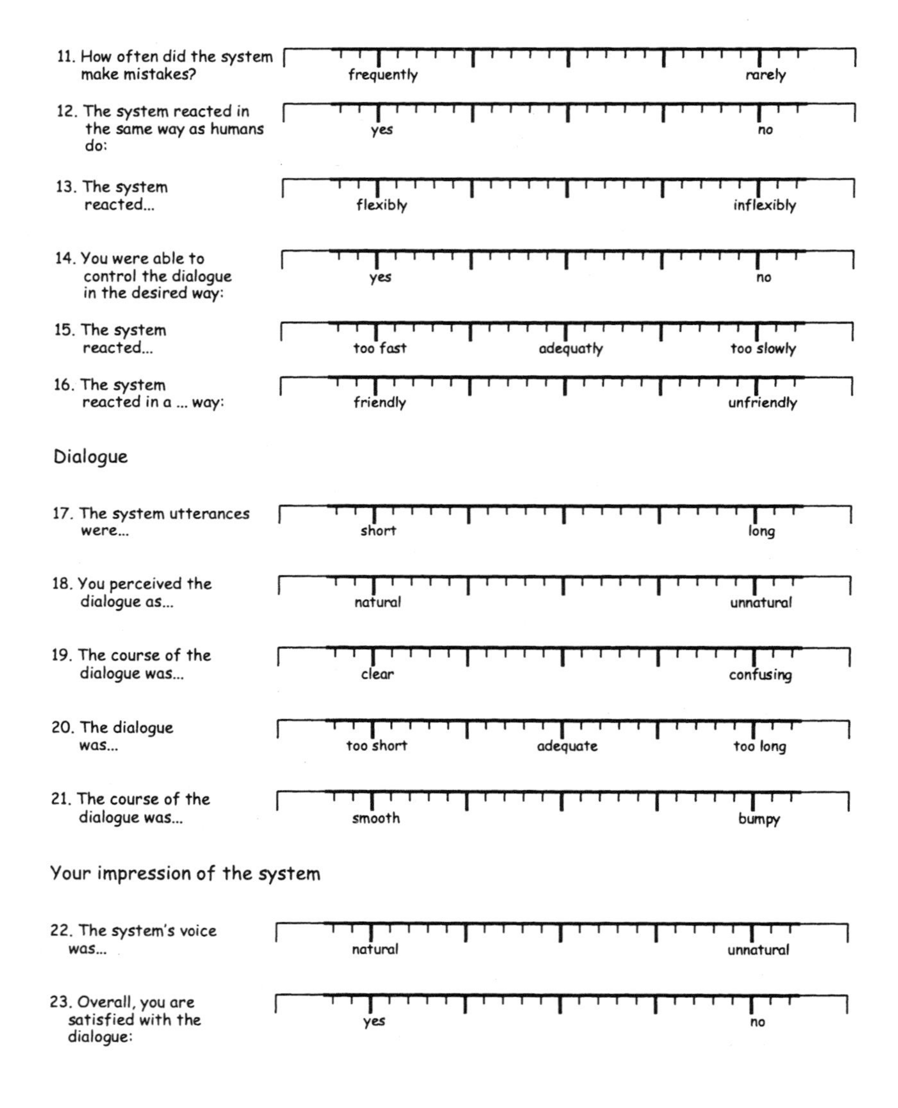

11. How often did the system make mistakes? — frequently … rarely

12. The system reacted in the same way as humans do: — yes … no

13. The system reacted... — flexibly … inflexibly

14. You were able to control the dialogue in the desired way: — yes … no

15. The system reacted... — too fast … adequatly … too slowly

16. The system reacted in a ... way: — friendly … unfriendly

Dialogue

17. The system utterances were... — short … long

18. You perceived the dialogue as... — natural … unnatural

19. The course of the dialogue was... — clear … confusing

20. The dialogue was... — too short … adequate … too long

21. The course of the dialogue was... — smooth … bumpy

Your impression of the system

22. The system's voice was... — natural … unnatural

23. Overall, you are satisfied with the dialogue: — yes … no

Personal impression

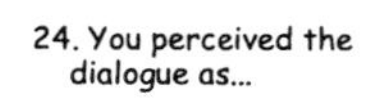

pleasant unpleasant

Part C

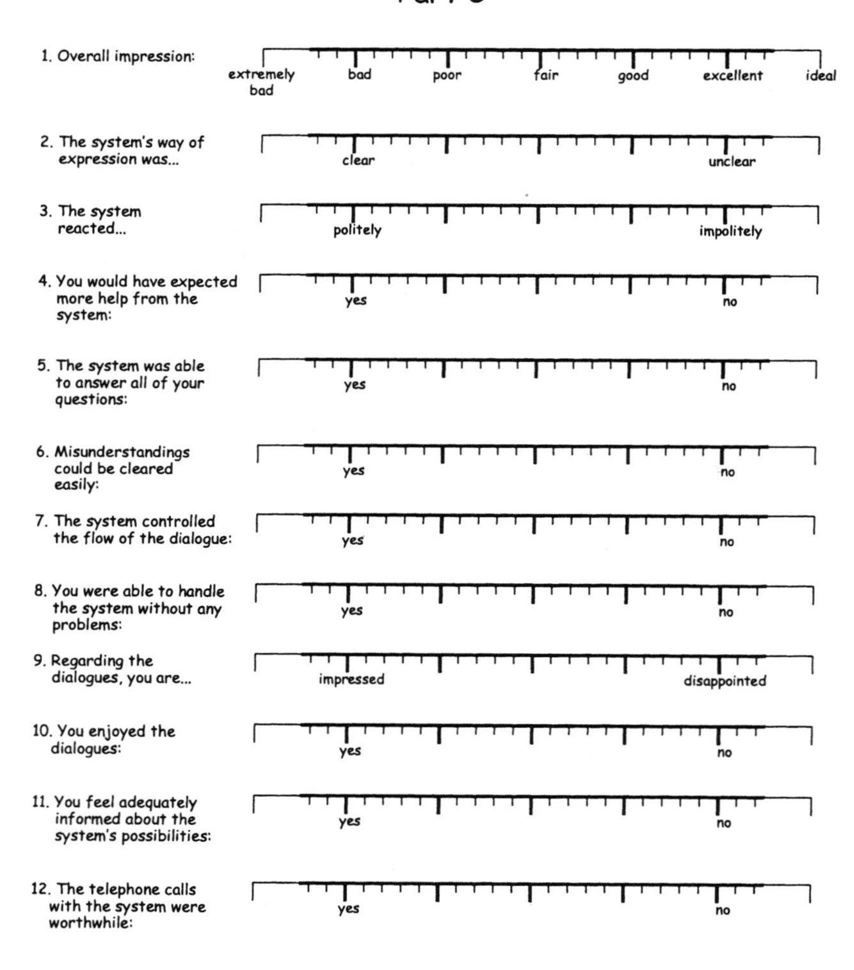

1. Overall impression: extremely bad — bad — poor — fair — good — excellent — ideal

2. The system's way of expression was... clear — unclear

3. The system reacted... politely — impolitely

4. You would have expected more help from the system: yes — no

5. The system was able to answer all of your questions: yes — no

6. Misunderstandings could be cleared easily: yes — no

7. The system controlled the flow of the dialogue: yes — no

8. You were able to handle the system without any problems: yes — no

9. Regarding the dialogues, you are... impressed — disappointed

10. You enjoyed the dialogues: yes — no

11. You feel adequately informed about the system's possibilities: yes — no

12. The telephone calls with the system were worthwhile: yes — no

13. You perceived this possibility for obtaining information as...
helpful — not helpful

14. You rate the system as...
reliable — unreliable

15. You prefer to use another source of information:
yes — no

16. The handling of the system was...
easy — complicated

17. You prefer a human operator:
yes — no

18. In the future, you would use the system again?
yes — no

19. Which characteristics of the system did you like best?

____

____

20. Which characteristics of the system disturbed you mostly?

____

____

21. Do you have any proposals for system improvements?

____

____

With your current experiences of using BoRIS, we ask you to answer the following questions once again.

22. If you would call a restaurant information system, how important is it...

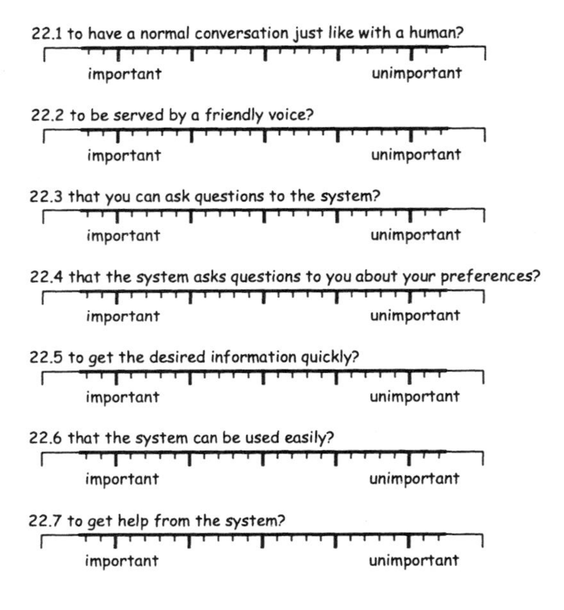

22.1 to have a normal conversation just like with a human?

important unimportant

22.2 to be served by a friendly voice?

important unimportant

22.3 that you can ask questions to the system?

important unimportant

22.4 that the system asks questions to you about your preferences?

important unimportant

22.5 to get the desired information quickly?

important unimportant

22.6 that the system can be used easily?

important unimportant

22.7 to get help from the system?

important unimportant

Thank you for your effort!

References

Abe, K., Kurokawa, K., Taketa, K., Ohno, S., and Fujisaki, H. (2000). A New Method for Dialogue Management in an Intelligent System for Information Retrieval. *In: Proc. 6th Int. Conf. on Spoken Language Processing (ICSLP 2000)*, 2:118–121, CHN–Beijing.

Allen, J., Ferguson, G., and Stent, A. (2001). An Architecture for More Realistic Conversational Systems. *In: Proc. of Intelligent User Interfaces 2001 (IUI-01)*, 1–8, USA–Santa Fe NM.

Amalberti, R., Carbonell, N., and Falzon, P. (1993). User Representations of Computer Systems in Human-Computer Speech Interaction. *International Journal on Man-Machine Studies*, 38:547–566.

Andernach, T., Deville, G., and Mortier, L. (1993). The Design of a Real World Wizard of Oz Experiment for a Speech Driven Telephone Directory Information System. *In: Proc. 3rd Europ. Conf. on Speech Communication and Technology (EUROSPEECH'93)*, 2:1165–1168, D–Berlin.

Antoine, J.-Y., Bousquet-Vernhettes, C., Goulian, J., Kurdi, M. Z., Rousset, S., Vigouroux, N., and Villaneau, J. (2002). Predictive and Objective Evaluation of Speech Understanding: The "Challenge" Evaluation Campaign of the I3 Speech Workgroup of the French CNRS. *In: Proc. 3rd Int. Conf. on Language Resources and Evaluation (LREC 2002)*, 2:529–536, ES–Las Palmas.

Antoine, J.-Y., Siroux, J., Caelen, J., Villaneau, J., Goulian, J., and Ahafhaf, M. (2000). Obtaining Predictive Results with an Objective Evaluation of Spoken Dialogue Systems: Experiments with the DCR Assessment Paradigm. *In: Proc. 2nd Int. Conf. on Language Resources and Evaluation (LREC 2000)*, 2:713–720, GR–Athens.

Antoniol, G., Fiutem, R., Lazzari, G., and de Mori, R. (1998). System Architectures and Applications. *Spoken Dialogues with Computers*, R. de Mori, ed., 583–609, Academic Press, UK–London.

Araki, M. and Doshita, S. (1997). Automatic Evaluation Environment for Spoken Dialogue Systems. *In: Dialogue Processing in Spoken Language Systems*, ECAI'96 Workshop, H–Budapest, E. Maier, M. Mast, and S. LuperFoy, eds., Lecture Notes in Artificial Intelligence No. 1236, 183–194, Springer, D–Berlin.

Arden, P. (1997). Subjective Assessment Methods for Text-To-Speech Systems. *In: Proc. Speech and Language Technology (SALT) Club Workshop on Evaluation in Speech and Language Technology*, 9–16, UK–Sheffield.

Atwell, E., Howarth, P., Souter, C., Baldo, P., Bisiani, R., Pezzotta, D., Bonaventura, P., Menzel, W., Herron, D., Morton, R., and Schmidt, J. (2000). User-Guided System Development in Interactive Spoken Language Education. *Natural Language Engineering*, 6(3-4):229–241.

Aust, H. and Oerder, M. (1995). Dialogue Control in Automatic Inquiry Systems. *In: Proc. ESCA Workshop on Spoken Dialogue Systems*, P. Dalsgaard, L.B. Larsen, L. Boves, and I. Thomsen, eds., 121–124, DK–Vigsø.

Aust, H., Oerder, M., Seide, F., and Steinbiss, V. (1995). The Philips Automatic Train Timetable Information System. *Speech Communication*, 17:249–262.

Baekgaard, A., Bernsen, O., Brøndsted, T., Dalsgaard, P., Dybkjær, H., Dybkjær, L., Kristiansen, J., Larsen, L. B., Lindberg, B., Maegaard, M., Music, B., Offersgaard, L., and Povlsen, C. (1995). The Danish Spoken Language Dialogue Project: A General Overview. *In: Proc. ESCA Workshop on Spoken Dialogue Systems*, P. Dalsgaard, L.B. Larsen, L. Boves, and I. Thomsen, eds., 89–92, DK–Vigsø.

Baekgaard, A. (1995). A Platform for Spoken Dialogue Systems. *In: Proc. ESCA Workshop on Spoken Dialogue Systems*, P. Dalsgaard, L.B. Larsen, L. Boves, and I. Thomsen, eds., 105–108, DK–Vigsø.

Baekgaard, A. (1996). Dialogue Management in a Generic Dialogue System. *In: Dialogue Management in Natural Language Systems. Proc. 11th Twente Workshop on Language Technology (TWLT 11)*, University of Twente, S. LuperFoy, A. Nijholt, and G. V. van Zanten, eds., 123–132, NL–Enschede.

Baggia, P. and Rullent, C. (1993). Partial Parsing as a Robust Parsing Strategy. *In: Proc. Int. Conf. Acoustics Speech and Signal Processing (ICASSP'93)*, 2:123–126, IEEE, USA–Piscataway NJ.

Baggia, P., Castagneri, G., and Danieli, M. (1998). Field Trials of the Italian ARISE Train Timetable System. *In: Proc. IEEE 4th Workshop Interactive Voice Technology for Telecommunications Applications (IVTTA'98)*, 97–102, I–Torino.

Baggia, P., Castagneri, G., and Danieli, M. (2000). Field Trials of the Italian ARISE Train Timetable System. *Speech Communication*, 31:355–367.

Baggia, P., Gerbino, E., Giachin, E., and Rullent, C. (1994). Experiences of Spontaneous Speech Interaction with a Dialogue System. *In: Progress and Prospects of Speech Research and Technology*, H. Niemann, R. de Mori, and G. Hanrieder, eds., 241–248, Infix, D–Sankt Augustin.

Balestri, M., Foti, E., Nebbia, L., Oreglia, M., Salza, P. L., and Sandri, S. (1992). Comparison of Natural and Synthetic Speech Intelligibility for a Reverse Telephone Directory Service. *In: Proc. 2nd Int. Conf. on Spoken Language Processing (ICSLP'92)*, 1:559–562, CND–Banff.

Bappert, V. and Blauert, J. (1994). Auditory Quality Evaluation of Speech-Coding Systems. *acta acustica*, 2:49–58.

Basson, S., Springer, S., Fong, C., Leung, H., Man, E., Olson, M., Pitrelli, J., Singh, R., and Wong, S. (1996). User Participation and Compliance in Speech Automated Telecommunications Applications. *In: Proc. 4th Int. Conf. on Spoken Language Processing (ICSLP'96)*, H.T. Bunnell and W. Idsardi, eds., 3:1680–1683, IEEE, USA–Piscataway NJ.

Bates, M. and Ayuso, D. (1991). A Proposal for Incremental Dialogue Evaluation. *In: Proc. DARPA Speech and Natural Language Workshop*, 319–322, USA–Pacific Grove CA.

Bates, M., Bobrwo, R., Fung, P., Ingria, R., Kubala, F., Makhoul, J., Nguyen, L., Schwartz, R., and Stallard, D. (1993). The BBN/HARC Spoken Language Understanding System. *In: Proc. Int. Conf. Acoustics Speech and Signal Processing (ICASSP'93)*, 2:111–114, IEEE, USA–Piscataway NJ.

Bates, M., Boisen, S., and Makhoul, J. (1990). Developing an Evaluation Methodology for Spoken Language Systems. *In: Proc. DARPA Speech and Natural Language Workshop*, 102–108, USA–Hidden Valley PA.

Baum, L. F. (1900). *The Wonderful Wizard of Oz.* Kansas Centennial Ed. (1999), University Press of Kansas, USA-Lawrence KS.

Beerends, J. G., Hekstra, A. P., Rix, A. W., and Hollier, M. P. (2002). Perceptual Evaluation of Speech Quality (PESQ) – The New ITU Standard for End-to-End Speech Quality Assessment. Part II – Psychoacoustic Model. *J. Audio Eng. Soc.*, 50(10):765–778.

Bel, N., Caminero, J., Hernández, L., Marimón, M., Morlesín, J. F., Otero, J. M., Relaño, J., Rodríguez, M. C., Ruz, P. M., and Tapias, D. (2002). Design and Evaluation of a SLDS for E-Mail Access Through the Telephone. *In: Proc. 3rd Int. Conf. on Language Resources and Evaluation (LREC 2002)*, 2:537–544, ES–Las Palmas.

Bellotti, V., MacLean, A., and Moran, T. (1991). *Generating Good Design Questions.* Technical Report EPC-1991-136, Rank Xerox Research Centre, Cambridge Laboratory, UK-Cambridge.

Bengler, K. (2000). Automotive Speech-Recognition – Success Conditions Beyond Recognition Rates. *In: Proc. 2nd Int. Conf. on Language Resources and Evaluation (LREC 2000)*, 3:1357–1359, GR–Athen.

Bennacef, S., Devillers, L., Rosset, S., and Lamel, L. (1996). Dialog in the RAILTEL Telephone-Based System. *In: Proc. 4th Int. Conf. on Spoken Language Processing (ICSLP'96)*, H.T. Bunnell and W. Idsardi, eds., 1:550–553, IEEE, USA–Piscataway NJ.

Benoît, C., Emerard, F., Schnabel, B., and Tseva, A. (1991). Quality Comparisons of Prosodic and of Acoustic Components of Various Synthesizers. *In: Proc. 2nd Europ. Conf. on Speech Communication and Technology (EUROSPEECH'91)*, 2:875–878, I–Genova.

Beranek, L. L. (1971). *Noise and Vibration Control.* McGraw-Hill, USA-New York NY.

Berger, J. (1998). *Instrumentelle Verfahren zur Sprachqualitätsschätzung – Modelle auditiver Tests.* Doctoral dissertation, Christian-Albrechts-Universität Kiel (Arbeiten über Digitale Signalverarbeitung No. 13, U. Heute, ed.), Shaker Verlag, D-Aachen.

Beringer, N., Kartal, U., Louka, K., Schiel, F., and Türk, U. (2002a). PROMISE – A Procedure for Multimodal Interactive System Evaluation. *In: Proc. LREC Workshop on Multimodal Resources and Multimodal Systems Evaluation*, 4 pages, ES–Las Palmas.

Beringer, N., Louka, K., Penide-Lopez, V., and Türk, U. (2002b). End-to-End Evaluation of Multimodal Dialogue Systems: Can we Transfer Established Methods? *In: Proc. 3rd Int. Conf. on Language Resources and Evaluation (LREC 2002)*, 2:558–563, ES–Las Palmas.

Bernsen, N. O. (1997). Towards a Tool for Predicting Speech Functionality. *Speech Communication*, 23:181–210.

Bernsen, N. O. (2003). On-Line User Modelling in a Mobile Spoken Dialogue System. *In: Proc. 8th Europ. Conf. on Speech Communication and Technology (EUROSPEECH 2003 – Switzerland)*, 2:737–740, CH–Geneva.

Bernsen, N. O. (2004). Measuring Relative Target User Group Success in Spoken Conversation for Edutainment. *In: Proc. LREC Workshop on Multimodal Corpora: Models of Human Behaviour for the Specification and Evaluation of Multimodal Input and Output Interfaces*, 17–20, P–Lisbon.

Bernsen, N. O., Dybkjær, H., and Dybkjær, L. (1996). Principles for the Design of Cooperative Spoken Human-Machine Dialogue. *In: Proc. 4th Int. Conf. on Spoken Language Processing (ICSLP'96)*, H.T. Bunnell and W. Idsardi, eds., 2:729–732, IEEE, USA–Piscataway NJ.

Bernsen, N. O., Dybkjær, H., and Dybkjær, L. (1998). *Designing Interactive Speech Systems: From First Ideas to User Testing.* Springer, D-Berlin.

Bernsen, N. O. and Dybkjær, L. (1997). The DISC Concerted Action. *In: Proc. Speech and Language Technology (SALT) Club Workshop on Evaluation in Speech and Language Technology*, 35–42, UK–Sheffield.

Bernsen, N. O. and Dybkjær, L. (1999). A Theory of Speech in Multimodal Systems. *In: Proc. ESCA Workshop on Interactive Dialogue in Multi-Modal Systems*, P. Dalsgaard, C.–H. Lee, P. Heisterkamp, and R. Cole, eds., 105–108, D–Kloster Irsee.

Bernsen, N. O. and Dybkjær, L. (2000). A Methodology for Evaluating Spoken Language Dialogue Systems and Their Components. *In: Proc. 2nd Int. Conf. on Language Resources and Evaluation (LREC 2000)*, 2:183–188, GR–Athens.

Bernsen, N. O., Dybkjær, L., and Kiilerich, S. (2004). Evaluating Conversation with Hans Christian Andersen. *In: Proc. 4th Int. Conf. on Language Resources and Evaluation (LREC 2004)*, 3:1011–1014, P–Lisbon.

Billi, R. and Lamel, L. F. (1997). RailTel: Railway Telephone Services. *Speech Communication*, 23:63–65.

Billi, R., Canavesio, F., and Rullent, C. (1998). Automation of Telecom Italia Directory Assistance Service: Field Trial Results. *In: Proc. IEEE 4th Workshop Interactive Voice Technology for Telecommunications Applications (IVTTA'98)*, 11–16, I–Torino.

Billi, R., Castagneri, G., and Danieli, M. (1996). Field Trial Evaluations of Two Different Information Inquiry Systems. *In: Proc. 3rd IEEE Workshop on Interactive Voice Technology for Telecommunications Applications (IVTTA'96)*, 129–134, USA–Basking Ridge NJ.

Bimbot, F. and Chollet, G. (1997). Assessment of Speaker Verification Systems. *In: Handbook on Standards and Resources for Spoken Language Systems*, D. Gibbon, R. Moore, and R. Winski, eds., 408–480, Mouton de Gruyter, D–Berlin.

Black, E. (1997). Evaluation of Broad-Coverage Natural-Language Parsers. *In: Survey of the State of the Art in Human Language Technology*, R. Cole, J. Mariani, A. Uszkoreit, H. Zaenen, and V. Zue, eds., 420–422, Cambridge University Press and Giardini Editori, I–Pisa.

Blauert, J. (1997). *Spatial Hearing: The Psychophysics of Human Sound Localization*. The MIT Press, USA-Cambridge MA.

Blauert, J. and Jekosch, U. (1996). Sound-Quality Evaluation – A Multilayered Problem. *In: Proc. EAA Tutorium on Aurally-Adequate Sound-Quality Evaluation*, 17 pages, B–Antwerp.

Blaylock, N., Allen, J., and Ferguson, G. (2002). Synchronization in an Asynchronous Agent-Based Architecture for Dialogue Systems. *In: Proc. Third SIGdial Workshop on Discourse and Dialogue*, 1–10, USA–Philadelphia PA.

Bodden, M. and Jekosch, U. (1996). *Entwicklung und Durchführung von Tests mit Versuchspersonen zur Verifizierung von Modellen zur Berechnung der Sprachübertragungsqualität*. Project report, Institut für Kommunikationsakustik, Ruhr-Universität, D-Bochum.

Böhm, A. (1993). *Maschinelle Sprachausgabe deutschen und englischen Textes*. Doctoral dissertation, Institut für Kommunikationsakustik, Ruhr-Universität Bochum, Shaker Verlag, D-Aachen.

Boite, R., Bourlard, H., Dutoit, T., Hancq, J., and Leich, H. (2000). *Traîtement de la parole*. Presses Polytechniques Universitaires Romandes, CH-Lausanne.

Bonneau-Maynard, H., Devillers, L., and Rosset, S. (2000). Predictive Performance of Dialogue Systems. *In: Proc. 2nd Int. Conf. on Language Resources and Evaluation (LREC 2000)*, 1:177–181, GR–Athen.

Borg, I. and Staufenbiel, T. (1993). *Theorien und Methoden der Skalierung: Eine Einführung*. Verlag Hans Huber, CH-Bern.

Boros, M., Eckert, W., Gallwitz, F., Görz, G., Hanrieder, G., and Niemann, H. (1996). Towards Understanding Spontaneous Speech: Word Accuracy vs. Concept Accuracy. *In: Proc. 4th Int. Conf. on Spoken Language Processing (ICSLP'96)*, H.T. Bunnell and W. Idsardi, eds., 2:1009–1012, IEEE, USA–Piscataway NJ.

Bortz, J. (1995). *Forschungsmethoden und Evaluation.* Springer, D-Berlin.

Bourlard, H. and Morgan, N. (1998). Hybrid HMM/ANN Systems for Speech Recognition: Overview and New Research Directions. *In: Adaptive Processing of Sequences and Data Structures*, Int. Summer School on Neural Networks, I–Vietri sul Mare, C. L. Giles and M. Gori, eds., Lecture Notes in Artificial Intelligence No. 1387, 389–417, Springer, D–Berlin.

Bourlard, H. and Wellekens, C. J. (1992). Links Between Markov Models and Multi-Layer Perceptron. *IEEE Trans. Pattern Analysis, Machine Intelligence*, 12:1167–1178.

Boves, L. and den Os, E. (1998). Speaker Verification in Telecom Applications. *In: Proc. IEEE 4th Workshop Interactive Voice Technology for Telecommunications Applications (IVTTA'98)*, 203–208, I–Torino.

Boyce, S. J. and Gorin, A. L. (1996). User Interface Issues for Natural Spoken Dialogue Systems. *In: Proc. 1996 Int. Symp. on Spoken Dialogue (ISSD 96)*, H. Fujisaki, ed., 65–68, USA–Philadelphia.

Bronkhorst, A. W., Bosman, A. J., and Smoorenburg, G. F. (1993). A Model for Context Effects in Speech Recognition. *J. Acoust. Soc. Am.*, 93(1):499–509.

Bruce, G., Granström, B., Gustafson, K., Horne, M., House, D., and Touati, P. (1995). Towards an Enhanced Prosodic Model Adapted to Dialogue Applications. *In: Proc. ESCA Workshop on Spoken Dialogue Systems*, P. Dalsgaard, L.B. Larsen, L. Boves, and I. Thomsen, eds., 201–204, DK–Vigsø.

Brüggen, M. (2001). *Klangverfärbungen durch Rückwürfe und ihre auditive und instrumentelle Kompensation.* Doctoral dissertation, Institut für Kommunikationsakustik, Ruhr-Universität Bochum, dissertation.de, D-Berlin.

Bub, T. and Schwinn, J. (1996). VERBMOBIL: The Evolution of a Complex Large Speech-to-Speech Translation System. *In: Proc. 4th Int. Conf. on Spoken Language Processing (ICSLP'96)*, H.T. Bunnell and W. Idsardi, eds., 4:2371–2374, IEEE, USA–Piscataway NJ.

Buntschuh, B., Kamm, C. A., di Fabbrizio, G., Abella, A., Mohri, M., Narayanan, S., Zeljkovic, I., Sharp, R. D., Wright, J. H., Marcus, S., Shaffer, J., Duncan, R., and Wilpon, J. G. (1998). VPQ: A Spoken Language Interface to Large Scale Directory Information. *In: Proc. 5th Int. Conf. on Spoken Language Processing (ICSLP'98)*, 7:2863–2866, AUS–Sydney.

Button, G. (1990). Going Up a Blind Alley: Conflating Conversation Analysis and Computational Modelling. *In: Computers and Conversation*, P. Luff and N. Gilbert, eds., 67–90, Academic Press, UK–London.

Caminero, J., González-Rodríguez, J., Ortega-García, J., Tapias, D., Ruz, P. M., and Solá, M. (2002). A Multilingual Speaker Verification System: Architecture and Performance Evaluation. *In: Proc. 3rd Int. Conf. on Language Resources and Evaluation (LREC 2002)*, 2:626–631, ES–Las Palmas.

Carletta, J. (1996). Assessing Agreement on Classification Tasks: The Kappa Statistics. *Computational Linguistics*, 22(2):249–254.

Carletta, J., Isard, A., Isard, S., Kowtko, J. C., Doherty-Sneddon, G., and Anderson, A. H. (1997). The Reliability of a Dialogue Structure Coding Scheme. *Computational Linguistics*, 23(1):13–31.

Carletta, J. C. (1992). *Risk Taking and Recovery in Task-Oriented Dialogue.* PhD thesis, University of Edinburgh, UK-Edinburgh.

Carroll, J., Briscoe, T., and Sanfilippo, A. (1998). Parser Evaluation: A Survey and a New Proposal. *In: Proc. 1st Int. Conf. on Language Resources and Evaluation (LREC'98)*, 1:447–454, ES–Granada.

Casali, S. P., Williges, B. H., and Dryden, R. D. (1990). Effects of Recognition Accuracy and Vocabulary Size of a Speech Recognition System on Task Performance and User Acceptance. *Human Factors*, 32(2):183–196.

Chang, H. M. (2000). Is ASR Ready for Wireless Primetime: Measuring the Core Technology for Selected Applications. *Speech Communication*, 31:293–307.

Charfuelán, M., Gómez, L. H., López, C. E., and Hemsen, H. (2002). A XML-Based Tool for Evaluation of SLDS. *In: Proc. 3rd Int. Conf. on Language Resources and Evaluation (LREC 2002)*, 2:551–557, ES–Las Palmas.

Chollet, G., Cochard, J.-L., Constantinescu, A., Jaboulet, C., and Langlais, P. (1996). *Swiss French PolyPhone and PolyVar: Telephone Speech Databases to Model Inter- and Intra-Speaker Variability*. Technical Report RR-96-01, IDIAP, CH-Martigny.

Chu, M. and Peng, H. (2001). An Objective Measure for Estimating MOS of Synthesized Speech. *In: Proc. 7th Europ. Conf. on Speech Communication and Technology (EUROSPEECH 2001 – Scandinavia)*, 3:2087–2090, DK–Aalborg.

Churcher, G. E., Atwell, E. S., and Souter, C. (1997a). *Dialogue Management Systems: A Survey and Overview*. Report 97.06, School of Computer Studies, University of Leeds, UK-Leeds.

Churcher, G. E., Atwell, E. S., and Souter, C. (1997b). A Generic Template to Evaluate Integrated Components in Spoken Dialogue Systems. *In: Proc. Speech and Language Technology (SALT) Club Workshop on Evaluation in Speech and Language Technology*, 51–58, UK–Sheffield.

Cochran, W. G. and Cox, G. M. (1992). *Experimental Designs*. John Wiley & Sons, Inc., USA-New York.

Cohen, P. and Oviatt, S. (1995). The Role of Voice in Human-Machine Communication. *Voice Communication Between Humans and Machines*, R. Roe and J. Wilpon, eds., 34–75, National Academy Press, USA–Washington DC.

Cohen, P. R. (1992). The Role of Natural Language in a Multimodal Interface. *In: Proc. of the ACM Symposium on User Interface Software and Technology (UIST'92)*, USA–Monterey CA, 143–149, ACM Press, USA–New York NY.

Cole, R., Novick, D. G., Fanty, M., Vermeulen, P., Sutton, S., Burnett, D., and Schalkwyk, J. (1994). A Prototype Voice-Response Questionnaire for the U.S. Census. *In: Proc. 3rd Int. Conf. on Spoken Language Processing (ICSLP'94)*, 2:683–686, JP–Yokohama.

Coleman, A. E., Gleiss, N., and Usai, P. (1988). A Subjective Testing Methodology for Evaluating Medium Rate Codecs for Digital Mobile Radio Applications. *Speech Communication*, 7:151–166.

Constantinides, P. C. and Rudnicky, A. I. (1999). Dialog Analysis in the Carnegie Mellon Communicator. *In: Proc. 6th Europ. Conf. on Speech Communication and Technology (EUROSPEECH'99)*, 1:243–246, H–Budapest.

Cookson, S. (1988). Final Evaluation of VODIS – Voice Operated Database Enquiry System. *In: Proc. of SPEECH'88, 7th FASE Symposium*, 4:1311–1320, UK–Edinburgh.

Cox, S. J., Linford, P. W., Hill, W. B., and Johnston, R. D. (1997). Towards a Rating System for Speech Recognizers. *In: Proc. Speech and Language Technology (SALT) Club Workshop on Evaluation in Speech and Language Technology*, 64–70, UK–Sheffield.

Dahlbäck, N. (1995). Kind of Agents and Types of Dialogues. *In: Corpus-Based Approaches to Dialogue Modelling. Proc. 9th Twente Workshop on Language Technology (TWLT 9)*, University of Twente, J. A. Andernach, S. P. van de Burgt, and G. F. van der Hoeven, eds., 1–11, NL–Enschede.

Dahlbäck, N. (1997). Towards a Dialogue Taxonomy. *In: Dialogue Processing in Spoken Language Systems*, ECAI'96 Workshop, H–Budapest, E. Maier, M. Mast, and S. LuperFoy, eds., Lecture Notes in Artificial Intelligence No. 1236, 29–40, Springer, D–Berlin.

Dahlbäck, N., Jönsson, A., and Ahrenberg, L. (1993). Wizard of Oz Studies – Why and How. *Knowledge-Based Systems*, 6(4):258–266.

Dalsgaard, P. and Baekgaard, A. (1994). Spoken Language Dialogue Systems. *In: Progress and Prospects of Speech Research and Technology*, H. Niemann, R. de Mori, and G. Hanrieder, eds., 178–191, Infix, D–Sankt Augustin.

Danieli, M. and Gerbino, E. (1995). Metrics for Evaluating Dialogue Strategies in a Spoken Language System. *In: Empirical Methods in Discourse Interpretation and Generation. Papers from the 1995 AAAI Symposium*, USA–Stanford CA, 34–39, AAAI Press, USA–Menlo Park CA.

Das, S., Lubensky, D., and Wu, C. (1999). Towards Robust Speech Recognition in the Telephony Network Environment – Cellular and Landline Conditions. *In: Proc. 6th Europ. Conf. on Speech Communication and Technology (EUROSPEECH'99)*, 5:1959–1962, H–Budapest.

de Ruyter, B. and Hoonhout, J. (2002). *Usage Scenarios, User Requirements and Functional Specifications.* Deliverable 1.1, IST project 2001-32746 INSPIRE (INfotainment management with SPeech Interaction via REmote-microphones and telephone interfaces), Philips Research, NL-Eindhoven.

Delogu, C., Conte, S., and Sementina, C. (1997). Cognitive Factors in the Evaluation of Synthetic Speech. *Speech Communication*, 24:153–168.

Delogu, C., Paoloni, P., Pocci, P., and Sementina, C. (1991). Quality Evaluation of Text-to-Speech Synthesizers Using Magnitude Estimation, Categorical Estimation, Pair Comparison and Reaction Time Methods. *In: Proc. 2nd Europ. Conf. on Speech Communication and Technology (EUROSPEECH'91)*, 1:353–355, I–Genova.

Delogu, C., di Carlo, A., Rotundi, P., and Sartori, D. (1998). A Comparison Between DTMF and ASR IVR Services Through Objective and Subjective Evaluation. *In: Proc. IEEE 4th Workshop Interactive Voice Technology for Telecommunications Applications (IVTTA'98)*, 145–150, I–Torino.

Delogu, C., di Carlo, A., Sementina, C., and Stecconi, S. (1993). A Methodology for Evaluating Human-Machine Spoken Language Interaction. *In: Proc. 3rd Europ. Conf. on Speech Communication and Technology (EUROSPEECH'93)*, 2:1427–1430, D–Berlin.

Delogu, C., Paoloni, A., Ridolfi, P., and Vagges, K. (1995). Intelligibility of Speech Produced by Text-to-Speech Systems in Good and Telephonic Conditions. *acta acustica*, 3:89–96.

den Os, E. and Bloothooft, G. (1998). Evaluating Various Spoken Dialogue Systems with a Single Questionnaire: Analysis of the ELSNET Olympics. *In: Proc. 1st Int. Conf. on Language Resources and Evaluation (LREC'98)*, 1:51–54, ES–Granada.

Devillers, L. and Bonneau-Maynard, H. (1998). Evaluation of Dialogue Strategies for a Tourist Information Retrieval System. *In: Proc. 5th Int. Conf. on Spoken Language Processing (ICSLP'98)*, 4:1187–1190, AUS–Sydney.

Devillers, L., Rosset, S., Bonneau-Maynard, H., and Lamel, L. (2002). Annotations for Dynamic Diagnosis of the Dialog State. *In: Proc. 3rd Int. Conf. on Language Resources and Evaluation (LREC 2002)*, 5:1594–1601, ES–Las Palmas.

di Fabbrizio, G., Dutton, D., Gupta, N., Hollister, B., Rahim, M., Riccardi, G., Schapire, R., and Schroeter, J. (2002). AT&T Help Desk. *In: Proc. 7th Int. Conf. on Spoken Language Processing (ICSLP-2002)*, 4:2681–2684, USA–Denver CO.

Dintruff, D. L., Grice, D. G., and Wang, T.-G. (1985). User Acceptance of Speech Technologies. *Speech Technology*, 2(4):16–21.

DISC Deliverable D2.7a (1999). *State-of-the-Art Survey of Dialogue Management Tools*. Esprit Long-Term Research Concerted Action No. 24823 DISC (Spoken Language Dialogue Systems and Components: Best Practice in Development and Evaluation), Natural Interactive Systems Laboratory, Odense University, DK-Odense, `http://www.disc2.dk`.

Doran, C., Aberdeen, J., Damianos, L., and Hirschman, L. (2001). Comparing Several Aspects of Human-Computer and Human-Human Dialogues. *In: Proc. 2nd SIGdial Workshop on Discourse and Dialogue*, J. van Kuppevelt and R. Smith, eds., 48–57, DK–Aalborg.

Dudda, C. (2001). *Evaluierung eines natürlichsprachlichen Dialogsystems für Restaurantauskünfte*. Diploma thesis (unpublished), Institut für Kommunikationsakustik, Ruhr-Universität, D-Bochum.

Duncanson, J. P. (1969). The Average Telephone Call Is Better than the Average Telephone Call. *The Public Opinion Quarterly*, 33(1):112–116.

Dutoit, T. (1997). *An Introduction to Text-to-Speech Synthesis*. Kluwer Academic Publ., NL-Dordrecht.

Dutton, R. T., Foster, J. C., and Jack, M. A. (1999). Please Mind the Doors – Do Interface Metaphors Improve the Usability of Voice Response Services? *BT Technology Journal*, 17(1):172–177.

Dybkjær, H., Bernsen, N. O., and Dybkjær, L. (1993). Wizard-of-Oz and the Trade-Off Between Naturalness and Recognizer Constraints. *In: Proc. 3rd Europ. Conf. on Speech Communication and Technology (EUROSPEECH'93)*, 2:947–950, D–Berlin.

Dybkjær, L., André, E., Minker, W., and Heisterkamp, P., editors (2002). *Proc. ISCA Tutorial and Research Workshop on Multi-Modal Dialogue in Mobile Environments*. SIGdial, Special Interest Group of the ISCA (International Speech Communication Association) and ACL (Association for Computational Linguistics), D-Kloster Irsee.

Dybkjær, L. and Bernsen, N. O. (2000). Usability Issues in Spoken Dialogue Systems. *Natural Language Engineering*, 6(3-4):243–271.

Dybkjær, L., Bernsen, N. O., and Dybkjær, H. (1995). Scenario Design for Spoken Language Dialogue Systems Development. *In: Proc. ESCA Workshop on Spoken Dialogue Systems*, P. Dalsgaard, L.B. Larsen, L. Boves, and I. Thomsen, eds., 93–96, DK–Vigsø.

Dybkjær, L., Bernsen, N. O., and Dybkjær, H. (1996). Evaluation of Spoken Dialogue Systems. *In: Dialogue Management in Natural Language Systems. Proc. 11th Twente Workshop on Language Technology (TWLT 11)*, University of Twente, S. LuperFoy, A. Nijholt, and G. V. van Zanten, eds., 15 pages, NL–Enschede.

Dybkjær, L., Bernsen, N. O., and Minker, W. (2004). Evaluation and Usability of Multimodal Spoken Language Dialogue Systems. *Speech Communication*, 43:33–54.

Dybkjær, L. and Dybkjær, H. (1993). *Wizard of Oz Experiments in the Development of the Dialogue Module for P1*. Report 3a, Spoken Language Dialogue Systems Program, Centre for Cognitive Informatics, Roskilde University, DK-Roskilde.

Ehrette, T., Chateau, N., d'Alessandro, C., and Maffiolo, V. (2003). Predicting the Perceptive Judgment of Voices in a Telecom Context: Selection of Acoustic Parameters. *In: Proc. 8th Europ. Conf. on Speech Communication and Technology (EUROSPEECH 2003 – Switzerland)*, 1:117–120, CH–Geneva.

Eichner, M., Wolff, M., Odenwald, S., and Hoffmann, R. (2001). Speech Synthesis Using Stochastic Markov Graphs. *In: Proc. Int. Conf. Acoustics Speech and Signal Processing (ICASSP 2001)*, 2:829–832, IEEE, USA–Piscataway NJ.

Elenius, K. (1999). Experiences from Building Two Large Telephone Speech Databases for Swedish. *Quarterly Progress and Status Report, KTH / DEF / Institutionen för Tal, Musik och Hörsel (TMH-QPSR)*, 1-2/1999:51–56.

Erbach, G. (2000). Sprachdialogsysteme für Telefondienste: Stand der Technik und zukünftige Entwicklungen. *In: Proc. Workshop Sprachtechnologie für eine dynamische Wirtschaft im Medienzeitalter*, D-Köln.

ETSI Standard ES 201 108 (2000). *Speech Processing, Transmission and Quality Aspects (STQ); Distributed Speech Recognition; Front-End Feature Extraction Algorithm; Compression Algorithms.* European Telecommunications Standards Institute, F-Sophia Antipolis, v1.1.2 edition.

ETSI Technical Report ETR 051 (1992). *Human Factors (HF); Usability Checklist of Telephones; Basic Requirements.* European Telecommunications Standards Institute, F-Sophia Antipolis.

ETSI Technical Report ETR 095 (1993). *Human Factors (HF); Guide for Usability Evaluations of Telecommunication Systems and Services.* European Telecommunications Standards Institute, F-Sophia Antipolis.

ETSI Technical Report ETR 147 (1994). *Human Factors (HF); Usability Checklist for Integrated Services Digital Network (ISDN) Telephone Terminal Equipment.* European Telecommunications Standards Institute, F-Sophia Antipolis.

ETSI Technical Report ETR 250 (1996). *Transmission and Multiplexing (TM); Speech Communication Quality from Mouth to Ear for 3,1 kHz Handset Telephony Across Networks.* European Telecommunications Standards Institute, F-Sophia Antipolis.

Euler, S. and Zinke, J. (1994). The Influence of Speech Coding Algorithms on Automatic Speech Recognition. *In: Proc. Int. Conf. Acoustics Speech and Signal Processing (ICASSP'94)*, 1:621–624, IEEE, USA–Piscataway NJ.

EURESCOM Project P.807 Deliverable 1 (1998). *Jupiter II - Usability, Performability and Interoperability Trials in Europe.* European Institute for Research and Strategic Studies in Telecommunications, D-Heidelberg.

Failenschmid, K. (1998). Spoken Dialogue System Design – The Influence of the Organisational Context on the Design Process. *In: Proc. IEEE 4th Workshop Interactive Voice Technology for Telecommunications Applications (IVTTA'98)*, 60–64, I–Torino.

Feldes, S., Fries, G., Hagen, E., and Wirth, A. (1998). A Design Environment for Acoustic Interfaces to Databases. *In: Proc. IEEE 4th Workshop Interactive Voice Technology for Telecommunications Applications (IVTTA'98)*, 103–106, I–Torino.

Fellbaum, K. and Ketzmerick, B. (2002). Über die Rolle der Audio-Komponente bei der Multimedia-Kommunikation. *Elektronische Sprachsignalverarbeitung, Studientexte zur Sprachkommunikation 24*, R. Hoffmann, ed., 331–340, w.e.b. Universitätsverlag, D–Dresden.

Fettke, K. (2001). Der Einsatz von Text to Speech in den Informationsdiensten der DTAG. *Elektronische Sprachsignalverarbeitung, Studientexte zur Sprachkommunikation 22*, W. Hess and K. Stöber, eds., 250–259, w.e.b. Universitätsverlag, D–Dresden.

Flammia, G. and Zue, V. (1995). A Graphical User Interface for Annotating Spoken Dialogue. *In: Empirical Methods in Discourse Interpretation and Generation. Papers from the 1995 AAAI Symposium*, USA–Stanford CA, 40–46, AAAI Press, USA–Menlo Park CA.

Foster, J. C., Dutton, R., Jack, M. A., Love, S., Nairn, I. A., Vergeynst, N., and Stentiford, F. W. M. (1993). Intelligent Dialogues in Automated Telephone Services. *In: Interactive Speech Technology: Human Factor Issues in the Application of Speech Input/Output to Computers*, C. Baber and J. M. Noyes, eds., 167–175, Taylor and Francis, UK–London.

Fox, B. A. (1987). *Discourse Structure and Anaphora.* Cambridge University Press, USA-Cambridge MA.

Francis, A. L. and Nusbaum, H. C. (1999). Evaluating the Quality of Synthetic Speech. *In: Human Factors and Voice Interactive Systems*, D. Gardner–Bonneau, ed., 63–97, Kluwer Academic Publ., USA–Boston MA.

Fraser, N. (1997). Assessment of Interactive Systems. *In: Handbook on Standards and Resources for Spoken Language Systems*, D. Gibbon, R. Moore, and R. Winski, eds., 564–615, Mouton de Gruyter, D–Berlin.

Fraser, N. M. (1995). Quality Standards for Spoken Language Dialogue Systems: A Report on Progress in EAGLES. *In: Proc. ESCA Workshop on Spoken Dialogue Systems*, P. Dalsgaard, L.B. Larsen, L. Boves, and I. Thomsen, eds., 157–160, DK–Vigsø.

Fraser, N. M. and Dalsgaard, P. (1996). Spoken Dialogue Systems: A European Perspective. *In: Proc. 1996 Int. Symp. on Spoken Dialogue (ISSD 96)*, H. Fujisaki, ed., 25–36, USA–Philadelphia.

Fraser, N. M. and Gilbert, G. N. (1991a). Effects of System Voice Quality on User Utterances in Speech Dialogue Systems. *In: Proc. 2nd Europ. Conf. on Speech Communication and Technology (EUROSPEECH'91)*, 1:57–60, I–Genova.

Fraser, N. M. and Gilbert, G. N. (1991b). Simulating Speech Systems. *Computer Speech and Language*, 5:81–99.

Fraser, N. M., Salmon, B., and Thomas, T. (1996). Call Routing by Name Recognition: Field Trial Results for the Operetta^{TM} System. *In: Proc. 3rd IEEE Workshop on Interactive Voice Technology for Telecommunications Applications (IVTTA-96)*, 101–104, USA–Basking Ridge NJ.

Fujisaki, H., Kameda, H., Ohno, S., Ito, T., Tajima, K., and Abe, K. (1997). An Intelligent System for Information Retrieval over the Internet Through Spoken Dialogue. *In: Proc. 5th Europ. Conf. on Speech Communication and Technology (EUROSPEECH'97)*, 3:1675–1678, GR–Rhodes.

Furui, S. (1996). An Overview of Speaker Recognition Technology. *In: Automatic Speech and Speaker Recognition*, C.–H. Lee, F. K. Soong, and K. K. Paliwal, eds., 31–56, Kluwer Academic Publ., USA–Boston.

Furui, S. (2001a). *Digital Speech Processing, Synthesis, and Recognition*. Marcel Dekker Inc., USA-New York NY.

Furui, S. (2001b). From Read Speech Recognition to Spontaneous Speech Understanding. *In: Proc. 6th Natural Language Processing Pacific Rim Symposium*, 19–25, JP–Tokyo.

Gallardo-Antolín, A., Peláez-Moreno, C., and Díaz-de-María, F. (2001). A Robust Front-End for ASR over IP and GSM Networks: An Integrated Scenario. *In: Proc. 7th Europ. Conf. on Speech Communication and Technology (EUROSPEECH 2001 – Scandinavia)*, 2:1103–1106, DK–Aalborg.

Gates, D., Lavie, A., Levin, L., Waibel, A., Gavaldà, M., Mayfield, L., Woszczyna, M., and Zhan, P. (1997). End-to-End Evaluation in JANUS: A Speech-to-Speech Translation System. *In: Dialogue Processing in Spoken Language Systems*, ECAI'96 Workshop, H–Budapest, E. Maier, M. Mast, and S. LuperFoy, eds., Lecture Notes in Artificial Intelligence No. 1236, 195–206, Springer, D–Berlin.

Gerbino, E., Baggia, P., Giachin, E., and Rullent, C. (1995). Analysis and Evaluation of Spontaneous Speech Utterances in Focussed Dialogue Contexts. *In: Proc. ESCA Workshop on Spoken Dialogue Systems*, P. Dalsgaard, L.B. Larsen, L. Boves, and I. Thomsen, eds., 185–188, DK–Vigsø.

Gerbino, E., Baggia, P., Ciaramella, A., and Rullent, C. (1993). Test and Evaluation of a Spoken Dialogue System. *In: Proc. Int. Conf. Acoustics Speech and Signal Processing (ICASSP'93)*, 2:135–138, IEEE, USA–Piscataway NJ.

Gibbon, D., Mertins, I., and Moore, R., editors (2000). *Handbook of Multimodal and Spoken Dialogue Systems: Resources, Terminology and Product Evaluation*. Kluwer Academic Publ., USA-Boston.

Gibbon, D., Moore, R., and Winski, R., editors (1997). *Handbook on Standards and Resources for Spoken Language Systems.* Mouton de Gruyter, D-Berlin.

Gilbert, N., Wooffitt, R., and Fraser, N. (1990). Organizing Computer Talk. *In: Computers and Conversation*, P. Luff, N. Gilbert and D. Frohlich, eds., 235–257, Academic Press, UK–London.

Gillick, L. and Cox, S. J. (1989). Some Statistical Issues in the Comparison of Speech Recognition Algorithms. *In: Proc. Int. Conf. Acoustics Speech and Signal Processing (ICASSP'89)*, 1:532–535, IEEE, USA–Piscataway NJ.

Giuliani, D., Matassoni, M., Omologo, M., and Svaizer, P. (1999). Training of HMM with Filtered Speech Material for Hands-Free Recognition. *In: Proc. Int. Conf. Acoustics Speech and Signal Processing (ICASSP'99)*, 1:449–452, IEEE, USA–Piscataway NJ.

Glass, J., Polifroni, J., Seneff, S., and Zue, V. (2000). Data Collection and Performance Evaluation of Spoken Dialogue Systems: The MIT Experience. *In: Proc. 6th Int. Conf. on Spoken Language Processing (ICSLP 2000)*, 4:1–4, CHN–Beijing.

Glass, J. and Weinstein, E. (2001). SpeechBuilder: Facilitating Spoken Dialogue System Development. *In: Proc. 7th Europ. Conf. on Speech Communication and Technology (EUROSPEECH 2001 – Scandinavia)*, 2:1335–1338, DK–Aalborg.

Gleiss, N. (1992). Usability – Concepts and Evaluation. *TELE (English edition)*, 2/92:24–30, Swedish Telecommunications Administration, S–Stockholm.

Goodine, D., Hirschman, L., Polifroni, J., Seneff, S., and Zue, V. (1992). Evaluating Interactive Spoken Language Systems. *In: Proc. 2nd Int. Conf. on Spoken Language Processing (ICSLP'92)*, 1:201–204, CND–Banff.

Gorin, A. L., Parker, B. A., Sachs, R. M., and Wilpon, J. G. (1996). How May I Help You? *In: Proc. 3rd IEEE Workshop on Interactive Voice Technology for Telecommunications Applications (IVTTA'96)*, 57–60, USA–Basking Ridge NJ.

Gorin, A. L., Riccardi, G., and Wright, J. H. (1997). How may I help you? *Speech Communication*, 23:113–127.

Grice, H. P. (1975). *Logic and Conversation*, Syntax and Semantics, Vol. 3: Speech Acts (P. Cole and J. L. Morgan, eds.), 41–58. Academic Press, USA-New York (NY).

Grosz, B. (1977). *The Representation and Uses of Focus in Dialogue Understanding.* PhD thesis, University of California, USA-Berkeley CA.

Grosz, B. J. and Sidner, C. L. (1986). Attention, Intentions, and the Structure of Discourse. *Computational Linguistics*, 12(3):175–204.

Guilford, J. P. (1954). *Psychometric Methods.* McGraw-Hill Book Company, USA-New York.

Guindon, R. (1988). A Multidisciplinary Perspective on Dialogue Structure in User-Advisory Dialogues. *In: Cognitive Science and its Application for Human-Computer Interaction*, R. Guindon, ed., Lawrence Erlbaum Publishers, USA-Hillsdale NJ.

Guindon, R., Shuldberg, K., and Connor, J. (1987). Grammatical and Ungrammatical Structures in User-Advisor Dialogues: Evidence for Sufficiency of Restricted Languages in Natural Language Interfaces to Advisory Systems. *In: Proc. 25th Ann. Meeting of the Association for Computational Linguistics*, 41–44, USA–Stanford CA.

Guindon, R., Sladky, P., Brunner, H., and Connor, J. (1986). The Structure of User-Advisor Dialogues: Is There Method in Their Madness? *In: Proc. 24th Ann. Meeting of the Association for Computational Linguistics*, 224–230, USA–New York NY.

Gupta, V., Robillard, S., and Pelletier, C. (1998). Automation of Locality Recognition in ADAS Plus. *In: Proc. IEEE 4th Workshop Interactive Voice Technology for Telecommunications Applications (IVTTA'98)*, 1–4, I–Torino.

Gustafson, J., Lundeberg, M., and Liljencrants, J. (1999). Experiences from the Development of August – a Multi-Modal Spoken Dialogue System. *In: Proc. ESCA Workshop on Interactive Dialogue in Multi-Modal Systems*, P. Dalsgaard, C.–H. Lee, P. Heisterkamp, and R. Cole, eds., 61–64, D–Kloster Irsee.

Hacioglu, K. and Ward, W. (2002). A Figure of Merit for the Analysis of Spoken Dialogue Systems. *In: Proc. 7th Int. Conf. on Spoken Language Processing (ICSLP-2002)*, 2:877–880, USA–Denver CO.

Hansen, M. (1998). *Assessment and Prediction of Speech Transmission Quality with an Auditory Processing Model.* Doctoral dissertation, Carl-von-Ossietzky-Universität, D-Oldenburg.

Hansen, M. and Kollmeier, B. (2000). Objective Modeling of Speech Quality with a Psychoacoustically Validated Auditory Model. *J. Audio Eng. Soc.*, 48(5):395–409.

Hardt, D., Fellbaum, K., Kapust, R., and Michael, K.-D. (1998). Einfluss der Sprachcodierung in der Telekommunikation auf die Qualität einer textabhängigen Sprecherverifizierung. *Sprachkommunikation, ITG-Fachbericht 152*, R. Hoffmann, ed., 93–96, VDE–Verlag GmbH, D–Berlin.

Hastie, H. W., Prasad, R., and Walker, M. (2002a). Automatic Evaluation: Using a DATE Dialogue Act Tagger for User Satisfaction and Task Completion Prediction. *In: Proc. 3rd Int. Conf. on Language Resources and Evaluation (LREC 2002)*, 2:641–648, ES–Las Palmas.

Hastie, H. W., Prasad, R., and Walker, M. (2002b). What's the Trouble: Automatically Identifying Problematic Dialogues in DARPA Communicator Dialogue Systems. *In: Proc. of the 40th Ann. Meeting of the Assoc. for Computational Linguistics*, 384–391, USA–Philadelphia PA.

Hauenstein, M. (1997). *Psychoakustisch motivierte Maße zur instrumentellen Sprachgütebeurteilung*. Doctoral dissertation, Christian-Albrechts-Universität Kiel (Arbeiten über Digitale Signalverarbeitung No. 10, U. Heute, ed.), Shaker Verlag, D-Aachen.

Heeman, P. A., Yang, F., and Strayer, S. E. (2002). DialogueView: An Annotation Tool for Dialogue. *In: Proc. Third SIGdial Workshop on Discourse and Dialogue*, 50–59, USA–Philadelphia PA.

Hellbrück, J., Fastl, H., and Keller, B. (2002). Effects of Meaning of Sound on Loudness Judgements. *In: Proc. 3rd European Congress on Acoustics (Forum Acusticum Sevilla 2002), Special Issue Revista de Acústica*, 33:6 pages, ES–Sevilla.

Hennebert, J., Melin, H., Petrivska, D., and Genoud, D. (2000). POLYCOST: A Telephone-Speech Database for Speaker Recognition. *Speech Communication*, 31:265–270.

Hermansky, H. (1990). Perceptual Linear Predictive (PLP) Analysis of Speech. *J. Acoust. Soc. Am.*, 87(4):1738–1752.

Hermansky, H. and Morgan, N. (1994). RASTA Processing of Speech. *IEEE Trans. Speech and Audio Processing*, 2(4):578–589.

Hermansky, H., Morgan, N., Bayya, A., and Kohn, P. (1991). Compensation for the Effect of the Communication Channel in Auditory-Like Analysis of Speech (RASTA-PLP). *In: Proc. 2nd Europ. Conf. on Speech Communication and Technology (EUROSPEECH'91)*, 3:1367–1370, I–Genova.

Higashinaka, R., Miyazaki, N., Nakano, M., and Aikawa, K. (2003). Evaluating Discourse Understanding in Spoken Dialogue Systems. *In: Proc. 8th Europ. Conf. on Speech Communication and Technology (EUROSPEECH 2003 – Switzerland)*, 3:1941–1944, CH–Geneva.

Hirsch, H.-G. (2001). HMM Adaptation for Applications in Telecommunication. *Speech Communication*, 34:127–139.

Hirsch, H.-G. (2002). The Influence of Speech Coding on Recognition Performance in Telecommunication Networks. *In: Proc. 7th Int. Conf. on Spoken Language Processing (ICSLP-2002)*, 3:1877–1880, USA–Denver CO.

Hirsch, H.-G. and Pearce, D. (2000). The AURORA Experimental Framework for the Performance Evaluation of Speech Recognition Systems Under Noisy Conditions. *In: Proc. ISCA Tutorial and Research Workshop on Automatic Speech Recognition: Challenges for the New Millenium (ASR2000)*, 8 pages, F–Paris.

Hirschberg, J., Litman, D., and Swerts, M. (2000). Generalizing Prosodic Prediction of Speech Recognition Errors. *In: Proc. 6th Int. Conf. on Spoken Language Processing (ICSLP 2000)*, 1:254–257, CHN–Beijing.

Hirschman, L. (1998). The Evolution of Evaluation: Lessons from the Message Understanding Conferences. *Computer Speech and Language*, 12:281–305.

Hirschman, L., Bates, M., Dahl, D., Fisher, W., Garofolo, J., Pallett, D., Hunicke-Smith, K., Price, P., Rudnicky, A., and Tzoukermann, E. (1993). Multi-Site Data Collection and Evaluation in Spoken Language Understanding. *In: Proc. DARPA Human Language Technology Workshop*, 19–24, USA–Princeton NJ.

Hirschman, L., Dahl, D. A., McKay, D. P., Norton, L. M., and Linebarger, M. C. (1990). Beyond Class A: A Proposal for Automatic Evaluation of Discourse. *In: Proc. DARPA Speech and Natural Language Workshop*, 109–113, USA–Hidden Valley PA.

Hirschman, L. and Pao, C. (1993). The Cost of Errors in a Spoken Language System. *In: Proc. 3rd Europ. Conf. on Speech Communication and Technology (EUROSPEECH'93)*, 2:1419–1422, D–Berlin.

Hirschman, L. and Thompson, H. S. (1997). Overview of Evaluation in Speech and Natural Language Processing. *In: Survey of the State of the Art in Human Language Technology*, R. Cole, J. Mariani, A. Uszkoreit, H. Zaenen, and V. Zue, eds., 409–414, Cambridge University Press and Giardini Editori, I–Pisa.

Hjalmarsson, A. (2002). *Evaluating AdApt, a Multi-Modal Conversational, Dialogue System, Using PARADISE.* Master thesis, Dept. of Speech, Music and Hearing, KTH, S-Stockholm.

Höge, H., Tropf, H. S., Winski, R., van der Heuvel, H., Haeb-Umbach, R., and Choukri, K. (1997). European Spech Database for Telephone Applications. *In: Proc. Int. Conf. Acoustics Speech and Signal Processing (ICASSP'97)*, 3:1771–1774, IEEE Sign. Proc. Soc., USA–Piscataway NJ.

Höge, H., Draxler, C., van der Heuvel, H., Johansen, F. T., Sanders, E., and Tropf, H. S. (1999). SpeechDat Multilingual Speech Databases for Teleservices: Across the Finish Line. *In: Proc. 6th Europ. Conf. on Speech Communication and Technology (EUROSPEECH'99)*, 1:2699–2702, H–Budapest.

Hone, K. S. and Graham, R. (2000). Towards a Tool for the Subjective Assessment of Speech System Interfaces (SASSI). *Natural Language Engineering*, 6(3-4):287–303.

Hone, K. S. and Graham, R. (2001). Subjective Assessment of Speech-System Interface Usability. *In: Proc. 7th Europ. Conf. on Speech Communication and Technology (EUROSPEECH 2001 – Scandinavia)*, 3:2083–2086, DK–Aalborg.

Hoth, D. F. (1941). Room Noise Spectra at Subscribers' Telephone Locations. *J. Acoust. Soc. Am.*, 12:499–504.

Howard-Jones, P. (1992). *Specification of Listener Dimensions.* Final Project Report, ESPRIT Project 2589 (SAM), Multilingual Speech Input/Output Assessment, Methodology and Standardization, University College, UK-London.

Hutchinson, B. (2001). A Functional Approach to Speech Recognition Evaluation. *In: Proc. 7th Europ. Conf. on Speech Communication and Technology (EUROSPEECH 2001 – Scandinavia)*, 3:1683–1686, DK–Aalborg.

IEC Standard 60268-16 (1998). *Sound System Equipment – Part 16: Objective Rating of Speech Intelligibility by Speech Transmission Index*. European Committee for Electrotechnical Standardization, B-Brussles.

ISO Standard ISO/IEC 9126-1 (2001). *Software Engineering – Product Quality – Part 1: Quality Model*. International Organization for Standardization/International Electrotechnical Commission, CH-Geneva.

ISO Technical Report ISO/TR 19358 (2002). *Ergonomics – Construction and Application of Tests for Speech Technology*. International Organization for Standardization, CH-Geneva.

Issar, S. and Ward, W. (1993). CMU's Robust Spoken Language Understanding System. *In: Proc. 3rd Europ. Conf. on Speech Communication and Technology (EUROSPEECH'93)*, 3:2147–2150, D–Berlin.

ITU-T Appendix I to Rec. G.113 (2002). *Provisional Planning Values for the Equipment Impairment Factor Ie and Packet-Loss Robustness Factor Bpl*. International Telecommunication Union, Geneva.

ITU-T Contribution COM 12-176 (1987). *Subjective Quality Assessment of Synthetic Speech*. Source: Sweden. International Telecommunication Union, CH-Geneva.

ITU-T Delayed Contribution D.108 (2003). *User's Perspective Performance Parameters for WEB Hosting, E-Mail and Streaming Media*. Source: AT&T (C. A. Dvorak). International Telecommunication Union, Study Group 12, CH-Geneva.

ITU-T Delayed Contribution D.29 (2001). *Derivation of Equipment Impairment Factors Using Instrumental Models – Test Results and Proposal for a New Recommendation P.DIEIM*. Source: Deutsche Telekom AG (S. Möller, J. Berger). International Telecommunication Union, Study Group 12, CH-Geneva.

ITU-T Delayed Contribution D.44 (2001). *Modelling Impairment Due to Packet Loss for Application in the E-Model*. Source: Deutsche Telekom AG (A. Raake). International Telecommunication Union, Study Group 12, CH-Geneva.

ITU-T Handbook on Telephonometry (1992). International Telecommunication Union, CH-Geneva.

ITU-T Rec. E.800 (1994). *Terms and Definitions Related to Quality of Service and Network Performance Including Dependability*. International Telecommunication Union, CH-Geneva.

ITU-T Rec. G.1000 (2001). *Communications Quality of Service: A Framework and Definitions*. International Telecommunication Union, CH-Geneva.

ITU-T Rec. G.107 (2003). *The E-Model, a Computational Model for Use in Transmission Planning*. International Telecommunication Union, CH-Geneva.

ITU-T Rec. G.108 (1999). *Application of the E-model: A Planning Guide*. International Telecommunication Union, CH-Geneva.

ITU-T Rec. G.109 (1999). *Definition of Categories of Speech Transmission Quality*. International Telecommunication Union, CH-Geneva.

ITU-T Rec. G.111 (1993). *Loudness Ratings (LRs) in an International Connection*. International Telecommunication Union, CH-Geneva.

ITU-T Rec. G.114 (2003). *One-Way Transmission Time*. International Telecommunication Union, CH-Geneva.

ITU-T Rec. G.121 (1993). *Loudness Ratings (LRs) of National Systems*. International Telecommunication Union, CH-Geneva.

ITU-T Rec. G.126 (1993). *Listener Echo in Telephone Networks.* International Telecommunication Union, CH-Geneva.

ITU-T Rec. G.131 (1996). *Control of Talker Echo.* International Telecommunication Union, CH-Geneva.

ITU-T Rec. G.711 (1988). *Pulse Code Modulation (PCM) of Voice Frequencies.* International Telecommunication Union, CH-Geneva.

ITU-T Rec. G.712 (2001). *Transmission Performance Characteristics of Pulse Code Modulation Channels.* International Telecommunication Union, CH-Geneva.

ITU-T Rec. G.722 (1988). *7 kHz Audio-Coding Within 64 kbit/s.* International Telecommunication Union, CH-Geneva.

ITU-T Rec. G.726 (1990). *40, 32, 24, 16 kbit/s Adaptive Differential Pulse Code Modulation (ADPCM).* International Telecommunication Union, CH-Geneva.

ITU-T Rec. G.728 (1992). *Coding of Speech at 16 kbit/s Using Low-Delay Code Excited Linear Prediction.* International Telecommunication Union, CH-Geneva.

ITU-T Rec. G.729 (1996). *Coding of Speech at 8 kbit/s Using Conjugate-Structure Algebraic-Code-Excited Linear-Prediction (CS-ACELP).* International Telecommunication Union, CH-Geneva.

ITU-T Rec. P.310 (2003). *Transmission Characteristics for Telephone-Band (300-3400 Hz) Digital Telephones.* International Telecommunication Union, CH-Geneva.

ITU-T Rec. P.340 (2000). *Transmission Characteristics and Speech Quality Parameters of Hands-Free Terminals.* International Telecommunication Union, CH-Geneva.

ITU-T Rec. P.48 (1988). *Specification for an Intermediate Reference System.* International Telecommunication Union, CH-Geneva.

ITU-T Rec. P.53 (1994). *Psophometer for Use on Telephone-Type Circuits.* International Telecommunication Union, CH-Geneva.

ITU-T Rec. P.56 (1993). *Objective Measurement of Active Speech Level.* International Telecommunication Union, CH-Geneva.

ITU-T Rec. P.561 (2002). *In-Service, Non-Intrusive Measurement Device – Voice Service Measurements.* International Telecommunication Union, CH-Geneva.

ITU-T Rec. P.562 (2004). *Analysis and Interpretation of INMD Voice-Services Measurements.* International Telecommunication Union, CH-Geneva.

ITU-T Rec. P.64 (1999). *Determination of Sensitivity / Frequency Characteristics of Local Telephone Systems.* International Telecommunication Union, CH-Geneva.

ITU-T Rec. P.79 (1999). *Calculation of Loudness Ratings for Telephone Sets.* International Telecommunication Union, CH-Geneva.

ITU-T Rec. P.800 (1996). *Methods for Subjective Determination of Transmission Quality.* International Telecommunication Union, CH-Geneva.

ITU-T Rec. P.810 (1996). *Modulated Noise Reference Unit (MNRU).* International Telecommunication Union, CH-Geneva.

ITU-T Rec. P.830 (1996). *Subjective Performance Assessment of Telephone-Band and Wideband Digital Codecs.* International Telecommunication Union, CH-Geneva.

ITU-T Rec. P.834 (2002). *Methodology for the Derivation of Equipment Impairment Factors from Instrumental Models.* International Telecommunication Union, Geneva.

ITU-T Rec. P.85 (1994). *A Method for Subjective Performance Assessment of the Quality of Speech Voice Output Devices.* International Telecommunication Union, CH-Geneva.

ITU-T Rec. P.862 (2001). *Perceptual Evaluation of Speech Quality (PESQ), an Objective Method for End-to-End Speech Quality Assessment of Narrowband Telephone Networks and Speech Codecs*. International Telecommunication Union, CH-Geneva.

ITU-T Suppl. 23 to P-Series Rec. (1998). *ITU-T Coded-Speech Database*. International Telecommunication Union, CH-Geneva.

ITU-T Suppl. 3 to P-Series Rec. (1993). *Models for Predicting Transmission Quality from Objective Measurements*. International Telecommunication Union, CH-Geneva.

Jack, M. A., Foster, J. C., and Stentiford, F. W. M. (1992). Intelligent Dialogues in Automated Telephone Services. *In: Proc. 2nd Int. Conf. on Spoken Language Processing (ICSLP'92)*, 1:715–718, CND–Banff.

Jack, M. A. and Lefèvre, J.-P. (1997). Usability Analysis of Spoken Dialogues for Automated Telephone Banking Services (OVID). *Human Comfort and Security of Information Systems*, K. Varghese and S. Pfleger, eds., 125–133, Springer, D–Berlin.

Jacquemin, C., Mariani, J., and Paroubek, P., editors (2000). *Using Evaluation Within HLT Programs: Results and Trends. Proc. of the CLASS Pre-Conference Workshop to LREC 2000*. GR-Athens, `http://www.class-tech.org/publications`.

Jarke, M., Turner, J. A., Stohr, E. A., Vassiliou, Y., White, N. H., and Michielsen, K. (1985). A Field Evaluation of Natural Language for Data Retrieval. *IEEE Trans. Software Engineering*, SE-11(1):97–113.

Jekosch, U. (2000). *Sprache hören und beurteilen: Ein Ansatz zur Grundlegung der Sprachqualitätsbeurteilung*. Habilitation thesis (unpublished), Universität/Gesamthochschule, D-Essen.

Jekosch, U. (2001). Projektionsmodell zu Sprachqualitätsmessungen. *In: Fortschritte der Akustik – DAGA 2001*, 596–597, DEGA e.V., D–Oldenburg.

Jekosch, U., Krause, S., and Mersdorf, J. (1997). *Evaluation der deutschsprachigen Synthese "Sprechmobil" im Verbmobilprojekt*. Project report, Institut für Kommunikationsakustik, Ruhr-Universität, D-Bochum.

Jekosch, U. and Pols, L. C. W. (1994). A Feature-Profile for Application-Specific Speech Synthesis Assessment and Evaluation. *In: Proc. 3rd Int. Conf. on Spoken Language Processing (ICSLP'94)*, 3:1319–1322, JP–Yokohama.

Johannesson, N. O. (1997). The ETSI Computation Model: A Tool for Transmission Planning of Telephone Networks. *IEEE Communications Magazine*, Jan.:70–79.

Johansen, F. T., Warakagoda, N., Lindberg, B., Lehtinen, G., Kačič, Z., Žgank, A., Elenius, K., and Salvi, G. (2000). The COST 249 SpeechDat Multilingual Reference Recognizer. *In: Proc. 2nd Int. Conf. on Language Resources and Evaluation (LREC 2000)*, 3:1351–1355, GR–Athen.

Johnsen, M. H., Svendsen, T., Amble, T., Holter, T., and Harborg, E. (2000). TABOR – A Norwegian Spoken Dialogue System for Bus Travel Information. *In: Proc. 6th Int. Conf. on Spoken Language Processing (ICSLP 2000)*, 3:1049–1052, CHN–Beijing.

Johnston, D. (2000). An Overview of the EURESCOM MIVA Project. *In: Multi-Lingual Interoperability in Speech Technology*, NATO Research and Technology Organization, F–Neuilly–sur–Seine, 7 pages.

Johnston, R. D. (1997). Towards a Reference System for Speech Technology. *In: Proc. Speech and Language Technology (SALT) Club Workshop on Evaluation in Speech and Language Technology*, 90–97, UK–Sheffield.

Jost, U. (1997). Evaluation in Verbmobil. *In: Proc. Speech and Language Technology (SALT) Club Workshop on Evaluation in Speech and Language Technology*, 98–105, UK–Sheffield.

Juang, B. H. (1991). Speech Recognition in Adverse Environments. *Computer Speech and Language*, 5:275–294.

Jurafski, D., Wooters, C., Tajchman, G., Segal, J., Stolcke, A., Fosler, E., and Morgan, N. (1994). The Berkeley Restaurant Project. *In: Proc. 3rd Int. Conf. on Spoken Language Processing (ICSLP'94)*, 4:2139–2142, JP–Yokohama.

Kamm, C. A. and Walker, M. A. (1997). Design and Evaluation of Spoken Dialogue Systems. *In: Proc. 1997 IEEE Workshop on Automatic Speech Recognition and Understanding*, USA–Santa Barbara CA.

Kamm, C., Narayanan, S., Dutton, D., and Ritenour, R. (1997a). Evaluating Spoken Dialogue Systems for Telecommunication Services. *In: Proc. 5th Europ. Conf. on Speech Communication and Technology (EUROSPEECH'97)*, 4:2203–2206, GR–Rhodes.

Kamm, C., Walker, M., and Rabiner, L. (1997b). The Role of Speech Processing in Human-Computer Intelligent Communication. *Speech Communication*, 23:263–278.

Kamm, C. A., Litman, D. J., and Walker, M. A. (1998). From Novice to Expert: The Effect of Tutorials on User Expertise with Spoken Dialogue Systems. *In: Proc. 5th Int. Conf. on Spoken Language Processing (ICSLP'98)*, 4:1211–1214, AUS–Sydney.

Karray, L., Ben Jelloun, A., and Mokbel, C. (1998). Solutions for Robust Recognition over the GSM Cellular Network. *In: Proc. Int. Conf. Acoustics Speech and Signal Processing (ICASSP'98)*, 1:261–264, IEEE, USA–Piscataway NJ.

Keeney, R. L. and Raiffa, H. (1993). *Decisions with Multiple Objectives: Preferences and Value Tradeoffs*. Cambridge University Press, USA-Cambridge MA.

Kellner, A., Rueber, B., Seide, F., and Tran, B.-H. (1997). PADIS – An Automatic Telephone Switchboard and Directory Information System. *Speech Communication*, 23:95–111.

Kennedy, A., Wilkes, A., Elder, L., and Murray, W. (1988). Dialogue with Machines. *Cognition*, 30:73–105.

Kiss, I. (2000). A Comparison of Distributed and Network Speech Recognition for Mobile Communication Systems. *In: Proc. 6th Int. Conf. on Spoken Language Processing (ICSLP 2000)*, 4:250–253, CHN–Beijing.

Klaus, H., Fellbaum, K., and Sotscheck, J. (1997). Auditive Bestimmung und Vergleich der Sprachqualität von Sprachsynthesesystemen für die deutsche Sprache. *ACUSTICA/acta acustica*, 83:124–136.

Klein, M., Bernsen, N. O., Davies, S., Dybkjær, L., Garrido, J., Kasch, H., Mengel, A., Pirelli, V., Poesio, M., Quazza, S., and Soria, C. (1998). *Supported Coding Schemes*. Telematics Project LE4-8370 (Multilevel Annotation, Tools Engineering), Natural Interactive Systems Laboratory, Odense University, DK-Odense, `http://mate.nis.sdu.dk`.

Klemmert, H., Brau, H., and Marzi, R. (2001). Erweiterung der heuristischen Evaluation für sprachgesteuerte Softwaresysteme – Ergebnisse der Evaluation. *Elektronische Sprachsignalverarbeitung, Studientexte zur Sprachkommunikation 22*, W. Hess and K. Stöber, eds., 53–60, w.e.b. Universitätsverlag, D–Dresden.

Koehler, J., Morgan, N., Hermansky, H., Hirsch, H. G., and Tong, G. (1994). Integrating RASTA-PLP into Speech Recognition. *In: Proc. Int. Conf. Acoustics Speech and Signal Processing (ICASSP'94)*, 1:421–424, IEEE, USA–Piscataway NJ.

Köster, S. (2003). *Modellierung von Sprechweisen für widrige Kommunikationsbedingungen mit Anwendung auf die Sprachsynthese*. Doctoral dissertation, Institut für Kommunikationsakustik, Ruhr-Universität Bochum, Shaker Verlag, D-Aachen.

Kraft, V. (1997). *Konkatenation natürlichsprachiger Bausteine in der Sprachsynthese: Anforderungen, Verfahren und Evaluierung*. Doctoral dissertation, Institut für Kommunikationsakustik, Ruhr-Universität Bochum, VDI-Verlag, D-Düsseldorf.

Kraft, V. and Portele, T. (1995). Quality Evaluation of Five German Speech Synthesis Systems. *acta acustica*, 3:351–365.

Krause, D. (1997). Using an Interpretation System – Some Observations in Hidden Operator Simulations of 'VERBMOBIL'. *In: Dialogue Processing in Spoken Language Systems*, ECAI'96 Workshop, H–Budapest, E. Maier, M. Mast, and S. LuperFoy, eds., Lecture Notes in Artificial Intelligence No. 1236, 41–54, Springer, D–Berlin.

Krause, J. and Hitzenberger, L. (1992). *Computer Talk.* Sprache und Computer No. 12, Olms, D-Hildesheim.

Kruschke, H. (2001). Simulation of Speaking Styles with Adapted Prosody. *In: Proc. 4th Int. Conf. on Text, Speech and Dialogue (TSD 2001)*, CZ–Zelezna Ruda, Lecture Notes in Artificial Intelligence No. 2166, 278–284, Springer, D–Berlin.

Kurdi, M.-Z. and Ahafhaf, M. (2002). Toward an Objective and Generic Method for Spoken Language Understanding Systems Evaluation: An Extension of the DCR Method. *In: Proc. 3rd Int. Conf. on Language Resources and Evaluation (LREC 2002)*, 2:545–550, ES–Las Palmas.

Lai, J. C. and Lee, K. M. (2002). Choosing Speech or Touchtone Modality for Navigation Within a Telephony Natural Language System. *In: Proc. 7th Int. Conf. on Spoken Language Processing (ICSLP-2002)*, 3:1509–1512, USA–Denver CO.

Lamel, L., Bennacef, S., Gauvain, J. L., Dartigues, H., and Temem, J. N. (1998a). User Evaluation of the MASK Kiosk. *In: Proc. 5th Int. Conf. on Spoken Language Processing (ICSLP'98)*, 7:2875–2878, AUS–Sydney.

Lamel, L., Bennacef, S., Gauvain, J. L., Dartigues, H., and Temem, J. N. (2002). User Evaluation of the MASK Kiosk. *Speech Communication*, 38:131–139.

Lamel, L., Rosset, S., Gauvain, J. L., Bennacef, S., Garnier-Rizet, M., and Prouts, B. (1998b). The LIMSI ARISE System. *In: Proc. IEEE 4th Workshop Interactive Voice Technology for Telecommunications Applications (IVTTA'98)*, 209–214, I–Torino.

Lamel, L., Rosset, S., Gauvain, J. L., Bennacef, S., Garnier-Rizet, M., and Prouts, B. (2000a). The LIMSI ARISE System. *Speech Communication*, 31:339–353.

Lamel, L. F., Bennacef, S. K., Rosset, S., Devillers, L., Foukia, S., Gangolf, J. J., and Gauvain, J. L. (1997). The LIMSI RailTel System: Field Trial of a Telephone Service for Rail Travel Information. *Speech Communication*, 23:67–82.

Lamel, L., Minker, W., and Paroubek, P. (2000b). Towards Best Practice in the Development and Evaluation of Speech Recognition Components of a Spoken Language Dialogue System. *Natural Language Engineering*, 6(3-4):305–322.

Lane, H. L., Catania, A. C., and Stevens, S. S. (1961). Voice Level: Autophonic Scale, Perceived Loudness, and Effects of Sidetone. *J. Acoust. Soc. Am.*, 33(2):160–167.

Lane, H. L., Tranel, B., and Sisson, C. (1970). Regulation of Voice Communication by Sensory Dynamics. *J. Acoust. Soc. Am.*, 47(2):618–624.

Larsen, L. B. (1999). Combining Objective and Subjective Data in Evaluation of Spoken Dialogues. *In: Proc. ESCA Workshop on Interactive Dialogue in Multi-Modal Systems*, P. Dalsgaard, C.–H. Lee, P. Heisterkamp, and R. Cole, eds., 89–92, D–Kloster Irsee.

Larsen, L. B. (2003). Assessment of Spoken Dialogue System Usability – What Are We Really Measuring? *In: Proc. 8th Europ. Conf. on Speech Communication and Technology (EUROSPEECH 2003 – Switzerland)*, 3:1945–1948, CH–Geneva.

Larsen, L. B. (2004). Usability Evaluation of Spoken Dialogue Systems. *In: Proc. 4th Int. Conf. on Language Resources and Evaluation (LREC 2004)*, 6:2151–2154, P–Lisbon.

Lavie, A., Waibel, A., Levin, L., Gates, D., Gavaldà, M., Zeppenfeld, T., Zhan, P., and Glickman, O. (1996). Translation of Conversational Speech with JANUS-II. *In: Proc. 4th Int. Conf. on Spoken Language Processing (ICSLP'96)*, H.T. Bunnell and W. Idsardi, eds., 4:2375–2378, IEEE, USA–Piscataway NJ.

Lee, C.-H., Lin, C.-H., and Juang, B.-H. (1991). A Study on Speaker Adaptation of the Parameters of Continuous Density Hidden Markov Models. *IEEE Trans. Signal Processing*, 39(4):806–814.

Lee, M. (1999). Implicit Goals in Indirect Replies. *In: Proc. ESCA Workshop on Interactive Dialogue in Multi-Modal Systems*, P. Dalsgaard, C.–H. Lee, P. Heisterkamp, and R. Cole, eds., 25–28, D–Kloster Irsee.

Leonard, R. G. and Doddington, G. R. (1991). *A Speaker-Independent Connected-Digit Database.* Texas Instruments Inc., Central Research Laboratories, USA-Dallas TX.

Levin, E., Narayanan, S., Pieraccini, R., Biatov, K., Bocchieri, E., di Fabbrizio, D., Eckert, W., Lee, S., Pokrovsky, A., Rahim, M., Ruscitti, P., and Walker, M. (2000). The AT&T-DARPA Communicator Mixed-Initiative Spoken Dialogue System. *In: Proc. 6th Int. Conf. on Spoken Language Processing (ICSLP 2000)*, 2:122–125, CHN–Beijing.

Lewandowski, T. (1994). *Linguistisches Wörterbuch.* Quelle und Meyer, D-Heidelberg.

Life, M. A., Lee, B. P., and Long, J. B. (1988). Assessing the Usability of Future Speech Technology: Towards a Method. *In: Proc. of SPEECH'88, 7th FASE Symposium*, 4:1297–1304, UK–Edinburgh.

Likert, R. (1932). A Technique for the Measurement of Attitudes. *Archives of Psychology*, 140:1–55.

Lilly, B. T. and Paliwal, K. K. (1996). Effect of Speech Coders on Speech Recognition Performance. *In: Proc. 4th Int. Conf. on Spoken Language Processing (ICSLP'96)*, H.T. Bunnell and W. Idsardi, eds., 4:2344–2347, IEEE, USA–Piscataway NJ.

Lindberg, B. (1994). Recognizer Response Modelling from Testing on Series of Minimal Word Pairs. *In: Proc. 3rd Int. Conf. on Spoken Language Processing (ICSLP'94)*, 3:1275–1278, JP–Yokohama.

Lippmann, R. P. (1997). Speech Recognition by Machines and Humans. *Speech Communication*, 22:1–15.

Litman, D., Hirschberg, J., and Swerts, M. (2001). Predicting User Reactions to System Error. *In: Proc. of the ACL/EACL 39th Ann. Meeting of the Assoc. for Computational Linguistics*, 362–369, F–Toulouse.

Litman, D. J. and Pan, S. (1999). Empirically Evaluating an Adaptable Spoken Dialogue System. *In: Proc. of the 7th Int. Conf. on User Modeling (UM99)*, J. Kay, ed., 55–64, Springer, A–Wien.

Litman, D. J., Pan, S., and Walker, M. A. (1998). Evaluating Response Strategies in a Web-Based Spoken Dialogue Agent. *In: Proc. of the 36th Ann. Meeting of the Assoc. for Computational Linguistics and 17th Int. Conf. on Computational Linguistics (COLING-ACL 98)*, 780–786, CAN–Montreal.

Loderer, G. (1998). *Evaluierung von Dialogstrategien eines natürlichsprachlichen Dialogsystems durch Wizard-of-Oz-Experimente.* Master thesis, Institut für med. Kybernetik und AI, Universität Wien, A-Wien.

Lombard, E. (1911). Le signe de l'elévation de la voix. *Ann. Maladies Oreille, Larynx, Nez, Pharynx*, 37:101–119.

López-Cózar, R., de la Torre, A., Segura, J. C., and Rubio, A. J. (2003). Assessment of Dialogue Systems by Means of a New Simulation Technique. *Speech Communication*, 40:387–407.

Love, S., Dutton, R. T., Foster, J. C., Jack, M. A., and Stentiford, F. W. M. (1994). Identifying Salient Usability Attributes for Automated Telephone Services. *In: Proc. 3rd Int. Conf. on Spoken Language Processing (ICSLP'94)*, 3:1307–1310, JP–Yokohama.

Ludwig, T. (2003). *Messung von Signaleigenschaften zur referenzfreien Qualitätsbewertung von Telefonbandsprache*. Doctoral dissertation, Christian-Albrechts-Universität Kiel (Arbeiten über Digitale Signalverarbeitung No. 23, U. Heute, ed.), Shaker Verlag, D-Aachen.

Luff, P., Gilbert, N., and Frohlich, D., editors (1990). *Computers and Conversation*. Academic Press, UK-London.

MacLean, A., Young, R., Bellotti, V., and Moran, T. (1991). Design Space Analysis: Bridging from Theory to Practice via Design Rationale. *In: Esprit '91: Proceedings of the Annual Esprit Conference*, 720–730.

Maier, E., Mast, M., and LuperFoy, S. (1997). Overview. *In: Dialogue Processing in Spoken Language Systems*, ECAI'96 Workshop, H–Budapest, E. Maier, M. Mast, and S. LuperFoy, eds., Lecture Notes in Artificial Intelligence No. 1236, 1–13, Springer, D–Berlin.

Malenke, M., Bäumler, M., and Paulus, E. (2000). Speech Recognition Performance Assessment. *In: Verbmobil: Foundations of Speech-to-Speech Translation*, W. Wahlster, ed., 583–591, Springer, D–Berlin.

Marcus, M., Santorini, S., and Marcinkiewicz, M. (1993). Building a Large Annotated Corpus of English: The Penn Treebank. *Computational Linguistics*, 19(2):313–330.

Mariani, J. (1998). The Aupelf-Uref Evaluation-Based Language Engineering Actions and Related Projects. *In: Proc. 1st Int. Conf. on Language Resources and Evaluation (LREC'98)*, 1:123–128, ES–Granada.

Mariani, J. and Lamel, L. (1998). An Overview of EU Programs Related to Conversational/Interactive Systems. *In: Proc. DARPA Broadcast News Transcription and Understanding Workshop*, 247–253, USA–Lansdowne VA.

Mariniak, A. and Mersdorf, J. (1994). Ein Ansatz zur Beurteilung der Intonation von synthetischer Sprache. *In: Fortschritte der Akustik – DAGA '94*, 1369–1372, DPG GmbH, D–Bad Honnef.

Marzi, R. and John, P. (2001). Sprachdialog für die Unterstützung bei der Fehlerdiagnose an CNC-Werkzeugmaschinen. *Elektronische Sprachsignalverarbeitung, Studientexte zur Sprachkommunikation 22*, W. Hess and K. Stöber, eds., 274–281, w.e.b. Universitätsverlag, D–Dresden.

Masuko, T., Tokuda, K., Kobayashi, T., and Imai, S. (1996). Speech Synthesis Using HMMs with Dynamic Features. *In: Proc. Int. Conf. Acoustics Speech and Signal Processing (ICASSP'96)*, 1:389–392, IEEE, USA–Piscataway NJ.

Matassoni, M., Omologo, M., and Svaizer, P. (2001). Use of Real and Contaminated Speech for Training of a Hands-Free In-Car Speech Recognizer. *In: Proc. 7th Europ. Conf. on Speech Communication and Technology (EUROSPEECH 2001 – Scandinavia)*, 3:1569–1572, DK–Aalborg.

Mazor, B. and Zeigler, B. L. (1995). The Design of Speech-Interactive Dialogs for Transaction-Automation Systems. *Speech Communication*, 17:313–320.

McDermott, B. J. (1969). Multidimensional Analysis of Circuit Quality Judgments. *J. Acoust. Soc. Am.*, 45:774–781.

McGee, V. E. (1964). Semantic Components of the Quality of Processed Speech. *J. Speech and Hear. Res.*, 7:310–323.

McInnes, F. R., Attwater, D. J., Edgington, M. D., Schmidt, M. S., and Jack, M. A. (1999). User Attitudes to Concatenated Natural Speech and Text-to-Speech Synthesis in an Automated

Information Service. *In: Proc. 6th Europ. Conf. on Speech Communication and Technology (EUROSPEECH'99)*, 2:831–834, H–Budapest.

McTear, M. F. (2002). Spoken Dialogue Technology: Enabling the Conversational Interface. *ACM Computing Surveys*, 34(1):90–169.

McTear, M. F., Allen, S., Clatworthy, L., Ellison, N., Lavelle, C., and McCaffery, H. (2000). Integrating Flexibility into a Structured Dialogue Model: Some Design Considerations. *In: Proc. 6th Int. Conf. on Spoken Language Processing (ICSLP 2000)*, 1:110–113, CHN–Beijing.

Melin, H. and Lindberg, J. (1996). Guidelines for Experiments on the POLYCOST Database. *In: Proc. COST250 Workshop on the Application of Speaker Recognition Technologies in Telephony*, 59–69, ES–Vigo. Updated versions available under `http://www.speech.kth.se/cost250/polycost/be/latest`.

Meng, H., Busayapongchai, S., Glass, J., Goddeau, D., Hetherington, L., Hurley, E., Pao, C., Polifroni, J., Seneff, S., and Zue, V. (1996). WHEELS: A Conversational System in the Automobile Classifieds Domain. *In: Proc. 4th Int. Conf. on Spoken Language Processing (ICSLP'96)*, H.T. Bunnell and W. Idsardi, eds., 1:542–545, IEEE, USA–Piscataway NJ.

Meng, H. M., Yip, W. L., Mok, O. Y., and Chan, S. F. (2003). Natural Language Response Generation in Mixed-Initiative Dialogs Using Task Goals and Dialogue Acts. *In: Proc. 8th Europ. Conf. on Speech Communication and Technology (EUROSPEECH 2003 – Switzerland)*, 3:1689–1692, CH–Geneva.

Mersdorf, J. (2001). *Sprecherspezifische Parametrisierung von Sprachgrundfrequenzverläufen: Analyse, Synthese und Evaluation*. Doctoral dissertation, Institut für Kommunikationsakustik, Ruhr-Universität Bochum, Shaker Verlag, D-Aachen.

Metze, F., McDonough, J., and Soltau, H. (2001). Speech Recognition over NetMeeting Connections. *In: Proc. 7th Europ. Conf. on Speech Communication and Technology (EUROSPEECH 2001 – Scandinavia)*, 4:2389–2392, DK–Aalborg.

Meyer, M. and Hild, H. (1997). Recognition of Spoken and Spelled Proper Names. *In: Proc. 5th Europ. Conf. on Speech Communication and Technology (EUROSPEECH'97)*, 3:1579–1582, GR–Rhodes.

Minker, W. (1998). Evaluation Methodologies for Interactive Speech Systems. *In: Proc. 1st Int. Conf. on Language Resources and Evaluation (LREC'98)*, 1:199–206, ES–Granada.

Minker, W. (2002). Overview on Recent Activities in Speech Understanding and Dialogue Systems Evaluation. *In: Proc. 7th Int. Conf. on Spoken Language Processing (ICSLP-2002)*, 1:265–268, USA–Denver CO.

Mokbel, C., Mauuary, L., Karray, L., Jouvet, D., Monné, J., Simonin, J., and Bartkova, K. (1997). Towards Improving ASR Robustness for PSN and GSM Telephone Applications. *Speech Communication*, 23:141–159.

Mokbel, C., Monné, J., and Jouvet, D. (1993). On-Line Adaptation of a Speech Recognizer to Variations in Telephone Line Conditions. *In: Proc. 3rd Europ. Conf. on Speech Communication and Technology (EUROSPEECH'93)*, 2:1247–1250, D–Berlin.

Mokbel, C., Monné, J., and Jouvet, D. (1995). Blind Equalization Using Adaptive Filtering for Improving Speech Recognition over the Telephone. *In: Proc. 4th Europ. Conf. on Speech Communication and Technology (EUROSPEECH'95)*, 3:1987–1990, ES–Madrid.

Mokbel, C., Jouvet, D., Monné, J., and de Mori, R. (1998). Robust Speech Recognition. *Spoken Dialogues with Computers*, 405–460, Academic Press, UK–London.

Möller, S. (2000). *Assessment and Prediction of Speech Quality in Telecommunications*. Kluwer Academic Publ., USA-Boston MA.

Möller, S. (2002a). A New Taxonomy for the Quality of Telephone Services Based on Spoken Dialogue Systems. *In: Proc. Third SIGdial Workshop on Discourse and Dialogue*, 142–153, USA–Philadelphia PA.

Möller, S. (2002b). Towards Quantifying the Influence of User Expectation on the Quality of Mobile Services. *In: Proc. ISCA Tutorial and Research Workshop on Multi-Modal Dialogue in Mobile Environments*, 15 pages, D–Kloster Irsee.

Möller, S. (2004). Telephone Transmission Impact on Synthesized Speech: Quality Assessment and Prediction. *Acta Acustica united with Acustica*, 90:121–136.

Möller, S. and Bourlard, H. (2000). Real-Time Telephone Transmission Simulation for Speech Recognizer and Dialogue System Evaluation and Improvement. *In: Proc. 6th Int. Conf. on Spoken Language Processing (ICSLP 2000)*, 1:750–753, CHN–Beijing.

Möller, S. and Bourlard, H. (2002). Analytic Assessment of Telephone Transmission Impact on ASR Performance Using a Simulation Model. *Speech Communication*, 38:441–459.

Möller, S., Jekosch, U., Mersdorf, J., and Kraft, V. (2001). Auditory Assessment of Synthesized Speech in Application Scenarios: Two Case Studies. *Speech Communication*, 34:229–246.

Möller, S. and Kavallieratou, E. (2002). Diagnostic Assessment of Telephone Transmission Impact on ASR Performance and Human-to-Human Speech Quality. *In: Proc. 3rd Int. Conf. on Language Resources and Evaluation (LREC 2002)*, 4:1177–1184, ES–Las Palmas.

Möller, S. and Raake, A. (2002). Telephone Speech Quality Prediction: Towards Network Planning and Monitoring Models for Modern Network Scenarios. *Speech Communication*, 38:47–75.

Möller, S. and Riedel, J. (1999). Expectation in Quality Assessment of Internet Telephony. *Joint Meeting ASA/EAA/DEGA, Forum Acusticum 1999, ACUSTICA/acta acustica*, 85:Suppl. 1, S48, D–Berlin.

Möller, S. and Skowronek, J. (2003a). Einfluss von Spracherkennung und Sprachsynthese auf die Qualität natürlichsprachlicher Dialogsysteme. *In: Fortschritte der Akustik – DAGA 2003*, 726–727, DEGA e.V., D–Oldenburg.

Möller, S. and Skowronek, J. (2003b). Quantifying the Impact of System Characteristics on Perceived Quality Dimensions of a Spoken Dialogue Service. *In: Proc. 8th Europ. Conf. on Speech Communication and Technology (EUROSPEECH 2003 – Switzerland)*, 3:1953–1956, CH–Geneva.

Moore, R. K. (1997). Users Guide. *In: Handbook on Standards and Resources for Spoken Language Systems*, D. Gibbon, R. Moore, and R. Winski, eds., 1–28, Mouton de Gruyter, D–Berlin.

Moore, R. K. (1998). Understanding Speech Understanding. *In: Proc. Int. Conf. Acoustics Speech and Signal Processing (ICASSP'98)*, 2:1049–1052, IEEE, USA–Piscataway NJ.

Moore, R. K. (1985). Evaluating Speech Recognizers. *IEEE Trans. Acoustics Speech and Signal Processing*, ASSP-25(2):178–183.

Moore, R. K. and Cutler, A. (2001). Constraints on Theories of Human vs. Machine Recognition of Speech. *In: Proc. Workshop on Speech Recognition as Pattern Classification (SPRAAC)*, Jonkerbosch Conf. Center, NL-Nijmegen.

Moreno, P. J. and Stern, R. M. (1994). Sources of Degradation of Speech Recognition in the Telephone Network. *In: Proc. 3rd Int. Conf. on Spoken Language Processing (ICSLP'94)*, 1:109–112, JP–Yokohama.

Moulines, E. and Charpentier, F. (1990). Pitch Synchronous Waveform Processing Techniques for Text-to-Speech Synthesis Using Diphones. *Speech Communication*, 9:453–467.

Naito, M., Kuroiwa, S., Takeda, K., Yamamoto, S., and Yato, F. (1995). A Real-Time Speech Dialogue System for a Voice-Activated Telephone Extension Service. *In: Proc. ESCA Workshop on Spoken Dialogue Systems*, P. Dalsgaard, L.B. Larsen, L. Boves, and I. Thomsen, eds., 129–132, DK–Vigsø.

Narayanan, S., Subramaniam, M., Stern, B., Hollister, B., and Lin, C.-M. (1998). Probing the Relationship between Qualitative and Quantitative Performance Measures for Voice-Enabled Telecommunication Services. *In: Proc. Int. Conf. Acoustics Speech and Signal Processing (ICASSP'98)*, 6:3769–3772, IEEE, USA–Piscataway NJ.

Nebbia, L., Quazza, S., and Salza, P. L. (1998). A Specialised Speech Synthesis Technique for Application to Automatic Reverse Directory Service. *In: Proc. IEEE 4th Workshop Interactive Voice Technology for Telecommunications Applications (IVTTA'98)*, 223–228, I–Torino.

Netter, K., Armstrong, S., Kiss, T., Klein, J., Lehmann, S., Milward, D., Regnier-Prost, S., Schäler, R., and Wegst, T. (1998). DiET – Diagnostic and Evaluation Tools for Natural Language Processing Applications. *In: Proc. 1st Int. Conf. on Language Resources and Evaluation (LREC'98)*, 1:573–579, ES–Granada.

Neumeyer, L. G., Digalakis, V. V., and Weintraub, M. (1994). Training Issues and Channel Equalization Techniques for the Construction of Telephone Acoustic Models Using a High-Quality Speech Corpus. *IEEE Trans. Speech and Audio Processing*, 2(4):590–597.

Niculescu, A. (2002). *Mensch-Maschine Kommunikation: Eine Rezipientenstudie zur Identifizierung und Bewertung qualitätsbeeinflussender Faktoren der Mensch-Maschine Kommunikation am Beispiel von BoRIS, einem automatischen sprachbasierten Restaurant-Auskunftssystem.* Magister thesis (unpublished), Fakultät für Philosophie und Publizistik, Ruhr-Universität, D-Bochum.

Nielsen, J. (1994). Heuristic Evaluation. *In: Usability Inspection Methods*, J. Nielsen and R. L. Mack, eds., 25–62, John Wiley & Sons, USA–New York NY.

Nielsen, J. and Mack, R. L., editors (1994). *Usability Inspection Methods.* John Wiley & Sons, USA-New York NY.

NIST HUB-5 Evaluation Plan (2000). *The 2000 NIST Evaluation Plan for Recognition of Conversational Speech over the Telephone.* National Institute of Standards and Technology, `http://www.nist.gov/speech/tests`, USA-Gaithersburg MD, edition 1.3.

NIST HUB-5 Evaluation Plan (2001). *The 2001 NIST Evaluation Plan for Recognition of Conversational Speech over the Telephone.* National Institute of Standards and Technology, `http://www.nist.gov/speech/tests`, USA-Gaithersburg MD, edition 1.1.

NIST Speech Recognition Scoring Toolkit (2001). *Speech Recognition Scoring Toolkit.* National Institute of Standards and Technology, `http://www.nist.gov/speech/tools`, USA-Gaithersburg MD.

Nöth, E., Batliner, A., Warnke, V., Haas, J., Boros, M., Buckow, J., Huber, R., Gallwitz, F., Nutt, M., and Niemann, H. (2002). On the Use of Prosody in Automatic Dialogue Understanding. *Speech Communication*, 36:45–62.

Nouza, J. and Holada, M. (1998). A City Information System Operating over the Phone. *In: Proc. IEEE 4th Workshop Interactive Voice Technology for Telecommunications Applications (IVTTA'98)*, 141–144, I–Torino.

Oglesby, J. (1995). What's a Number? Moving Beyond the Equal Error Rate. *Speech Communication*, 17:193–208.

Oria, D. and Koskinen, E. (2002). E-Mail Goes Mobile: The Design and Implementation of a Spoken Language Interface to E-Mail. *In: Proc. 7th Int. Conf. on Spoken Language Processing (ICSLP-2002)*, 4:2697–2700, USA–Denver CO.

Osgood, C. E., Suci, G., and Tannenbaum, P. (1957). *The Measurement of Meaning*. University of Illinois Press, USA-Urbana IL.

O'Shaughnessy, D. (2000). *Speech Communications: Human and Machine*. IEEE Press, USA-New York NY.

Paek, T. (2001). Empirical Methods for Evaluating Dialogue Systems. *In: Proc. 2nd SIGdial Workshop on Discourse and Dialogue*, J. van Kuppevelt and R. Smith, eds., 100–107, DK–Aalborg.

Pallett, D. S., Fisher, W. M., Fiscus, J. G., and Garofolo, J. S. (1990a). DARPA ATIS Test Results. *In: Proc. DARPA Speech and Natural Language Workshop*, 114–121, USA–Hidden Valley PA.

Pallett, D. S. (1985). Performance Assessment of Automatic Speech Recognizers. *J. of Research of the National Bureau of Standards*, 90(5):371–387.

Pallett, D. S. (1998). The NIST Role in Automatic Speech Recognition Benchmark Tests. *In: Proc. 1st Int. Conf. on Language Resources and Evaluation (LREC'98)*, 1:327–330, ES–Granada.

Pallett, D. S., Fiscus, J. G., Fisher, W. M., and Garofolo, J. S. (1993). Benchmark Tests for the DARPA Spoken Language Program. *In: Proc. DARPA Human Language Technology Workshop*, 7–18, USA–Princeton NJ.

Pallett, D. S., Fisher, W. M., and Fiscus, J. G. (1990b). Tools for the Analysis of Benchmark Speech Recognition Tests. *In: Proc. Int. Conf. Acoustics Speech and Signal Processing (ICASSP'90)*, 1:97–100, IEEE, USA–Piscataway NJ.

Pallett, D. S. and Fourcin, A. (1997). Speech Input: Assessment and Evaluation. *In: Survey of the State of the Art in Human Language Technology*, R. Cole, J. Mariani, A. Uszkoreit, H. Zaenen, and V. Zue, eds., 425–429, Cambridge University Press and Giardini Editori, I–Pisa.

Pavlovic, C., Sorin, C., Roumiguiere, J. P., and Lucas, J. P. (1990). Cross Validation Between a Magnitude-Estimation Technique and a Pair Comparison Technique for Assessing Text-to-Speech Synthesis Systems. *J. d'Acoustique*, 3:75–83.

Peckham, J. (1989). VODIS – A Voice Operated Database Inquiry System. *In: Recent Developments and Applications of Natural Language Processing*, J. Peckham, ed., 117–128, Kogan Page, UK–London.

Peckham, J. (1991). Speech Understanding and Dialogue over the Telephone: An Overview of the ESPRIT SUNDIAL Project. *In: Proc. of the DARPA Workshop on Speech and Language*, 14–27, USA–Pacific Grove CA.

Peckham, J. and Fraser, N. M. (1994). Spoken Language Dialogue over the Telephone. *In: Progress and Prospects of Speech Research and Technology*, H. Niemann, R. de Mori, and G. Hanrieder, eds., 192–203, Infix, D–Sankt Augustin.

Peckham, J., Thomas, T., and Frangoulis, E. (1990). Recognizer Sensitivity Analysis: A Method for Assessing the Performance of Speech Recognizers. *Speech Communication*, 9:317–327.

Peláez-Moreno, C., Gallardo-Antolín, A., and Díaz-de-María, F. (2001). Recognizing Voice over IP: A Robust Front-End for Speech Recognition on the World Wide Web. *IEEE Trans. Multimedia*, 3(2):209–218.

Pellegrini, C. S. (2003). Quality Evaluation of Bochum Restaurant Information System: A Spoken Dialogue System. *In: Proc. ISCA Tutorial and Research Workshop on Auditory Quality of Systems (AQS 2003)*, U. Jekosch and S. Möller, eds., 134–138, D–Mont–Cenis.

Picone, J., Doddington, G. R., and Pallett, D. S. (1990). Phone-Mediated Word Alignment for Speech Recognition Evaluation. *IEEE Trans. Acoustics, Speech, and Signal Processing*, 38(3):559–562.

Picone, J., Goudie-Marshall, K. M., Doddington, G. R., and Fisher, W. (1986). Automatic Text Alignment for Speech System Evaluation. *IEEE Trans. Acoustics, Speech, and Signal Processing*, 34(4):780–784.

Pirker, H., Loderer, G., and Trost, H. (1999). Thus Spoke the User to the Wizard. *In: Proc. 6th Europ. Conf. on Speech Communication and Technology (EUROSPEECH'99)*, 3:1171–1174, H–Budapest.

Polifroni, J., Hirschman, L., Seneff, S., and Zue, V. (1992). Experiments in Evaluating Interactive Spoken Language Systems. *In: Proc. DARPA Speech and Natural Language Workshop*, 28–33, USA–Harriman CA.

Polifroni, J. and Seneff, S. (2000). Galaxy-II as an Architecture for Spoken Dialogue Evaluation. *In: Proc. 2nd Int. Conf. on Language Resources and Evaluation (LREC 2000)*, 2:725–730, GR–Athens.

Polifroni, J., Seneff, S., Glass, J., and Hazen, T. J. (1998). Evaluation Methodology for a Telephone-Based Conversational System. *In: Proc. 1st Int. Conf. on Language Resources and Evaluation (LREC'98)*, 1:43–49, ES–Granada.

Popovici, C. and Baggia, P. (1997). Specialized Language Models Using Dialogue Predictions. *In: Proc. Int. Conf. Acoustics Speech and Signal Processing (ICASSP'97)*, 2:815–818, IEEE Sign. Proc. Soc., USA–Piscataway NJ.

Portele, T., Goronzy, S., Emele, M., Kellner, A., Torge, S., and te Vrugt, J. (2003). SmartKom-Home – An Advanced Multi-Modal Interface to Home Entertainment. *In: Proc. 8th Europ. Conf. on Speech Communication and Technology (EUROSPEECH 2003 – Switzerland)*, 3:1897–1900, CH–Geneva.

Prasad, R. and Walker, M. (2002). Training a Dialogue Act Tagger for Human-Human and Human-Computer Travel Dialogues. *In: Proc. Third SIGdial Workshop on Discourse and Dialogue*, 162–173, USA–Philadelphia PA.

Price, P. J. (1990). Evaluation of Spoken Language Systems: The ATIS Domain. *In: Proc. DARPA Speech and Natural Language Workshop*, 91–95, USA–Hidden Valley PA.

Price, P. J., Hirschman, L., Shriberg, E., and Wade, E. (1992). Subject-Based Evaluation Measures for Interactive Spoken Language Systems. *In: Proc. DARPA Speech and Natural Language Workshop*, 34–39, USA–Harriman CA.

Puel, J.-B. and André-Obrecht, R. (1997). Cellular Phone Speech Recognition: Noise Compensation vs. Robust Architectures. *In: Proc. 5th Europ. Conf. on Speech Communication and Technology (EUROSPEECH'97)*, 3:1151–1154, GR–Rhodes.

Raake, A. (2003). Speech Quality of Heterogeneous Networks Involving VoIP: Are Time Varying Impairments Additive to Classical Stationary Ones? *In: Proc. ISCA Tutorial and Research Workshop on Auditory Quality of Systems (AQS 2003)*, U. Jekosch and S. Möller, eds., 63–70, D–Mont–Cenis.

Raake, A. (2004). *Assessment and Parametric Modelling of Speech Quality in Voice-over-IP Networks*. Doctoral dissertation, Institut für Kommunikationsakustik, Ruhr-Universität, D-Bochum.

Raake, A. and Möller, S. (1999). *Analysis and Verification of the Tektronix M366 GSM Network QoS Analyser's Measurements and Quality Predictions*. Final report (unpublished), Institut für Kommunikationsakustik, Ruhr-Universität, D-Bochum.

Rabiner, L. R., Juang, B.-H., and Lee, C.-H. (1996). An Overview of Automatic Speech Recognition. *In: Automatic Speech and Speaker Recognition*, C.–H. Lee, F. K. Soong, and K. K. Paliwal, eds., 1–30, Kluwer Academic Publ., USA–Boston.

Rahim, M., Pieaccini, R., Eckert, W., Levin, E., di Fabbrizio, G., Riccardi, G., Kamm, C., and Narayanan, S. (2000). A Spoken Dialogue System for Conference/Workshop Services. *In:*

Proc. 6th Int. Conf. on Spoken Language Processing (ICSLP 2000), 3:1041–1044, CHN–Beijing.

Rajman, M., Rajman, A., Seydoux, F., and Trutnev, A. (2003). Assessing the Usability of a Dialogue Management System Designed in the Framework of a Rapid Dialogue Prototyping Methodology. *In: Proc. ISCA Tutorial and Research Workshop on Auditory Quality of Systems (AQS 2003)*, U. Jekosch and S. Möller, eds., 126–133, D–Mont–Cenis.

Ramshaw, L. A. and Boisen, S. (1990). An SLS Answer Comparator. *SLS Note 7*, BBN Systems and Technologies Corp., USA-Cambridge MA.

Rao, A., Roth, B., Nagesha, V., McAllaster, D., Liberman, N., and Gillick, L. (2000). Large Vocabulary Continuous Speech Recognition of Read Speech over Cellular and Landline Networks. *In: Proc. 6th Int. Conf. on Spoken Language Processing (ICSLP 2000)*, 4:402–405, CHN–Beijing.

Rehmann, S. (1999). *Entwurf und Implementierung einer Versuchsumgebung zum Testen von Sprachsynthese und Spracherkennung in einer Dialogsituation.* Study project report (unpublished), Institut für Kommunikationsakustik, Ruhr-Universität, D-Bochum.

Rehmann, S. (2001). *Sprachqualität von Voice-over-Internet-Protocol-(VoIP)-Verbindungen bei zeitlich instationären Störungen.* Diploma thesis (unpublished), Institut für Kommunikationsakustik, Ruhr-Universität, D-Bochum.

Rehmann, S., Raake, A., and Möller, S. (2002). Parametric Simulation of Impairments Caused by Telephone and Voice over IP Network Transmission. *In: Proc. 3rd European Congress on Acoustics (Forum Acusticum Sevilla 2002), Special Issue Revista de Acústica*, 33:6 pages, ES–Sevilla.

Reilly, R. (1987). Ill-Formedness and Mis-Communication in Person-Machine Dialogue. *Information and Software Technology*, 29:69–74.

Relaño Gil, J., Tapias, D., Villar, J. M., Gancedo, M., and Hernández, L. A. (1999). Flexible Mixed-Initiative Dialogue for Telephone Services. *In: Proc. 6th Europ. Conf. on Speech Communication and Technology (EUROSPEECH'99)*, 3:1179–1182, H–Budapest.

Richards, D. L. (1973). *Telecommunication by Speech.* Butterworths, UK-London.

Richards, M. A. and Underwood, K. M. (1984). Talking to Machines. How are People Naturally Inclined to Speak? *In: Contemporary Ergonomics. Proc. of the Ergonomics Society's Conference*, E. D. Megaw, ed., 62–67, Taylor and Francis, UK–London.

Rix, A. W., Berger, J., and Beerends, J. G. (2003). Perceptual Quality Assessment of Telecommunications Systems Including Terminals. *In: Preprints of the 114th Convention of the Audio Eng. Soc.*, paper 5724, NL–Amsterdam, 22–25 March.

Rix, A. W., Hollier, M. P., Hekstra, A. P., and Beerends, J. G. (2002). Perceptual Evaluation of Speech Quality (PESQ) – The New ITU Standard for End-to-End Speech Quality Assessment. Part I – Time-Delay Compensation. *J. Audio Eng. Soc.*, 50(10):755–764.

Roe, D. B. and Riley, M. D. (1994). Prediction of Word Confusabilities for Speech Recognition. *In: Proc. 3rd Int. Conf. on Spoken Language Processing (ICSLP'94)*, 1:227–230, JP–Yokohama.

Rosset, S., Bennacef, S., and Lamel, L. (1999). Design Strategies for Spoken Language Dialogue Systems. *In: Proc. 6th Europ. Conf. on Speech Communication and Technology (EUROSPEECH'99)*, 4:1535–1538, H–Budapest.

Rothkrantz, L. J. M., van Vark, R. J., and Koppelaar, H. (1997). Corpus-Based Test System for an Automated Speech Processing System. *In: Proc. Speech and Language Technology (SALT) Club Workshop on Evaluation in Speech and Language Technology*, 164–171, UK–Sheffield.

Rudnicky, A. I. (2002). The DARPA Communicator Programme. *elsnews*, 10.4:13.

Salza, P. L., Foti, E., Nebbia, L., and Oreglia, M. (1996). MOS and Pair Comparison Combined Methods for Quality Evaluation of Text-to-Speech Systems. *ACUSTICA/acta acustica*, 82:650–656.

San-Segundo, R., Montero, J. M., Colás, J., Gutiérrez, J., Ramos, J. M., and Pardo, J. M. (2001a). Methodology for Dialogue Design in Telephone-Based Spoken Dialogue Systems: A Spanish Train Information System. *In: Proc. 7th Europ. Conf. on Speech Communication and Technology (EUROSPEECH 2001 – Scandinavia)*, 3:2165–2168, DK–Aalborg.

San-Segundo, R., Montero, J. M., Gutiérrez, J., Gallardo, A., Romeral, J. D., and Pardo, J. M. (2001b). A Telephone-Based Railway Information System for Spanish: Development of a Methodology for Spoken Dialogue Design. *In: Proc. 2nd SIGdial Workshop on Discourse and Dialogue*, J. van Kuppevelt and R. Smith, eds., 140–148, DK–Aalborg.

Sanderman, A., Sturm, J., den Os, E., Boves, L., and Cremers, A. (1998). Evaluation of the Dutch Train Timetable Information System Developed in the ARISE Project. *In: Proc. IEEE 4th Workshop Interactive Voice Technology for Telecommunications Applications (IVTTA'98)*, 91–96, I–Torino.

Sanders, G. A., Le, A. N., and Garofolo, J. S. (2002). Effects of Word Error Rate in the DARPA Communicator Data During 2000 and 2001. *In: Proc. 7th Int. Conf. on Spoken Language Processing (ICSLP-2002)*, 1:277–280, USA–Denver CO.

Schaden, S. (2003). Rule-Based Lexical Modelling of Foreign-Accented Pronunciation Variants. *In: Proc. of the 10th Conference of the European Chapter of the Association for Computational Linguistics (EACL '03)*, H–Budapest.

Searle, J. R. (1969). *Speech Acts. An Essay in the Philosophy of Language*. Cambridge University Press, USA-Cambridge MA.

Searle, J. R. (1979). *Expression and Meaning. Studies in the Theory of Speech Acts*. Cambridge University Press, USA-New York NY.

Seide, F. and Kellner, A. (1997). Toward an Automated Directory Information System. *In: Proc. 5th Europ. Conf. on Speech Communication and Technology (EUROSPEECH'97)*, 3:1327–1330, GR–Rhodes.

Seneff, S. (1998). Galaxy-II: A Reference Architecture for Conversational System Development. *In: Proc. 5th Int. Conf. on Spoken Language Processing (ICSLP'98)*, 3:931–934, AUS–Sydney.

Shin, J., Narayanan, S., Gerber, L., Kazemzadeh, A., and Byrd, D. (2002). Analysis of User Behavior Under Error Conditions in Spoken Dialogs. *In: Proc. 7th Int. Conf. on Spoken Language Processing (ICSLP-2002)*, 3:2069–2072, USA–Denver CO.

Shneiderman, B. (1992). *Designing the User Interface: Strategies for Effective Human-Computer Interaction*. Addison Wesley, USA-Reading MA.

Shriberg, E., Wade, E., and Price, P. (1992). Human-Machine Problem Solving Using Spoken Language Systems (SLS): Factors Affecting Performance and User Satisfaction. *In: Proc. DARPA Speech and Natural Language Workshop*, 49–54, USA–Harriman CA.

Siegel, S. and Castellan, N. J. (1988). *Nonparametric Statistics for the Behavioral Sciences*. McGraw-Hill, USA-New York NY.

Sikorski, T. and Allen, J. F. (1997). A Task-Based Evaluation of the TRAINS-95 Dialogue System. *In: Dialogue Processing in Spoken Language Systems*, ECAI'96 Workshop, H–Budapest, E. Maier, M. Mast, and S. LuperFoy, eds., Lecture Notes in Artificial Intelligence No. 1236, 207–220, Springer, D–Berlin.

Silverman, K., Basson, S., and Levas, S. (1990). Evaluating Synthesiser Performance: Is Segmental Intelligibility Enough? *In: Proc. 1st Int. Conf. on Spoken Language Processing (ICSLP'90)*, 2:981–984, JP–Kobe.

Simpson, A. and Fraser, N. M. (1993). Black Box and Glass Box Evaluation of the SUNDIAL System. *In: Proc. 3rd Europ. Conf. on Speech Communication and Technology (EUROSPEECH'93)*, 2:1423–1426, D–Berlin.

Skowronek, J. (2002). *Entwicklung von Modellierungsansätzen zur Vorhersage der Dienstequalität bei der Interaktion mit einem natürlichsprachlichen Dialogsystem.* Diploma thesis (unpublished), Institut für Kommunikationsakustik, Ruhr-Universität, D-Bochum.

Smeele, P. M. T. and Waals, J. A. J. S. (2003). Evaluation of a Speech-Driven Telephone Information Service Using the PARADISE Framework: A Closer Look at Subjective Measures. *In: Proc. 8th Europ. Conf. on Speech Communication and Technology (EUROSPEECH 2003 – Switzerland)*, 3:1949–1952, CH–Geneva.

Smith, R. W. and Gordon, S. A. (1997). Effects of Variable Initiative on Linguistic Behavior in Human-Computer Spoken Natural Language Dialogue. *Computational Linguistics*, 23(1):141–168.

Sonntag, G. P. and Portele, T. (1997). A Method for Prosody Evaluation. *In: Proc. Speech and Language Technology (SALT) Club Workshop on Evaluation in Speech and Language Technology*, 188–194, UK–Sheffield.

South, C. and Usai, P. (1992). Subjective Performance of CCITT's 16 kbit/s LD-CELP Algorithm with Voice Signals. *In: Communication for Global Users, IEEE Global Telecommunications Conf. GLOBECOM '92*, USA–Orlando FL, 1709–1714, USA–Piscataway NJ.

Souvignier, B., Kellner, A., Rueber, B., Schramm, H., and Seide, F. (2000). The Thoughtful Elephant: Strategies for Spoken Dialog Systems. *IEEE Trans. Speech and Audio Processing*, 8(1):51–62.

Sparck-Jones, K. and Gallier, J. R. (1996). *Evaluating Natural Language Processing Systems: An Analysis and Review*. Lecture Notes in Artificial Intelligence No. 1083, Springer, D-Berlin.

Spiegel, M. F., Altom, M. J., Macchi, M. J., and Wallace, K. L. (1990). Comprehensive Assessment of the Telephone Intelligibility of Synthesized and Natural Speech. *Speech Communication*, 9:279–291.

Sproat, R., editor (1997). *Multilingual Text-to-Speech Synthesis: The Bell Labs Approach.* Kluwer Academic Publ., USA-Boston MA.

Steeneken, H. J. M. and van Leeuwen, D. A. (1995). Multi-Lingual Assessment of Speaker Independent Large Vocabulary Speech-Recognition Systems: The SQALE-Project. *In: Proc. 4th Europ. Conf. on Speech Communication and Technology (EUROSPEECH'95)*, 2:1271–1274, ES–Madrid.

Steeneken, H. J. M. and van Velden, J. G. (1989a). Objective and Diagostic Assessment of (Isolated) Word Recognizers. *In: Proc. Int. Conf. Acoustics Speech and Signal Processing (ICASSP'89)*, 1:540–543, IEEE, USA–Piscataway NJ.

Steeneken, H. J. M. and van Velden, J. G. (1989b). RAMOS: Recognizer Assessment by Means of Manipulation Of Speech. *In: Proc. Europ. Conf. on Speech Communication and Technology (EUROSPEECH 89)*, 2:316–319, F–Paris.

Steeneken, H. J. M. and Varga, A. (1993). Assessment for Automatic Speech Recognition: I. Comparison of Assessment Methods. *Speech Communication*, 12:241–246.

Steffens, J. and Paulus, E. (2000). Speech Synthesis Quality Assessment. *In: Verbmobil: Foundations of Speech-to-Speech Translation*, W. Wahlster, ed., 592–596, Springer, D–Berlin.

Stent, A., Walker, M., Whittaker, S., and Maloor, P. (2002). User-Tailored Generation for Spoken Dialogue: An Experiment. *In: Proc. 7th Int. Conf. on Spoken Language Processing (ICSLP-2002)*, 2:1281–1284, USA–Denver CO.

Strik, H., Cucchiarini, C., and Kessens, J. M. (2000). Comparing the Recognition Performance of CSRs: In Search of an Adequate Metric and Statistical Significance Test. *In: Proc. 6th Int. Conf. on Spoken Language Processing (ICSLP 2000)*, 4:740–743, CHN–Beijing.

Strik, H., Cucchiarini, C., and Kessens, J. M. (2001). Comparing the Performance of Two CSRs: How to Determine the Significance Level of the Differences. *In: Proc. 7th Europ. Conf. on Speech Communication and Technology (EUROSPEECH 2001 – Scandinavia)*, 3:2091–2094, DK–Aalborg.

Sturm, J., Bakx, I., Cranen, B., Terken, J., and Wang, F. (2002a). The Effect of Prolonged Use of Multimodal Interaction. *In: Proc. ISCA Workshop on Multi-Modal Dialogue in Mobile Environments*, L. Dybkjaer, E. André, W. Minker, and P. Heisterkamp, eds., 1–15, D–Kloster Irsee.

Sturm, J., Bakx, I., Cranen, B., Terken, J., and Wang, F. (2002b). Usability Evaluation of a Dutch Multimodal System for Train Timetable Information. *In: Proc. 3rd Int. Conf. on Language Resources and Evaluation (LREC 2002)*, 1:255–261, ES–Las Palmas.

Sturm, J., den Os, E., and Boves, L. (1999). Issues in Spoken Dialogue Systems: Experiences with the Dutch ARISE System. *In: Proc. ESCA Workshop on Interactive Dialogue in Multi-Modal Systems*, P. Dalsgaard, C.–H. Lee, P. Heisterkamp, and R. Cole, eds., 1–4, D–Kloster Irsee.

Suhm, B. (2003). Towards Best Practices for Speech User Interface Design. *In: Proc. 8th Europ. Conf. on Speech Communication and Technology (EUROSPEECH 2003 – Switzerland)*, 3:2217–2220, CH–Geneva.

Sundheim, B. (2001). Assessment of Text Analysis Technologies: How "Message Understanding" Came to Mean "Information Extraction". *Presentation at the International Bullet Course on Speech and Language Engineering Evaluation*, July 2–3, LIMSI–CNRS, F–Paris.

Sutton, S., Cole, R., de Villiers, J., Schalkwyk, J., Vermeulen, P., Macon, M., Yan, Y., Kaiser, E., Rundle, B., Shobaki, K., Hosom, P., Kain, A., Wouters, J., Massaro, M., and Cohen, M. (1998). Universal Speech Tools: The CSLU Toolkit. *In: Proc. 5th Int. Conf. on Spoken Language Processing (ICSLP'98)*, 7:3221–3224, AUS–Sydney.

Sutton, S., Hansen, B., Lander, T., Novick, D. G., and Cole, R. (1995). Evaluating the Effectiveness of Dialogue for an Automated Spoken Questionnaire. *In: Empirical Methods in Discourse Interpretation and Generation. Papers from the 1995 AAAI Symposium, USA-Stanford CA*, 156–161, AAAI Press, USA–Menlo Park CA.

Sutton, S., Novick, D. G., Cole, R., Vermeulen, P., de Villiers, J., Schalkwyk, J., and Fanty, M. (1996). Building 10,000 Spoken Dialogue Systems. *In: Proc. 4th Int. Conf. on Spoken Language Processing (ICSLP'96)*, H.T. Bunnell and W. Idsardi, eds., 2:709–712, IEEE, USA–Piscataway NJ.

Tamura, M., Masuko, T., Tokuda, K., and Kobayashi, T. (2001). Adaptation of Pitch and Spectrum for HMM-Based Speech Synthesis Using MLLR. *In: Proc. Int. Conf. Acoustics Speech and Signal Processing (ICASSP 2001)*, 2:805–808, IEEE, USA–Piscataway NJ.

Tarcisio, C., Daniele, F., Roberto, G., and Marco, O. (1999). Use of Simulated Data for Robust Telephone Speech Recognition. *In: Proc. 6th Europ. Conf. on Speech Communication and Technology (EUROSPEECH'99)*, 6:2825–2828, H–Budapest.

Tatham, M. and Morton, K. (1995). Speech Synthesis in Dialogue Systems. *In: Proc. ESCA Workshop on Spoken Dialogue Systems*, P. Dalsgaard, L.B. Larsen, L. Boves, and I. Thomsen, eds.:221–224, DK–Vigsø.

Thyfault, M. E. (1999). Voice Gets Reliable. *Information Week Online*, Feb. 22.

Trancoso, I., Ribeiro, C., Rodrigues, R., and Rosa, M. (1995). Issues in Speech Recognition Applied to Directory Listing Retrieval. *In: Proc. ESCA Workshop on Spoken Dialogue Systems*, P. Dalsgaard, L.B. Larsen, L. Boves, and I. Thomsen, eds., 49–52, DK–Vigsø.

Traum, D. R. (1994). *A Computational Theory of Grounding in Natural Language Conversation.* PhD thesis, University of Rochester, USA-Rochester NY.

Traum, D. R., Robinson, S., and Stephan, J. (2004). Evaluation of Multi-Party Virtual Reality Dialogue Interaction. *In: Proc. 4th Int. Conf. on Language Resources and Evaluation (LREC 2004)*, 5:1699–1702, P–Lisbon.

Trias-Sanz, R. and Mariño, J. B. (2002). Basurde[lite], a Machine-Driven Dialogue System for Accessing Railway Timetable Information. *In: Proc. 7th Int. Conf. on Spoken Language Processing (ICSLP-2002)*, 4:2685–2688, USA–Denver CO.

Trutnev, A., Ronzenknop, A., and Rajman, M. (2004). Speech Recognition Simulation and its Application for Wizard-of-Oz Experiments. *In: Proc. 4th Int. Conf. on Language Resources and Evaluation (LREC 2004)*, 2:611–614, P–Lisbon.

Tukey, J. W. (1977). *Exploratory Data Analysis.* Addison-Wesley, USA-Reading MA.

Turing, A. M. (1950). Computing Machinery and Intelligence. *Mind*, 59:433–460.

Turunen, J. and Vlaj, D. (2001). A Study of Speech Coding Parameters in Speech Recognition. *In: Proc. 7th Europ. Conf. on Speech Communication and Technology (EUROSPEECH 2001 – Scandinavia)*, 4:2363–2366, DK–Aalborg.

van Bezooijen, R. and Pols, L. C. W. (1990). Evaluating Text-to-Speech Systems: Some Methodological Aspects. *Speech Communication*, 9:263–270.

van Bezooijen, R. and van Heuven, V. (1997). Assessment of Synthesis Systems. *In: Handbook on Standards and Resources for Spoken Language Systems*, D. Gibbon, R. Moore, and R. Winski, eds., 481–563, Mouton de Gruyter, D–Berlin.

van Leeuwen, D. and Steeneken, H. (1997). Assessment of Recognition Systems. *In: Handbook on Standards and Resources for Spoken Language Systems*, D. Gibbon, R. Moore, and R. Winski, eds., 381–407, Mouton de Gruyter, D–Berlin.

van Leeuwen, D. A. (2003). Speaker Verification Systems and Security Considerations. *In: Proc. 8th Europ. Conf. on Speech Communication and Technology (EUROSPEECH 2003 – Switzerland)*, 3:1661–1664, CH–Geneva.

van Leeuwen, D. A. and de Louwere, M. (1999). Objective and Subjective Evaluation of the Acoustic Models of a Continuous Speech Recognition System. *In: Proc. 6th Europ. Conf. on Speech Communication and Technology (EUROSPEECH'99)*, 2:1915–1918, H–Budapest.

van Santen, J. P. H. (1993). Perceptual Experiments for Diagnostic Testing of Text-to-Speech Systems. *Computer Speech and Language*, 7:49–100.

van Santen, J. P. H., Sproat, R. W., Olive, J. P., and Hirschberg, J. (1997). *Progress in Speech Synthesis.* Springer, USA-New York NY.

Vary, P., Heute, U., and Hess, W. (1998). *Digitale Sprachsignalverarbeitung.* Teubner, D-Stuttgart.

Västfjäll, D. (2003). Tapping into the Personal Experience of Quality: Expectation-Based Sound Quality Evaluation. *In: Proc. ISCA Tutorial and Research Workshop on Auditory Quality of Systems (AQS 2003)*, U. Jekosch and S. Möller, eds., 24–28, D–Mont–Cenis.

Veldhuijzen van Zanten, G. (1998). Adaptive Mixed-Initiative Dialogue Management. *In: Proc. IEEE 4th Workshop Interactive Voice Technology for Telecommunications Applications (IVTTA '98)*, 65–70, I–Torino.

Veldhuijzen van Zanten, G. (1999). User Modelling in Adaptive Dialogue Management. *In: Proc. 6th Europ. Conf. on Speech Communication and Technology (EUROSPEECH'99)*, 3:1183–1186, H–Budapest.

Viikki, O., editor (2001). *Speech Communication. Special Issue on Noise Robust ASR*. 34(1-2), Elsevier, NL-Amsterdam.

Wahlster, W., editor (2000). *Verbmobil: Foundations of Speech-to-Speech Translation*. Springer, D-Berlin.

Wahlster, W. and Kobsa, A. (1989). User Models in Dialog Systems. *User Models in Dialog Systems*, A. Kobsa and W. Wahlster, eds., 4–34, Springer, D–Berlin.

Wahlster, W., Reithinger, N., and Blocher, A. (2001). SmartKom: Multimodal Communication with a Life-Like Character. *In: Proc. 7th Europ. Conf. on Speech Communication and Technology (EUROSPEECH 2001 – Scandinavia)*, 3:1547–1550, DK–Aalborg.

Walker, M. A., Litman, D. J., Kamm, C. A., and Abella, A. (1997). PARADISE: A Framework for Evaluating Spoken Dialogue Agents. *In: Proc. of the ACL/EACL 35th Ann. Meeting of the Assoc. for Computational Linguistics*, 271–280, ES–Madrid.

Walker, M., Kamm, C., and Litman, D. (2000a). Towards Developing General Models of Usability with PARADISE. *Natural Language Engineering*, 6(3-4):363–377.

Walker, M., Kamm, C., and Boland, J. (2000b). Developing and Testing General Models of Spoken Dialogue System Performance. *In: Proc. 2nd Int. Conf. on Language Resources and Evaluation (LREC 2000)*, 1:189–196, GR–Athens.

Walker, M. A. (1993). *Informational Redundancy and Resource Bounds in Dialogue*. PhD thesis, University of Pennsylvania, USA-Pennsylvania PA.

Walker, M. A. (1994). Experimentally Evaluating Communicative Strategies: The Effect of the Task. *In: Proc. Conf. Am. Assoc. Artificial Intelligence (AAAI'94)*, 86–93, Assoc. for Computing Machinery (ACM), USA–New York NY.

Walker, M. A. (2004). Can We Talk? Prospects for Automatically Training Spoken Dialogue Systems. *In: Proc. 4th Int. Conf. on Language Resources and Evaluation (LREC 2004)*, 3:XVII, P–Lisbon.

Walker, M. A., Fromer, J., di Fabbrizio, G., Mestel, C., and Hindle, D. (1998a). What Can I Say?: Evaluating a Spoken Language Interface to Email. *In: Human Factors in Computing Systems. CHI'98 Conference Proc.*, USA–Los Angeles CA, 582–589, Assoc. for Computing Machinery (ACM), USA–New York NY.

Walker, M. A., Litman, D. J., Kamm, C. A., and Abella, A. (1998b). Evaluating Spoken Dialogue Agents with PARADISE: Two Case Studies. *Computer Speech and Language*, 12(3):317–347.

Walker, M. A., Passonneau, R., and Boland, J. E. (2001). Quantitative and Qualitative Evaluation of DARPA Communicator Spoken Dialogue Systems. *In: Proc. of the ACL/EACL 39th Ann. Meeting of the Assoc. for Computational Linguistics*, 515–522.

Walker, M. A., Rudnicky, A., Aberdeen, J., Bratt, E. O., Garofolo, J., Hastie, H., Le, A., Pellom, B., Potamianos, A., Passonneau, R., Prasad, R., Roukos, S., Sanders, G., Seneff, S., and Stallard, D. (2002a). DARPA Communicator Evaluation: Progress from 2000 to 2001. *In: Proc. 7th Int. Conf. on Spoken Language Processing (ICSLP-2002)*, 1:273–276, USA–Denver CO.

Walker, M. A., Rudnicky, A., Prasad, R., Aberdeen, J., Bratt, E. O., Garofolo, J., Hastie, H., Le, A., Pellom, B., Potamianos, A., Passonneau, R., Roukos, S., Sanders, G., Seneff, S., and Stallard, D. (2002b). DARPA Communicator: Cross System Results for the 2001 Evaluation. *In: Proc. 7th Int. Conf. on Spoken Language Processing (ICSLP-2002)*, 1:269–272, USA–Denver CO.

Whittaker, S. and Stenton, P. (1988). Cues and Control in Expert-Client Dialogues. *In: Proc. of the 26th Ann. Meeting of the Assoc. for Computational Linguistics*, 123–130.

Whittaker, S., Walker, M., and Maloor, P. (2003). Should I Tell All?: An Experiment on Conciseness in Spoken Dialogue. *In: Proc. 8th Europ. Conf. on Speech Communication and Technology (EUROSPEECH 2003 – Switzerland)*, 3:1685–1688, CH–Geneva.

Whittaker, S. and Attwater, D. (1995). Advanced Speech Applications – The Integration of Speech Technology into Complex Devices. *In: Proc. ESCA Workshop on Spoken Dialogue Systems*, P. Dalsgaard, L.B. Larsen, L. Boves, and I. Thomsen, eds., 113–116, DK–Vigsø.

Whittaker, S., Walker, M., and Moore, J. (2002). Fish or Fowl: A Wizard of Oz Evaluation of Dialogue Strategies in the Restaurant Domain. *In: Proc. 3rd Int. Conf. on Language Resources and Evaluation (LREC 2002)*, 5:1602–1609, ES–Las Palmas.

Whittaker, S. J. and Attwater, D. J. (1996). The Design of Complex Applications Using Large Vocabulary Speech Technology. *In: Proc. 4th Int. Conf. on Spoken Language Processing (ICSLP'96)*, H.T. Bunnell and W. Idsardi, eds., 2:705–708, IEEE, USA–Piscataway NJ.

Williams, J. D., Shaw, A. T., Piano, L., and Abt, M. (2003). Preference, Perception, and Task Completion of Open, Menu-Based, and Directed Prompts for Call Routing: A Case Study. *In: Proc. 8th Europ. Conf. on Speech Communication and Technology (EUROSPEECH 2003 – Switzerland)*, 3:2209–2212, CH–Geneva.

Williams, S. (1996). Dialogue Management in a Mixed-Initiative, Cooperative, Spoken Language System. *In: Dialogue Management in Natural Language Systems. Proc. 11th Twente Workshop on Language Technology (TWLT 11)*, University of Twente, S. LuperFoy, A. Nijholt, and G. V. van Zanten, eds., 199–208, NL–Enschede.

Witt, S. M. and Williams, J. D. (2003). Two Studies of Open vs. Directed Dialog Strategies in Spoken Dialog Systems. *In: Proc. 8th Europ. Conf. on Speech Communication and Technology (EUROSPEECH 2003 – Switzerland)*, 1:589–592, CH–Geneva.

Wyard, P. (1993). The Relative Importance of the Factors Affecting Recognizer Performance with Telephone Speech. *In: Proc. 3rd Europ. Conf. on Speech Communication and Technology (EUROSPEECH'93)*, 3:1805–1808, D–Berlin.

Xu, W. and Rudnicky, A. (2000). Language Modeling for Dialogue System. *In: Proc. 6th Int. Conf. on Spoken Language Processing (ICSLP 2000)*, 1:118–121, CHN–Beijing.

Young, S., Adda-Decker, M., Aubert, X., Dugast, C., Gauvain, J. L., Kershaw, D. J., Lamel, L., Leeuwen, D. A., Pye, D., Robinson, A. J., Steeneken, H. J. M., and Woodland, P. C. (1997a). Mulitilingual Large Vocabulary Speech Recognition: The European SQALE Project. *Computer Speech and Language*, 11:73–89.

Young, S. J. (1997). Speech Recognition Evaluation: A Review of the ARPA CSR Programme. *In: Proc. Speech and Language Technology (SALT) Club Workshop on Evaluation in Speech and Language Technology*, 197–205, UK–Sheffield.

Young, S., Kershaw, D., Odell, J., Ollason, D., Valtchev, V., and Woodland, P. (2000). *The HTK Book (for HTK Version 3.0)*. Microsoft Corporation, UK-Cambridge.

Young, S., Ollason, D., Valtchev, V., and Woodland, P. (1997b). *The HTK Book (for HTK Version 2.1)*. Entropic Cambridge Research Laboratory, UK-Cambridge.

Zhan, P., Ries, K., Gavaldà, M., Gates, D., Lavie, A., and Waibel, A. (1996). JANUS-II: Towards Spontaneous Spanish Speech Recognition. *In: Proc. 4th Int. Conf. on Spoken Language Processing (ICSLP'96)*, H.T. Bunnell and W. Idsardi, eds., 4:2285–2288, IEEE, USA–Piscataway NJ.

Zoltan-Ford, E. (1991). How to Get People to Say and Type What Computers Can Understand. *International Journal on Man-Machine Studies*, 34:527–547.

Zue, V., Seneff, S., Glass, J. R., Polifroni, J., Pao, C., Hazen, T. J., and Hetherington, L. (2000). JUPITER: A Telephone-Based Conversational Interface for Weather Information. *IEEE Trans. Speech and Audio Processing*, 8(1):85–96.

Zue, V., Seneff, S., Polifroni, J., Phillips, M., Pao, C., Goodine, D., Goddeau, D., and Glass, J. (1994). PEGASUS: A Spoken Language Interface for On-Line Air Travel Planning. *Speech Communication*, 15:331–340.

About the Author

Sebastian Möller was born in 1968 and studied electrical engineering at the universities of Bochum (Germany), Orléans (France) and Bologna (Italy). Since 1994, he has held the position of a scientific researcher at the Institute of Communication Acoustics (IKA), Ruhr-University Bochum, and works on speech signal processing, speech technology, communication acoustics, as well as on speech communication quality aspects.

He received a Doctor-of-Engineering degree at Ruhr-University Bochum in 1999 for his work on the assessment and prediction of speech quality in telecommunications. In 2000, he was a guest scientist at the Institut dalle Molle d'Intélligence Artificielle Perceptive (IDIAP) in Martigny (Switzerland) where he worked on the quality of speech recognition systems. With the present book, he gained the qualification needed to be a professor (venia legendi) at the Faculty of Electrical Engineering and Information Technology at Ruhr-University Bochum.

Sebastian Möller was awarded the GEERS prize in 1998 for his interdisciplinary work on the analysis of infant cries for early hearing-impairment detection, the ITG prize of the German Association for Electrical, Electronic & Information Technologies (VDE) in 2001, and the Lothar-Cremer prize of the German Acoustical Association (DEGA) in 2003. Since 1997, he has taken part in the standardization activities of the International Telecommunication Union (ITU-T) on transmission performance of telephone networks and terminals.

Index